Probability and Mathematical Statistics (C

WILLIAMS • Diffusions, Markov Processes, and Martingales,
Foundations

ZACKS • Theory of Statistical Inference

Applied Probability and Statistics

ANDERSON, AUQUIER, HAUCK, OAKES, VANDAELE, and WEISBERG • Statistical Methods for Comparative Studies

ARTHANARI and DODGE • Mathematical Programming in Statistics

BAILEY • The Elements of Stochastic Processes with Applications to the Natural Sciences

BAILEY • Mathematics, Statistics and Systems for Health

BARNETT and LEWIS • Outliers in Statistical Data

BARTHOLOMEW • Stochastic Models for Social Processes, *Second Edition*

BARTHOLOMEW and FORBES • Statistical Techniques for Manpower Planning

BECK and ARNOLD • Parameter Estimation in Engineering and Science

BELSLEY, KUH, and WELSCH • Regression Diagnostics: Identifying Influential Data and Sources of Collinearity

BENNETT and FRANKLIN • Statistical Analysis in Chemistry and the Chemical Industry

BHAT • Elements of Applied Stochastic Processes

BLOOMFIELD • Fourier Analysis of Time Series: An Introduction

BOX • R. A. Fisher, The Life of a Scientist

BOX and DRAPER • Evolutionary Operation: A Statistical Method for Process Improvement

BOX, HUNTER, and HUNTER • Statistics for Experimenters: An Introduction to Design, Data Analysis, and Model Building

BROWN and HOLLANDER • Statistics: A Biomedical Introduction

BROWNLEE • Statistical Theory and Methodology in Science and Engineering, *Second Edition*

BURY • Statistical Models in Applied Science

CHAMBERS • Computational Methods for Data Analysis

CHATTERJEE and PRICE • Regression Analysis by Example

CHERNOFF and MOSES • Elementary Decision Theory

CHOW • Analysis and Control of Dynamic Economic Systems

CLELLAND, BROWN, and deCANI • Basic Statistics with Business Applications, *Second Edition*

COCHRAN • Sampling Techniques, *Third Edition*

COCHRAN and COX • Experimental Designs, *Second Edition*

CONOVER • Practical Nonparametric Statistics, *Second Edition*

CORNELL • Experiments with Mixtures: Designs, Models and The Analysis of Mixture Data

COX • Planning of Experiments

DANIEL • Biostatistics: A Foundation for Analysis in the Health Sciences, *Second Edition*

DANIEL • Applications of Statistics to Industrial Experimentation

DANIEL and WOOD • Fitting Equations to Data: Computer Analysis of Multifactor Data, *Second Edition*

DAVID • Order Statistics, *Second Edition*

DEMING • Sample Design in Business Research

DODGE and ROMIG • Sampling Inspection Tables, *Second Edition*

DRAPER and SMITH • Applied Regression Analysis, *Second Edition*

DUNN • Basic Statistics: A Primer for the Biomedical Sciences, *Second Edition*

DUNN and CLARK • Applied Statistics: Analysis of Variance and Regression

ELANDT-JOHNSON • Probability Models and Statistical Methods in Genetics

ELANDT-JOHNSON and JOHNSON • Survival Models and Data Analysis

continued on back

Mathematical Programming in Statistics

Mathematical Programming in Statistics

T. S. Arthanari

Yadolah Dodge

JOHN WILEY & SONS New York • Chichester • Brisbane • Toronto

Library of Congress Cataloging in Publication Data:

Arthanari, T S 1946–
Mathematical programming in statistics.

(Wiley series in probability and mathematical statistics)
Includes index.
1. Mathematical statistics. 2. Programming (Mathematics) I. Dodge, Yadolah, 1944– joint author. II. Title.

QA276.A75 519.5 80-21637
ISBN 0-471-08073-X

Printed in the United States of America

10 9 8 7 6 5 4 3 2 1

Dedicated to C. R. Rao

for his great contribution
to the field of statistics

Foreword

All statistical procedures are, in the ultimate analysis, solutions to suitably formulated optimization problems. Whether it is designing a scientific experiment, or planning a large-scale survey for collection of data, or choosing a stochastic model to characterize observed data, or drawing inference from available data, such as estimation, testing of hypotheses, and decision making, one has to choose an objective function and minimize or maximize it subject to given constraints on unknown parameters and inputs such as the costs involved. The classical optimization methods based on differential calculus are too restrictive, and are either inapplicable or difficult to apply in many situations that arise in statistical work. This, together with the lack of suitable numerical algorithms for solving optimizing equations, has placed severe limitations on the choice of objective functions and constraints and led to the development and use of some inefficient statistical procedures.

Attempts have therefore been made during the last three decades to find other optimization techniques that have wider applicability and can be easily implemented with the available computing power. One such technique that has the potential for increasing the scope for application of efficient statistical methodology is mathematical programming. Although endowed with a vast literature, this method has not come into regular use in statistical practice mainly because of lack of good expositions integrating the techniques of mathematical programming with statistical concepts and procedures. Dr. Arthanari and Dr. Dodge have done an excellent job in this direction by writing a book on mathematical programming with special reference to statistical theory and methods.

Mathematical Programming in Statistics contains extensive discussions of problems in the mainstream of research in statistics and is thus a valuable reference book for both applied and theoretical statisticians. The book is written in a simple style without sacrificing rigor and contains numerous illustrations using live data, which makes it an ideal textbook for one-semester courses on mathematical programming in statistics.

The authors, Dr. Arthanari and Dr. Dodge, deserve to be congratulated for the valuable service they have done to the statistical profession by writing a full-length monograph on tools and techniques that are expected to play a significant role in the development and use of statistical methods in the future.

C. R. Rao

Pittsburgh
April 21, 1980

Preface

This book is an attempt to bring together most of the available results on applications of mathematical programming in statistics.

In the development of the theory underlying statistical methods, one is often faced with an optimization problem. The techniques for solving such problems can be broadly classified as classical, numerical, variational methods, and mathematical programming. The emphasis of this book is on mathematical programming.

The fundamental paper by Charnes, Cooper, and Ferguson (1955) introduced the application of mathematical programming to statistics. As an alternative to the least-squares approach to linear regression, they chose to minimize the sum of the absolute deviations (MINMAD), and showed the equivalence between the MINMAD problem and a linear-programming (LP) problem. Wagner (1959) suggested solving the problem through the dual approach. An efficient modification of the simplex method by Barrodale and Roberts (1973) increased the possibility of using MINMAD regression as an alternative to classical regression.

Other areas of applications of mathematical programming in statistics developed simultaneously. Lee, Judge, and Zellner (1968) applied it in maximum-likelihood and Bayes estimation of transition probabilities in Markov chains; Raj (1956), Folks and Antle (1965), Pfanzagl (1966), Kohan and Khan (1967), Bruvold and Murphy (1978), and Rao (1979) in sampling; Jensen (1969), Vinod (1969), Rao (1971), and Liittschwager and Wang (1978) in cluster analysis; Sedransk (1967) in designing some multifactor of survey data; Neuhardt and Bradley (1971) in selection of multifactor experiments with resource constraints; Foody and Hedayat (1977) in the construction of BIB designs with repeated blocks; Neuhardt, Bradley, and Henning (1973) in optimal design of multifactor experiments; Dantzig and Wald (1951), Barankin (1951), Francis and Wright (1969), Krafft (1970), Meeks and Francis (1973), and Pulkelsheim (1978) in testing statistical hypotheses. In 1971 there was enough interest that a seminar in this area was organized at the symposium on "Optimization Methods in

Statistics" at Ohio State University. Recently an international conference was held at the Indian Institute of Technology, Bombay, during December 1977.

The material in this book is based on research done during the last three decades by several research workers, along with results obtained recently by the authors. The vastness of the literature on mathematical programming and on statistical methods has made it impossible for us to cover all aspects of the possible applications, but we are sure this work can accelerate interest in this area among students and research workers in both fields.

Chapter 1 introduces the subject matter of the book and points out the optimizing nature of statistical problems.

In Chapter 2 the problem of regression analysis is discussed and the different optimizing criteria are first introduced, in the simple context of a single independent variable. Multiple regression is then considered. The theory of the simplex method in LP is developed and a special algorithm for the MINMAD problem is then explained. The bounded-variable method is developed for the dual formulation. Unbiased estimation of the parameters is then considered. The MINMAXAD criterion and the geometrical properties of the problem are discussed, with the use of the various LP formulations. Other new estimators that can be obtained with LP are discussed. Convex-programming theory is developed for solving restricted least-squares problems. Also, as a compromise between least squares and MINMAD, a convex combination of the two is considered. Obtaining best sets of independent variables is first approached through the branch-and-bound procedure; then the problem is posed as a cardinality-constrained linear-programming (CCLP) problem.

Computations of generalized inverses and their connection with the simplex method are brought out in Chapter 3.

In Chapter 4 of this book, the Neyman-Pearson problem in the theory of testing statistical hypotheses is posed as a bounded-variable LP problem. First the finite-sample-space case with simple null and simple alternate hypotheses is considered. Duality in the Neyman-Pearson problem is explained, and then the cases with composite null and composite alternate hypotheses are considered. On this basis a generalized version of the Neyman-Pearson problem is given, which requires duality results in the space of Lebesgue measurable functions.

Some problems in the area of sampling that are mathematical-programming problems are considered in Chapter 5. Optimal allocation of sample size in stratified sampling is shown as a dynamic programming problem. In the multivariate case the underlying convex-programming problem is identified and methods for solving are indicated. Integration of sample surveys with prescribed probabilities of selection of samples is then considered. Necessary theory from transportation problems of LP is also developed. This chapter also deals with the estimation of population proportions with restrictions on marginals. A convex cost-transportation formulation and a separable programming approximation are obtained. Maximum-likelihood estimation and its connection with minimum χ^2 estimation are brought out with a real-life case.

In Chapter 6 MINMAD estimates for a two-way classification model are considered and the problem is shown as one of transportation. The special structure of the two-way model as well as the MINMAD criterion are exploited in development of the method. Construction of BIB designs and their connection to certain 0–1 integer programming problems are studied. Minimal support size for BIB designs with repeated block is defined, and when $k=3$, a special procedure is given for obtaining such designs. Connected designs and transportation matrices are brought together to obtain minimal connected designs in the case of 2^n factorial designs.

The problem of clustering N elements into m clusters according to a given criterion is basically a combinatorial-optimization problem. In Chapter 7 different criteria are considered, and the different mathematical-programming problems arising therefrom. The Lagrangian relaxation method and subgradient optimization are discussed to develop a method for solving the cluster-median problem. Minimizing within cluster-sum of squares in the one-dimensional case is then considered, and a dynamic recursive method is suggested. Use of dynamic programming, LP, and the branch-and-bound approach are discussed at the end of Chapter 7.

Selection problems, optimal design problems, and obtaining nonnegative estimates in variance-components analysis and in Rao's MINQUE are discussed briefly in Chapter 8, as well as the need for learning programming in abstract spaces and exploiting the structure of the problem for effective application of mathematical programming.

There is enough material in this book for a one-term course in mathematical programming in statistics. The main mathematical prerequisites are matrix algebra and introductory mathematical statistics, along with a basic understanding of designs and linear models. Some of the mathematical-programming requirements are explained in the main body of the book as they are first applied to specific problems. However, basic knowledge of convex sets and functions seems essential.

In our opinion there is a need for developing mathematical programming in the context of statistical problems, in order to appreciate the potential of such applications. We deliberately avoided including applications of stochastic models of operation research. Throughout the book, except for Chapters 1 and 8, references and bibliographical discussions appear at the end of each chapter.

It gives us great pleasure to thank V. V. Iyer, J. V. Krishna, N. V. R. Mahadev, and P. Pathak for reading parts of the manuscript and suggesting useful improvements. We are very grateful to M. Lal, V. P. Sharma, M. Airut, U. Heldt, M. Lopez, and C. Redfern, who took the trouble to type the manuscript. To the School of Planning and Computer Applications in Tehran, which provided us with facilities during the four years of the preparation of the manuscript, go our sincere thanks. We also thank Professor A. Hedayat, Professor L. Kish, and Dr. F. Mehran for inspiring us with some problems. To D. Javan and Dr. J. Hogarth, who helped us in programming some of our results, go our special thanks. We thank Professor C. R. Prasad, Head of SQC and OR Division of the Indian Statistical Institute and Professor L. D. Calvin, Head of the Department of Statistics at Oregon State University, Corvallis, Oregon, for providing us with the facilities to work on the final draft. We are also grateful to V. A. Narashimhan of Air India, Tehran, for his help. To an unknown referee for his valuable suggestions and comments we are also thankful. Professor C. R. Rao gave us encouragement throughout the preparation of the manuscript. It was only after his visit to the School of Planning and Computer Applications in Tehran during the first Iranian statistics seminar (April 1976) that we did the initial writing. We are proud to dedicate this little work to Professor C. R. Rao's great and immeasurable contribution to the world of statistics.

T. S. ARTHANARI — *Madras, India*

YADOLAH DODGE — *Corvallis, Oregon*

July 1980

Contents

CHAPTER 1

Prologue

1.1 THE NATURE OF STATISTICAL PROBLEMS

Statistical technology plays an indispensable role in almost every possible sphere of human activity in the modern world. The inferential aspects of statistical methods have made them essential to the toolkit of anyone engaged in scientific enquiry. Regression analysis, estimation, testing of statistical hypotheses, design and analysis of experiments, sample surveys, data classification and grouping, and time-series analysis are most of the major statistical methods that have found many applications in various fields. Such applications have also contributed to the growth of the theory and methods of inference based on data.

In the development of the underlying theories in these statistical methods, more often than not one finds that the statistician is trying to solve an optimization problem and its related questions, such as existence or uniqueness of the solution. Although this statement may sound too general to be true, the discussion that follows can throw more light on it.

Estimation Problem

The general problem of estimation is one of choosing a density function belonging to a specified family of density functions, on the basis of observed data. For this purpose a function of the observations called "estimator" is defined so that the value of the estimator for a given observed datum is the estimate of the unknown density function. When interested only in estimating certain parameters of the density function, we might not estimate the entire density function. Such problems are indeed optimization problems, as Rao (1973) says in *Linear Statistical Inference and Its Applications*:

> To determine an estimator we need a set of criteria by which its performance can be judged....

Intuitively, by an estimator of a parameter θ we mean a function T of the observations $(x_1,\ldots,x_n)$ which is closest to the true value in some sense. In laying down criteria of estimation one attempts to provide a measure of closeness of an estimator to the true value of a parameter and to impose suitable restriction on the class of estimators. An optimum estimator in the restricted class is determined by minimizing the measure of closeness. Some restrictions on the class of estimators seem to be necessary to avoid trivial estimators....

Testing of Statistical Hypotheses

The problem of testing statistical hypotheses was considered originally by Neyman and Pearson (1936). Connections between the well-known Neyman-Pearson Lemma for constructing the uniformly most powerful test of a simple hypothesis having a single alternative, and optimization with linear models can be seen from the following passage from Dantzig's *Linear Programming and Extensions* (1963):

> In 1936 J. Neyman and E. S. Pearson clarified the basic concepts for validating statistical tests and estimating underlying parameters of a distribution for given observations (Neyman and Pearson (1936)). They used what is now the well-known Neyman-Pearson Lemma for constructing the best test of a simple hypothesis having a single alternative. For a more general class of hypotheses they showed that if a test existed satisfying a generalized form of their lemma, it would be optimal. In 1939 (and as a part of his doctoral thesis, (1946)), the author first showed that under very general conditions such a test always exists. This work was later published jointly with A. Wald, who independently reached the same result around 1950 (Dantzig and Wald (1951)). This effort constitutes not only an early proof of one form of the important duality theorem of linear programming, but one given for an infinite (denumerable) number of variables or (through the use of integrals) a nondenumerable number of variables. These are referred to by Duffin as infinite programs (Duffin (1956)). It is interesting to note that the conditions of the general Neyman-Pearson Lemma are in fact *the conditions that a solution to a bounded variable linear programming problem be optimal*. The author's research on this problem formed a background for his later research on linear programming.

Regression Analysis

Regression analysis is concerned with the problem of predicting a variable, called "dependent variable," on the basis of information provided by certain other variables, called "independent variables." A function $f(X_1,\ldots,X_p)$ of the independent variables $\mathbf{X}=(X_1,\ldots,X_p)$ is called a predictor of a dependent variable $\mathbf{Y}$ that is considered. Here again, different criteria are considered for optimization. For instance, the classical approach is to minimize the mean square error, that is, to minimize $E[\mathbf{Y}-f(\mathbf{X})]^2$.

Sample Survey Methodology

Since the need for reliable data to understand better the world in which we live is basic, statisticians have to devise methods of collecting such data. Information on a population may be collected either by complete enumeration or by sample enumeration. The cost of conducting sample enumeration is in general less than that of complete enumeration. On the other hand, precision suffers when too small a sample is considered. Thus the fundamental problem in sample survey is to choose a sampling design that either assures the maximum precision for a given cost of the survey or assumes the minimum cost for a given level of precision. Thus at the root of sample-survey methodology lies an optimization problem of considerable importance. Similarly, the consideration behind the statistical design of an experiment can be traced to the need for optimal balance of cost versus efficiency of the design chosen.

These examples show why optimization methods and theory are needed in statistics.

1.2 CLASSICAL OPTIMIZATION

Over 200 years ago, differential calculus was first used to solve certain optimization problems originating in geometry and physics. These methods have come to be known as classical optimization methods. The problem discussed below is one of classical optimization.

Let $f(\mathbf{x})$ be a real-valued function defined on R^n. We assume that f is differentiable. We wish to find an $\mathbf{x}^* \in R^n$ such that $f(\mathbf{x}^*) \leqslant f(\mathbf{x}), \mathbf{x} \in R^n$. $\mathbf{x}^*$ is called a global minimum of f. The function $f(\mathbf{x})$ is said to have a local minimum at $\mathbf{x}_0$ if there exists an ε, $\varepsilon > 0$, such that, for all $\mathbf{x}$ in the ε-neighborhood of $\mathbf{x}_0$, $f(\mathbf{x}) \geqslant f(\mathbf{x}_0)$. The classical approach provides a means for determining the local minima, using the partial derivatives of f. From differential calculus we have: if f is differentiable and $f(\mathbf{x})$ has a local minimum at the point $\mathbf{x}_0$, then $\mathbf{x}_0$ must be a solution to the set of n equations

$$\frac{\partial f(\mathbf{x})}{\partial x_j} = 0, \qquad j = 1, \ldots, n. \tag{1.2.1}$$

With a similar definition of local maximum, this result is true for local maximum as well.

However, it is not true that if $\mathbf{x}_0$ satisfies (1.2.1) it is a local minimum or maximum. Still this result is of use, as, resolving all the solutions to (1.2.1), we can find all the local minima of $f(\mathbf{x})$ over R^n. And consequently, if a global minimum exists, we can also find $\mathbf{x}^* \in R^n$ such that $f(\mathbf{x}^*) \leqslant f(\mathbf{x})$, for all $\mathbf{x}$ satisfying (1.2.1).

Finding the global optimal solution to the following problem was also considered by classical mathematicians and physicists.

$$\text{Find } \mathbf{x}^* \in R^n, \text{ such that } f(\mathbf{x}^*) \leqslant f(\mathbf{x}), \qquad \text{for all } \mathbf{x} \in \mathscr{F}$$

$$\text{where } \mathscr{F} = \{\mathbf{x} \mid g_i(\mathbf{x}) = b_i,\ i = 1, \ldots, m;\ m \leqslant n\} \subset R^n$$

where g_i's are real-valued differentiable functions defined on R^n.

This problem is approached through Lagrangian multipliers. We write the Lagrangian function

$$F(\mathbf{x}, \boldsymbol{\lambda}) = f(\mathbf{x}) + \sum_{i=1}^{m} \lambda_i [b_i - g_i(\mathbf{x})] \tag{1.2.2}$$

where $\boldsymbol{\lambda} = (\lambda_1, \ldots, \lambda_m)$. The unconstrained method is to find necessary conditions for a local minimum at $\mathbf{x}_0 \in \mathscr{F}$, considering $F(\mathbf{x}, \boldsymbol{\lambda})$. If $f(\mathbf{x})$ attains a local minimum at $\mathbf{x}_0$ for $\mathbf{x} \in \mathscr{F}$, then there exists a $\boldsymbol{\lambda}$ such that

$$\frac{\partial f(\mathbf{x}_0)}{\partial x_j} - \sum_{i=1}^{m} \lambda_i \frac{\partial g_i(\mathbf{x}_0)}{\partial x_j} = 0, \qquad j = 1, \ldots, n$$

and

$$g_i(\mathbf{x}_0) = b_i, \qquad i = 1, \ldots, m. \tag{1.2.3}$$

Hadley (1964) discusses the difficulties of using such approaches to problems in which $\mathbf{x}^*$ should be a global optimum over the nonnegative orthand, or should satisfy some inequality restrictions.

EXAMPLE 1.2.1 Consider a steel plant engaged in producing steel billets, using as input different types of scrap materials. The metallurgist in charge is interested in predicting the yield of steel produced as a function of the input scrap of different types. Let Y_i be the tons of steel produced on the ith melting and let X_{ij} be the tons of scrap of jth type used in the ith melting, $i = 1, \ldots, n$, and $j = 1, \ldots, p$. From experience, the metallurgist knows that it is enough to find the yield as a linear function of the inputs. Thus he assumes a linear model for prediction,

$$\mathbf{Y}=\mathbf{X}\boldsymbol{\beta}+\boldsymbol{\varepsilon},$$

where $\mathbf{Y}'=(Y_1,\dots,Y_n)$, $\mathbf{X}=((X_{ij}))$, $\boldsymbol{\varepsilon}$ is the error, and $\boldsymbol{\beta}$ gives the unknown parameters to be estimated. Assume that he resorts to estimating $\boldsymbol{\beta}$ using the least-squares approach, that is, he wishes to minimize

$$f(\boldsymbol{\beta})=(\mathbf{Y}-\mathbf{X}\boldsymbol{\beta})'(\mathbf{Y}-\mathbf{X}\boldsymbol{\beta}). \tag{1.2.4}$$

Notice that this function is differentiable, so we can resort to the classical unconstrained minimization method, and hence form (1.2.1). We obtain the well-known normal equations of least-squares regression analysis

$$-2\mathbf{X}'\mathbf{Y}+2\mathbf{X}'\mathbf{X}\boldsymbol{\beta}=\mathbf{0}$$

or

$$\mathbf{X}'\mathbf{X}\boldsymbol{\beta}=\mathbf{X}'\mathbf{Y}.$$

If we assume $\mathbf{X}$ is of full rank, then $\boldsymbol{\beta}^*$ is the unique solution to (1.2.4), or $\boldsymbol{\beta}^*=(\mathbf{X}'\mathbf{X})^{-1}\mathbf{X}'\mathbf{Y}$. However, when we find $\boldsymbol{\beta}^*$ and present this to the metallurgist, he refuses to accept this $\boldsymbol{\beta}^*$ as anything meaningful, because some of the β_j^*'s are negative. In his mind he had certain restrictions on $\boldsymbol{\beta}$; in fact, he tacitly assumed that β_j's would turn out to be proportions. Usually the output tonnage is less than or equal to the input tonnage, as there is some loss while melting. We need to introduce these additional restrictions on $\boldsymbol{\beta}$, that is,

$$1\geqslant\beta_j\geqslant 0, \qquad j=1,2,\dots,p.$$

Suppose we wish to adopt the classical approach to solve the problem with these constraints. We have the problem:

$$\text{Find } \boldsymbol{\beta}, \text{ such that } f(\boldsymbol{\beta}^*)\leqslant f(\boldsymbol{\beta}) \qquad \text{for all } \boldsymbol{\beta}\in\mathfrak{F}$$

$$\text{where } \mathfrak{F}=\{\boldsymbol{\beta}\,|\,0\leqslant\beta_j\leqslant 1, j=1,\dots,p\}. \tag{1.2.5}$$

We proceed as follows.

If an unconstrained minimum of $f(\boldsymbol{\beta})$ is such that it is an interior point of $\mathfrak{F}$, then it is also a global minimum for problem (1.2.5). However, global minimum for (1.2.5) need not be an interior point of $\mathfrak{F}$. Therefore we have to consider the boundaries of $\mathfrak{F}$ as well. We must first find all solutions to (1.2.1) lying in the interior of the nonnegative unit hypercube. Systematically, we consider the boundaries of the hypercube by enumerating all possible boundaries. Corresponding to each j, we have two problems depending on $\beta_j=0$ or $\beta_j=1$, which constitute the two boundaries.

Consider any boundary of the hypercube. Some of the β_j's are at 0 level, some are at 1 level, and the rest assume values between 0 and 1. Fixing the β_j's at 0 and 1 levels, correspondingly, we have an unconstrained problem that can be approached through (1.2.1). There are 3^p-1 possibilities in all. For each possibility we consider the β_j's that are neither at 0 level nor at 1 level; we solve the corresponding unconstrained problem and find all solutions to (1.2.1) lying in the interior of the hypercube in the appropriate dimension. This can be done systematically by considering the increasing order of the number of β_j's that are at the 0 and 1 levels.

Then the global minimum of $f(\boldsymbol{\beta})$ is obtained as the smallest of all the values of $f(\boldsymbol{\beta})$ obtained in these problems.

This discussion throws light on the computational burden of introducing nonnegativity contraints and upper-bound constraints into the classical unconstrained problem. Now we may consider more efficient optimization approaches.

1.3 MATHEMATICAL-PROGRAMMING PROBLEMS

A *mathematical-programming problem* can be stated as that of finding an $\mathbf{x}^*$ such that

$$f(\mathbf{x}^*) \leqslant f(\mathbf{x}) \qquad \text{for all } \mathbf{x} \in \mathfrak{F}. \tag{1.3.1}$$

Here $\mathfrak{F}$ is called the feasible set. $\mathfrak{F}$ can be any set not necessarily a subset of R^n; f is a real-valued function, defined on $\mathfrak{F}$. Existence of such an $\mathbf{x}^*$ in general is not guaranteed. However, solving this problem can be understood as that of finding an $\mathbf{x}^*$ or showing the nonexistence of such an $\mathbf{x}^*$.

When $\mathfrak{F}$ is given as a subset of R^n, we have the *nonlinear programming problem*, stated as follows:

$$\text{Find an } \mathbf{x}^* \in \mathfrak{F} \text{ such that } f(\mathbf{x}^*) \leqslant f(\mathbf{x}), \qquad \text{for all } \mathbf{x} \in \mathfrak{F} \tag{1.3.2}$$

where $\mathfrak{F} = \{\mathbf{x} \mid g_i(\mathbf{x}) \leqslant b_i,\ i=1,\ldots,m;\ \mathbf{x} \geqslant \mathbf{0}\}$.

Here f is called the objective function, $g_i(\mathbf{x}) \leqslant b_i$ are known as the constraints of the problem where g_i is a real-valued function defined on $\mathcal{F}$, and the restriction $\mathbf{x} \geqslant \mathbf{0}$ is known as the nonnegative restriction. This problem is usually stated as:

$$\begin{aligned} &\text{Minimize} && f(\mathbf{x}) \\ &\text{Subject to} && g_i(\mathbf{x}) \leqslant b_i, \qquad i=1,\ldots,m. \\ &&& \mathbf{x} \geqslant \mathbf{0}. \end{aligned}$$

REMARK 1.3.1 It is easy to handle an unrestricted variable, x_j, which can be any real number, by replacing x_j with the difference of two nonnegative variables. Also, equality constraints can be replaced by two inequality constraints of the kind used in the problem. Classical unconstrained problems and equality-constrained problems can be viewed as special cases of the problem above. Problems that seek integer solutions come under mathematical programming, as nonlinear programming problems make use of the compactness of the feasible set.

REMARK 1.3.2 When the objective function $f(\mathbf{x})$ and the constraints $g_i(\mathbf{x})$ are linear we have the celebrated linear-programming problem. Further, if we restrict the solutions to the linear-programming problem to integers we have the integer-programming problem.

Mathematical-programming problems have received the attention of researchers in mathematics, economics, and operations research for over three decades. Since the development of the simplex method for efficiently solving the linear-programming problem, both the theory and the methods of mathematical programming have seen unprecedented growth. Also the emphasis has turned for solving certain problems, toward finding efficient methods suitable for computers.

However, the general problem of mathematical programming or, for that matter, nonlinear programming, even today does not have an efficient solution procedure. Fortunately, certain special problems introducing linearity or convexity in the objective function or constraints have been studied completely, and reasonably efficient methods are available today. Also, problems in which the solutions are further restricted to having only integer components have received considerable attention from the researchers.

Thus it is encouraging to attempt to use the available literature on mathematical programming to formulate and solve several of the statistical problems that might not be solved efficiently with classical methods alone.

1.4 MATHEMATICAL PROGRAMMING IN STATISTICS

This section, which bears the title of the book, shows the need for using mathematical programming in statistics. Many problems in regression analysis, sample surveys, cluster analysis, construction of designs, estimation, decision theory, and so on can be viewed as mathematical-programming problems.

Mathematical programming illuminates problems already solved by the classical methods, and also provides ways to deal with possible additional restrictions.

For instance, the least-squares regression problem with additional linear inequality restrictions or nonnegative restriction on $\boldsymbol{\beta}$ can be handled by treating the problem as a nonlinear-programming problem with quadratic objective function and linear restrictions. Construction of BIB designs can be subjected to efficiently programmed methods of finding solutions to certain special integer-programming problems. The use of L_1-norm in estimation and regression problems has certain advantages over the least-squares approach, when the underlying assumption of normality is violated. These problems can be brought under a linear-programming setup and solved efficiently.

Certain cost considerations in selecting predictor (independent) variables in regression problems can be handled with mathematical programming. Partial search procedures or discrete programming can be used to find the best few independent variables in a regression model according to certain criteria other than cost.

The basic problem of cluster analysis naturally leads to mathematical-programming formulations of the problem. The maximum-likelihood estimation and minimum χ^2 estimation under restrictions on marginals can be successfully solved with convex transportation models. Further, there is a connection between geometric programming and maximum-likelihood estimation.

Generalized versions of the Neyman-Pearson Lemma are obtained through duality results in mathematical programming. Mathematical programming can be applied to ensure nonnegative estimates of Rao's MINQUE approach. Mathematical-programming application is also found in the theory of optimal designs. These are a few of the applications of mathematical programming in statistics.

The aim of this book is to give enough of a sample from the abundance of literature available on the applications of mathematical programming in statistics to create interest in this area, among the students of both fields.

Some of these mathematical-programming approaches are explained in detail and applied to specific problems in different areas. We use the approach of developing the mathematical-programming theory and method for the first problem in each area and then bringing out generalizations, limitations of the proofs and methods, and what remains to be done for a general application. Thus these methods are applied to a problem with which the students of statistics are familiar. We are also able, in a limited way, to specialize the proofs and methods, making complete use of the structure of the particular problem considered.

As the stress of mathematical programming provides better ways of solving the problems considered, computational aspects are emphasized. However, we do not want to undermine the fruitfulness of the use of mathematical programming in these problems by occasionally making discouraging remarks about the computational efficiencies of certain other methods available today.

For this reason we have chosen and describe in detail those methods that are easy to follow from the theory developed in the book, rather than trying to give the best available method, description of which may be involved. But we have taken care to list references to the other methods for the interested reader.

REFERENCES

Dantzig, G. B. (1963). *Linear Programming and Extensions*. Princeton Univ. Press, Princeton, N.J.

Dantzig, G. B., and Wald, A. (1951). "On the Fundamental Lemma of Neyman and Pearson." *Ann. Math. Stat.* **22**, 87–93.

Duffin, R. J. (1956). "Infinite Programs." In *Linear Inequalities and Related Systems. Annals of Mathematics Study* (Vol. 38). H. W. Kuhn and A. W. Tucker, Eds. Princeton Univ. Press, Princeton, N.J., pp. 157–170.

Hadley, G. (1964). *Nonlinear and Dynamic Programming*. Addison-Wesley, Reading, Mass.

Krafft, O. (1970). "Programming Methods in statistics and probability theory," In *Nonlinear Programming*. J. B. Rosen, O. L. Mangasarian, and K. Ritter, Eds. Academic Press, New York, pp. 425–446.

Neyman, J., and Pearson, E. S. (1936). "Contributions to the Theory of Testing Statistical Hypothesis." *Stat. Res. Mem.* I, 1–37 (1936) and II, 25–57 (1938).

Rao, C. R. (1973). *Linear Statistical Inference and Its Applications*. Wiley, New York.

Rustagi, J. S., Ed. (1971). *Optimizing Methods in Statistics*. Academic Press, New York.

Rustagi, J. S. (1976). *Variational Methods in Statistics*. Academic Press, New York.

Rustagi, J. S. Ed. (1979). *Optimizing Methods in Statistics. II*. Academic Press, New York.

Zanakis, S. H. (1975). "Optimization in Statistics." *Interfaces*. Vol. 6, 84.

CHAPTER 2

Linear Regression Analysis

2.1 INTRODUCTION

One of the most extensively and exhaustively discussed methods among the statistical tools available for analysis of data is *regression*. Theory of regression deals with the prediction of one or more variables, called "dependent (response) variables" on the basis of other variables called "independent variables." (The dependent variable is also called "criterion variable." For independent variables names such as "predictor" or "explanatory" variables are also common.) Such problems are encountered in almost every branch of experimental science and technology.

When the model used to explain the dependent variable in terms of independent variables assumes a linear relationship between them we have a linear regression model. Otherwise we have a nonlinear regression model.

In the classical approach to the regression problem, the objective is to minimize the sum of squared deviations from the observed and the predicted values of the dependent variable. This method is known as the *least-squares method*; it uses classical optimization methods and generalized inverses.

Another method used is minimizing mean absolute deviation from the predicted and observed values of the dependent variable. This problem is known as L_1-norm minimization.

A third method considered in the literature is the Chebyshev criterion of minimizing the maximum of the absolute deviations from the observed and the predicted values of the dependent variable.

In this chapter we highlight the use of mathematical programming in regression analysis. First, for a complete understanding of the problem, we present the classical least-squares approach for the problem in brief, for the simple case when there is only one independent variable, along with some of the other methods used.

Next, we consider the multiple-regression problem, where we have more than one independent variable. This is followed by mathematical-programming approaches for the methods explained earlier. We consider the minimization of the mean absolute deviation from the observed and the predicted values of the dependent variable and also minimization of the maximum of the deviations. In each of the situations we show that the corresponding problem is a linear-programming problem. We then show that the classical least-squares problem with additional restrictions to obtain nonnegative estimates can be formulated as a linear complementary programming problem.

Some of the properties of these prediction methods and some generalizations of the regression problem are also considered. Two more criteria for finding linear-regression lines are proposed. The problem of selecting optimally a subset of independent variables from the available set of independent variables is considered at the end.

2.2 SIMPLE LINEAR REGRESSION

In many statistical analyses we wish to investigate how changes in one variable affect another variable. For example, height and weight, income and amount of food consumed, gross national product and change in money supply, registration procedures and voting turnout, yield and fertilizer used are related in such a fashion. Suppose we have n pairs of observations (X_i, Y_i), $(i=1,2,\dots,n)$, we can plot these points and try to fit a smooth curve through them so that the points are as close as possible to the fitted curve. Such a diagram is called a scatter or dot-diagram and is shown in Fig. 2.2.1. Even though some variables do have an exact relationship, plotted points would not fall exactly on the line but somewhat around it, because of measurement errors. Often linear relation is not exact, like that between height and weight, but a meaningful relation does exist between them. For any given weight there exists a range of height, and vice versa. But the average observed weight increases as height increases.

Usually the type of curve to be fitted is suggested by empirical evidence or theoretical arguments. When there are no such bases it is difficult to decide what type of curve should be fitted. One such case is shown in Fig. 2.2.2 in which both the straight line and the dotted curve seem to be appropriate.

In the first place we start with a *model*. Then at a later stage we must analyze our model very carefully to see whether it is adequate.

Suppose we assume that the relationship is linear between the two variables. That is, the functional relationship of Y and X is of the following

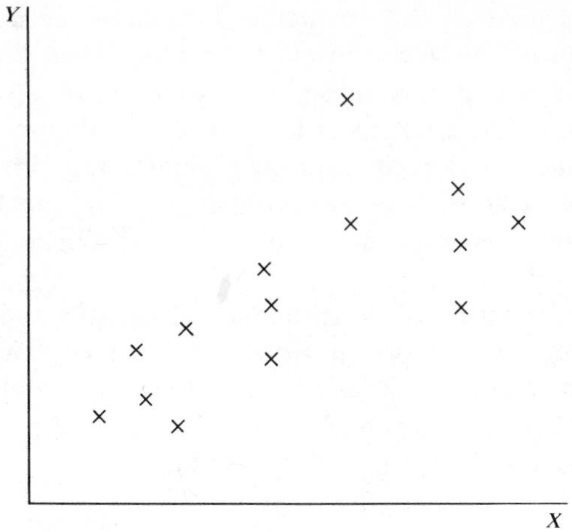

Figure 2.2.1 Scatter diagram between X and Y

form

$$Y = \beta_0 + \beta_1 X + \varepsilon, \tag{2.2.1}$$

which is known as simple linear regression of Y on X. β_0 and β_1 are called *parameters*, and should be found. Equation 2.2.1 means that for a given X_i, a corresponding Y_i consists of $\beta_0 + \beta_1 X_i$ and an ε_i by which an observation may fall off the true regression line (see Fig. 2.2.3). On the basis of the information available from the observations we would like to find β_0 and β_1. The term ε is a random variable and is called "error term." By (2.2.1)

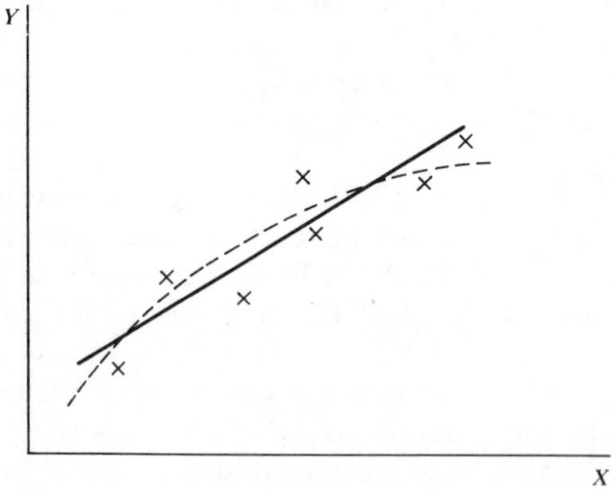

Figure 2.2.2 Two different curves fitted to the same data.

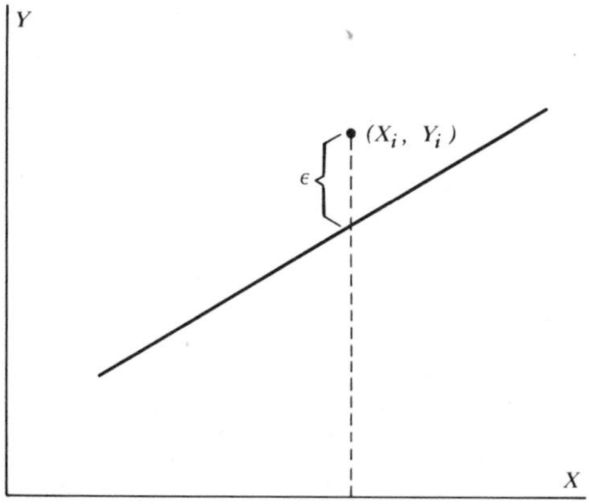

Figure 2.2.3 An observation off the true regression line.

we can write

$$Y_i - \beta_0 - \beta_1 X_i = \varepsilon_i. \tag{2.2.2}$$

Finding β_0 and β_1 from (X_i, Y_i), $i = 1, 2, \ldots, n$ is called *estimation* of the parameters. There are different methods of obtaining such estimates. In the sections that follow we will consider methods that find the parameters by minimizing (A) sum of squared deviations, (B) mean absolute deviation, and (C) maximum of absolute deviations.

A. Minimizing Sum of Squared Deviations (Least-Squares Regression)

This method is based on choosing β_0 and β_1 so as to minimize the sum of squares of the vertical deviations of the data points from the fitted line.

The sum of squares of deviations (SSD) from the line is

$$SSD = \sum_{i=1}^{n} \varepsilon_i^2 = \sum_{i=1}^{n} (Y_i - \beta_0 - \beta_1 X_i)^2. \tag{2.2.3}$$

Then we would choose the estimates β_0 and β_1 so that if we substitute them in (2.2.3) the sum of squared deviations is minimum. Differentiating (2.2.3) with respect to β_0 and β_1 and setting the resultant partial derivatives to zero, we have

$$\frac{\partial SSD}{\partial \beta_0} = -2\sum_{i=1}^{n}(Y_i - \beta_0 - \beta_1 X_i)$$

$$\frac{\partial SSD}{\partial \beta_1} = -2\sum_{i=1}^{n}X_i(Y_i - \beta_0 - \beta_1 X_i). \tag{2.2.4}$$

And hence,

$$\sum_{i=1}^{n}(Y_i - \beta_0 - \beta_1 X_i) = 0$$

$$\sum_{i=1}^{n}X_i(Y_i - \beta_0 - \beta_1 X_i) = 0. \tag{2.2.5}$$

From (2.2.5) we have

$$\beta_0 n + \beta_1 \sum_{i=1}^{n} X_i = \sum_{i=1}^{n} Y_i$$

$$\beta_0 \sum_{i=1}^{n} X_i + \beta_1 \sum_{i=1}^{n} X_i^2 = \sum_{i=1}^{n} X_i Y_i. \tag{2.2.6}$$

Equations 2.2.6 are called *normal equations*. From (2.2.6) we have

$$\hat{\beta}_1 = \frac{\Sigma X_i Y_i - (\Sigma X_i)(\Sigma Y_i)/n}{\Sigma X_i^2 - (\Sigma X_i)^2/n}$$

and $\hat{\beta}_0 = \overline{Y} - \hat{\beta}_1 \overline{X}$, where $\overline{Y}$ and $\overline{X}$ are $\Sigma_{i=1}^{n} Y_i/n$ and $\Sigma_{i=1}^{n} X_i/n$, respectively. $\hat{\beta}_0$ and $\hat{\beta}_1$ obtained in this fashion are called *least-squares estimates* of β_0 and β_1, respectively. Thus, we can write our estimated regression equation as

$$\hat{Y}=\hat{\beta}_0+\hat{\beta}_1 X,$$

which is called the *prediction* equation.

EXAMPLE 2.2.1 Consider the following data:

Observation Number	X	Y
1	50	22
2	54	25
3	56	34
4	59	28
5	60	26
6	61	32
7	62	30
8	65	30
9	67	28
10	71	34
11	71	36
12	74	40

We find that

$$n=12 \qquad \bar{X}=62.5 \qquad \Sigma X_i^2=47470$$

$$\Sigma Y_i=365 \qquad \bar{Y}=30.4 \qquad (\Sigma X_i)(\Sigma Y_i)=273750$$

$$\Sigma X_i=750 \qquad \Sigma X_i Y_i=23134$$

$$\hat{\beta}_1=\frac{23134-22812.5}{47470-46875}=0.54$$

and

$$\hat{\beta}_0=30.4-0.54(62.5)=-3.35.$$

The fitted equation is

$$\hat{Y}=-3.35+0.54X.$$

Table 2.2.1

Observation Number	Y_i	$\hat{Y}_i$	$Y_i - \hat{Y}_i$
1	22	23.63	−1.63
2	25	25.81	−0.81
3	34	26.89	7.11
4	28	28.51	−0.51
5	26	29.05	−3.05
6	32	29.59	2.41
7	30	30.13	−0.13
8	30	31.75	−1.75
9	28	32.83	−4.83
10	34	34.99	−0.99
11	36	35.01	0.99
12	40	36.61	3.39

For each value of X_i, $i=1,\ldots,n$, we can find $d_i = Y_i - \hat{Y}_i$. d_i is also called the *residual*. As a consequence of the first normal equation the sum of residuals (deviations) in any regression problem when the β_0 term is in the model is always zero. In Table 2.2.1 we find the fitted value $\hat{Y}_i$, and the deviations for the data given in Ex. 2.2.1.
Here, we see that $\Sigma_{i=1}^{n} d_i = 0.20$, and is not equal to zero as it should be. Because of rounding off, this error is introduced.

B. Minimizing Mean Absolute Deviations (MINMAD* Regression)

For the simple linear regression model, namely,

$$Y = \beta_0 + \beta_1 X + \varepsilon,$$

we have observed data on X and Y given by (X_i, Y_i), $i = 1, 2, \ldots, n$. We are interested in finding the coefficients β_0 and β_1 such that

$$\frac{1}{n} \sum_{i=1}^{n} |Y_i - \beta_0 - \beta_1 X_i| \tag{2.2.7}$$

is minimized. The expression in (2.2.7) is known as the mean absolute

*Certain researchers have chosen to call this the Least Absolute Value estimator, LAV estimator. LAV, pronounced "love," is a kind of MADness anyway.

deviation from the observed and the predicted values of the dependent variable. It can easily be seen that minimizing (2.2.7) is the same as minimizing

$$\sum_{i=1}^{n} |Y_i - \beta_0 - \beta_1 X_i|, \tag{2.2.8}$$

which is the sum of absolute deviations. First we consider the problem with an additional restriction that β_0 and β_1 as well as minimizing the sum of the absolute deviations should be such that

$$Y_0 = \beta_0 + \beta_1 X_0$$

for a given pair (X_0, Y_0).

This restriction simplifies the analysis and so enhances the understanding of the approach. Later we consider the problem without this restriction. Given (X_0, Y_0) we transform the given data as follows:

$$x_i = X_i - X_0$$

$$y_i = Y_i - Y_0.$$

Now the problem can be stated as that of finding a β that minimizes

$$\sum_{i=1}^{n} |y_i - \beta x_i|.$$

Before attacking this problem, we consider the minimum of $|y_i - \beta x_i|$ for any i (Fig. 2.2.4). It consists of two straight lines with a minimum at $(y_i/x_i, 0)$ and slopes $-|x_i|$ and $|x_i|$.

Now consider the data

i	x_i	y_i
1	1	3
2	1	1
3	2	4

We can draw the graph of $|y_i - \beta x_i|$ for $i = 1, 2, 3$ and $\Sigma|y_i - \beta x_i|$ (Fig. 2.2.5). In this example $\Sigma|y_i - \beta x_i|$ is a piecewise linear convex function as seen in Fig. 2.2.4. In general we can show that $\Sigma|y_i - \beta x_i|$ is a piecewise linear convex function, as given by Result 2.2.1.

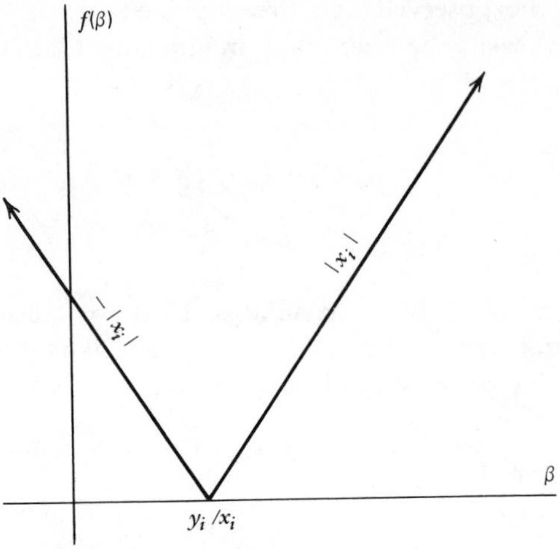

Figure 2.2.4 The graph of $f(\beta)=|y_i-\beta x_i|$.

RESULT 2.2.1 The function $f(\beta)=\Sigma_{i=1}^{n}|y_i-\beta x_i|$ for given (x_i, y_i), $i=1,2,\ldots,n$ is a piecewise linear convex function.

Proof We shall show that for $\beta'<\beta''$ and $1\geqslant\lambda\geqslant 0$, and $\beta=\lambda\beta'+(1-\lambda)\beta''$,

$$f(\beta)\leqslant\lambda f(\beta')+(1-\lambda)f(\beta'').$$

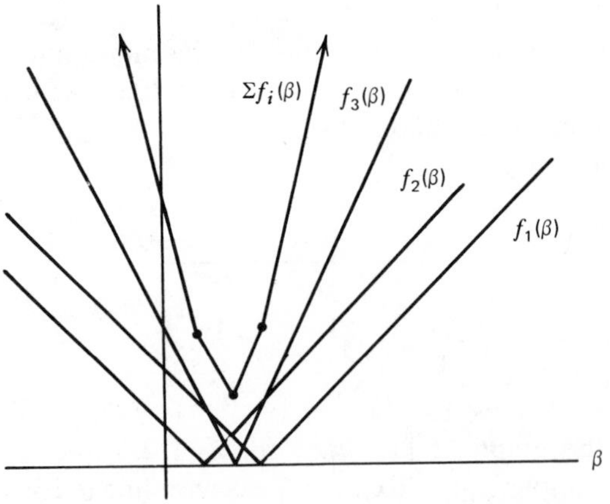

Figure 2.2.5 Graph showing the functions $f_1(\beta)=|3-\beta|$, $f_2(\beta)=|1-\beta|$, $f_3(\beta)=|4-2\beta|$ and $\Sigma f_i(\beta)$.

This will prove f is convex. If we denote by

$$f_i(\beta)=|y_i-\beta x_i|$$

then

$$f_i(\beta)=f_i(\lambda\beta'+(1-\lambda)\beta'')=|y_i-\lambda\beta' x_i-(1-\lambda)\beta'' x_i|$$

$$=|\lambda(y_i-\beta' x_i)+(1-\lambda)(y_i-\beta'' x_i)|$$

$$\leqslant\lambda|y_i-\beta' x_i|+(1-\lambda)|y_i-\beta'' x_i|$$

$$=\lambda f_i(\beta')+(1-\lambda)f_i(\beta'').$$

We now use the fact that the sum of two convex functions is convex repeatedly and we have $f(\beta)$ is convex.

Observe that $f_i(\beta)$ is piecewise linear. As the sum of piecewise linear functions is piecewise linear, we have the result.

RESULT 2.2.2 The function $f(\beta)$ has the following properties:

1 The slope of the extreme left-hand linear piece equals $-\Sigma|x_i|$, and similarly the slope of the extreme right-hand linear piece equals $\Sigma|x_i|$.

2 The grid points of $f(\beta)$ are of the form (y_i/x_i), when y_i/x_i is the minimum point for $f_i(\beta)$; if $(i_1,\ldots,i_n)$ is such that $y_{i_1}/x_{i_1}\leqslant\cdots\leqslant y_{i_n}/x_{i_n}$, then the slope of $f(\beta)$ increases at each grid point $\beta_{(k)}$ by $2|x_{i_k}|$, where $\beta_{(k)}=y_{i_k}/x_{i_k}$.

Proof **(1)** Notice that for all i, $f_i(\beta)$ has slope $-|x_i|$, for $\beta\leqslant\beta_{(1)}$. Therefore $f(\beta)$ has slope $-\Sigma_{i=1}^n|x_i|$ for $\beta\leqslant\beta_{(1)}$. Similarly, $f(\beta)$ has slope

$$\sum_{i=1}^{n}|x_i| \qquad \text{for } \beta\geqslant\beta_{(n)}.$$

(2) Consider $\beta_{(2)}\geqslant\beta>\beta_{(1)}$. We can observe that the slope of $f(\beta)$ remains the same in this interval, for $f_{i_1}(\beta)$ has slope $|x_{i_1}|$ in this interval and all other $f_i(\beta)$'s have slope $-|x_i|$, respectively, and so $f(\beta)$ has slope $-\Sigma_{i=1}^n|x_i|+2|x_{i_1}|$. In fact, proceeding this way we can see that the slope changes only at the points $\beta_{(1)},\beta_{(2)},\ldots,\beta_{(n)}$. Hence they are the grid points of $f(\beta)$ and there are no other grid points.

Thus from the above result we have a procedure to find the minimum of $f(\beta)$. The minimum is attained for $\beta_{(r)}$ such that

$$-\sum_{i=1}^{n}|x_i|+2\sum_{k=1}^{r-1}|x_{i_k}| \quad \text{is negative,}$$

but

$$-\sum_{i=1}^{n}|x_i|+2\sum_{k=1}^{r}|x_{i_k}| \quad \text{is nonnegative.}$$

If $-\Sigma_{i=1}^{n}|x_i|+2\Sigma_{k=1}^{r}|x_{i_k}|=0$, then $\beta_{(r)}\leqslant\beta\leqslant\beta_{(r+1)}$ are all optimal. Then we can choose $\beta_{(r)}$ or $\beta_{(r+1)}$ with equal probability.

We get $\hat{\beta}_0$ and $\hat{\beta}_1$ for the original model, as

$$\hat{\beta}_1=\frac{y_r}{x_r}$$

$$\hat{\beta}_0=Y_0-(y_r/x_r)X_0$$

if we choose $\beta_{(r)}$ as the solution.

EXAMPLE 2.2.2 Consider the data Y_i, X_i given in Table 2.3.1. Let Y_0, X_0 be $\overline{Y}$, $\overline{X}$ where $\overline{Y}$, $\overline{X}$ are the averages of Y_i's and X_i's, respectively. We have $\overline{Y}=30.4$, $\overline{X}=62.5$. Hence we have $\Sigma_{i=1}^{12}|x_i|=71.0$. The slopes for the different linear pieces from left to right are obtained by adding $2|x_{i_k}|$ to S_{k-1}, where $S_{k-1}=-\Sigma_{i=1}^{12}|x_i|+2\Sigma_{l=1}^{k-1}|x_{i_l}|$ for $k=1,2,\ldots,r$ until S_r is nonnegative. We have

l	0	1	2	3	4	5	6	7
S_l	−71.0	−68.0	−55.0	−46.0	−41.0	−24.0	−7.0	10.0
$2\|x_{i_l}\|$	–	3.0	13.0	9.0	5.0	17.0	17.0	17.0

Thus, $y_{i_7}/x_{i_7}=0.659$ is the optimal value of β, and the regression line is given by

$$\hat{Y}=-10.8+0.659X \tag{2.2.9}$$

Next we consider the problem without the restriction that the fitted line should pass through a specified point. So we have to find both β_0 and β_1

Table 2.3.1

i	Y_i	X_i	y_i	x_i	y_i/x_i	Rank
1	22	50	−8.4	−12.5	0.672	8
2	25	54	−5.4	−8.5	0.635	6
3	34	56	3.6	−6.5	−.554	2
4	28	59	−2.4	−3.5	0.686	9
5	26	60	−4.4	−2.5	1.760	12
6	32	61	1.6	−1.5	−1.067	1
7	30	62	−0.4	−0.5	0.800	10
8	30	65	−0.4	2.5	−0.160	4
9	28	67	−2.4	4.5	−0.534	3
10	34	71	3.6	8.5	0.423	5
11	36	71	5.6	8.5	0.659	7
12	40	74	9.6	11.5	0.835	11
Σ	365	750				

such that

$$Y=\beta_0+\beta_1 X$$

and

$$f(\beta)=\sum_{i=1}^{n}|Y_i-\beta_0-\beta_1 X_i|$$

is minimized.

Let

$$f_i(\beta)=|Y_i-\beta_0-\beta_1 X_i|. \tag{2.2.10}$$

Observe that $f_i(\beta)$ is a convex function. So $\Sigma_{i=1}^{n} f_i(\beta)=f(\beta)$ is convex.

Earlier we found that, given any point $z=(z_1, z_2)\in R^2$, the best line passing through z also passes through one of the given points (X_i, Y_i). Let $V_i=\{z\,|\,z\in R^2$ and the best line minimizing the mean absolute deviation passing through z passes through $(X_i, Y_i)\}$. Thus

$$R^2=\bigcup_{i=1}^{n} V_i.$$

Therefore, the best line for the given data, without any restriction that it should pass through any specified point, is obtained when we find z^* such that

$$f_{z^*} = \min_{z \in R^2} f_z$$

where f_z is equal to the optimal value of $f(\beta)$ corresponding to the best line passing through z, i.e.,

$$f_{z^*} = \min_{1 \leqslant i \leqslant n} \left\{ \min_{z \in V_i} f_z \right\}.$$

As for any $z \in V_i$ the best line passes through (X_i, Y_i), we can find the best line passing through (X_i, Y_i) for each i and choose the best line over all.

We can in fact do as follows:

1 Choose any point of the given data, say (X_1, Y_1). Find optimal $\beta_0^{(1)}$ and $\beta_1^{(1)}$, such that the best line is $Y = \beta_0^{(1)} + \beta_1^{(1)} X$. Let the corresponding $f(\beta)$ be f_1.

2 This line passes through another point of the given data, say (X_2, Y_2). Using (X_2, Y_2) we can find optimal $\beta_0^{(2)}$, $\beta_1^{(2)}$ and the corresponding f_2 will be smaller than or equal to f_1, as $Y = \beta_0^{(1)} + \beta_1^{(1)} X$ is one of the lines passing through (X_2, Y_2) but $Y = \beta_0^{(2)} + \beta_1^{(2)} X$ is the best line.

We repeat this procedure until a point (X_t, Y_t) is reached such that the optimal line passes through (X_{t-1}, Y_{t-1}). This is the best line over all, as the function $f(\beta)$ is convex.

EXAMPLE 2.2.3 Consider the example discussed earlier. We find that starting with $(X_1, Y_1) = (50, 22)$, we set the line $Y - 22 = 0.667(X - 50)$ as the best line. This line passes through the points $(59, 28), (62, 30), (71, 36)$ of the given data. Taking $(X_2, Y_2) = (59, 28)$, we find that the new line obtained is $(Y - 28) = 0.667(X - 59)$, which is the same as $(Y - 22) = 0.667(X - 50)$ and passes through $(50, 22)$.

So we stop. The MINMAD line is given by

$$\hat{Y} = -11.35 + 0.667X.$$

C. Minimizing Maximum of Absolute Deviations (MINMAXAD Regression)

We here consider estimating the parameters β_0, β_1 by minimizing the maximum of absolute deviations. Under this criterion we have the objective of finding β_0, β_1 such that β_0, β_1 is a solution to

$$\underset{(\beta_0,\beta_1)}{\text{Minimize}}\left[\max_{1 \leqslant i \leqslant n} |Y_i - (\beta_0 + X_i\beta_1)|\right].$$

First we discuss the problem without the β_0 term, i.e., $Y = \beta X + \varepsilon$. As shown in Section 2.2B, above, $f_1(\beta) = |Y_i - X_i\beta|$ for any i, consists of two straight lines with minimum at $(Y_i/X_i, 0)$ and slopes $-|X_i|, |X_i|$.

Now consider the data given below

i	Y_i	X_i
1	3	1
2	1	1
3	4	2

We can draw the graphs of $f_i(\beta)$ for $i = 1,2,3$ and $\max_i f_i(\beta) = g(\beta)$ as shown in Fig. 2.2.6. It is seen that in this example, $g(\beta)$ is a piecewise linear convex function. In general we can show that $g(\beta)$ is such a function, as given by Result 2.2.3, proof of which is similar to that of Result 2.2.1.

RESULT 2.2.3 The function $g(\beta) = \max_i |Y_i - X_i\beta|$ for given (X_i, Y_i), $i = 1, \ldots, n$ is a piecewise linear convex function.

Observe that the grid points of $g(\beta)$ are not necessarily the points Y_i/X_i, as can be seen from Fig. 2.2.7. However, the grid points of $g(\beta)$ are

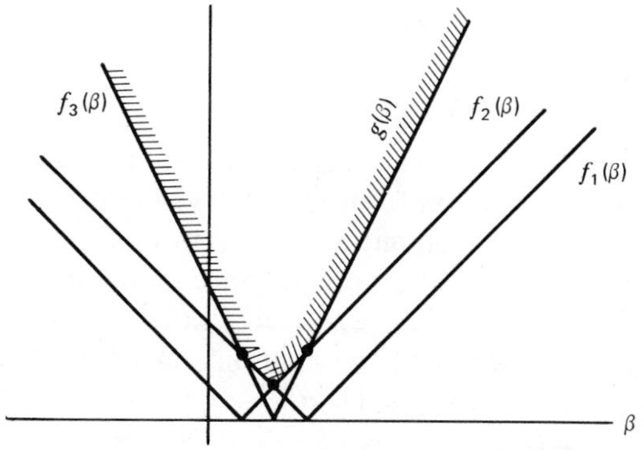

Figure 2.2.6 Graph showing $f_1(\beta) = |3 - \beta|$, $f_2(\beta) = |1 - \beta|$, $f_3(\beta) = |4 - 2\beta|$ and max $f_i(\beta)$ $= g(\beta)$.

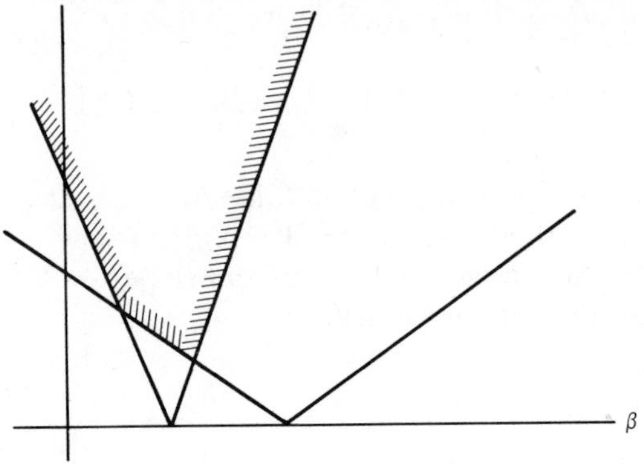

Figure 2.2.7 Graph of a linear convex function different from that shown in Fig. 2.2.6 (see text).

the β coordinates of the intersections of some of the lines

$$g(\beta)=Y_i+X_i\beta \quad \text{or} \quad g(\beta)=-(Y_i+X_i\beta)$$

$$g(\beta)=Y_j+X_j\beta \quad \text{or} \quad g(\beta)=-(Y_j+X_j\beta)$$

for some i and j, $i\neq j$.

Therefore, it is possible to find all such intersections and choose the one for which $g(\beta)$ is minimum.

The geometric interpretation of MINMAXAD estimate and the solution procedure when both β_0 and β_1 terms are in the model are discussed in Section 2.17, where we consider the regression model with several β_j's.

2.3 OTHER ESTIMATORS

We have seen the least-squares regression criterion used to minimize $\Sigma_{i=1}^{n} d_i^2$, and d_i is the deviation from the observed and predicted values of the dependent variable corresponding to the ith observation. Now notice that $\Sigma_{i=1}^{n} d_i^2$ can be thought of as the variance of the deviations, namely, $1/n\Sigma(d_i-\bar{d})^2$, where $\bar{d}=1/n\Sigma_{i=1}^{n} d_i=1/n\Sigma_{i=1}^{n}(Y_i-\hat{Y}_i)$. But $\bar{d}=0$, for the least-squares regression line as it passes through the point $(\bar{X},\bar{Y})$. $\Sigma(d_i-\bar{d})^2$ can be equivalently written as $(1/n)\Sigma_{i<j}(d_i-d_j)^2$. Thus minimizing $\Sigma_{i<j}(d_i-d_j)^2$ is a possible criterion for finding a regression line. In case $\bar{d}=0$ this is equivalent to the least-squares regression line. Replacing $(d_i-d_j)^2$ by $|d_i-d_j|$ we obtain another criterion, namely, the sum of the absolute difference between deviations. We consider minimization of this

sum in the following section. Along the same lines we also consider minimization of the sum of absolute differences between absolute deviations.

A. Minimizing Sum of Absolute Differences Between Deviations (MINSADBED Regression)

We here consider the estimation of the parameters β_0, β_1, minimizing the sum of absolute differences between deviations; that is,

$$\text{Minimize} \sum_{i<j} |d_i - d_j|.$$

From the expression for d_i, d_j we have

$$\sum_{i<j} |d_i - d_j| = \sum_{i<j} |(Y_i - \beta_0 - \beta_1 X_i) - (Y_j - \beta_0 - \beta_1 X_j)|$$

$$= \sum_{i<j} |(Y_i - Y_j) - \beta_1 (X_i - X_j)|$$

Let $Y_{ij} = Y_i - Y_j$ and $X_{ij} = X_i - X_j$, $i<j$. Then we have

$$\sum_{i<j} |d_i - d_j| = \sum_{i<j} |Y_{ij} - \beta_1 X_{ij}|.$$

Then the objective function in this case is similar to that of the MINMAD regression, with β_1 alone to be estimated, as the β_0 term cannot be estimated using this method. So we can apply the procedure developed for MINMAD with $n(n-1)/2$ data points Y_{ij}, X_{ij} to solve this problem. One way of obtaining an estimate of β_0 is to force the line to pass through $(\overline{X}, \overline{Y})$ and take the corresponding constant term as an estimate of β_0; that is,

$$\hat{\beta}_0 = \overline{Y} - \overline{X}\hat{\beta}_1.$$

There are other ways of estimating β_0. One such estimate is

$$\hat{\beta}_0 = \underset{i<j}{\text{median}}\, \tfrac{1}{2}(Y_i + Y_j).$$

EXAMPLE 2.3.1 Consider the data given below.

Observation Number	Y	X
1	2	5
2	5	4
3	4	6
4	8	9
5	3	7

From these data we calculate $Y_{ij}=Y_i-Y_j$ and $X_{ij}=X_i-X_j$ for $i<j, 1 \leqslant i, j \leqslant 5$. Also we calculate Y_{ij}/X_{ij} and rank them, as shown in Table 2.3.1. We have $\Sigma|X_{ij}|=24.00$. The rest of the computations are summarized in Table 2.3.2. Thus $Y_{24}/X_{24}=0.60$ is the optimal value of β_1; that is, $\hat{\beta}_1=0.60$. Now we choose $\hat{\beta}_0$ according to

$$\hat{\beta}_0=\underset{i<j}{\text{median}}\left\{\tfrac{1}{2}(Y_i+Y_j)\right\}$$

as shown in Table 2.3.3. We find $\hat{\beta}_0=4.25$. Hence, the MINSADBED regression line is given by

$$\hat{Y}=4.25+0.60X.$$

Table 2.3.1.

i, j	Y_{ij}	X_{ij}	Y_{ij}/X_{ij}	Rank
1,2	−3	1	−3.00	1
1,3	−2	−1	2.00	9
1,4	−6	−4	1.50	8
1,5	−1	−2	0.50	5
2,3	1	−2	−0.50	4
2,4	−3	−5	0.60	6
2,5	2	−3	−0.67	3
3,4	−4	−3	1.33	7
3,5	1	−1	−1.00	2
4,5	5	2	2.50	10

Table 2.3.2

l	0	1	2	3	4	5	6
S_l	−24.0	−22.0	−20.0	−14.0	−10.0	−6.0	4.0
$2\|X_{(ij)_l}\|$	–	2	2	6	4	4	10

Table 2.3.3

i, j	1,2	1,3	1,4	1,5	2,3	2,4	2,5	3,4	3,5	4,5
$\frac{1}{2}(Y_i+Y_j)$	3.5	3.0	5.0	2.5	4.5	6.5	4.0	6.0	3.5	5.5
Rank	3	2	7	1	6	10	5	9	4	8

B. Minimizing Sum of Absolute Differences Between Absolute Deviations (MINSADBAD Regession)

We consider the estimation of parameter β_0, and β_1 in the simple regression by minimizing the sum of absolute differences between absolute deviations.

$$\underset{\beta_0, \beta_1}{\text{Minimize}} \sum_{i<j} ||d_i|-|d_j||$$

where

$$d_i = Y_i - (\beta_0 + \beta_1 X_i).$$

In Section 2.27 we give procedures for obtaining this kind of estimate in a more general setup.

So far we have considered some of the estimation procedures for the simple regression model. When we relate the dependent variable to more than one independent variable we have the *multiple regression* model. In the subsequent sections we consider such a model and discuss different criteria for estimation of the parameters.

2.4 MULTIPLE LINEAR REGRESSION

Consider the model

$$Y = \beta_0 + \beta_1 X_1 + \cdots + \beta_{p-1} X_{p-1} + \varepsilon \tag{2.4.1}$$

where $X_1, X_2, \ldots, X_{p-1}$ are known constants and β_j's are unknown parameters to be estimated and ε is the error term. If the X_j's are varied and n values of Y are observed, denoted by

$$\mathbf{Y}' = (Y_1, Y_2, \ldots, Y_n),$$

then we have

$$\mathbf{Y}=\mathbf{X}\boldsymbol{\beta}+\boldsymbol{\varepsilon} \tag{2.4.2}$$

where

$$\mathbf{X}'=(\mathbf{X}'_1,\mathbf{X}'_2,\ldots,\mathbf{X}'_n) \qquad \text{and} \quad \mathbf{X}_i=(1, X_{i1},\ldots, X_{ip-1})$$

corresponds to the ith choice of the variables $\mathbf{X}_1,\ldots,\mathbf{X}_{p-1}$,

$$\boldsymbol{\beta}'=(\beta_0,\ldots,\beta_{p-1})$$

and

$$\boldsymbol{\varepsilon}'=(\varepsilon_1,\ldots,\varepsilon_n).$$

As in Section 2.2, the least-squares method of estimating $\boldsymbol{\beta}$ consists of minimizing $\Sigma_i\varepsilon_i^2$ with respect to $\boldsymbol{\beta}$; that is, we minimize $\boldsymbol{\varepsilon}'\boldsymbol{\varepsilon}=\|\mathbf{Y}-\mathbf{X}\boldsymbol{\beta}\|^2$ with respect to $\boldsymbol{\beta}$. Now

$$\begin{aligned}\boldsymbol{\varepsilon}'\boldsymbol{\varepsilon}&=(\mathbf{Y}-\mathbf{X}\boldsymbol{\beta})'(\mathbf{Y}-\mathbf{X}\boldsymbol{\beta})\\&=\mathbf{Y}'\mathbf{Y}-2\boldsymbol{\beta}'\mathbf{X}'\mathbf{Y}+\boldsymbol{\beta}'\mathbf{X}'\mathbf{X}\boldsymbol{\beta}.\end{aligned}$$

Differentiating $\boldsymbol{\varepsilon}'\boldsymbol{\varepsilon}$ with respect to $\boldsymbol{\beta}$ and equating $\partial\boldsymbol{\varepsilon}'\boldsymbol{\varepsilon}/\partial\boldsymbol{\beta}$ to zero, we get

$$-2\mathbf{X}'\mathbf{Y}+2\mathbf{X}'\mathbf{X}\boldsymbol{\beta}=\mathbf{0},$$

or

$$\mathbf{X}'\mathbf{X}\boldsymbol{\beta}=\mathbf{X}'\mathbf{Y}. \tag{2.4.3}$$

Equation 2.4.3 is called the *normal equation(s)*. If $\mathbf{X}$ is of rank p then $\mathbf{X}'\mathbf{X}$ is positive definite, and so, nonsingular. Hence we have a unique solution to (2.4.3). Thus

$$\hat{\boldsymbol{\beta}}=(\mathbf{X}'\mathbf{X})^{-1}\mathbf{X}'\mathbf{Y}. \tag{2.4.4}$$

Then for any $\boldsymbol{\beta}$,

$$\begin{aligned}(\mathbf{Y}-\mathbf{X}\boldsymbol{\beta})'(\mathbf{Y}-\mathbf{X}\boldsymbol{\beta})&=\left[\mathbf{Y}-\mathbf{X}\boldsymbol{\beta}+\mathbf{X}(\hat{\boldsymbol{\beta}}-\hat{\boldsymbol{\beta}})\right]'\left[\mathbf{Y}-\mathbf{X}\boldsymbol{\beta}+\mathbf{X}(\hat{\boldsymbol{\beta}}-\hat{\boldsymbol{\beta}})\right]\\&=(\mathbf{Y}-\mathbf{X}\hat{\boldsymbol{\beta}})'(\mathbf{Y}-\mathbf{X}\hat{\boldsymbol{\beta}})+(\hat{\boldsymbol{\beta}}-\boldsymbol{\beta})'\mathbf{X}'\mathbf{X}(\hat{\boldsymbol{\beta}}-\boldsymbol{\beta})\\&\geqslant(\mathbf{Y}-\mathbf{X}\hat{\boldsymbol{\beta}})'(\mathbf{Y}-\mathbf{X}\hat{\boldsymbol{\beta}}),\end{aligned}$$

which shows that the minimum of $(\mathbf{Y}-\mathbf{X}\boldsymbol{\beta})'(\mathbf{Y}-\mathbf{X}\boldsymbol{\beta})$ is $(\mathbf{Y}-\mathbf{X}\hat{\boldsymbol{\beta}})'(\mathbf{Y}-\mathbf{X}\hat{\boldsymbol{\beta}})$ and is attained at $\boldsymbol{\beta}=\hat{\boldsymbol{\beta}}$.

This solution is shown to minimize $\boldsymbol{\varepsilon}'\boldsymbol{\varepsilon}$.

EXAMPLE 2.4.1. Consider the following data given in Table 2.4.1, where we have four independent variables and thirteen observations:

Table 2.4.1

Observation Number	Y	X_0	X_1	X_2	X_3	X_4
1	78.5	1.0	7.0	26.0	6.0	60.0
2	74.3	1.0	1.0	29.0	15.0	52.0
3	104.3	1.0	11.0	56.0	8.0	20.0
4	87.6	1.0	11.0	31.0	8.0	47.0
5	95.4	1.0	7.0	52.0	6.0	33.0
6	109.2	1.0	11.0	55.0	9.0	22.0
7	102.7	1.0	3.0	71.0	17.0	6.0
8	72.5	1.0	1.0	31.0	22.0	44.0
9	93.1	1.0	2.0	54.0	18.0	22.0
10	115.9	1.0	21.0	47.0	4.0	26.0
11	83.8	1.0	1.0	40.0	23.0	34.0
12	113.3	1.0	11.0	66.0	9.0	12.0
13	109.4	1.0	1.0	68.0	8.0	12.0

Here

$$\mathbf{Y}=\begin{bmatrix}78.5\\74.3\\ \vdots\\109.4\end{bmatrix},\qquad \mathbf{X}=\begin{bmatrix}1 & 7.0 & 26.0 & 6.0 & 60.0\\ 1 & 1.0 & 29.0 & 15.0 & 52.0\\ \vdots & \vdots & \vdots & \vdots & \vdots\\ 1 & 1.0 & 68.0 & 8.0 & 12.0\end{bmatrix}$$

and $\boldsymbol{\beta}'=(\beta_0,\beta_1,\beta_2,\beta_3,\beta_4)$. We find

$$\mathbf{X}'\mathbf{X}=\begin{bmatrix}13 & 97 & 626 & 153 & 390\\ & 1139 & 4922 & 769 & 2620\\ & & 33050 & 7201 & 15739\\ & & & 2293 & 4628\\ & & & & 15062\end{bmatrix},\qquad \mathbf{X}'\mathbf{Y}=\begin{bmatrix}1240.5\\10032.0\\62027.8\\13981.5\\34733.3\end{bmatrix}$$

As $\mathbf{X}'\mathbf{X}$ is symmetric we have shown only the upper triangular portion of the matrix. With (2.4.4), the estimates of the parameters are found to be

$$\hat{\beta}_0=62.14,\ \hat{\beta}_1=1.55,\ \hat{\beta}_2=0.51,\ \hat{\beta}_3=0.10,\ \text{and}\ \hat{\beta}_4=-0.14.$$

Thus the prediction equation has the following form:

$$\hat{Y}=62.14+1.55X_1+0.51X_2+0.1X_3-0.14X_4.$$

2.5 LEAST-SQUARES ESTIMATION WITH CONSISTENT LINEAR RESTRICTIONS

Suppose we wish to estimate $\boldsymbol{\beta}$ using the least-squares method where $\boldsymbol{\beta}$ is subjected to consistent linear-equality restrictions. Then we have the problem

$$\text{Minimize} \quad \boldsymbol{\varepsilon}'\boldsymbol{\varepsilon}$$

$$\text{subject to} \quad \mathbf{A}\boldsymbol{\beta}=\mathbf{C} \tag{2.5.1}$$

where $\mathbf{A}$ is a known $q \times p$ matrix of rank r, $\mathbf{C}$ is a known column vector, and $\boldsymbol{\varepsilon}'\boldsymbol{\varepsilon}$ is defined as before. This problem is approached using Lagrangian multipliers.

The Lagrangian function is

$$\mathbf{L}=\boldsymbol{\varepsilon}'\boldsymbol{\varepsilon}+(\boldsymbol{\beta}'\mathbf{A}'-\mathbf{C}')\boldsymbol{\lambda}. \tag{2.5.2}$$

Now, equating $\partial\mathbf{L}/\partial\boldsymbol{\beta}=0$ and $\partial\mathbf{L}/\partial\boldsymbol{\lambda}=0$, we obtain

$$\begin{aligned} -2\mathbf{X}'\mathbf{Y}+2\mathbf{X}'\mathbf{X}\boldsymbol{\beta}+\mathbf{A}'\boldsymbol{\lambda}&=0 \\ \mathbf{A}\boldsymbol{\beta}&=\mathbf{C}. \end{aligned} \tag{2.5.3}$$

Let the solution of (2.5.3) be $\hat{\boldsymbol{\beta}}_R$ and $\hat{\boldsymbol{\lambda}}_R$, the subscript R denoting the restricted problem.

$$\begin{aligned} \hat{\boldsymbol{\beta}}_R&=(\mathbf{X}'\mathbf{X})^{-1}\mathbf{X}'\mathbf{Y}-\tfrac{1}{2}(\mathbf{X}'\mathbf{X})^{-1}\mathbf{A}'\boldsymbol{\lambda}_R \\ &=\hat{\boldsymbol{\beta}}-\tfrac{1}{2}(\mathbf{X}'\mathbf{X})^{-1}\mathbf{A}'\hat{\boldsymbol{\lambda}}_R \end{aligned} \tag{2.5.4}$$

As $\hat{\boldsymbol{\beta}}_R$ satisfies $\mathbf{A}\boldsymbol{\beta}=\mathbf{C}$

$$\mathbf{C}=\mathbf{A}\hat{\boldsymbol{\beta}}_R$$
$$=\mathbf{A}\hat{\boldsymbol{\beta}}-\tfrac{1}{2}\mathbf{A}(\mathbf{X}'\mathbf{X})^{-1}\mathbf{A}'\hat{\boldsymbol{\lambda}}_R.$$

We have assumed that $\mathbf{A}$ is of rank r; $(\mathbf{X}'\mathbf{X})^{-1}$ is positive definite as $(\mathbf{X}'\mathbf{X})$ is positive definite. And so $\mathbf{A}(\mathbf{X}'\mathbf{X})^{-1}\mathbf{A}'$ is also positive definite. Hence

$$-\tfrac{1}{2}\hat{\boldsymbol{\lambda}}_R=\left[\mathbf{A}(\mathbf{X}'\mathbf{X})^{-1}\mathbf{A}'\right]^{-1}(\mathbf{C}-\mathbf{A}\hat{\boldsymbol{\beta}}). \tag{2.5.5}$$

Thus $\hat{\boldsymbol{\beta}}_R$ can be obtained from (2.5.4) using (2.5.5), as

$$\hat{\boldsymbol{\beta}}_R=\hat{\boldsymbol{\beta}}+(\mathbf{X}'\mathbf{X})^{-1}\mathbf{A}'\left[\mathbf{A}(\mathbf{X}'\mathbf{X})^{-1}\mathbf{A}'\right]^{-1}(\mathbf{C}-\mathbf{A}\hat{\boldsymbol{\beta}}). \tag{2.5.6}$$

It can be shown that $\hat{\boldsymbol{\beta}}_R$ actually is optimal.

It is known that $\hat{\boldsymbol{\beta}}$ and $\hat{\boldsymbol{\beta}}_R$ are unbiased estimates of $\boldsymbol{\beta}$, if $E(\boldsymbol{\varepsilon})=\mathbf{0}$. Further, if the variance-covariance matrix of $\boldsymbol{\varepsilon}$ is given by $\sigma^2 I_n$, then for any linear combination, $\mathbf{a}'\hat{\boldsymbol{\beta}}$ is the minimum variance unbiased linear estimate of $\mathbf{a}'\boldsymbol{\beta}$ for every $\mathbf{a}$. Moreover, if the $\boldsymbol{\varepsilon}\sim N(\mathbf{0},\sigma^2 I_n)$, maximizing the likelihood function is equivalent to minimizing the quantity $\boldsymbol{\varepsilon}'\boldsymbol{\varepsilon}$. So in this case least squares estimate $\hat{\boldsymbol{\beta}}$ is also the maximum likelihood estimate.

2.6 BIASED ESTIMATORS

Given the general model

$$\mathbf{Y}=\mathbf{X}\boldsymbol{\beta}+\boldsymbol{\varepsilon}$$

with $\boldsymbol{\varepsilon}$ following $N(\mathbf{0},\sigma^2\mathbf{I}_n)$, $\mathbf{a}'\hat{\boldsymbol{\beta}}$ is the minimum-variance unbiased estimate of $\mathbf{a}'\boldsymbol{\beta}$. If $\mathbf{X}'\mathbf{X}$ is near singular, then the total variance may be too large for practical purposes. With such ill-conditioned $\mathbf{X}$, we use certain estimators known as *ridge estimators*, given by

$$\hat{\boldsymbol{\beta}}^*_{(k)}=(\mathbf{X}'\mathbf{X}+k\mathbf{I}_n)^{-1}\mathbf{X}'\mathbf{Y}, \qquad (1\leqslant k<\infty). \tag{2.6.1}$$

Since

$$\hat{\boldsymbol{\beta}}^*_{(k)}=(\mathbf{X}'\mathbf{X}+k\mathbf{I}_n)^{-1}\mathbf{X}'\mathbf{X}\hat{\boldsymbol{\beta}}$$
$$=\left[\mathbf{I}_n+k(\mathbf{X}'\mathbf{X})^{-1}\right]^{-1}\hat{\boldsymbol{\beta}}.$$

Then $\hat{\boldsymbol{\beta}}^*_{(k)}$ is a *biased* estimate of $\boldsymbol{\beta}$, for $k>0$. There always exists a $k>0$ such that the total mean square error for $\hat{\boldsymbol{\beta}}^*_{(k)}$ is less than that of $\hat{\boldsymbol{\beta}}$.

Another class of biased estimators is of the form $\lambda\hat{\boldsymbol{\beta}}$ for $0<\lambda\leqslant 1$; these are known as *shrunken estimators*. The general form of these estimators is

$$\boldsymbol{\beta}^{**}=(\mathbf{X}'\mathbf{X}+\mathbf{C})^{-1}\mathbf{X}'\mathbf{Y} \tag{2.6.2}$$

where $\mathbf{C}$ is a positive definite matrix that commutes with $\mathbf{X}$. If $\mathbf{C}=(1/\lambda-1)\mathbf{X}'\mathbf{X}$ we have the shrunken estimator, where λ is close enough to unity, and if $\mathbf{C}=k\mathbf{I}_n$ we have the ridge estimators, for suitable k.

2.7 MINIMIZING MEAN ABSOLUTE DEVIATIONS (MINMAD Regression)

Consider the problem of minimizing $\Sigma|d_i|$ with respect to $\boldsymbol{\beta}$ where d_i is the deviation from the observed, and predicted values of Y_i the ith observation. The problem is the same as minimizing mean absolute deviation; it is alternatively known as the L_1-norm minimization problem.

The problem can be stated as follows:

$$\begin{aligned} &\text{Minimize} \quad \Sigma|d_i| \\ &\text{subject to} \quad \mathbf{X}\boldsymbol{\beta}+\mathbf{d}=\mathbf{Y} \\ &\mathbf{d},\boldsymbol{\beta} \quad \text{unrestricted in sign.} \end{aligned} \tag{2.7.1}$$

Noting the fact that $|d_i|=d_{1i}+d_{2i}$ where d_{1i} and d_{2i} are nonnegative, and $d_i=d_{1i}-d_{2i}$, we can reformulate the problem as:

$$\begin{aligned} &\text{Minimize} \quad \Sigma d_{1i}+\Sigma d_{2i} \\ &\text{subject to} \quad \mathbf{X}\boldsymbol{\beta}+\mathbf{d}_1-\mathbf{d}_2=\mathbf{Y} \\ &\boldsymbol{\beta} \text{ unrestricted in sign} \\ &\mathbf{d}_1,\mathbf{d}_2\geqslant\mathbf{0} \end{aligned} \tag{2.7.2}$$

We now proceed to prove certain results that are used in developing a method for solving this problem.

DEFINITION 2.7.1 Any $(\boldsymbol{\beta}, \mathbf{d}_1, \mathbf{d}_2)$ satisfying $\mathbf{X}\boldsymbol{\beta}+\mathbf{I}\mathbf{d}_1-\mathbf{I}\mathbf{d}_2=\mathbf{Y}$ is called a *solution* to (2.7.2).

Let $(\mathbf{X}, \mathbf{I}, -\mathbf{I})$ be denoted by the matrix $\mathbf{A}$ of order $n \times p+2n$ and $(\boldsymbol{\beta}, \mathbf{d}_1, \mathbf{d}_2)$ be denoted by $\mathbf{W}$. Any column of $\mathbf{A}$ is denoted by $\mathbf{a}_j$. Thus any $\mathbf{W}$ satisfying $\mathbf{AW}=\mathbf{Y}$ is a *solution* to (2.7.2). Let $\mathbf{C}'$ be the vector $(\mathbf{0}, \mathbf{e}', \mathbf{e}')$ where $\mathbf{0}$ is a $1 \times p$ vector and $\mathbf{e}'=(1,\ldots,1)$ a $1 \times n$ vector. Then $\mathbf{C}'\mathbf{W}$ is called the *objective function* of the Problem 2.7.2.

DEFINITION 2.7.2 Any solution $\mathbf{W}$ to (2.7.2), if it further satisfies

$$W_j \geqslant 0, \qquad j=p+1,\ldots,p+2n,$$

we call it a *feasible solution* to the problem.

REMARK 2.7.1 If there is a solution $\mathbf{W}$ to the problem, we can find a corresponding feasible solution $\mathbf{W}^*$ as follows:

$$W_j^* = \begin{cases} W_j,\ j=1,\ldots,p & \\ W_{p+r}-W_{p+n+r} \text{ if } W_{p+r}-W_{p+n+r}>0 & \text{for } j=p+r, r=1,\ldots,n \\ 0 \quad \text{otherwise} & \\ -W_{p+r}+W_{p+n+r} \text{ if } W_{p+r}-W_{p+n+r}<0 & \text{for } j=p+n+r, \\ & r=1,\ldots,n \\ 0 \quad \text{otherwise} & \end{cases}$$

Now $W_j^* \geqslant 0$, $j=p+1,\ldots,p+2n$ and

$$\begin{aligned} \mathbf{AW}^* &= [\mathbf{X}, \mathbf{I}, -\mathbf{I}]\mathbf{W}^* \\ &= \sum_{l=1}^{p} X_l W_l + \sum_{r=1}^{n} e_r W_{p+r}^* - \sum_{r=1}^{n} e_r W_{p+n+r}^* \end{aligned}$$

As only one of W_{p+r}^* or W_{p+n+r}^* can be positive by definition of W^*, we get,

$$\mathbf{AW}^* = \mathbf{AW} = \mathbf{Y}$$

Or, $\mathbf{W}^*$ is a feasible solution to the problem.

Unless otherwise specified, "solution" and "feasible solution" refer to (2.7.2).

DEFINITION 2.7.3 Any feasible solution $\mathbf{W}$ is called a *basic feasible solution* if, the columns of $\mathbf{A}$ corresponding to the nonzero components of $\mathbf{W}$ form a linearly independent set of vectors in R^n.

REMARK 2.7.2 If the rank of $[\mathbf{a}_1,\ldots,\mathbf{a}_p]$ is r we have at most r columns from $[\mathbf{a}_1,\ldots,\mathbf{a}_p]$ that are linearly independent in any basic feasible solution to the problem. Therefore, at most r of these p variables W_j will be nonzero in any basic feasible solution.

REMARK 2.7.3 Define

$$W_{p+r}=\begin{cases} Y_r & \text{if } \; Y_r>0 \\ 0 & \text{otherwise} \end{cases}$$

and

$$W_{p+n+r}=\begin{cases} -Y_r & \text{if } \; Y_r<0 \\ 0 & \text{otherwise} \end{cases}$$

and the rest of the W_j's are zero. $\mathbf{W}$ as defined above is a basic feasible solution to the problem. At this point, we can discuss the geometry of the set of feasible solutions to the problem.

RESULT 2.7.1

$$\mathcal{F}=\{\mathbf{W}|\mathbf{AW}=\mathbf{Y}, W_j\geqslant 0, \qquad j=p+1,\ldots,p+2n\}$$

is a convex set.

Proof We shall show that for $\mathbf{W}^1,\mathbf{W}^2\in\mathcal{F}$ and $1\geqslant\lambda\geqslant 0, \mathbf{W}=\lambda\mathbf{W}^1+(1-\lambda)\mathbf{W}^2$ is in $\mathcal{F}$.

As $\lambda\geqslant 0, 1-\lambda>0$,

$$W_{p+r}=\lambda W_{p+r}^1+(1-\lambda)W_{p+r}^2, \qquad r=1,\ldots,2n$$

are nonnegative. Moreover,

$$\begin{aligned}\mathbf{AW}&=\mathbf{A}(\lambda\mathbf{W}^1+(1-\lambda)\mathbf{W}^2)\\ &=\lambda\mathbf{AW}^1+(1-\lambda)\mathbf{AW}^2\\ &=\lambda\mathbf{Y}+(1-\lambda)\mathbf{Y}=\mathbf{Y}\end{aligned}$$

hence the result.

DEFINITION 2.7.4 A point $\mathbf{W}\in\mathcal{F}$ is called an extreme point of $\mathcal{F}$ if there is no λ such that $1>\lambda>0$ and $\mathbf{W}^1,\mathbf{W}^2\in\mathcal{F}$ such that $\mathbf{W}^1\neq\mathbf{W}^2$

$$\mathbf{W}=\lambda\mathbf{W}^1+(1-\lambda)\mathbf{W}^2.$$

We are interested in extreme points as they are related to basic feasible solutions.

RESULT 2.7.2 $\mathbf{W}\in\mathcal{F}$ is an extreme point of $\mathcal{F}$ then the columns of $\mathbf{A}$ that correspond to nonzero components of $\mathbf{W}$ are linearly independent.

Proof Let $\mathbf{W}$ be an extreme point of $\mathcal{F}$. Without loss of generality, we may assume that first s components of $\mathbf{W}$ are nonzero since we can rename the variables, some of which are of course restricted in sign. If s is zero the corresponding set of columns is empty and therefore trivially linearly independent.

If the columns $(\mathbf{a}_1,\ldots,\mathbf{a}_s)$ are linearly dependent, then there exist real numbers $\alpha_1,\ldots,\alpha_s$, not all zero, such that

$$\sum_{l=1}^{s}\alpha_l\mathbf{a}_l=\mathbf{0}.$$

CASE 1 If none of these $\mathbf{a}_l$ corresponds to restricted W_j's, we observe that

$$W_j\pm\alpha_j,\qquad j=1,\ldots,s,\qquad \text{and}\quad W_j=0\quad \text{otherwise}$$

satisfies $\mathbf{AW}=\mathbf{Y}$, that is,

$$\sum_{j=1}^{s}(W_j\pm\alpha_j)\mathbf{a}_j=\sum_{j=1}^{s}W_j\mathbf{a}_j\pm\sum_{j=1}^{s}\alpha_j\mathbf{a}_j$$

$$=\mathbf{Y}+\mathbf{0}=\mathbf{Y}$$

and is feasible, as all the restricted W_j's are zeroes, and

$$W_j=\tfrac{1}{2}(W_j+\alpha_j)+\tfrac{1}{2}(W_j-\alpha_j).$$

Then $\mathbf{W}$ is not an extreme point of $\mathcal{F}$, leading to a contradiction.

CASE 2 If one or more of these $\mathbf{a}_l$ correspond to restricted W_j's we observe that there exists a sufficiently small $\delta > 0$ such that

$$W_j \pm \delta\alpha_j > 0 \text{ for the restricted } W_j\text{'s.}$$

Also,

$$\sum_{j=1}^{s} (W_j \pm \delta\alpha_j)\mathbf{a}_j = \sum_{j=1}^{s} W_j\mathbf{a}_j \pm \delta \sum_{j=1}^{s} \alpha_j \mathbf{a}_j.$$

Let

$$W_j^1 = \begin{cases} W_j + \delta\alpha_j & \text{for } j = 1, 2, \ldots, s \\ 0 & \text{otherwise} \end{cases}$$

and

$$W_j^2 = \begin{cases} W_j - \delta\alpha_j & \text{for } j = 1, 2, \ldots, s \\ 0 & \text{otherwise.} \end{cases}$$

Hence $\mathbf{W}^1$ and $\mathbf{W}^2$ are feasible solution to the problem. Moreover,

$$\mathbf{W} = \tfrac{1}{2}\mathbf{W}^1 + \tfrac{1}{2}\mathbf{W}^2,$$

which leads to contradiction as $\mathbf{W}$ is an extreme point of $\mathcal{F}$. Hence the result.

COUNTER-EXAMPLE Consider $\mathbf{X}, \mathbf{Y}$, given below

$$\mathbf{X} = \begin{bmatrix} 1 & 2 & 4 \\ 1 & 1 & 3 \\ 1 & 2 & 5 \end{bmatrix}; \qquad \mathbf{Y} = \begin{bmatrix} 4 \\ 2 \\ 4 \end{bmatrix}.$$

Notice that $\mathbf{W} = (-1, 0, 1, 1, 0, 0, 0, 0, 0)$ satisfies $(\mathbf{X}, \mathbf{I}_3, -\mathbf{I}_3)\mathbf{W} = \mathbf{Y}$ and the columns $(\mathbf{a}_1, \mathbf{a}_3, \mathbf{a}_4)$ are linearly independent. But $\mathbf{W}$ is not an extreme point of $\mathcal{F}$ as

$$\mathbf{W} = \tfrac{1}{2}\mathbf{W}^1 + \tfrac{1}{2}\mathbf{W}^2,$$

where

$$\mathbf{W}^1 = \left(-\tfrac{1}{2}, 1, \tfrac{1}{2}, \tfrac{1}{2}, 0, 0, 0, 0, 0\right)$$

$$\mathbf{W}^2 = \left(-\tfrac{3}{2}, -1, \tfrac{3}{2}, \tfrac{3}{2}, 0, 0, 0, 0, 0\right).$$

Where both $\mathbf{W}^1$ and $\mathbf{W}^2$ are feasible solutions to the problem.

If all the variables are restricted in sign, then the converse is also true.

As a corollary to this result, we observe that if $\mathbf{W}$ is an extreme point of $\mathcal{F}$, then $\mathbf{W}$ has at most n nonzero components. The remaining are zeroes.

RESULT 2.7.3 $\mathcal{F}$ contains only finitely many basic feasible solutions.

Proof In each basic feasible solution the nonzero components are determined uniquely by a corresponding set of columns from $\mathbf{A}$. But there are only finitely many subsets of linearly independent column vectors of size less than or equal to n. Hence the result. As a corollary, we observe that $\mathcal{F}$ has finitely many extreme points.

RESULT 2.7.4 If $\mathcal{F}$ is not empty, then the set of extreme points of $\mathcal{F}, \mathcal{F}^0$ also is not empty.

Proof For any $\mathbf{W} \in \mathcal{F}$ we define a function $\rho(\mathbf{W})$ by the number of nonzero components of $\mathbf{W}$. We have $0 \leqslant \rho(\mathbf{W}) \leqslant p+2n$. If $\mathcal{F}$ is not empty, the function attains its minimum, ρ_0, on $\mathcal{F}$. Suppose $\rho(\overline{\mathbf{W}}) = \rho_0$. We shall show that $\overline{\mathbf{W}}$ is an extreme point of $\mathcal{F}$. If $\rho_0 = 0$, then $\overline{\mathbf{W}} = 0$, and by definition $\overline{\mathbf{W}}$ is an extreme point, since the set of columns corresponding to nonzero components is empty and therefore linearly independent.

If $\rho_0 > 0$, we may assume $\overline{\mathbf{W}} = (\overline{W}_1, \ldots, \overline{W}_{\rho_0}, 0, 0, \ldots, 0)$. We suppose $\overline{\mathbf{W}}$ is not an extreme point of $\mathcal{F}$ and get a contradiction. Then the columns $\mathbf{a}_1, \ldots, \mathbf{a}_{\rho_0}$ are linearly dependent, and there exists $\alpha_1, \ldots, \alpha_{\rho_0}$, not all zero, such the $\sum_{j=1}^{\rho_0} \alpha_j \mathbf{a}_j = 0$. For those indices for which $\alpha_j \neq 0$, consider $\overline{W}_j / |\alpha_j|$ and find the smallest of the numbers over those j's corresponding to restricted variables. Let λ be this number and let $\lambda = \overline{W}_{j_0} / |\alpha_{j_0}|$. There is no loss of generality in assuming $\alpha_{j_0} > 0$. The point given by

$$\overline{\overline{W}}_j = \begin{cases} \overline{W}_j - \lambda \alpha_j & j = 1, \ldots, \rho_0 \\ 0 & \text{otherwise,} \end{cases}$$

belongs to $\mathcal{F}$ because,

$$\mathbf{A}\overline{\overline{\mathbf{W}}} = \mathbf{A}\overline{\mathbf{W}} - \lambda \sum_{j=1}^{\rho_0} \alpha_j \mathbf{a}_j = \mathbf{Y}$$

and $\overline{\overline{W}}_j$'s corresponding to the restricted variables are nonnegative by the choice of λ. Also $\overline{\overline{\mathbf{W}}}$ has fewer than ρ_0 components because $\overline{W}_{j_0}-\lambda\alpha_{j_0}=0$, contradicting the fact that ρ_0 is the minimum of $\rho(\mathbf{W})$ over $\mathcal{F}$.

Therefore $\overline{\mathbf{W}}$ is an extreme point.

When $\mathcal{F}$ is *bounded* it is called a convex *polyhedron*.* We state the following result for convex *polyhedrons* without proof.

RESULT 2.7.5 If $\mathcal{F}$ is a convex polyhedron, then every point $\mathbf{W}$ of $\mathcal{F}$ can be written as a convex combination of the extreme points of $\mathcal{F}$.

DEFINITION 2.7.5 Any $\mathbf{W}\in\mathcal{F}$ minimizing $\mathbf{C}'\mathbf{W}$ is called an optimal solution to the problem.

RESULT 2.7.6 If $\mathcal{F}$ is a convex polyhedron, then $\mathbf{C}'\mathbf{W}$ attains its minimum at, at least one of the extreme points of $\mathcal{F}$.

Proof As $\mathcal{F}$ is a bounded convex set, $\mathbf{C}'\mathbf{W}$ attains its minimum at some $\mathbf{W}^0\in\mathcal{F}$. If $\mathbf{W}^0$ is an extreme point, there is nothing to prove. $\mathcal{F}$ has only a finite number of extreme points, say, $\mathbf{W}^1,\mathbf{W}^2,\ldots,\mathbf{W}^k$. From Result 2.7.5 there exists $\lambda_j\geqslant 0$ such that

$$\mathbf{W}^0=\sum_{j=1}^{k}\lambda_j\mathbf{W}^j,\qquad \text{with}\quad \sum_{j=1}^{k}\lambda_j=1.$$

Now

$$\mathbf{C}'\mathbf{W}^0=\sum_{j=1}^{k}\lambda_j\mathbf{C}'\mathbf{W}^j.$$

Let $\mathbf{W}^{j_0}$ be such that $\mathbf{C}'\mathbf{W}^{j_0}=\min_{1<j<k}\mathbf{C}'\mathbf{W}^j$, therefore

$$\mathbf{C}'\mathbf{W}^0=\sum_{j=1}^{k}\lambda_j\mathbf{C}'\mathbf{W}^j\geqslant\sum_{j=1}^{k}\lambda_j\mathbf{C}'\mathbf{W}^{j_0}=\mathbf{C}'\mathbf{W}^{j_0}$$

as $\lambda_j\geqslant 0$ and $\Sigma_{j=1}^{k}\lambda_j=1$, which implies $\mathbf{C}'\mathbf{W}^0\geqslant\mathbf{C}'\mathbf{W}^{j_0}$. On the other hand

*Some authors use "polytope" in place of "polyhedron." For them a polyhedron means a region formed by the intersection of closed half-spaces, and a polytope is a polyhedron that is bounded.

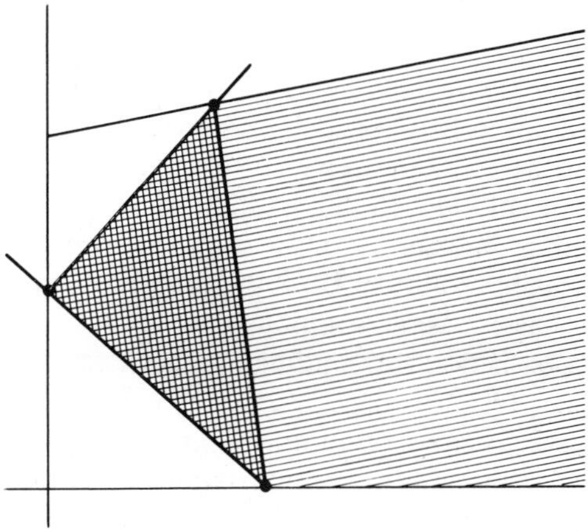

Figure 2.7.1 A counter example to Result 2.7.5 when $\mathcal{F}$ is not bounded.

$\mathbf{C'W}^0 \leqslant \mathbf{C'W}$ *for* $\mathbf{W} \in \mathcal{F}$. *So*

$$\mathbf{C'W} \leqslant \mathbf{C'W}^{j_0}.$$

Therefore $\mathbf{C'W}^0 = \mathbf{C'W}^{j_0}$. Hence the result.

In the case in which $\mathcal{F}$ is not bounded, a result similar to Result 2.7.5 is not true (see Fig. 2.7.1). However, we have the following result which we state without proof.

RESULT 2.7.7 If $\mathcal{F}$ is unbounded and $\mathbf{C'W}$ attains its minimum on $\mathcal{F}$, then at least one extreme point of $\mathcal{F}$ is an optimal solution to the problem. We can conclude that while searching for an optimal solution, basic feasible solutions may be considered, as they include the extreme points of $\mathcal{F}$.

2.8 SIMPLEX PROCEDURE FOR MINMAD REGRESSION

Now we develop a method known as the *simplex* procedure to obtain an optimal basic feasible solution if one exists. For Problem 2.7.2 an optimal solution always exists, as we find a minimum of a continuous function that is bounded below by zero over a closed convex set.

DEFINITION 2.8.1 A set of n linearly independent columns from the matrix **A**, is called a *basis* of **A**.

Corresponding to these columns $\mathbf{B}=[\mathbf{a}_{i_1},\ldots,\mathbf{a}_{i_n}]$ there are $n+p$ other columns which are not in **B**. These columns are called *non*basic columns. If we can find a feasible solution, in which the $W_j=0$ for these columns, we have a basic feasible solution. In this solution some of the W_j's corresponding to basic columns may be zero. In such case we call the solution a *degenerate* basic feasible solution. If all the n basic W_j are nonzero we call it a *nondegenerate* basic feasible solution.

Starting with a basis **B** corresponding to a basic feasible solution we try to find a better basic feasible solution if possible. Let **B** be a basis with the corresponding basic feasible solution **W**. Let $\mathbf{W}_B$ denote the vector of variables W_j corresponding to the basic columns in **B**. Let the corresponding coefficients C_j be denoted by $\mathbf{C}_\mathbf{B}$.

Observe that any vector not in the basis can be expressed as a linear combination of vectors in the basis. Let these coefficients be denoted by α_{ij}, which expresses $\mathbf{a}_j$ as a linear combination of $(\mathbf{a}_{\mathbf{B}_1},\ldots,\mathbf{a}_{\mathbf{B}_n})$, $\mathbf{B}_i$ denoting the ith basic column, that is,

$$\mathbf{a}_j=\sum_{i=1}^{n}\alpha_{ij}\mathbf{a}_{\mathbf{B}_i}$$

or

$$\mathbf{B}^{-1}\mathbf{a}_j=\boldsymbol{\alpha}_j.$$

If we wish to consider a new basis formed from **B** by including one nonbasic column $\mathbf{a}_j$ into **B** removing some column $\mathbf{B}_r$ from **B** in turn, so as to insure getting a corresponding basic feasible solution, we can not do so arbitrarily. Here are the conditions for ensuring feasibility in such an exchange.

We have

$$\sum_{i=1}^{n}W_{B_i}\mathbf{B}_i=\mathbf{Y}$$

and $W_{B_i}\geqslant 0$ for those B_i's with the corresponding variables that are restricted in sign. Let R_B denote the indices of those restricted variables in the basis. Notice that **B** is of rank n and there are only p unrestricted variables in the model, $p<n$; hence R_B will be always nonempty. However, developing the theory with the possibility of R_B's being empty helps in generalizing the theory.

Now $\mathbf{a}_j$ can replace any vector $\mathbf{B}_r$ for which $\alpha_{rj} \neq 0$, and the new set of vectors will form a linearly independent set of n vectors. We select an $\mathbf{a}_j$ such that at least one $\alpha_{ij} \neq 0$, and insert it into the basis.

Then

$$\mathbf{B}_r = \frac{1}{\alpha_{rj}}\mathbf{a}_j - \sum_{\substack{i=1 \\ i \neq r}}^{n} \frac{\alpha_{ij}}{\alpha_{rj}}\mathbf{B}_i.$$

Replacing $\mathbf{B}_r$ in the expression of $\mathbf{Y}$ in terms of $\mathbf{B}_i$'s, we get

$$\sum_{\substack{i=1 \\ i \neq r}}^{n} \left(W_{B_i} - W_{B_r}\frac{\alpha_{ij}}{\alpha_{rj}} \right)\mathbf{B}_i + \frac{W_{B_r}}{\alpha_{rj}}\mathbf{a}_j = \mathbf{Y}.$$

We require the new solution to be feasible, that is $W_{B_i} - W_{B_r}\alpha_{ij}/\alpha_{rj} \geqslant 0$ for $i \in R_B$, $i \neq r$; also we may require $W_{B_r}/\alpha_{rj} \geqslant 0$ if W_j is restricted in sign.

CASE 1 W_j is unrestricted in sign. Then W_{B_r}/α_{rj} need not be nonnegative and we require only

$$W_{B_i} - \frac{W_{B_r}}{\alpha_{rj}}\alpha_{ij} \geqslant 0, \qquad i \in R_B, i \neq r \tag{2.8.1}$$

Notice that when R_B is not empty, there is a W_{B_r} restricted in sign, that is, $W_{B_r} \geqslant 0$. Therefore (2.8.1) is satisfied for all $\alpha_{ij} \leqslant 0$ when $\alpha_{rj} > 0$, and for all $\alpha_{ij} \geqslant 0$ when $\alpha_{rj} < 0$. Therefore if we choose r either by

$$\frac{W_{B_r}}{\alpha_{rj}} = \min_{\substack{i \in R_B \\ \alpha_{ij} > 0}} \left[\frac{W_{B_i}}{\alpha_{ij}} \right]$$

or

$$\frac{W_{B_r}}{\alpha_{rj}} = \max_{\substack{i \in R_B \\ \alpha_{ij} < 0}} \left[\frac{W_{B_i}}{\alpha_{ij}} \right]$$

(2.8.1) is satisfied.

When R_B is empty, (2.8.1) is vacuously satisfied. So any $\mathbf{B}_r$ can be removed from the basis.

CASE 2 W_j is restricted in sign. Then W_{B_r}/α_{rj} must be greater than or equal to zero. Also we require (2.8.1). Thus we require,

$$W_{B_i} - \frac{W_{B_r}}{\alpha_{rj}}\alpha_{ij} \geqslant 0, \qquad \underset{i\neq r}{i \in R_B}$$

$$\frac{W_{B_r}}{\alpha_{rj}} \geqslant 0. \tag{2.8.2}$$

If $r \in R_B$, then $W_{B_r} \geqslant 0$. Therefore, α_{rj} needs to be positive. Then, as in Case 1, we choose r such that

$$\frac{W_{B_r}}{\alpha_{rj}} = \min_{\substack{i\in R_B \\ \alpha_{ij}>0}} \left[\frac{W_{B_i}}{\alpha_{ij}} \right].$$

On the other hand, if R_B is empty then we can choose any r such that $(W_{B_r}/\alpha_{rj}) \geqslant 0$. Next we attempt to choose the column $\mathbf{a}_j$ in such a manner that we improve the objective function. Let $Z_j = \Sigma_{i=1}^{n} C_{B_i}\alpha_{ij}$. Let $Z(\mathbf{B})$ be equal to $\Sigma_{i=1}^{n} C_{B_i} W_{B_i}$ and $Z(\hat{\mathbf{B}})$ denote the corresponding objective function value of $\mathbf{W}_{\hat{B}}$ when $\hat{\mathbf{B}} = (\mathbf{B}_1, \ldots, \mathbf{B}_{r-1}, \mathbf{a}_j, \mathbf{B}_{r+1} \cdots \mathbf{B}_n)$ is the basis. We require $Z(\hat{\mathbf{B}}) \leqslant Z(\mathbf{B})$. Consider

$$Z(\hat{\mathbf{B}}) = \sum_{\substack{i=1 \\ i\neq r}}^{n} C_{B_i}\left(W_{B_i} - \frac{W_{B_r}}{\alpha_{rj}}\alpha_{ij} \right) + C_j \frac{W_{B_r}}{\alpha_{rj}}$$

$$= \sum_{i=1}^{n} C_{B_i}\left(W_{B_i} - \frac{W_{B_r}}{\alpha_{rj}}\alpha_{ij} \right) + C_j \frac{W_{B_r}}{\alpha_{rj}}$$

$$= \sum_{i=1}^{n} C_{B_i} W_{B_i} + \frac{W_{B_r}}{\alpha_{rj}}\left[-\sum_{i=1}^{n} C_{B_i}\alpha_{ij} + C_j \right].$$

Therefore $Z(\hat{\mathbf{B}}) \leqslant Z(\mathbf{B})$ if

$$\frac{W_{B_r}}{\alpha_{rj}}\left[C_j - Z_j \right] \leqslant 0 \tag{2.8.3}$$

CASE 1 $C_j - Z_j < 0$.

If W_j is unrestricted in sign then W_{B_r}/α_{rj} can be any real number. But to satisfy (2.8.3) we require $W_{B_r}/\alpha_{rj} \geqslant 0$. Therefore if R_B is nonempty, $\alpha_{rj} > 0$.

If W_j is restricted in sign then we require $W_{B_r}/\alpha_{rj} \geqslant 0$. Therefore if R_B is nonempty, $\alpha_{rj} > 0$. Unless $W_{B_r}/\alpha_{rj} = 0$.

CASE 2 $C_j - Z_j > 0$.

If W_j is unrestricted in sign, then to satisfy (2.8.3) we require $W_{B_r}/\alpha_{rj} \leqslant 0$. Therefore if R_B is nonempty, $\alpha_{rj} < 0$.

If W_j is restricted in sign, W_{B_r}/α_{rj} has to be greater than or equal to zero to satisfy (2.8.3) which is impossible, unless $W_{B_r}/\alpha_{rj} = 0$. Then the new solution is just as good as the previous solution. But we know R_B is nonempty for the problem 2.7.2.

Given a basic feasible solution $\mathbf{W}_B = \mathbf{B}^{-1}\mathbf{Y}$ to the set of constraints $\mathbf{AW} = \mathbf{Y}$ for the problem 2.7.2 with the value of the objective function for this solution being $Z = \mathbf{C}_B\mathbf{W}_B$, for any column $\mathbf{a}_j$ in $\mathbf{A}$, but not in $\mathbf{B}$, the condition $C_j - Z_j < 0$ holds. If at least one $\alpha_{ij} > 0$, $i \in R_B$, then it is possible to obtain a new basic feasible solution by replacing one of the columns of $\mathbf{B}$ by $\mathbf{a}_j$, and the new value of the objective function $\hat{Z}$ satisfies $\hat{Z} \leqslant Z$. On the other hand, if the condition $C_j - Z_j > 0$ is satisfied and if $\mathbf{a}_j$ corresponds to an unrestricted variable W_j, with at least one $\alpha_{ij} < 0$, $i \in R_B$, then it is possible to obtain a new basic feasible solution by replacing one of the columns of $\mathbf{B}$ by $\mathbf{a}_j$, and the new value of the objective function $\hat{Z}$ satisfies $\hat{Z} \leqslant Z$.

If we continue this process of exchanging one vector at a time in the basis we go through a sequence of bases, improving the objective function at each step if possible. We will have strict improvement if

$$\frac{W_{B_r}}{\alpha_{rj}}\left[C_j - Z_j\right] < 0,$$

i.e., if W_{B_r}/α_{rj} is not zero whenever $C_j - Z_j$ is nonzero. But $\alpha_{rj} \neq 0$. Therefore, if no $W_{B_i} = 0$, $i \in R_B$, we can have strict improvement in the objective function. Thus, if all the restricted variables in the basis are at positive level, in each step we improve the objective function. As there are only a finite number of bases of this kind, we come to a stage where no more improvement is possible. We next characterize the conditions for optimality in such an eventuality.

The process of basis changing comes to an end when the following case arises: $C_j - Z_j \geqslant 0$ for all positively restricted nonbasic variables W_j, and $C_j - Z_j = 0$ for all nonbasic unrestricted variables W_j.

Of course, for (2.7.2) it is not possible to end up in a case in which we have either (1) a nonbasic unrestricted variable with $C_j - Z_j > 0$ (< 0) and none of the α_{ij}'s, $i \in R_B$ being < 0 (> 0), or (2) a nonbasic restricted variable with $C_j - Z_j < 0$ and $\alpha_{ij} \leqslant 0$ for all $i \in R_B$.

Suppose we have Case (1) above. The implication is that we have an $\mathbf{a}_j$ corresponding to an unrestricted W_j with $C_j - Z_j > 0$, and that none of the α_{ij}, $i \in R_B$ is less than zero.

Consider

$$\sum_{i=1}^{n} W_{B_i} \mathbf{B}_i = \mathbf{Y}$$

with $Z = \mathbf{C}_B \mathbf{W}_B$. Suppose we add and subtract $\theta \mathbf{a}_j$, θ any scalar, to obtain

$$\sum_{i=1}^{n} W_{B_i} \mathbf{B}_i + \theta \mathbf{a}_j - \theta \mathbf{a}_j = \mathbf{Y}$$

but

$$-\theta \mathbf{a}_j = -\theta \sum_{i=1}^{n} \alpha_{ij} \mathbf{B}_i .$$

Substituting this in the previous equation we obtain.

$$\sum_{i=1}^{n} (W_{B_i} - \theta \alpha_{ij}) \mathbf{B}_i + \theta \mathbf{a}_j = \mathbf{Y}$$

when $\theta < 0$; then $W_{B_i} - \theta \alpha_{ij} \geqslant 0$, since $\alpha_{ij} \geqslant 0$, $i \in R_B$; therefore $(W_{B_i} - \theta \alpha_{ij})$, $i = 1, \ldots, n$ and θ is a feasible solution to the problem 2.7.2 in which $n+1$ variables can be nonzero.

Look at the objective function value of this solution.

$$\hat{Z} = \sum_{i=1}^{n} C_{B_i}(W_{B_i} - \theta \alpha_{ij}) + C_j \theta = Z + \theta (C_j - Z_j).$$

But $C_j - Z_j > 0$ and $\theta < 0$, therefore, choosing θ arbitrarily small, $\hat{Z}$ can be made arbitrarily small. This contradicts the fact that $\mathbf{C'W}$ is bounded below by zero. Hence this case cannot arise. Similarly we can prove the impossibility of Case 2 above.

However for a problem of this type, in general with arbitrary $\mathbf{A}$ and $\mathbf{Y}$, such eventualities are possible as the objective function may not be bounded from below.

We are now in a position to describe the algorithm based on the above discussions.

Algorithm 2.8.1

Step 1 As mentioned in Remark 2.7.3, start with the basic feasible solution.

$$W_{p+r}=\begin{cases} Y_r & \text{if } Y_r>0 \\ 0 & \text{otherwise}, \quad 1\leqslant r\leqslant n \end{cases}$$

$$W_{p+n+r}=\begin{cases} -Y_r & \text{if } Y_r<0 \\ 0 & \text{otherwise}, \quad 1\leqslant r\leqslant n \end{cases}$$

and

$$W_j=0, \qquad j=1,\ldots,p.$$

We present the corresponding α_{ij}'s in the form of a table. Notice that the basis $\mathbf{B}$ corresponding to this solution contains exactly one of the columns $\mathbf{e}_r$ or $-\mathbf{e}_r$ for all $r=1,\ldots,n$, and the inverse of $\mathbf{B}$ is itself—i.e., $\mathbf{B}^{-1}=\mathbf{B}$. Hence $\boldsymbol{\alpha}_j=\mathbf{B}^{-1}\mathbf{a}_j$ is obtained by multiplying the rth row of $\mathbf{A}$ by $+1$ if W_{p+r} is in the basis and by -1 if W_{p+n+r} is in the basis, and $\mathbf{B}^{-1}\mathbf{Y}=\mathbf{W}_B$ gives the corresponding basic feasible solution. We also note down the vectors in the basis and their C_j's. The first table thus obtained is Table 2.8.1.

C_j-Z_j is obtained by subtracting $\Sigma_{i=1}^{n}C_{B_i}\alpha_{ij}$ from C_j for each j. Notice that $C_j-Z_j=0$ for the $\mathbf{a}_j$'s in the basis.

Step 2 Choose an $\mathbf{a}_j$ not in the basis, for exchanging with a vector in the basis as follows:

(a) Let

$$C_{j_1}-Z_{j_1}=\max_{\substack{k \\ C_k-Z_k>0 \\ W_k \text{ unrestricted variable}}} C_k-Z_k$$

Table 2.8.1

$\mathbf{C}_B$	Vector in the Basis	$\mathbf{W}_B$	$\boldsymbol{\alpha}_1$	$\boldsymbol{\alpha}_2$	$\cdots$	$\boldsymbol{\alpha}_j$	$\cdots$	$\boldsymbol{\alpha}_{2n+p}$
1	$\mathbf{a}_{p+1}$ or $\mathbf{a}_{p+n+1}$	$\lvert Y_1\rvert$	α_{11}	α_{12}	$\cdots$	α_{1j}	$\cdots$	$\alpha_{1,2n+p}$
$\vdots$	$\vdots$	$\vdots$	$\vdots$	$\vdots$		$\vdots$		$\vdots$
1	$\mathbf{a}_{p+n}$ or $\mathbf{a}_{p+2n}$	$\lvert Y_n\rvert$	α_{n1}	α_{n2}	$\cdots$	α_{nj}	$\cdots$	$\alpha_{n,2n+p}$
	C_k-Z_k	$Z=\sum_{l=1}^{n}\lvert Y_l\rvert$				C_j-Z_j		

(b) Let

$$|C_{j_2}-Z_{j_2}|=\max_{\substack{k\\ C_k-Z_k<0}}|C_k-Z_k|.$$

Choose j, as

$$|C_j-Z_j|=\max\left[C_{j_1}-Z_{j_1},|C_{j_2}-Z_{j_2}|\right].$$

This step gives the criterion for choosing the vector to enter the basis. If both j_1 and j_2 cannot be found go to Step 5; otherwise, go to Step 3.

Step 3 If $C_j-Z_j>0$, choose r as follows:

$$\frac{W_{B_r}}{\alpha_{rj}}=\max_{i\in R_B}\left[\frac{W_{B_i}}{\alpha_{ij}},\alpha_{ij}<0\right].$$

If $C_j-Z_j<0$, choose r as follows:

$$\frac{W_{B_r}}{\alpha_{rj}}=\min_{i\in R_B}\left[\frac{W_{B_i}}{\alpha_{ij}},\alpha_{ij}>0\right]$$

This step gives the criterion for choosing the vector to leave the basis. Go to Step 4.

Step 4 Form the new table corresponding to the new basis **B** as follows:

$$\hat{W}_{B_r}=\frac{W_{B_r}}{\alpha_{rj}}$$

$$\hat{\alpha}_{rl}=\frac{\alpha_{rl}}{\alpha_{rj}}$$

and

$$\hat{W}_{B_i}=W_{B_i}-\frac{W_{B_r}}{\alpha_{rj}}\alpha_{ij},\qquad i\neq r,\qquad i=1,\ldots,n$$

$$\hat{\alpha}_{il}=\alpha_{il}-\frac{\alpha_{ij}\cdot\alpha_{rl}}{\alpha_{rj}},\qquad i\neq r,\qquad i=1,\ldots,n,\qquad l=1,\ldots,p+2n$$

$$\hat{Z}=Z+(C_j-Z_j)\cdot\frac{W_{B_r}}{\alpha_{rj}}$$

$$\hat{C}_l-\hat{Z}_l=(C_l-Z_l)-(C_j-Z_j)\cdot\frac{\alpha_{rl}}{\alpha_{rj}},\qquad l=1,\ldots,p+2n.$$

This transformation is called the ring-around-rosy method. Return to Step 2 and continue with $\mathbf{B}=\hat{\mathbf{B}}$.

Step 5 Stop. The current basis is optimal. $W_1,\ldots,W_p$ give the parameters $\beta_0,\ldots,\beta_{p-1}$, and $W_{p+r}+W_{p+n+r}$ gives the absolute error in the rth observation. Notice that if we wish to use this algorithm in general for solving a problem whose objective function is not bounded from below, we may end up in showing that the problem has no finite minimum. This will be indicated by the inability to find an r in Step 3.

We illustrate Algorithm 2.8.1 by solving MINMAD regression problem, with the data given in the following example.

EXAMPLE 2.8.1 For the given data, find a MINMAD regression equation $\hat{Y}=\hat{\beta}_0+\hat{\beta}_1X_1+\hat{\beta}_2X_2$ estimating $\beta_0, \beta_1, \beta_2$ by applying Algorithm 2.8.1.

$$\mathbf{X}=\begin{bmatrix}1 & 1 & 3\\ 1 & 2 & 2\\ 1 & 3 & 1\\ 1 & 4 & 2\\ 1 & 5 & 3\end{bmatrix}, \qquad \mathbf{Y}=\begin{bmatrix}4\\ 3\\ 4\\ 5\\ 5\end{bmatrix}.$$

Step 1 We find that the initial basis is formed by the columns corresponding to $d_{11},\ldots,d_{15}$. We have the initial table as given by Table 2.8.2.

Step 2 C_k-Z_k for $k=1,2,3$ are negative, and so Step 2a does not yield any j_1. We go to Step 2b.

$$|C_2-Z_2|=15=\max[|-5|,|-15|,|-11|];$$

hence, $j=2$ and $C_j-Z_j<0$.

Step 3 We choose $r=5$ by forming the ratios

$$W_{B_i}/\alpha_{i2}, \alpha_{i2}>0 \text{ and finding}$$

$$W_{B_5}/\alpha_{52}=5/5=\min[4/1,3/2,4/3,5/4,5/5].$$

Thus $\mathbf{a}_2$ replaces $\mathbf{a}_8$ in the basis. The new table is achieved as per Step 4. And we have Table 2.8.3. For instance $\hat{\alpha}_{23}=2-2\times 3/5=4/5$. Going back to Step 2, we find

$$|C_1-Z_1|=\max[|-2|,|-2|,|-1|].$$

Therefore $j=1$ and $C_j-Z_j<0$.

Table 2.8.2

C_B	Vectors in the Basis	$\mathbf{W}_B$	α_1	α_2	α_3	α_4	α_5	α_6	α_7	α_8	α_9	α_{10}	α_{11}	α_{12}	α_{13}
1	$\mathbf{a}_4$	4	1	1	3	1	0	0	0	0	−1	0	0	0	0
1	$\mathbf{a}_5$	3	1	2	2	0	1	0	0	0	0	−1	0	0	0
1	$\mathbf{a}_6$	4	1	3	1	0	0	1	0	0	0	0	−1	0	0
1	$\mathbf{a}_7$	5	1	4	2	0	0	0	1	0	0	0	0	−1	0
1	$\mathbf{a}_8$	5	1	5^a	3	0	0	0	0	1	0	0	0	0	−1
	$C_k - Z_k$	$Z=21$	−5	−15	−11	0	0	0	0	0	2	2	2	2	2

[a]The pivotal element α_{rj}.

In Step 3, we find $r=2$, as $W_{B_2}/\alpha_{21}=5/3=\min[15/4, 5/3, 5/2, 5/1, 5/1]$. Thus $\mathbf{a}_1$ replaces $\mathbf{a}_5$ in the basis. The new table is obtained as per Step 4. We have Table 2.8.4. Going back to Step 2, from step 2a, we find $j_1=3$ as $C_3-Z_3=2/3=\max C_k-Z_k$, over $C_k-Z_k>0, W_k$ unrestricted in sign, and Step 2b yields $j_2=10$ and $|C_{10}-Z_{10}|=4/3$. Thus $j=10$, as $\max[2/3,4/3]=4/3$ and $C_{10}-Z_{10}<0$. In Step 3, we find $r=3$, as $W_{B_3}/\alpha_{310}=1/2=\min[5/4,1/2,2/2]$. Thus $\mathbf{a}_{10}$ replaces $\mathbf{a}_6$ in the basis. The new table is obtained as per Step 4 (Table 2.8.5.)

Now we find $j=3$ in Step 2, as $\mathbf{a}_3$ is the only eligible vector to enter the basis. Step 3 yields $r=1$. Replacing $\mathbf{a}_4$ by $\mathbf{a}_3$ we get Table 2.8.6. In Table 2.8.6 we find there is no j eligible for entry. So we go to Step 5 and we stop.

The estimates of β_0, β_1, and β_2 are read off from the Table 2.8.6, corresponding to the basic rows, 2, 5, and 1, respectively. We have:

$$\hat{Y}=3+\tfrac{1}{4}X_1+\tfrac{1}{4}X_2.$$

Also notice that Observations 4 and 5 lie on this MINMAD regression hyperplane. From the tables we observe the following:

1 The $(C_{p+i}-Z_{p+i})+(C_{p+n+1}-Z_{p+n+1})$ always remains 2, for $i=1,2,\ldots,n$. This result is usual, not peculiar to this example. For, given any basis $\mathbf{B}$,

$$\mathbf{B}^{-1}\mathbf{a}_{p+i}=-\mathbf{B}^{-1}\mathbf{a}_{p+n+i} \qquad \text{as } \mathbf{a}_{p+i}=\mathbf{e}_i=-\mathbf{a}_{p+n+i}$$

that is

$$\boldsymbol{\alpha}_{p+i}=-\boldsymbol{\alpha}_{p+n+i}.$$

(This result is also seen in Tables 2.8.1 through 2.8.5.)

Therefore

$$C_{p+i}-Z_{p+i}=1-\mathbf{C}_B\boldsymbol{\alpha}_{p+i}=1+\mathbf{C}_B\boldsymbol{\alpha}_{p+n+i}.$$

Thus

$$(C_{p+i}-Z_{p+i})+(C_{p+n+i}-Z_{p+n+i})=1+\mathbf{C}_B\boldsymbol{\alpha}_{p+n+i}+1-\mathbf{C}_B\boldsymbol{\alpha}_{p+n+i}=2.$$

If we know $C_{p+i}-Z_{p+i}$ or $C_{p+n+i}-Z_{p+n+i}$ we can find the other without computing it by using Step 4, while using the algorithm.

2 If $\mathbf{a}_{p+i}$ is in the basis then $\mathbf{B}^{-1}\mathbf{a}_{p+i}=\boldsymbol{\alpha}_{p+i}=\mathbf{e}_i$. Therefore $\boldsymbol{\alpha}_{p+n+i}=-\mathbf{e}_i$. Hence we need not compute these two columns in any of the tables,

Table 2.8.3

$\mathbf{C}_B$	Vectors in the Basis	$\mathbf{W}_B$	α_1	α_2	α_3	α_4	α_5	α_6	α_7	α_8	α_9	α_{10}	α_{11}	α_{12}	α_{13}
1	$\mathbf{a}_4$	3	4/5	0	12/5	1	0	0	0	−1/5	−1	0	0	0	1/5
1	$\mathbf{a}_5$	1	3/5[a]	0	4/5	0	1	0	0	−2/5	0	−1	0	0	2/5
1	$\mathbf{a}_6$	1	2/5	0	−4/5	0	0	1	0	−3/5	0	0	−1	0	3/5
1	$\mathbf{a}_7$	1	1/5	0	−2/5	0	0	0	1	−4/5	0	0	0	−1	4/5
0	$\mathbf{a}_2$	1	1/5	1	3/5	0	0	0	0	1/5	0	0	0	0	−1/5
	$C_k - Z_k$	6	−2	0	−2	0	0	0	0	3	2	2	2	2	−1

[a]The pivotal element α_{rj}.

Table 2.8.4

$\mathbf{C}_B$	Vectors in the Basis	$\mathbf{W}_B$	α_1	α_2	α_3	α_4	α_5	α_6	α_7	α_8	α_9	α_{10}	α_{11}	α_{12}	α_{13}
1	$\mathbf{a}_4$	5/3	0	0	4/3	1	−4/3	0	0	1/3	−1	4/3	0	0	−1/3
0	$\mathbf{a}_1$	5/3	1	0	4/3	0	5/3	0	0	−2/3	0	−5/3	0	0	2/3
1	$\mathbf{a}_6$	1/3	0	0	−4/3	0	−2/3	1	0	−1/3	0	2/3[a]	−1	0	1/3
1	$\mathbf{a}_7$	2/3	0	0	−2/3	0	−1/3	0	1	−2/3	0	1/3	0	−1	2/3
0	$\mathbf{a}_2$	3/3	0	1	1/3	0	−1/3	0	0	1/3	0	1/3	0	0	−1/3
	$C_k - Z_k$	8/3	0	0	2/3	0	10/3	0	0	5/3	2	−4/3	2	2	1/3

[a]The pivotal element α_{rj}.

Table 2.8.5

C_B	Vectors in the Basis	W_B	α_1	α_2	α_3	α_4	α_5	α_6	α_7	α_8	α_9	α_{10}	α_{11}	α_{12}	α_{13}
1	a_4	1	0	0	4^a	1	0	−2	0	1	−1	0	2	0	−1
0	a_1	5/2	1	0	−2	0	0	5/2	0	−3/2	0	0	−5/2	0	3/2
1	a_{10}	1/2	0	0	−2	0	−1	3/2	0	−1/2	0	1	−3/2	0	1/2
1	a_7	1/2	0	0	0	0	0	−1/2	1	−1/2	0	0	1/2	−1	1/2
0	a_2	1/2	0	1	1	0	0	−1/2	0	1/2	0	0	1/2	0	−1/2
	$C_k - Z_k$	2	0	0	−2	0	2	2	0	1	2	0	0	2	1

[a]The pivotal element α_{rj}.

Table 2.8.6

C_B	Vectors in the Basis	W_B	α_1	α_2	α_3	α_4	α_5	α_6	α_7	α_8	α_9	α_{10}	α_{11}	α_{12}	α_{13}
0	a_3	1/4	0	0	1	1/4	0	−1/2	0	1/4	−1/4	0	1/2	0	−1/4
0	a_1	3	1	0	0	1/2	0	3/2	0	−1	−1/2	0	−3/2	0	1
1	a_{10}	1	0	0	0	1/2	−1	1/2	0	0	−1/2	1	−1/2	0	0
1	a_7	1/2	0	0	0	0	0	−1/2	1	−1/2	0	0	1/2	−1	1/2
0	a_2	1/4	0	1	0	−1/4	0	0	0	1/4	1/4	0	0	0	−1/4
	$C_k - Z_k$	3/2	0	0	0	1/2	2	1	0	3/2	3/2	0	1	2	1/2

as $\boldsymbol{\alpha}_{p+i}$ ($\boldsymbol{\alpha}_{p+n+i}$) is either $\mathbf{e}_i$ or $-\mathbf{e}_i$, depending on whether $\mathbf{a}_{p+i}$ ($\mathbf{a}_{p+n+i}$) is in the basis or outside the basis. In any case both cannot be in the basis, as they are linearly dependent.

3 If both $\mathbf{a}_{p+i}$ and $\mathbf{a}_{p+n+i}$ are nonbasic columns, then it is enough to have $\boldsymbol{\alpha}_{p+i}$ or $\boldsymbol{\alpha}_{p+n+i}$ computed.

If we replace a basic $\mathbf{a}_{p+i}$ ($\mathbf{a}_{p+n+i}$) by $\mathbf{a}_j$, in the subsequent table we need only the columns $\hat{\boldsymbol{\alpha}}_k$'s for the nonbasic k's, such that k corresponds to β_k, or k corresponds to a d_{1i} or d_{2i} when both are nonbasic. Hence at each iteration, we can store in the column entering the basis, the $\boldsymbol{\alpha}_{p+r}$ or $\boldsymbol{\alpha}_{p+n+r}$, depending on whether d_{1r} or d_{2r} is leaving the basis. We have

$$\hat{\boldsymbol{\alpha}}_{p+r} \quad \text{or} \quad \hat{\boldsymbol{\alpha}}_{p+n+r} = \begin{bmatrix} -\alpha_{1j}/\alpha_{rj} \\ \vdots \\ 1/\alpha_{rj} \\ \vdots \\ -\alpha_{nj}/\alpha_{rj} \end{bmatrix}$$

Therefore we can conclude that we can suppress $2n$ columns out of the total $2n+p$ columns in each table corresponding to the $\boldsymbol{\alpha}_k$'s—that is, the columns corresponding to basic $\boldsymbol{\alpha}_k$'s (say s in number, $s \leqslant p$); both d_{1i} and d_{2i}, if one of them is basic ($2[n-s]$ in number); and either d_{1i} or d_{2i}, if both are nonbasic (s in number). (The number of columns suppressed therefore is $s+2n-2s+s=2n$.) This conclusion implies tremendous computational improvement.

Further we observe that whenever we replace a basic d_{1r} or d_{2r} by d_{2r} or d_{1r} (which is nonbasic), respectively, the only changes are that the ith row of the basis corresponding to d_{1r} or d_{2r} is multiplied by -1, and the $C_k - Z_k$ row of the table is transformed as follows:

$$\hat{C}_k - \hat{Z}_k = (C_k - Z_k) + 2\cdot\alpha_{rk}$$

and

$$\hat{Z} = Z - 2W_{B_r}.$$

At this point we introduce two heuristics which have been found to reduce the number of iterations required to obtain an optimal solution while using the simplex procedure.

1 In Step 2 of the algorithm, we consider only nonbasic β_j's for entry in the first few iterations. When no more nonbasic β_j can replace basic d_{1i} or d_{2i}, we consider other nonbasic variables for entry.

2 Corresponding to a variable β_j to enter the basis we decrease or increase its present value depending on $C_j - Z_j > 0$ or $C_j - Z_j < 0$.

We know that if β_j is decreased or increased (as the case may be) beyond W_{B_r}/α_{rj} some of the basic variables restricted in sign will turn out to be negative. (So in the algorithm, we do not go beyond $W_{\hat{B}_r} = W_{B_r}/\alpha_{rj}$— see Step 3.)

From the last observation, if a d_{1r} or d_{2r} becomes negative, the corresponding d_{2r} or d_{1r}, respectively, can be brought into the basis and the corresponding variable can be made positive; this would affect the $C_k - Z_k$ row of the table and we would multiply the rth row by -1.

From what we have discussed so far, we can modify the vector to leave the basis as follows: Choose r as per Step 3 of the algorithm. Now compute $\hat{C}_j - \hat{Z}_j = C_j - Z_j + 2\alpha_{rj}$. If $C_j - Z_j$ and $\hat{C}_j - \hat{Z}_j$ are of opposite signs, we introduce the corresponding variable and remove $\mathbf{B}_r$ from the basis. Otherwise, we change all $C_k - Z_k$ to $C_k - Z_k + 2\alpha_{rk}$ and Z to $Z - 2W_{B_r}$. And we replace the vector $\mathbf{a}_{p+r}$ or $\mathbf{a}_{p+n+r}$ in the basis, in Row r, by $\mathbf{a}_{p+n+r}$ or $\mathbf{a}_{p+r}$, respectively. That is, the basic variable d_{1r} or d_{2r} is replaced by d_{2r} or d_{1r}, respectively. Multiply the rth row by -1. Discard $\mathbf{B}_r$ from consideration for removal. Find the next possible vector for removal as per Step 3, and repeat these steps. We do the necessary transformations in the table to get the next table. The termination of the process is as in the algorithm.

An example illustrates these modifications in the algorithm.

EXAMPLE 2.8.2 Find the MINMAD regression line for the data given below:

$$\mathbf{X} = \begin{bmatrix} 1 & 1 \\ 1 & 2 \\ 1 & 3 \\ 1 & 4 \\ 1 & 5 \end{bmatrix}, \qquad \mathbf{Y} = \begin{bmatrix} 1 \\ 1 \\ 2 \\ 3 \\ 2 \end{bmatrix}.$$

We have the initial table without suppressing any column (Table 2.8.7.)

We choose as per Step 2 of the algorithm β_2 to become basic. We find $r=5$, with $W_{B_5}/\alpha_{52} = \min[W_{B_i}/\alpha_{i2}, \alpha_{i2} > 0] = \frac{2}{5}$. The table after suppression

Table 2.8.7

C_B	Vectors in the Basis	W_B	α_1	α_2	α_3	α_4	α_5	α_6	α_7	α_8	α_9	α_{10}	α_{11}	α_{12}
1	$\mathbf{a}_3$	1	1	1	1	0	0	0	0	-1	0	0	0	0
1	$\mathbf{a}_4$	1	1	2	0	1	0	0	0	0	-1	0	0	0
1	$\mathbf{a}_5$	2	1	3	0	0	1	0	0	0	0	-1	0	0
1	$\mathbf{a}_6$	3	1	4	0	0	0	1	0	0	0	0	-1	0
1	$\mathbf{a}_7$	2	1	5	0	0	0	0	1	0	0	0	0	-1
	$C_k - Z_k$	9	-5	-15	0	0	0	0	0	2	2	2	2	2

becomes

C_B	Vectors in the Basis	W_B	α_1	α_2
1	$\mathbf{a}_3$	1	1	1
1	$\mathbf{a}_4$	1	1	2
1	$\mathbf{a}_5$	2	1	3
1	$\mathbf{a}_6$	3	1	4
1	$\mathbf{a}_7$	2	1	5
	$C_k - Z_k$	9	-5	-15

Note that $\mathbf{a}_7$ has to be replaced by $\mathbf{a}_2$, as per the algorithm. As per the modification in Step 3, we calculate

$$\hat{C}_2 - \hat{Z}_2 = -15 + 2 \times 5 = -5.$$

We consider increasing β_2 beyond $\frac{2}{5}$. Replace $\mathbf{a}_7$ by $\mathbf{a}_{12}$, multiply Row 5 by -1, add $2\alpha_{52}$ to all $C_k - Z_k$, and subtract $2W_{B_5}$ from Z. This gives the new last row as

$C_k - Z_k$	$Z=5$	-3	-5

Next, excluding Row 5, we consider an r for removal. As per Step 3 we get $r=2$ and $W_{B_2}/\alpha_{22} = \frac{1}{2}$. Calculate

$$\hat{C}_2 - \hat{Z}_2 = -5 + 2 \times 2 = -1.$$

We consider increasing β_2 beyond $\frac{1}{2}$. Replace $\mathbf{a}_4$ by $\mathbf{a}_9$, multiply Row 2 by

-1, and change the last row. We get the new last row as

$C_k - Z_k$	$Z=3$	-1	-1

The next r turns out to be $r=3$ and $W_{B_r}/\alpha_{r2}=\frac{2}{3}$. Calculate

$$\hat{C}_2 - \hat{Z}_2 = -1 + 2\times 3 = 5.$$

Hence we find that $\hat{C}_2 - \hat{Z}_2$ has become positive. This means that we introduce $\mathbf{a}_2$ and remove $\mathbf{a}_5$ from the basis. The transformed table (Table 2.8.8) is given below.

Notice that in the new table, new $\boldsymbol{\alpha}_5$ replaces old $\boldsymbol{\alpha}_2$. In the next iteration, we choose β_1 to enter the basis. $C_1 - Z_1 = -\frac{2}{3} < 0$, so we find in Step 3 that $r=1, W_{B_1}/\alpha_{11} = \frac{1}{2}$. Calculate

$$\hat{C}_1 - \hat{Z}_1 = -\tfrac{2}{3} + 2\times\tfrac{2}{3} = \tfrac{2}{3}.$$

Hence we introduce $\mathbf{a}_1$ and remove $\mathbf{a}_3$ from the basis. The next table after transformation is Table 2.8.9.

Now all the $C_j - Z_j$'s for the nonbasic variables are positive, including the suppressed ones. For the nonbasic $\mathbf{a}_4$, $\mathbf{a}_{11}$, and $\mathbf{a}_7$ the corresponding $C_k - Z_k$'s are 2's, and those of the nonbasic $\mathbf{a}_8$ and $\mathbf{a}_{10}$ are 1 and 2, respectively. And so we terminate the process, as the optimality conditions

Table 2.8.8

C_B	Vectors in the Basis	$\mathbf{W}_B$	$\boldsymbol{\alpha}_1$	$\boldsymbol{\alpha}_5$
1	$\mathbf{a}_3$	$\frac{1}{3}$	$\frac{2}{3}$	$-\frac{1}{3}$
1	$\mathbf{a}_9$	$\frac{1}{3}$	$-\frac{1}{3}$	$\frac{2}{3}$
0	$\mathbf{a}_2$	$\frac{2}{3}$	$\frac{1}{3}$	$\frac{1}{3}$
1	$\mathbf{a}_6$	$\frac{1}{3}$	$-\frac{1}{3}$	$-\frac{4}{3}$
1	$\mathbf{a}_{12}$	$\frac{4}{3}$	$\frac{2}{3}$	$\frac{5}{3}$
	$C_k - Z_k$	$Z=\frac{7}{3}$	$-\frac{2}{3}$	$\frac{1}{3}$

Table 2.8.9

C_B	Vectors in the Basis	W_B	α_3	α_5
0	$\mathbf{a}_1$	$\frac{1}{2}$	$\frac{3}{2}$	$-\frac{1}{2}$
1	$\mathbf{a}_9$	$\frac{1}{2}$	$\frac{1}{2}$	$\frac{1}{2}$
0	$\mathbf{a}_2$	$\frac{1}{2}$	$-\frac{1}{2}$	$\frac{1}{2}$
1	$\mathbf{a}_6$	$\frac{1}{2}$	$\frac{1}{2}$	$-\frac{3}{2}$
1	$\mathbf{a}_{12}$	1	-1	2
	$C_k - Z_k$	$Z=2$	1	0

are satisfied as per Step 5. The MINMAD regression line obtained is

$$\hat{Y} = \tfrac{1}{2} + \tfrac{1}{2}X$$

This example illustrates the modifications in the simplex procedure and also the obvious reduction in the regular simplex computations while transforming one table to the other. The modified algorithm is known as Barrodale and Roberts' algorithm.

2.9 GENERAL LINEAR PROGRAMMING PROBLEM

In general, problems similar to (2.7.2) in form can be stated as follows:

$$\begin{aligned} \text{Minimize} \quad & \mathbf{C}'\mathbf{W} \\ \text{Subject to} \quad & \mathbf{A}^1\mathbf{W} \leqslant \mathbf{b}^1 \\ & \mathbf{A}^2\mathbf{W} \geqslant \mathbf{b}^2 \\ & \mathbf{A}^3\mathbf{W} = \mathbf{b}^3 \end{aligned} \tag{2.9.1}$$

and $W_1,\ldots,W_p$ unrestricted in sign; $W_{p+1} \geqslant 0,\ldots,W_n \geqslant 0$. Where $\mathbf{A}^k$ is an $r_k \times n$ matrix, $k = 1,2,3$, and

$$\mathbf{A} = \begin{bmatrix} \mathbf{A}^1 \\ \mathbf{A}^2 \\ \mathbf{A}^3 \end{bmatrix}$$

is an $m \times n$ matrix, where $m = r_1 + r_2 + r_3$. Similarly, $\mathbf{b}^k$ is an $r_k \times 1$ vector and $\mathbf{C}'$ is an $n \times 1$ vector. Without loss of generality we can assume $\mathbf{b}^k \geqslant \mathbf{0}$. Such problems are known as linear-programming problems.

For such general problems we may not have any feasible solutions ($\mathcal{F} = \varnothing$). Even if $\mathcal{F} \neq \varnothing$, we may not have an optimal solution, as in general $\mathbf{C}'\mathbf{W}$ need not be bounded below. Therefore, we first try to find whether $\mathcal{F} = \varnothing$ or not, by transforming the problem to an equivalent problem in which all constraints are equalities.

Corresponding to the equalities we add what are known as *artificial* variables.

$$\sum_{j=1}^{n} a_{ij} W_j + W_{a_i} = b_i$$

$$W_{a_i} \geqslant 0 \qquad \text{with} \quad C_{a_i} = M > 0$$

where M is a large positive constant.

The inequality constraints $\mathbf{A}^1\mathbf{W} \leqslant \mathbf{b}^1$ are converted to equality constraints by adding a nonnegative variable W_{s_i} to the left-hand side of each of these constraints. Similarly the inequality constraints $\mathbf{A}^2\mathbf{W} \geqslant \mathbf{b}^2$ are converted to equality constraints by subtracting a nonnegative variable W_{s_i} from the left-hand side of each of these constraints. These variables are called *slack* and *surplus* variables, respectively. The objective function coefficients for these variables are zero.

Now we have the problem:

$$\text{Minimize} \qquad \mathbf{C}'\mathbf{W} + M \sum_{i=1}^{r_3} W_{a_i}$$

$$\text{subject to} \begin{bmatrix} \mathbf{A}^1 & \mathbf{I} & \mathbf{0} & \mathbf{0} \\ \mathbf{A}^2 & \mathbf{0} & -\mathbf{I} & \mathbf{0} \\ \mathbf{A}^3 & \mathbf{0} & \mathbf{0} & \mathbf{I} \end{bmatrix} \begin{bmatrix} \mathbf{W} \\ \mathbf{W}_s \\ \mathbf{W}_a \end{bmatrix} = \mathbf{b}$$

$$W_i \qquad \text{unrestricted in sign} \quad i = 1, \ldots, p$$

$$W_i \geqslant 0, \qquad i = p+1, \ldots, n$$

$$W_{s_i} \geqslant 0, \qquad W_{a_i} \geqslant 0 \qquad (2.9.2)$$

where $\mathbf{W}_s$ is an $(r_1+r_2)\times 1$ vector giving the slack and surplus variables and $\mathbf{W}_a$ in an $r_3\times 1$ vector.

This problem now can be solved using Algorithm 2.8.1. In this problem we have a basis formed by the identity matrix corresponding to slack and surplus and artificial variables. This serves as an initial basis **B**. At the end of the simplex process there are three possibilities:

1 Optimality conditions in the algorithm are satisfied and some $W_{a_i}>0$ in the optimal table.

2 Optimality conditions in the algorithm are satisfied and none of the $W_{a_i}>0$ in the optimal table.

3 We are not able to find an r satisfying the conditions of Step 3.

In Case 1 we have no feasible solution to the original problem. If $\mathcal{F}\neq\varnothing$, then the objective function value of a $\mathbf{W}\in\mathcal{F}$ will be $\mathbf{C}'\mathbf{W}$ and all $W_{a_i}=0$. And this is also true for the enlarged problem. This finding leads to a contradiction as some of the W_{a_i}'s are >0 with coefficient M, very large positive, in the optimal solution found for the enlarged problem.

In Case 2 we have an optimal solution to the problem, given by the last table's W_{B_i}'s. If some of the artificial variables in the basis cannot be replaced by the original variables or slack and surplus variables, the implication is that the corresponding row of the matrix **A** is redundant, and the rank of **A** is less than m.

Case 3 indicates that there is no finite minimum to the problem (see the note following Algorithm 2.8.1).

This method of solving a linear-programming problem is called the Charnes M-method. Another method for a linear-programming problem is to solve the problem in two phases. In Phase 1 we consider the problem of minimizing $\Sigma_{i=1}^{r_3}W_{a_i}$ subject to the restriction of Problem 2.9.2. As all the $W_{a_i}\geqslant 0$, the minimum of $\Sigma W_{a_i}=0$. If $\Sigma W_{a_i}>0$, there is no feasible solution to the original problem.

At the end of Phase 1, if $W_{a_i}=0$ for all artificial variables and if some of them are in the basis at zero level, $\mathcal{F}\neq\varnothing$. However, there may be redundancy in the rows of A. If $W_{a_i}=0$ for all artificial variables and all artificial variables are nonbasic, we have $\mathcal{F}\neq\varnothing$ and there is no redundancy, and rank of $\mathbf{A}=m$.

Once the objective function value in Phase 1 is zero, we go to Phase 2 with the original objective function of Problem 2.9.1, namely, $\mathbf{C}'\mathbf{W}$, and proceed with the simplex exchanges. Either we get an optimal solution to

Problem 2.9.1 or we have an indication that there is no finite optimum for the problem. This method is known as the two-phase method.

Another version of the method arises from the observation that if we have the inverse of the present basis at any of the iterations we can compute the α_{ij}'s for the vector that is selected for entry. So there is no need to compute all the α_{ij}'s. This way we can reduce a considerable amount of the computational burden. The revised simplex method makes use of this observation.

2.10 DUALITY IN MINMAD REGRESSION

Given the problem,

$$\begin{aligned} &\text{Minimize} && \mathbf{C}'\mathbf{W} \\ &\text{subject to} && \mathbf{AW}=\mathbf{Y} \\ &&& W_{p+r} \geqslant 0, \quad r=1,\ldots,2n \\ &&& W_1,\ldots,W_p \quad \text{unrestricted in sign,} \end{aligned}$$

we define the *dual* of this problem to be

$$\text{Maximize } \mathbf{Y}'\boldsymbol{\lambda}$$

$$\text{subject to} \qquad \mathbf{A}'\boldsymbol{\lambda}=\left\{\begin{bmatrix}\mathbf{X}'\\ \mathbf{I}\\ -\mathbf{I}\end{bmatrix}\boldsymbol{\lambda}\right\} \begin{matrix}=\\ \leqslant\\ \leqslant\end{matrix} \left\{\mathbf{C}=\begin{bmatrix}\mathbf{0}\\ \mathbf{e}'\\ \mathbf{e}'\end{bmatrix}\right.$$

$$\boldsymbol{\lambda} \text{ unrestricted in sign,} \tag{2.10.1}$$

where $\mathbf{0}$ is a $p\times 1$ vector, $\mathbf{e}$ is a $1\times n$ vector $(1,\ldots,1)$, and $\boldsymbol{\lambda}$ is an $n\times 1$ vector. We now have n variables $\lambda_1,\ldots,\lambda_n$ and $2n+p$ constraints; p of them are equations and the rest are inequalities.

EXAMPLE 2.10.1 The dual of the problem considered in Example 2.8.1 turns out to be

$$\text{Maximize } 4\lambda_1 + 3\lambda_2 + 4\lambda_3 + 5\lambda_4 + 5\lambda_5$$

$$\text{subject to } \begin{bmatrix} 1 & 1 & 1 & 1 & 1 \\ 1 & 2 & 3 & 4 & 5 \\ 3 & 2 & 1 & 2 & 3 \end{bmatrix} \begin{bmatrix} \lambda_1 \\ \lambda_2 \\ \lambda_3 \\ \lambda_4 \\ \lambda_5 \end{bmatrix} = \begin{bmatrix} 0 \\ 0 \\ 0 \end{bmatrix} \qquad (1)$$

$$\begin{bmatrix} 1 & 0 & 0 & 0 & 0 \\ 0 & 1 & 0 & 0 & 0 \\ 0 & 0 & 1 & 0 & 0 \\ 0 & 0 & 0 & 1 & 0 \\ 0 & 0 & 0 & 0 & 1 \end{bmatrix} \begin{bmatrix} \lambda_1 \\ \lambda_2 \\ \lambda_3 \\ \lambda_4 \\ \lambda_5 \end{bmatrix} \leqslant \begin{bmatrix} 1 \\ 1 \\ 1 \\ 1 \\ 1 \end{bmatrix} \qquad (2)$$

$$\begin{bmatrix} -1 & 0 & 0 & 0 & 0 \\ 0 & -1 & 0 & 0 & 0 \\ 0 & 0 & -1 & 0 & 0 \\ 0 & 0 & 0 & -1 & 0 \\ 0 & 0 & 0 & 0 & -1 \end{bmatrix} \begin{bmatrix} \lambda_1 \\ \lambda_2 \\ \lambda_3 \\ \lambda_4 \\ \lambda_5 \end{bmatrix} \leqslant \begin{bmatrix} 1 \\ 1 \\ 1 \\ 1 \\ 1 \end{bmatrix} \qquad (3)$$

$$\lambda_1, \ldots, \lambda_5 \text{ unrestricted in sign.}$$

The second and the third sets of constraints can be stated equivalently as $-1 \leqslant \lambda_i \leqslant 1$.

First we establish the relationship between this problem (2.10.1) and the problem of MINMAD regression (2.7.2), and then explain how to reduce the size of the dual problem.

RESULT 2.10.1 For any feasible solution to (2.7.2) and any solution to the dual, we have

$$\mathbf{Y}'\boldsymbol{\lambda} \leqslant \mathbf{C}'\mathbf{W}.$$

Proof As $\boldsymbol{\lambda}$ is a solution to (2.10.1) we have

$$\mathbf{A}'\boldsymbol{\lambda} \begin{Bmatrix} = \\ \leqslant \\ \leqslant \end{Bmatrix} \mathbf{C} \qquad (2.10.2)$$

Multiplying both sides of (2.10.2) by $\mathbf{W}'$ does not reverse the inequalities, since we have $W_{P+r} \geqslant 0$, $r=1,\ldots,2n$ and corresponding to $W_1,\ldots,W_p$ we have equalities. We get from the fact that $\mathbf{W}$ is feasible for (2.7.2) $\mathbf{W}'\mathbf{A}'=\mathbf{Y}'$. Hence $\mathbf{Y}'\boldsymbol{\lambda} \leqslant \mathbf{C}'\mathbf{W}$. We know that we have an optimal solution to the MINMAD regression problem. We shall now give the conditions for any $\mathbf{W}^*$, $\boldsymbol{\lambda}^*$ to be optimal for Problem 2.7.2 and for the dual problem.

RESULT 2.10.2 If $\mathbf{W}^*$ is feasible for (2.7.2) and $\boldsymbol{\lambda}^*$ any solution to the dual and $\mathbf{C}'\mathbf{W}^*=\mathbf{Y}'\boldsymbol{\lambda}^*$, then $\boldsymbol{\lambda}^*$ is optimal for the dual problem and $\mathbf{W}^*$ optimal for (2.7.2).

Proof From Result 2.10.1 we know that $\mathbf{C}'\mathbf{W}^* \geqslant \mathbf{Y}'\boldsymbol{\lambda}$ for any $\boldsymbol{\lambda}$ solution to (2.10.1). But $\mathbf{C}'\mathbf{W}^*=\mathbf{Y}'\boldsymbol{\lambda}^*$; hence $\mathbf{Y}'\boldsymbol{\lambda}^* \geqslant \mathbf{Y}'\boldsymbol{\lambda}$ or $\boldsymbol{\lambda}^*$ is optimal for the dual problem. Similarly we can show that $\mathbf{W}^*$ is optimal. The converse of this result is obtained in the proof of the following result, which states that both these problems have optimal solutions and that the objective function values agree.

RESULT 2.10.3 Consider an optimal basic feasible solution $\mathbf{W}_B$ to Problem 2.7.2. Then $\boldsymbol{\lambda}'=\mathbf{C}_B\mathbf{B}^{-1}$ is an optimal solution to the dual problem, and the objective function values agree.

Proof Because $\mathbf{W}_B$ is optimal for (2.7.2), we have $C_j-Z_j=0$ for all unrestricted variables and $C_j-Z_j \geqslant 0$ for all positively restricted variables. Then $Z_j=\mathbf{C}_B\mathbf{B}^{-1}\mathbf{a}_j=C_j$ for all unrestricted variables and $C_j=0$, for $j=1,\ldots,p$. Also

$$\mathbf{C}_B\mathbf{B}^{-1}\mathbf{a}_j=\mathbf{C}_B\mathbf{B}^{-1}\mathbf{e}_r \leqslant C_j=1, \qquad j=p+r, \qquad r=1,\ldots,n$$

and

$$\mathbf{C}_B\mathbf{B}^{-1}\mathbf{a}_j=\mathbf{C}_B\mathbf{B}^{-1}(-\mathbf{e}_r) \leqslant C_j=1, \qquad j=p+n+r, \qquad r=1,\ldots,n.$$

Then $\boldsymbol{\lambda}'=\mathbf{C}_B\mathbf{B}^{-1}$ satisfies all the constraints of Problem 2.10.1.

Now we consider the dual objective function,

$$\mathbf{Y}'\boldsymbol{\lambda}=\boldsymbol{\lambda}'\mathbf{Y}=\mathbf{C}_B\mathbf{B}^{-1}\mathbf{Y}.$$

But $\mathbf{B}^{-1}\mathbf{Y}=\mathbf{W}_B$. Therefore $\mathbf{Y}'\boldsymbol{\lambda}=\mathbf{C}_B\mathbf{W}_B=Z$. From Result 2.10.2, $\boldsymbol{\lambda}$ is optimal for the Problem 2.10.1. As for Result 2.10.3, we can solve the dual of Problem 2.7.2 and find an optimal solution. This proof is left as an exercise, as we are not going to solve the dual as it is.

2.11 DUALITY IN LINEAR PROGRAMMING

Given the linear programming problem as below,

$$\begin{aligned} \text{Minimize} \quad & \mathbf{C}'\mathbf{W} \\ \text{subject to} \quad & \mathbf{A}\mathbf{W} \geqslant \mathbf{b} \\ & \mathbf{W} \geqslant \mathbf{0}. \end{aligned} \tag{2.11.1}$$

Let us call this problem the *primal* problem. The dual of this problem is defined as

$$\begin{aligned} \text{Maximize} \quad & \mathbf{b}'\mathbf{U} \\ \text{subject to} \quad & \mathbf{A}'\mathbf{U} \leqslant \mathbf{C} \\ & \mathbf{U} \geqslant \mathbf{0}. \end{aligned} \tag{2.11.2}$$

REMARK 2.11.1 For each primal constraint we have defined a variable in the dual problem and for each variable in the primal problem we have a dual constraint. The inequalities in primal problem are $\geqslant$ type; in dual, $\leqslant$ type. In both the problems we have nonnegative variables.

REMARK 2.11.2 In a linear-programming problem, some constraints like

$$a_{i1}W_1 + \cdots + a_{in}W_n \leqslant b_i$$

are equivalent to

$$-a_{i1}W_1 - \cdots - a_{in}W_n \geqslant -b_i.$$

Similarly, some constraints that are equalities, such as

$$a_{i1}W_1 + \cdots + a_{in}W_n = b_i$$

are equivalent to the two constraints,

$$a_{i1}W_1 + \cdots + a_{in}W_n \geqslant b_i$$

$$-a_{i1}W_1 - \cdots - a_{in}W_n \geqslant -b_i.$$

Also if the objective function $\mathbf{C}'\mathbf{W}$ is maximized instead of minimized we consider minimizing $-\mathbf{C}'\mathbf{W}$ in its place.

If some variable W_j is unrestricted in sign, we can replace it by $W_j = W_j^1 - W_j^2$, where $W_j^1 \geqslant 0, W_j^2 \geqslant 0$.

Thus any linear-programming problem like (2.9.1) can be recast as a Problem of the type 2.11.1. However, it is possible to write the dual of any linear-programming problem without putting it in the form 2.11.1, with the following results.

RESULT 2.11.1 The dual of the dual Problem 2.11.2 is the primal Problem 2.11.1.

Proof We cast the dual problem (2.11.2) as:

$$\begin{aligned} \text{Minimize} \quad & -\mathbf{b}'\mathbf{U} \\ \text{subject to} \quad & -\mathbf{A}'\mathbf{U} \geqslant -\mathbf{C} \\ & \mathbf{U} \geqslant \mathbf{0}. \end{aligned}$$

And now the dual of this problem:

$$\begin{aligned} \text{Maximize} \quad & -\mathbf{C}'\mathbf{W} \\ \text{subject to} \quad & -\mathbf{A}\mathbf{W} \leqslant -\mathbf{b} \\ & \mathbf{W} \geqslant \mathbf{0}, \end{aligned}$$

which is equivalent to Problem 2.11.1.

RESULT 2.11.2 If the ith constraint in the primal is an equality, then the ith dual variable is unrestricted in sign.

Proof Corresponding to the equation

$$a_{i1}W_1 + \cdots + a_{in}W_n = b_i$$

we have

$$a_{i1}W_1 + \cdots + a_{in}W_n \geqslant b_i$$

$$-a_{i1}W_1 - \cdots - a_{in}W_n \geqslant -b_i$$

when the equation is replaced by two inequalities, to cast the problem in the form of (2.11.1). Now we have two variables corresponding to these inequalities in the dual; let these variables be denoted by U_i^+ and U_i^-. The dual column corresponding to U_i^+ is the transpose of the ith row in the primal, and that of U_i^- is the negative of the transpose of the ith row in the primal. Also the objective function coefficient of U_i^+ is b_i and that of U_i^- is $-b_i$. Hence, when $U_i^+ - U_i^-$ is replaced by U_i, U_i is unrestricted in sign, and the column corresponding to U_i is the transpose of the ith row of the primal problem, also b_i is its objective function coefficient.

Hence the variable corresponding to the ith row, which is an equality in the primal, is unrestricted in sign in the dual.

From Results 2.11.1 and 2.11.2 we have the following.

COROLLARY 2.11.1 If the jth variable in the primal is unrestricted in sign, then the jth constraint in the dual is an equality.

Now going back to the dual problem (2.10.1) of the MINMAD Regression problem (2.7.2), we notice that the dual is in accordance with the general definition of the dual of a linear-programming problem.

Results 2.10.1–3 for Problems 2.7.2 and 2.10.1 are in general true, and we state them without proof as the proofs are similar to that above.

RESULT 2.11.3 If **W** is a feasible solution to Problem 2.11.1 and **U** is a feasible solution to Problem 2.11.2, then

$$\mathbf{C}'\mathbf{W} \geqslant \mathbf{b}'\mathbf{U}$$

RESULT 2.11.4 If $\mathbf{W}^*$ is feasible for (2.11.1) and $\mathbf{U}^*$ is feasible for (2.11.2), and $\mathbf{C}'\mathbf{W}^* = \mathbf{b}'\mathbf{U}^*$, then $\mathbf{W}^*$ and $\mathbf{U}^*$ are optimal for their respective problems.

RESULT 2.11.5 If either Problem 2.11.1 or Problem 2.11.2 has an optimal solution, the other also has an optimal solution and the objective function values are equal.

REMARK 2.11.3 In the proof that Result 2.10.3 corresponds to Result 2.11.5, we show that corresponding to a basic feasible solution $\mathbf{W}_B$ that is optimal to the primal, $\mathbf{C}_B\mathbf{B}^{-1}$ gives an optimal solution to the dual problem. From Result 2.11.1, the same is true, corresponding to a dual optimal basis.

This observation is useful in obtaining an optimal solution to the MINMAD problem, through solving its dual. In general we noticed earlier that a linear-programming problem need not have a finite optimum. In such cases we say that the problem has an unbounded solution. We have the following result for such cases.

RESULT 2.11.6 If the primal (dual) problem has an unbounded solution, the dual (primal) problem has no feasible solution. The proof follows from Result 2.10.1.

REMARK 2.11.4 Notice that both the problems may not have feasible solutions, as illustrated by the following problems:

$$\begin{aligned} \text{Minimize} \quad & -CX_1 - CX_2 \\ \text{subject to} \quad & aX_1 - aX_2 \geqslant b \\ & -aX_1 + aX_2 \geqslant b \\ & X_1, X_2 \geqslant 0, \end{aligned}$$

and

$$\begin{aligned} \text{Maximize} \quad & bU_1 + bU_2 \\ \text{subject to} \quad & aU_1 - aU_2 \leqslant -C \\ & -aU_1 + aU_2 \leqslant -C \\ & U_1, U_2 \geqslant 0. \end{aligned}$$

Obviously both the problems have no feasible solution, and they are dual to each other.

A very useful property satisfied by the optimal solutions to the primal and dual problems is the *complementarity* between primal surplus variables and dual variables, and primal variables and dual slack variables. In order to observe this complementarity we add slack and surplus variables to the dual and primal problems. We then have

$$\begin{aligned} \text{Minimize} \quad & \mathbf{C}'\mathbf{W} \\ \text{subject to} \quad & [\mathbf{A} \quad -\mathbf{I}]\begin{bmatrix} \mathbf{W} \\ \mathbf{W}_s \end{bmatrix} = \mathbf{b} \\ & \mathbf{W}, \mathbf{W}_s \geqslant \mathbf{0} \end{aligned} \tag{2.11.3}$$

and

$$\begin{aligned} \text{Maximize} \quad & \mathbf{b}'\mathbf{U} \\ \text{subject to} \quad & [\mathbf{A}' \quad \mathbf{I}]\begin{bmatrix} \mathbf{U} \\ \mathbf{U}_s \end{bmatrix} = \mathbf{C} \\ & \mathbf{U}, \mathbf{U}_s \geqslant \mathbf{0}. \end{aligned} \tag{2.11.4}$$

RESULT 2.11.7 If the surplus variable W_{s_i} added to the constraint of Problem 2.11.3 is positive in any optimal solution to Problem 2.11.3, then in Problem 2.11.4 the ith variable U_i is zero in every optimal solution to (2.11.4). On the other hand, if U_i is positive in any optimal solution to (2.11.4), then $W_{s_i}=0$ in every optimal solution to (2.11.3).

Proof Consider $\mathbf{AW}-\mathbf{IW}_s=\mathbf{b}$. Premultiplying both sides by $\mathbf{U}'\geqslant\mathbf{0}$, we obtain

$$\mathbf{U}'\mathbf{AW}+\mathbf{U}'\mathbf{IW}_s=\mathbf{U}'\mathbf{b}=\mathbf{b}'\mathbf{U}. \tag{2.11.5}$$

Also we have, $\mathbf{C}'\mathbf{W}\geqslant\mathbf{U}'\mathbf{AW}\geqslant\mathbf{U}'\mathbf{b}$ from the constraints of Problems 2.11.1 and 2.11.2 and the fact that $\mathbf{W},\mathbf{U}\geqslant\mathbf{0}$. Suppose $\mathbf{W}$ and $\mathbf{U}$ are optimal for (2.11.3) and (2.11.4), respectively. Then, $\mathbf{C}'\mathbf{W}=\mathbf{U}'\mathbf{AW}=\mathbf{U}'\mathbf{b}$. Therefore (2.11.5) implies that $\mathbf{U}'\mathbf{IW}_s=\mathbf{U}'\mathbf{W}_s=\mathbf{0}$. As $\mathbf{U},\mathbf{W}\geqslant\mathbf{0}$, both U_i and W_{s_i} cannot be simultaneously positive. Therefore, we can also state this by $W_{s_i}U_i=0$, for all i, $\mathbf{W}$, and $\mathbf{U}$ optimal for the respective problems. Similarly, we can show that $W_jU_{s_j}=0$ for all j corresponding to optimal $\mathbf{W}$ and $\mathbf{U}$. Hence the result.

This property is known as the *complementary slackness property* of dual linear programming problems.

Going back to the discussion on MINMAD regression problem we find that the dual problem has more rows than Problem 2.7.2. Hence we need to consider a bigger basis while solving the dual. Also we need to add $2n$ more variables restricted in sign to make the inequality constraints in the dual into equalities. However, we can reduce the amount of computation involved by solving the problem as follows.

Let $\mathbf{V}=\boldsymbol{\lambda}+\mathbf{e}$. Then $\mathbf{Y}'\boldsymbol{\lambda}=\mathbf{Y}'\mathbf{V}-\mathbf{Y}'\mathbf{e}$ and $\mathbf{Y}'\mathbf{e}$ is a constant, so minimizing $\mathbf{Y}'\boldsymbol{\lambda}$ is the same as minimizing $\mathbf{Y}'\mathbf{V}$. Now,

$$\mathbf{A}'\boldsymbol{\lambda}=\mathbf{A}'(\mathbf{V}-\mathbf{e})$$

$$=\begin{bmatrix}\mathbf{X}'\mathbf{V}-\mathbf{X}'\mathbf{e}\\ \mathbf{V}-\mathbf{e}\\ -\mathbf{V}+\mathbf{e}\end{bmatrix}.$$

Thus the constraints become,

$$\mathbf{X}'\mathbf{V}=\mathbf{X}'\mathbf{e}$$

$$\mathbf{V}\leqslant\mathbf{e}+\mathbf{e}$$

$$-\mathbf{V}\leqslant\mathbf{e}-\mathbf{e}$$

or

$$\mathbf{X'V} = \mathbf{X'e}$$
$$\mathbf{0} \leqslant \mathbf{V} \leqslant 2\mathbf{e}.$$

Thus we have the problem,

$$\begin{aligned} \text{Maximize} \quad & \mathbf{Y'V} \\ \text{subject to} \quad & \mathbf{X'V} = \mathbf{X'e} \\ & \mathbf{0} \leqslant \mathbf{V} \leqslant 2\mathbf{e}. \end{aligned} \tag{2.11.6}$$

Now there are only p linear equations, as $\mathbf{X}'$ is a $p \times n$ matrix, but we do now have the restriction that $\mathbf{V}$ must be bounded above by $2\mathbf{e}$. Fortunately we can modify the simplex procedure developed earlier so that these upper bound constraints are taken care of, without explicitly adding them as constraints.

EXAMPLE 2.11.1 Consider Example 2.8.1. Now after transforming $\lambda_1, \ldots, \lambda_5$ we get in terms of $V_1, \ldots, V_5$ with $\mathbf{V} = \boldsymbol{\lambda} + \mathbf{e}$,

$$\text{Maximize} \quad 4V_1 + 3V_2 + 4V_3 + 5V_4 + 5V_5$$

$$\text{subject to} \quad \begin{bmatrix} 1 & 1 & 1 & 1 & 1 \\ 1 & 2 & 3 & 4 & 5 \\ 3 & 2 & 1 & 2 & 3 \end{bmatrix} \begin{bmatrix} V_1 \\ V_2 \\ V_3 \\ V_4 \\ V_5 \end{bmatrix} = \begin{bmatrix} 5 \\ 15 \\ 11 \end{bmatrix}$$

$$0 \leqslant V_i \leqslant 2.$$

2.12 THE BOUNDED-VARIABLE METHOD

Consider solving Problem 2.11.6. Contrary to the definition of nonbasic variables in the discussion of the method developed for solving (2.7.2), we now allow the nonbasic variables to be either 0 or the upper bound for any variable, namely, 2. Also we assume for the time being the rank of $\mathbf{X}'$ to be equal to p. Later we indicate how to handle the case when the rank of $\mathbf{X}'$ is less than p. Let any column of $\mathbf{X}'$ be denoted by $\mathbf{X}'_l$.

Now consider a feasible basis $\mathbf{B}$ for $\mathbf{X}'$. Let $\mathbf{V}_B$ denote coefficients of $\mathbf{B}$ while expressing $\mathbf{X'e}$ as a linear combination of $\mathbf{B}$ and the nonbasic vectors corresponding to the variables at Level 2. Let $N_2(B)$ denote the indices of the nonbasic variables at Level 2, corresponding to the basis **B**. Therefore

$$\mathbf{B}\mathbf{V}_B + \sum_{l \in N_2(B)} 2\mathbf{X}'_l = \mathbf{X}'\mathbf{e} \triangleq \mathbf{b}$$

$$\Rightarrow \mathbf{V}_B = \mathbf{B}^{-1}\mathbf{X}'\mathbf{e} - 2 \sum_{l \in N_2(B)} \mathbf{B}^{-1}\mathbf{X}'_l.$$

As earlier defined, let $\boldsymbol{\alpha}_j = \mathbf{B}^{-1}\mathbf{X}'_j$. Hence

$$\mathbf{V}_B = \mathbf{B}^{-1}\mathbf{X}'\mathbf{e} - 2 \sum_{l \in N_2(B)} \boldsymbol{\alpha}_l.$$

Also we have $C_j = Y_j$.

Therefore $Y_j - Z_j = Y_j - \mathbf{C}_B\mathbf{B}^{-1}\mathbf{X}'_j = Y_j - \mathbf{C}_B\boldsymbol{\alpha}_j$. Assume that $\mathbf{V}_B$ is feasible; that is, $2 \geqslant V_{B_i} \geqslant 0$, $i = 1, \ldots, p$. Now let us discuss what happens when a nonbasic V_j is chosen to increase its value from 0 to $\theta > 0$. That is, $V_j = \theta > 0$ is the new solution. Then

$$\mathbf{B}\hat{\mathbf{V}}_B + \theta\mathbf{X}'_j = \mathbf{b} - 2 \sum_{l \in N_2(B)} \mathbf{X}'_l$$

where $\hat{\mathbf{V}}_B$ is the vector of coefficience of **B** in the new solution. Note that the "^" is on $\mathbf{V}_B$ and not on **B**. In fact $V_{\hat{B}}$ refers to the coefficient of the columns of $\hat{\mathbf{B}} = (\mathbf{B}_1, \ldots, \mathbf{B}_{r-1}, \mathbf{X}'_j, \ldots, \mathbf{B}_p)$. Now replacing $\mathbf{X}'_j$ by $\sum_{i=1}^{p} \alpha_{ij}\mathbf{B}_i$, we get

$$\mathbf{B}\hat{\mathbf{V}}_B + \theta \sum_{i=1}^{p} \alpha_{ij}\mathbf{B}_i = \mathbf{b} - 2 \sum_{l \in N_2(B)} \mathbf{X}'_l$$

or

$$\mathbf{B}(\hat{\mathbf{V}}_B + \theta\boldsymbol{\alpha}_j) = \mathbf{b} - 2 \sum_{l \in N_2(B)} \mathbf{X}'_l.$$

But $\mathbf{V}_B$ is the vector of coefficients of **B**, while expressing

$$\mathbf{b} - 2 \sum_{l \in N_2(B)} \mathbf{X}'_l.$$

Therefore

$$\mathbf{V}_B = \hat{\mathbf{V}}_B + \theta \boldsymbol{\alpha}_j$$

or

$$\hat{\mathbf{V}}_B = \mathbf{V}_B - \theta \boldsymbol{\alpha}_j$$

$$\Rightarrow \hat{V}_{B_i} = V_{B_i} - \theta \alpha_{ij}, \qquad i = 1, \ldots, p.$$

But we require $2 \geqslant \hat{V}_{B_i} \geqslant 0$, for all $i = 1, \ldots, p$. To have $\hat{V}_{B_i} \geqslant 0$, we require $V_{B_i} - \theta \alpha_{ij} \geqslant 0$. Notice that as θ is positive, when $\alpha_{ij} \leqslant 0$ this requirement is automatically satisfied. Therefore for $\alpha_{ij} > 0$ we require $(V_{B_i}/\alpha_{ij}) \geqslant \theta$. Or if

$$\theta_1 = \min\left[\frac{V_{B_i}}{\alpha_{ij}}, \alpha_{ij} > 0\right]$$

then $\theta \leqslant \theta_1$.

Now we also require θ to satisfy $\theta \leqslant 2$ as $V_j = \theta$, and V_j should be less than or equal to 2. Finally, to have $\hat{V}_{B_i} \leqslant 2$, we require $V_{B_i} - \theta \alpha_{ij} \leqslant 2$. This requirement is satisfied automatically for $\alpha_{ij} \geqslant 0$. Therefore for $\alpha_{ij} < 0$, we need $2 - V_{B_i} \geqslant -\theta \alpha_{ij}$, or

$$\frac{2 - V_{B_i}}{-\alpha_{ij}} \geqslant \theta, \text{ for } i \text{ such that } \alpha_{ij} < 0.$$

Let

$$\theta_2 = \min\left[\frac{2 - V_{B_i}}{-\alpha_{ij}}, \alpha_{ij} < 0\right].$$

Then we require $\theta \leqslant \theta_2$.

Thus θ can be at most equal to

$$\min[\theta_1, \theta_2, 2]. \tag{2.12.1}$$

As long as the minimum in (2.12.1) is θ_1 we can continue the process of vector exchange steps as in Algorithm 2.8.1 with the difference that now there are no unrestricted variables in the problem. We choose a $\mathbf{B}_r$ to leave the basis such that

$$\theta_1 = \frac{V_{B_r}}{\alpha_{rj}} = \min\left(\frac{V_{B_i}}{\alpha_{ij}}, \alpha_{ij} > 0\right)$$

and V_j is in the new basis with the value $\theta = \theta_1$. If θ_1 is not the minimum in (2.12.1), then in the succeeding step one or more variables will reach their upper bounds. As soon as a variable outside the basis reaches its upper bound we mark "2" on the corresponding column in the table. (In fact all the nonbasic columns with the corresponding $V_j = 2$ are so marked.) If $\theta = 2$, then V_j reaches its upper bound from its previous value, 0. So we simply mark the column $\mathbf{X}'_j$ with a "2." Now we shall not change the basis, as we are not introducing the vector $\mathbf{X}'_j$ into the basis. But this will change the values $\mathbf{V}_B$ as

$$\hat{\mathbf{V}}_B = \mathbf{V}_B - 2\boldsymbol{\alpha}_j$$

as $\theta = 2$. Thus we subtract $2\boldsymbol{\alpha}_j$ from $\mathbf{V}_B$ if V_j becomes 2 from level 0. If the minimum in (2.12.1) is θ_2, the implication is that as V_j enters the basis, some variable V_{B_r} will go to its upper bound, and

$$\hat{V}_{B_i} = V_{B_i} - \left(\frac{2 - V_{B_r}}{-\alpha_{rj}}\right)\alpha_{ij}, \qquad i = 1, \dots, p$$

$$= V_{B_i} + \frac{2 - V_{B_r}}{\alpha_{rj}}\alpha_{ij}.$$

Therefore

$$\hat{V}_{B_r} = V_{B_r} + \frac{2 - V_{B_r}}{\alpha_{rj}}\alpha_{rj} = 2$$

and

$$\hat{V}_j = \frac{2 - V_{B_r}}{-\alpha_{rj}} = \theta.$$

Now suppose V_q is the variable corresponding to $\mathbf{B}_r$. Then $\alpha_{iq}=0,\ i\neq r$, and $\alpha_{rq}=1$. Notice that

$$\hat{V}_{B_i}=\left(V_{B_i}-\frac{V_{B_r}}{\alpha_{rj}}\alpha_{ij}\right)+2\frac{\alpha_{ij}}{\alpha_{rj}},\qquad i\neq r$$

$$=\left(V_{B_i}-\frac{V_{B_r}}{\alpha_{rj}}\alpha_{ij}\right)-2\left(\alpha_{iq}-\frac{\alpha_{ij}\alpha_{rq}}{\alpha_{rj}}\right)$$

by using $\alpha_{iq}=0,\ i\neq r$, and $\alpha_{rq}=1$. Recall that $V_{B_i}-(V_{B_r}/\alpha_{rj})\alpha_{ij}$ is the transformed value of V_{B_i} when $\mathbf{X}'_j$ replaces $\mathbf{B}_r$, or $V_{\hat{B}_i}=V_{B_i}-(V_{B_r}/\alpha_{rj})\alpha_{ij}$, $i\neq r$, where $\hat{\mathbf{B}}=[\mathbf{B}_1\cdots\mathbf{B}_{r-1},\mathbf{X}'_j,\mathbf{B}_{r+1}\cdots\mathbf{B}_p]$ and $(\alpha_{iq}-(\alpha_{ij}\alpha_{rq}/\alpha_{rj}))$ is the transformed value of α_{iq}; that is, $\hat{\alpha}_{iq}$ corresponding to $\hat{\mathbf{B}}$, the new basis. We have

$$V_{\hat{B}_r}=\frac{V_{B_r}}{\alpha_{rj}}.$$

Therefore

$$\hat{V}_j=\frac{V_{B_r}}{\alpha_{rj}}-2\frac{\alpha_{rq}}{\alpha_{rj}}=V_{\hat{B}_r}-2\hat{\alpha}_{rq}.$$

Hence we do the transformation of the table, introducing $\mathbf{X}'_j$ and removing $\mathbf{B}_r$ corresponding to

$$\theta_2=\min\left[\frac{2-V_{B_i}}{-\alpha_{ij}}\right]=\frac{2-V_{B_r}}{-\alpha_{rj}}.$$

In the transformed table we subtract $2\hat{\boldsymbol{\alpha}}_q$ from $\mathbf{V}_{\hat{B}}$, to obtain the new solution. Subtracting $2\hat{\boldsymbol{\alpha}}_q$ adjusts for the $\mathbf{B}_r$ column, leaving the basis at Level 2.

There is also the case in which some nonbasic variables V_j at Level 2 are chosen for *reducing* this value from 2 in the next step. Let $\mathbf{B}$ be the current basis. Let this amount of reduction be θ. Then θ has to be less than or equal to 2 for nonnegativity of V_j. Now

$$\mathbf{B}\hat{\mathbf{V}}_B-\theta\mathbf{X}'_j=\mathbf{b}-2\sum_{l\in N_2(B)}\mathbf{X}'_l$$

$$\hat{\mathbf{V}}_B=\mathbf{V}_B+\theta\boldsymbol{\alpha}_j,$$

now we require $2 \geqslant \hat{V}_{B_i} \geqslant 0$. Nonnegativity of $\hat{V}_{B_i}$ will require for $\alpha_{ij} < 0$,

$$V_{B_i} \geqslant -\theta\alpha_{ij}, \qquad i=1,\ldots,p.$$

Or

$$\frac{V_{B_i}}{-\alpha_{ij}} \geqslant \theta.$$

Let

$$\theta_1 = \min\left[\frac{V_{B_i}}{-\alpha_{ij}}, \alpha_{ij} < 0\right];$$

then $\theta \leqslant \theta_1$.

Upper bound on $\hat{\mathbf{V}}_B$ will imply for $\alpha_{ij} > 0$,

$$\theta \leqslant \frac{2 - V_{B_i}}{\alpha_{ij}}$$

or $\theta \leqslant \theta_2, \theta_2$ as defined by

$$\theta_2 = \min\left[\frac{2 - V_{B_i}}{\alpha_{ij}}, \alpha_{ij} > 0\right].$$

Hence in this case also $\theta \leqslant \min[\theta_1, \theta_2, 2]$ with a slight modification of θ_1 and θ_2, now the minimum taken over negative and positive α_{ij}'s, respectively.

If the minimum of $(\theta_1, \theta_2, 2)$ is 2, then V_j is reduced by 2 or it goes from Level 2 to Level 0. The basis is not changed and the new solution is given by $\hat{\mathbf{V}}_B$, which is obtained by adding $2\boldsymbol{\alpha}_j$ to $\mathbf{V}_B$.

If the minimum of $(\theta_1, \theta_2, 2)$ is θ_2 then some variable V_q corresponding to a basic column $\mathbf{B}_r$ reaches its upper bound when V_j replaces it in the basis. So we insert $\mathbf{X}'_j$ into the basis and remove $\mathbf{X}'_q$. We transform the table and mark "2" on the corresponding column $\mathbf{X}'_q$.

The new solution is given by

$$\hat{V}_{B_i} = \begin{cases} V_{\hat{B}_i} - 2\hat{\alpha}_{iq} & i \neq r \\ & i = 1,\ldots,p \\ V_{\hat{B}_r} + 2 - 2\hat{\alpha}_{rq} & i = r. \end{cases}$$

This case can be proved similar to that of the earlier case when θ_2 was minimum and $\mathbf{V}_j$ was at Level 0. If the minimum of $(\theta_1, \theta_2, 2)$ is θ_1, then $\mathbf{X}'_j$ replaces $\mathbf{B}_r$ in the basis for which $\theta_1 = (V_{B_r}/\alpha_{rj})$, $\alpha_{rj} < 0$. We transform the table and remove the mark on Column $\mathbf{X}'_j$, and the new solution is given by

$$\hat{V}_{B_i} = \begin{cases} V_{\hat{B}_i} + 2 & i = r \\ V_{\hat{B}_i} & i \neq r \; i = 1, \ldots, p. \end{cases}$$

This is the result because $\mathbf{X}'_j$ is in the rth row of the transformed table and as it is in the basis the new $\hat{\boldsymbol{\alpha}}_j$ will be a unit vector with 1 in the rth place and zeroes elsewhere. Thus we have explained the changes in the solution corresponding to the six possibilities, when V_j is chosen to change its value.

Next we consider the choice of column to enter the basis. There are two cases, depending on whether the vector chosen is at Level 0 or at Level 2. Let the objective function value corresponding to the present solution be

$$Z = \sum_{i=1}^{p} Y_{B_i} V_{B_i} + 2 \sum_{l \in N_2(B)} Y_l.$$

CASE 1 $V_j = 0$. The new solution is given by

$$\hat{V}_{B_i} = V_{B_i} - \theta \alpha_{ij}, \qquad i = 1, \ldots, p$$

and

$$\hat{V}_j = \theta.$$

Let the corresponding objective function value be denoted by $\hat{Z}$,

$$\hat{Z} = \sum_{i=1}^{p} Y_{B_i}(V_{B_i} - \theta\alpha_{ij}) + Y_j\theta + 2 \sum_{l \in N_2(B)} Y_l$$

$$= \sum_{i=1}^{p} Y_{B_i} V_{B_i} - \theta\left[-Y_j + \sum_{i=1}^{p} Y_{B_i}\alpha_{ij} \right] + 2 \sum_{l \in N_2(B)} Y_l$$

$$= Z + \theta[Y_j - Z_j].$$

So if $Y_j - Z_j > 0$, we have an improvement in the objective function.

Therefore we may choose

$$Y_j - Z_j = \max_{\substack{K \\ \text{nonbasic} \\ k \notin N_2(B)}} [Y_k - Z_k, Y_k - Z_k > 0].$$

CASE 2 $V_j = 2$. The new solution given by

$$\hat{V}_{B_i} = V_{B_i} + \theta\alpha_{ij}, \qquad i = 1, \ldots, p,$$

and V_j is reduced by θ.

Let the corresponding objective function value be $\hat{Z}$. Then

$$\hat{Z} = \sum_{i=1}^{p} Y_{B_i}(V_{B_i} + \theta\alpha_{ij}) - \theta Y_j + 2 \sum_{l \in N_2(B)} Y_l$$

$$= Z + \theta[Z_j - Y_j],$$

or if $Y_j - Z_j < 0$ there is an improvement in the objective function.

Observe that in the MINMAD problem we choose $Y_j - Z_j < 0$ for the restricted variables for entry into the basis. Here the sign is reversed as we are maximizing the objective function instead of minimizing as it was the case with Problem 2.7.2.

Thus we find the variable to be changed from its level by choosing V_j such that

$$Y_j - Z_j = \max[Y_{j_1} - Z_{j_1}, |Y_{j_2} - Z_{j_2}|],$$

where

$$Y_{j_1} - Z_{j_1} = \max_{\substack{Y_k - Z_k > 0 \\ k \notin N_2(B)}} [Y_k - Z_k]$$

and

$$|Y_{j_2} - Z_{j_2}| = \max_{\substack{Y_k - Z_k < 0 \\ k \in N_2(B)}} |Y_k - Z_k|.$$

From this we obtain the optimality conditions that $Y_j - Z_j \geqslant 0$ for all j nonbasic and $j \in N_2(B)$, and $Y_j - Z_j \leqslant 0$ for all j nonbasic and $j \notin N_2(B)$.

We can summarize the method for the bounded-variable problems as follows:

Let $\mathbf{B}, 2\mathbf{e} \geqslant \mathbf{V}_B \geqslant \mathbf{0}$, $N_2(B)$ be such that

$$\mathbf{B}\mathbf{V}_B + 2 \sum_{l \in N_2(B)} \mathbf{X}'_l = \mathbf{X}'\mathbf{e}.$$

Let

$$Y_j - Z_j = Y_j - \sum_{i=1}^{p} Y_{B_i} \alpha_{ij}$$

and $\boldsymbol{\alpha}_j = \mathbf{B}^{-1}\mathbf{X}'_j$.

Step 1 Compute $Y_j - Z_j = \min[Y_{j_1} - Z_{j_1}, |Y_{j_2} - Z_{j_2}|]$ where

$$Y_{j_1} - Z_{j_1} = \max_{\substack{Y_k - Z_k > 0 \\ k \notin N_2(B)}} [Y_k - Z_k]$$

$$|Y_{j_2} - Z_{j_2}| = \max_{\substack{Y_k - Z_k < 0 \\ k \in N_2(B)}} [|Y_k - Z_k|].$$

If no such j exists, go to Step 3. Otherwise go to Step 2. Note whether $j = j_1$ or j_2.

Step 2 Compute $\theta = \min[\theta_1, \theta_2, 2]$ where

$$\theta_1 = \min_i \left[\frac{V_{B_i}}{|\alpha_{ij}|} \right]$$

where i runs over $\alpha_{ij} > 0$ if $j = j_1$ and i runs over $\alpha_{ij} < 0$ if $j = j_2$, and

$$\theta_2 = \min_i \left[\frac{2 - V_{B_i}}{|\alpha_{ij}|} \right]$$

where i runs over $\alpha_{ij} < 0$ if $j = j_1$ and i runs over $\alpha_{ij} > 0$ if $j = j_2$.

There are six cases to be discussed.

CASE 1 $j = j_1$

(i) $\theta = \theta_1$ and column r of $\mathbf{B}$ is to be replaced. Insert $\mathbf{X}'_j$ into Column r of $\mathbf{B}$ and transform the table using the ring-around-the-rosy method.* Go to Step 1.

*The method of transforming the table explained under Algorithm 2.8.1.

(ii) $\theta=\theta_2$. Then $\mathbf{X}'_j$ replaces Column r of $\mathbf{B}$ for which $\theta_2=(2-V_{B_r}/-\alpha_{rj})$, which contains $\mathbf{X}'_q$. Transform the table using ring-around-rosy method; subtract $2\boldsymbol{\alpha}_q$ from $\mathbf{V}_B$. Go to Step 1. Mark Column $\mathbf{X}'_q$ by "2."

(iii) $\boldsymbol{\theta}=2$. Then mark Column $\mathbf{X}'_j$ to indicate that V_j is at its upper bound. $\mathbf{B}$ is unchanged. Subtract $2\boldsymbol{\alpha}_j$ from $\mathbf{V}_B$. Go to Step 1.

CASE 2 $j=j_2$

(iv) $\theta=\theta_1$, and Column r of $\mathbf{B}$ is to be replaced. Transform the table as usual and add 2 to the new V_{B_r}. Remove the mark from Column $\mathbf{X}'_j$, indicating that V_j is no longer at its upper bound. Go to Step 1.

(v) $\theta=\theta_2$, and the minimum is taken on at $i=r$. Replace Column r, say $\mathbf{X}'_q$, by $\mathbf{X}'_j$ in the basis, and place "2" on Column $\mathbf{X}'_q$. Add 2 to the new V_{B_r} and subtract $2\boldsymbol{\alpha}_q$ from the $\mathbf{V}_B$, so obtained. Go to Step 1.

(vi) $\theta=2$. Remove the mark from Column $\mathbf{X}'_j$ and add $2\boldsymbol{\alpha}_j$ to $\mathbf{V}_B$. Go to Step 1.

Step 3 When all $Y_j-Z_j\leqslant 0,\ j\notin N_2(B)$, and $Y_j-Z_j\geqslant 0,\ j\in N_2(B)$, an optimal solution to the problem has been obtained. The vector $\mathbf{V}_B$ gives the values of V_j in $\mathbf{B}$. Each marked column has $V_j=2$.

The same algorithm can be used for general linear-programming problems with bounded variables where the bounds are not the same for all variables, by replacing the upper bounds in place of 2's appropriately.

EXAMPLE 2.12.1 Consider the data given in Example 2.8.2. We have

$$\mathbf{X}'=\begin{bmatrix}1 & 1 & 1 & 1 & 1\\ 1 & 2 & 3 & 4 & 5\end{bmatrix}$$

$$\mathbf{Y}'=\begin{bmatrix}1 & 1 & 2 & 3 & 2\end{bmatrix}.$$

Suppose we use the bounded variable method to solve the problem

$$\begin{aligned}&\text{Maximize} && \mathbf{Y}'\mathbf{V}\\ &\text{subject to} && \mathbf{X}'\mathbf{V}=\mathbf{X}'\mathbf{e}\\ & && \mathbf{0}\leqslant\mathbf{V}\leqslant 2\mathbf{e}.\end{aligned}$$

Let the initial basis be given by

$$\mathbf{B}=[\mathbf{X}_1',\mathbf{X}_2'] \qquad \text{and} \quad N_2(B)=\{5\}.$$

That is, $V_5=2$ and nonbasic.

Now

$$\mathbf{B}\mathbf{V}_B=2\sum_{l\in N_2(B)}\mathbf{X}_l'=\mathbf{X}'\mathbf{e}=\begin{bmatrix}5\\15\end{bmatrix}$$

$$\mathbf{B}\mathbf{V}_B=\begin{bmatrix}5\\15\end{bmatrix}-2\begin{bmatrix}1\\5\end{bmatrix}=\begin{bmatrix}3\\5\end{bmatrix}.$$

We have

$$\mathbf{B}^{-1}=\begin{bmatrix}2&-1\\-1&1\end{bmatrix}.$$

Therefore

$$\mathbf{V}_B=\begin{bmatrix}1\\2\end{bmatrix}.$$

Let $\mathbf{B}^{-1}\mathbf{X}_j'=\boldsymbol{\alpha}_j$ and $\mathbf{Y}_B\mathbf{B}^{-1}\mathbf{X}_j'=Z_j$. The initial table is shown in Table 2.12.1.

Table 2.12.1

$Y_j\rightarrow$			1	1	2	3	2
Mark→							"2"
$\mathbf{Y}_B$	Vectors in the Basis	$\mathbf{V}_B$	$\boldsymbol{\alpha}_1$	$\boldsymbol{\alpha}_2$	$\boldsymbol{\alpha}_3$	$\boldsymbol{\alpha}_4$	$\boldsymbol{\alpha}_5$
1	$\mathbf{X}_1'$	1	1	0	-1	-2	-3
1	$\mathbf{X}_2'$	2	0	1	2	3	4
			0	0	1	2	1

We go to Step 1 of the method.

Step 1

$$\max_{\substack{Y_k - Z_k > 0 \\ k \notin N_2(B)}} = \max[1,2] = 2 = Y_4 - Z_4.$$

Therefore, $j_1 = 4$.

$$\max_{\substack{Y_k - Z_k < 0 \\ k \in N_2(B)}} [|Y_k - Z_k|]$$

is not defined as $Y_5 - Z_5 > 0$ and $N_2(B) = \{5\}$. Thus $j = j_1$. We go to Step 2.

Step 2

$$\theta_1 = \min_{\alpha_{ij} > 0} \left[\frac{V_{B_i}}{|\alpha_{ij}|} \right] = \min\left[\tfrac{2}{3}\right] = \tfrac{2}{3}.$$

$$\theta_2 = \min_{\alpha_{ij} < 0} \left[\frac{2 - V_{B_i}}{|\alpha_{ij}|} \right] = \min\left[\frac{2-1}{2} \right] = \tfrac{1}{2}.$$

Hence $\theta = \min[\frac{2}{3}, \frac{1}{2}, 2] = \frac{1}{2} = \theta_2$. We have case 1.(ii). $\mathbf{X}_4'$ replaces $\mathbf{X}_1'$. We transform the table and subtract $2\boldsymbol{\alpha}_1$ from $\mathbf{V}_B$. We mark the column $\mathbf{X}_1'$ by "2." This is shown in Table 2.12.2.

We find, after transforming,

$$\mathbf{V}_B = \begin{bmatrix} -\frac{1}{2} \\ \frac{7}{2} \end{bmatrix}.$$

Table 2.12.2

$Y_j \rightarrow$			1	1	2	3	2
Mark→			'2'				"2"
$\mathbf{Y}_B$	Vectors in the Basis	$\mathbf{V}_B$	$\boldsymbol{\alpha}_1$	$\boldsymbol{\alpha}_2$	$\boldsymbol{\alpha}_3$	$\boldsymbol{\alpha}_4$	$\boldsymbol{\alpha}_5$
3	$\mathbf{X}_4'$	$\frac{1}{2}$	$-\frac{1}{2}$	0	$\frac{1}{2}$	1	$\frac{3}{2}$
1	$\mathbf{X}_2'$	$\frac{1}{2}$	$\frac{3}{2}$	1	$\frac{1}{2}$	0	$-\frac{1}{2}$
			1	0	0	0	-2

But

$$2\boldsymbol{\alpha}_1 = 2\begin{bmatrix} -\frac{1}{2} \\ \frac{3}{2} \end{bmatrix} = \begin{bmatrix} -1 \\ 3 \end{bmatrix}.$$

Thus finally subtracting $2\boldsymbol{\alpha}_1$ from $\mathbf{V}_B$ we obtain the adjusted

$$\mathbf{V}_B = \begin{bmatrix} -\frac{1}{2} \\ \frac{7}{2} \end{bmatrix} - \begin{bmatrix} -1 \\ 3 \end{bmatrix} = \begin{bmatrix} \frac{1}{2} \\ \frac{1}{2} \end{bmatrix}$$

as shown in Table 2.12.2. We go to Step 1.

Step 1 j_1 is not defined as $N_2(B) = \{1, 5\}$ and outside $N_2(B)$ no

$Y_k - Z_k > 0.$

$$j_2 = 5 \text{ and } \max_{\substack{Y_k - Z_k < 0 \\ k \in N_2(B)}} |Y_k - Z_k| = 2 = |Y_5 - Z_5|.$$

Therefore $j = j_2 = 5$. We go to step 2.

Step 2

$$\theta_1 = \min_{\alpha_{ij} < 0} \left[\frac{V_{B_i}}{|\alpha_{ij}|} \right] = 1, \qquad \text{for } r = 2.$$

$$\theta_2 = \min_{\alpha_{ij} > 0} \left[\frac{2 - V_{B_i}}{|\alpha_{ij}|} \right] = \left[\frac{2 - \frac{1}{2}}{\frac{3}{2}} \right] = 1, \qquad \text{for } r = 1.$$

Therefore $\theta = \min[1, 1, 2] = 1 = \theta_2$ (say). This is Case 2(iv).

We introduce $\mathbf{X}'_5$ into the basis and remove $\mathbf{X}'_4$ from the basis. We transform the table. $\mathbf{X}'_4$ is now at its upper bound, namely 2. That column is marked by '2', and the mark in column $\mathbf{X}'_5$ is removed. Adding 2 to the new V_{B_1} and subtracting $2\boldsymbol{\alpha}_4$ from the $\mathbf{V}_B$ so obtained we get the adjusted $\mathbf{V}_B$. This is shown in Table 2.12.3.

Here, transformed

$$\mathbf{V}_B = \begin{bmatrix} \frac{1}{3} \\ \frac{2}{3} \end{bmatrix}.$$

Table 2.12.3

$Y_j \rightarrow$			1	1	3	2	2
Mark$\rightarrow$			"2"			"2"	
$\mathbf{Y}_B$	Vectors in the Basis	$\mathbf{V}_B$	α_1	α_2	α_3	α_4	α_5
2	$\mathbf{X}'_5$	1	$-\frac{1}{3}$	0	$\frac{1}{3}$	$\frac{2}{3}$	1
1	$\mathbf{X}'_2$	0	$\frac{8}{6}$	1	$\frac{2}{3}$	$\frac{2}{6}$	0
			$\frac{1}{3}$	0	$\frac{2}{3}$	$\frac{4}{3}$	0

Then the adjusted

$$\mathbf{V}_B = \begin{bmatrix} \frac{1}{3} \\ \frac{2}{3} \end{bmatrix} + \begin{bmatrix} 2 \\ 0 \end{bmatrix} - 2\begin{bmatrix} \frac{2}{3} \\ \frac{2}{6} \end{bmatrix} = \begin{bmatrix} 1 \\ 0 \end{bmatrix},$$

as shown in Table (2.12.3). We go to Step 1.

Step 1 $j_1 = 3$ and j_2 is not defined. Therefore $j = j_1$. We go to Step 2.

Step 2

$$\theta_1 = \min_{\alpha_{ij}>0} \frac{V_{B_i}}{|\alpha_{ij}|} = \min[3,0] = 0.$$

θ_2 need not be calculated as $\theta = \min[0, \theta_2, 2] = 0$. This is Case 1(i). We transform the table replacing $\mathbf{X}'_2$ by $\mathbf{X}'_3$. The new table is shown in Table 2.12.4.

Table 2.12.4

$Y_j \rightarrow$			1	1	2	3	2
Mark$\rightarrow$			'2'			'2'	
$\mathbf{Y}_B$	Vectors in the Basis	$\mathbf{V}_B$	α_1	α_2	α_3	α_4	α_5
2	$\mathbf{X}'_5$	1	-1	$-\frac{1}{2}$	0	$\frac{1}{2}$	1
2	$\mathbf{X}'_3$	0	2	$\frac{3}{2}$	1	$\frac{1}{2}$	0
			-1	-1	0	1	0

We go to Step 1.

Step 1 We find no $Y_k - Z_k > 0$, for $k \notin N_2(B)$ so j_1 is not defined. But $j_2 = 1$ as $Y_1 - Z_1 < 0$ and $1 \in N_2(B)$.

Step 2

$$\theta_1 = \min_{\alpha_{ij} < 0} \frac{V_{B_i}}{|\alpha_{ij}|} = 1, \qquad r = 1$$

$$\theta_2 = \min_{\alpha_{ij} > 0} \frac{2 - V_{B_i}}{|\alpha_{ij}|} = \left[\tfrac{2}{2}\right] = 1, \qquad r = 2.$$

Therefore $\theta = \min[\theta_1, \theta_2, 2] = 1 = \theta_2$ (say). This is Case 2(v). We replace $\mathbf{X}'_3$ by $\mathbf{X}'_1$. We remove the mark of $\mathbf{X}'_1$ and place a mark on $\mathbf{X}'_3$. And do the necessary adjustments in $\mathbf{V}_B$ as done earlier. The new table obtained is shown in Table 2.12.5. We go to Step 1.

Step 1 We find all $Y_k - Z_k \leqslant 0$ for $k \notin N_2(B)$ and $Y_k - Z_k > 0$ for $k \in N_2(B)$. So no j exists such that $j = j_1$ or j_2 as defined in Step 1. We go to Step 3 and stop. The solution at hand is optimal. We have $V_1 = 1, V_2 = 0, V_3 = 2, V_4 = 2$, and $V_5 = 0$. With the objective function value equal to $1 \times V_1 + 3 \times V_3 + 2 \times V_4 = 11$. Here

$$\mathbf{B} = \begin{bmatrix} 1 & 1 \\ 5 & 1 \end{bmatrix} \quad \text{or} \quad \mathbf{B}^{-1} = \begin{bmatrix} -\frac{1}{4} & \frac{5}{4} \\ \frac{1}{4} & -\frac{1}{4} \end{bmatrix}.$$

Hence,

$$\mathbf{C}_B \mathbf{B}^{-1} = [2, 1] \begin{bmatrix} -\frac{1}{4} & \frac{5}{4} \\ \frac{1}{4} & -\frac{1}{4} \end{bmatrix} = \begin{bmatrix} \frac{3}{4} \\ \frac{1}{4} \end{bmatrix}$$

Table 2.12.5

$Y_j \rightarrow$			1	1	2	3	2
Mark$\rightarrow$					'2'	'2'	
$\mathbf{Y}_B$	Vectors in the Basis	$\mathbf{V}_B$	$\boldsymbol{\alpha}_1$	$\boldsymbol{\alpha}_2$	$\boldsymbol{\alpha}_3$	$\boldsymbol{\alpha}_4$	$\boldsymbol{\alpha}_5$
2	$\mathbf{X}'_5$	0	0	$\frac{1}{4}$	$\frac{1}{2}$	$\frac{3}{4}$	1
1	$\mathbf{X}'_1$	1	1	$\frac{3}{4}$	$\frac{1}{2}$	$\frac{1}{4}$	0
			0	$-\frac{1}{4}$	$\frac{1}{2}$	$\frac{5}{4}$	0

or $\beta_0=\frac{3}{4}$ and $\beta_1=\frac{1}{4}$ are the estimates of the parameters. However, we obtained a different solution in Example 2.8.2. This is so because the dual optimal solution is degenerate. So the primal problem has alternate optimum.

In the linear-programming approach, addition of new independent variables or addition of new rows corresponding to additional data can be handled without solving the problem all over from scratch.

Adding New Variables β_p to the Model Consider the MINMAD regression problem. Suppose we wish to introduce a new parameter (variable) into the model after having obtained an optimal solution to the existing setup. From the final table of the simplex process, we have an optimal basis $\mathbf{B}$ and its inverse $\mathbf{B}^{-1}$. Let the new variable β_p have the corresponding column of observations $\mathbf{X}_{p+1}$, with $C_{p+1}=0$. Here $\boldsymbol{\beta}$ is unrestricted in sign.

Calculate

$$C_{p+1}-Z_{p+1}=0-\mathbf{C}_B\mathbf{B}^{-1}\mathbf{X}_{p+1}.$$

If $C_{p+1}-Z_{p+1}=0$, the present basic solution is optimal with $\beta_p=0$. We stop. Otherwise, we introduce a new column in the final table corresponding to β_p whose first n rows are given by $\boldsymbol{\alpha}_{p+2n+1}=\mathbf{B}^{-1}\mathbf{X}_{p+1}$ and the last row by $C_{p+1}-\mathbf{C}_B\mathbf{B}^{-1}\mathbf{X}_{p+1}$. Then we proceed with β_p as the variable to enter the basis, and continue. While solving the problem through its dual, adding a new variable to the primal corresponds to adding a constraint in the dual.

If the dual solution $\boldsymbol{\lambda}^*$ satisfies the new constraint

$$\mathbf{X}'_{p+1}\boldsymbol{\lambda}^*=0$$

we are through. The present optimal solution is optimal for the expanded system as well. We stop. This also implies that $\mathbf{X}'_{p+1}\mathbf{V}^*=\mathbf{X}'_{p+1}\mathbf{e}$ is satisfied as $\mathbf{V}^*=\boldsymbol{\lambda}^*+\mathbf{e}$, with $\boldsymbol{\lambda}=\boldsymbol{\lambda}^*$, corresponding to Problem 2.11.1. In the case of $\mathbf{X}'_{p+1}\mathbf{V}^*\neq\mathbf{X}'_{p+1}\mathbf{e}$, this new constraint is not satisfied by the optimal solution $\mathbf{V}^*$. We proceed as follows:

Calculate $x=-(\mathbf{X}'_{p+1})_B\mathbf{B}^{-1}\mathbf{X}'\mathbf{e}+\mathbf{X}'_{p+1}\mathbf{e}$, where $(\mathbf{X}'_{p+1})_B$ is that portion of $\mathbf{X}'_{p+1}$ corresponding to the optimal basis $\mathbf{B}$, for the dual problem. (We are using $\mathbf{B}$ for the basis of the primal as well as dual; this should not create any trouble.) If $x<0$, we introduce an artificial variable $V_{a_{p+1}}\geqslant 0$ and have the constraint

$$\mathbf{X}'_{p+1}\mathbf{V}-V_{a_{p+1}}=\mathbf{X}'_{p+1}\mathbf{e}.$$

Notice that $\mathbf{X}$ cannot be equal to zero, as that would mean $\mathbf{X}'_{p+1}\mathbf{V}^* = \mathbf{X}'_{p+1}\mathbf{e}$ as $\mathbf{V}^*_B = \mathbf{B}^{-1}\mathbf{X}'\mathbf{e}$. Thus the artificial variable $V_{a_{p+1}}$ has a column with $+1$ or -1 in the last row and zeroes elsewhere, depending on whether $x>0$ or $x<0$, respectively. So the initial basis for the expanded system is given by

$$\overline{\mathbf{B}} = \begin{bmatrix} \mathbf{B} & \mathbf{0} \\ (\mathbf{X}'_{p+1})_B & \pm 1 \end{bmatrix}$$

and its inverse is given by

$$\overline{\mathbf{B}}^{-1} = \begin{pmatrix} \mathbf{B}^{-1} & \mathbf{0} \\ \mp (\mathbf{X}'_{p+1})_B \mathbf{B}^{-1} & \pm 1 \end{pmatrix}.$$

Here $\overline{\mathbf{B}} = [\overline{\mathbf{B}}_1, \ldots, \overline{\mathbf{B}}_p, \overline{\mathbf{B}}_{p+1}]$ with $\overline{\mathbf{B}}'_{p+1} = (0, \ldots, \pm 1)$ and $\overline{\mathbf{B}}_i = \mathbf{B}_i$, $i = 1, \ldots, p$. Observe that

$$V_{\overline{B}_i} = \begin{cases} V_{B_i}, & i = 1, \ldots, p \\ \mp (\mathbf{X}'_{p+1})_B \mathbf{V}_B \pm \mathbf{X}'_{p+1}\mathbf{e}, & i = p+1. \end{cases}$$

Also, we have the $\overline{\boldsymbol{\alpha}}_l$, corresponding to any V_l of the expanded table, for the basis $\overline{\mathbf{B}}$ is given by

$$\overline{\boldsymbol{\alpha}}_l = \begin{pmatrix} \boldsymbol{\alpha}_l \\ \mp (\mathbf{X}'_{p+1})_B \boldsymbol{\alpha}_l \pm \mathbf{X}_{lp+1} \end{pmatrix}$$

where $\boldsymbol{\alpha}_l$ is corresponding V_l in the optimal table.

Let the objective function coefficient for $V_{a_{p+1}}$ be $-M$ with M, a large positive real number. The $Y_l - \overline{Z}_l$ corresponding V_l in the expanded table is then,

$$Y_l - \overline{Z}_l = Y_l - \mathbf{Y}_{\overline{B}} \overline{\boldsymbol{\alpha}}_l$$

where $\mathbf{Y}_{\overline{B}}$, the vector, gives the cost coefficients corresponding to $\overline{\mathbf{B}}$. Therefore

$$\mathbf{Y}_{\overline{B}} = (\mathbf{Y}_B, -M)$$

Hence $Y_l - \overline{Z}_l = Y_l - \mathbf{Y}_B \boldsymbol{\alpha}_l + M \overline{\alpha}_{p+1l}$ where

$$\overline{\alpha}_{p+1l} = \mp (\mathbf{X}'_{p+1})_B \boldsymbol{\alpha}_l \pm X_{lp+1}.$$

With this expanded table, we go through bounded-variable simplex steps. Notice that $V_{a_{p+1}}$ has no upper bound. Hence we do the necessary modifications. Eventually either (1) $V_{a_{p+1}}$ is driven to zero, or (2) $V_{a_{p+1}}$ is positive and the optimality conditions are satisfied. Eventuality (2) implies that there is no feasible solution to the problem when the new constraint is added. However, eventuality (2) is impossible, as the MINMAD regression problem has a nonempty feasible region and the objective function is bounded below by zero, which is not altered by the addition of the variable β_p into the model; hence the dual problem must have a feasible solution.

EXAMPLE 2.12.2 Consider Example 2.8.1. Suppose we want to add β_{new} to the model with $\mathbf{X}_{\text{new}}$ given by $\mathbf{X}'_{\text{new}} = (4,3,1,3,2)$. The $C_k - Z_k$ corresponding to this new variable is calculated as follows: We reproduce here the final table, Table 2.8.6, for convenience. The columns corresponding to the initial basis, namely $\mathbf{a}_4$, $\mathbf{a}_5$, $\mathbf{a}_6$, $\mathbf{a}_7$, and $\mathbf{a}_8$, contain the $\mathbf{B}^{-1}$ in the final table (Table 2.12.6).
We find

$$\mathbf{B}^{-1} = \begin{bmatrix} \frac{1}{4} & 0 & -\frac{1}{2} & 0 & \frac{1}{4} \\ \frac{1}{2} & 0 & \frac{3}{2} & 0 & -1 \\ \frac{1}{2} & -1 & \frac{1}{2} & 0 & 0 \\ 0 & 0 & -\frac{1}{2} & 1 & -\frac{1}{2} \\ -\frac{1}{4} & 0 & 0 & 0 & \frac{1}{4} \end{bmatrix}.$$

Table 2.12.6

C_B	Vectors in the Basis	W_B	α_1	α_2	α_3	α_4	α_5	α_6	α_7	α_8	α_9	α_{10}	α_{11}	α_{12}	α_{13}
0	$\mathbf{a}_3$	$\frac{1}{4}$	0	0	1	$\frac{1}{4}$	0	$-\frac{1}{2}$	0	$\frac{1}{4}$	$-\frac{1}{4}$	0	$\frac{1}{2}$	0	$-\frac{1}{4}$
0	$\mathbf{a}_1$	3	1	0	0	$\frac{1}{2}$	0	$\frac{3}{2}$	0	-1	$-\frac{1}{2}$	0	$-\frac{3}{2}$	0	1
1	$\mathbf{a}_{10}$	1	0	0	0	$\frac{1}{2}$	-1	$\frac{1}{2}$	0	0	$-\frac{1}{2}$	1	$-\frac{1}{2}$	0	0
1	$\mathbf{a}_7$	$\frac{1}{2}$	0	0	0	0	0	$-\frac{1}{2}$	1	$-\frac{1}{2}$	0	0	$\frac{1}{2}$	-1	$\frac{1}{2}$
0	$\mathbf{a}_2$	$\frac{1}{4}$	0	1	0	$-\frac{1}{4}$	0	0	0	$\frac{1}{4}$	$\frac{1}{4}$	0	0	0	$-\frac{1}{4}$
		$Z = \frac{3}{2}$	0	0	0	$\frac{1}{2}$	2	1	0	$\frac{3}{2}$	$\frac{3}{2}$	0	1	2	$\frac{1}{2}$

Therefore

$$C_k - Z_k = 0 - \mathbf{C}_B \mathbf{B}^{-1} \mathbf{a}_k.$$

Let $k = p + 2n + 1 = 14$ and $\mathbf{a}'_k = \mathbf{X}'_{\text{new}}$

$$C_k - Z_k = 0 - (0,0,1,1,0) \begin{bmatrix} 1 \\ \frac{3}{2} \\ -\frac{1}{2} \\ \frac{3}{2} \\ -\frac{1}{2} \end{bmatrix} = -1.$$

Hence we introduce $\mathbf{a}_{14}$ as basic and remove $\mathbf{a}_7$. We include Column $\boldsymbol{\alpha}_{14}$ corresponding to the new β as the last column and transform the table, into Table 2.12.7.

We find that the optimality conditions are satisfied and we have the estimates,

$$\hat{\beta}_0 = \tfrac{5}{2}, \qquad \hat{\beta}_1 = \tfrac{5}{12}, \qquad \hat{\beta}_2 = -\tfrac{1}{12} \qquad \text{and} \quad \hat{\beta}_{\text{new}} = \tfrac{1}{3}.$$

Notice that we could do this with the suppressed tables as well.

Adding a New Observation to the Model Suppose we wish to add a new observation, to the set of n observations already existing in the model. Let the data corresponding to this $(n+1)$st observation be given by $(X_{n+1,1}, \ldots, X_{n+1p})$ and Y_{n+1}. Adding this observation also introduces two

Table 2.12.7

C_B	Vectors in the Basis	$\mathbf{W}_B$	$\boldsymbol{\alpha}_1$	$\boldsymbol{\alpha}_2$	$\boldsymbol{\alpha}_3$	$\boldsymbol{\alpha}_4$	$\boldsymbol{\alpha}_5$	$\boldsymbol{\alpha}_6$	$\boldsymbol{\alpha}_7$	$\boldsymbol{\alpha}_8$	$\boldsymbol{\alpha}_9$	$\boldsymbol{\alpha}_{10}$	$\boldsymbol{\alpha}_{11}$	$\boldsymbol{\alpha}_{12}$	$\boldsymbol{\alpha}_{13}$	$\boldsymbol{\alpha}_{14}$
	$\mathbf{a}_3$	$-\frac{1}{12}$	0	0	1	$\frac{1}{4}$	0	$-\frac{1}{6}$	$-\frac{2}{3}$	$\frac{7}{12}$	$-\frac{1}{4}$	0	$\frac{1}{6}$	$\frac{2}{3}$	$-\frac{7}{12}$	0
	$\mathbf{a}_1$	$\frac{5}{2}$	1	0	0	$\frac{1}{2}$	0	2	-1	$-\frac{1}{2}$	$-\frac{1}{2}$	0	-2	1	$\frac{1}{2}$	0
	$\mathbf{a}_{10}$	$\frac{7}{6}$	0	0	0	$\frac{1}{2}$	-1	$\frac{2}{6}$	$\frac{1}{3}$	$-\frac{1}{6}$	$-\frac{1}{2}$	1	$-\frac{2}{6}$	$-\frac{1}{3}$	$\frac{1}{6}$	0
	$\mathbf{a}_{14}$	$\frac{1}{3}$	0	0	0	0	0	$-\frac{1}{3}$	$\frac{2}{3}$	$-\frac{1}{3}$	0	0	$\frac{1}{3}$	$-\frac{2}{3}$	$\frac{1}{3}$	1
	$\mathbf{a}_2$	$\frac{5}{12}$	0	1	0	$-\frac{1}{4}$	0	$-\frac{1}{6}$	$\frac{1}{3}$	$\frac{1}{12}$	$\frac{1}{4}$	0	$\frac{1}{6}$	$-\frac{1}{3}$	$-\frac{1}{12}$	0
	$C_k - Z_k$	$\frac{7}{6}$	0	0	0	$\frac{1}{2}$	2	$\frac{2}{3}$	$\frac{2}{3}$	$\frac{7}{6}$	$\frac{3}{2}$	0	$\frac{4}{3}$	$\frac{4}{3}$	$\frac{5}{6}$	0

more variables, $d_{n+1}^1, d_{n+1}^2 \geqslant 0$, with the objective function coefficients at 1 for both. Thus we have the new constraint

$$X_{n+11}\beta_0 + \cdots + X_{n+1p}\beta_{p-1} + d_{n+1}^1 - d_{n+1}^2 = Y_{n+1}.$$

From the optimal solution $\mathbf{W}^* = (\boldsymbol{\beta}^*, \mathbf{d}^{1*}, \mathbf{d}^{2*})$, if we find $X_{n+1,1}\beta_0^* + \cdots + X_{n+1p}\beta_{p-1}^* = Y_{n+1}$. It is obvious that the present optimal $\mathbf{W}^*$ is optimal for the expanded problem with $d_{n+1}^1 = d_{n+1}^2 = 0$. Otherwise the objective function value is altered. Consider the initial basis $\overline{\mathbf{B}}$ for the expanded problem. Let $(\mathbf{X}^{n+1})_B$ denote the components corresponding to the optimal basis in the row $\mathbf{X}^{n+1}$.

$$\overline{\mathbf{B}} = \begin{pmatrix} \mathbf{B} & \mathbf{0} \\ (\mathbf{X}^{n+1})_B & \pm 1 \end{pmatrix}$$

The new column added to the basis corresponds to d_{n+1}^1 or d_{n+1}^2, depending on whether $[Y_{n+1} - (X_{n+1,1}\beta_0^* + \cdots + X_{n+1p}\beta_{p-1}^*)]$ is positive or negative. Then

$$\overline{\mathbf{B}}^{-1} = \begin{pmatrix} \mathbf{B}^{-1} & \mathbf{0} \\ \mp (\mathbf{X}^{n+1})_B \mathbf{B}^{-1} & \pm 1 \end{pmatrix}.$$

$\mathbf{W}_{\overline{B}}, \overline{\boldsymbol{\alpha}}_j, C_j - \overline{Z}_j$, corresponding to the expanded problem, can be worked out as in the preceding discussion, and simplex steps can be carried out until optimality is achieved.

However, if we are interested in solving the MINMAD regression problem (2.7.2) through solving its dual, let us see what happens to the dual when a new constraint with the corresponding variables is added to Problem 2.7.2. Notice that adding the $(n+1)$st constraint to Problem 2.7.2 introduces a new variable, V_{n+1}, into the dual problem, and the new variables added, d_{n+1}^1 and d_{n+1}^2, introduce the constraints, which imply $0 \leqslant V_{n+1} \leqslant 2$. We calculate

$$Y_{n+1} - Z_{n+1} = Y_{n+1} - \mathbf{C}_B \mathbf{B}^{-1} \mathbf{X}^{n+1}$$

where $\mathbf{B}$ is the optimal basis for the dual problem. Notice that $(Y_l - Z_l)$'s are not altered by introducing V_{n+1} into the model. With the new column corresponding to V_{n+1} annexed to the optimal table, we proceed with the bounded-variable simplex process.

EXAMPLE 2.12.3 Consider Example 2.8.1. Suppose we have a new observation to be added to the data, given by

$$\beta_0 + 3\beta_1 + 5\beta_2 + d_6^1 - d_6^2 = 6.$$

We have from the final table (Table 2.8.6) (and also from Table 2.12.6)

$$(\hat{\beta}_0, \hat{\beta}_1, \hat{\beta}_2) = (3, \tfrac{1}{4}, \tfrac{1}{4}).$$

Therefore, $\hat{Y}_6 = 5$. As $Y_6 = 6$, $d_6^1 = 1$. We have to find an optimal solution to the expanded problem including Row 6 in the existing final basis, consisting of vectors $\bar{\mathbf{a}}_3$, $\bar{\mathbf{a}}_1$, $\bar{\mathbf{a}}_{10}$, $\bar{\mathbf{a}}_7$, and $\bar{\mathbf{a}}_2$, where the "$-$" denotes the expanded vectors, adding the coordinates corresponding to the new row.

$$(\mathbf{X}_6)_B = (5,1,0,0,3)$$

$$\bar{\mathbf{B}}^{-1} = \begin{pmatrix} \mathbf{B}^{-1} & \mathbf{0} \\ \mp(\mathbf{X}_6)_B\mathbf{B}^{-1} & 1 \end{pmatrix} = \begin{bmatrix} \frac{1}{4} & 0 & -\frac{1}{2} & 0 & \frac{1}{4} & 0 \\ \frac{1}{2} & 0 & \frac{3}{2} & 0 & -1 & 0 \\ \frac{1}{2} & -1 & \frac{1}{2} & 0 & 0 & 0 \\ 0 & 0 & -\frac{1}{2} & 1 & -\frac{1}{2} & 0 \\ -\frac{1}{4} & 0 & 0 & 0 & \frac{1}{4} & 0 \\ -1 & 0 & 1 & 0 & -1 & 1 \end{bmatrix}.$$

We need only to calculate the $\bar{\alpha}_{n+1,k}$'s, using

$$\bar{\alpha}_{n+1,k} = -(\mathbf{X}_6)_B\mathbf{B}^{-1}\mathbf{a}_k + a_{n+1,k}, \qquad k = 1,\ldots,2(n+1)+p$$

and

$$\bar{C}_k - \bar{Z}_k = C_k - Z_k - \bar{\alpha}_{n+1,k}C_{\bar{B}_{n+1}} = C_k - Z_k - \bar{\alpha}_{n+1,k}.$$

Table 2.12.8 is the expanded table.

The only candidate for entry is $\bar{\mathbf{a}}_{13}$. With $\bar{C}_{13} - \bar{Z}_{13} = -\frac{1}{2}$, $\bar{\mathbf{a}}_{14}$ can be removed from the basis. Transforming the table we find $\bar{C}_k - \bar{Z}_k$ satisfies optimality conditions and the estimates are found to be

$$\hat{\beta}_0 = 2, \qquad \hat{\beta}_1 = \tfrac{1}{2}, \qquad \hat{\beta}_2 = \tfrac{1}{2}$$

with Observations 2 and 5 falling off the fitted line by one unit each.

Table 2.12.8

C_B	Vectors in the Basis	W_B	α_1	α_2	α_3	α_4	α_5	α_6	α_7	α_8	α_9	α_{10}	α_{11}	α_{12}	α_{13}	α_{14}	α_{15}
0	$\bar{\mathbf{a}}_3$	$\frac{1}{4}$	0	0	1	$\frac{1}{4}$	0	$-\frac{1}{2}$	0	$\frac{1}{4}$	$-\frac{1}{4}$	0	$\frac{1}{2}$	0	$-\frac{1}{4}$	0	0
0	$\bar{\mathbf{a}}_1$	3	1	0	0	$\frac{1}{2}$	0	$\frac{3}{2}$	0	-1	$-\frac{1}{2}$	0	$-\frac{3}{2}$	0	1	0	0
1	$\bar{\mathbf{a}}_{10}$	1	0	0	0	$\frac{1}{2}$	-1	$\frac{1}{2}$	0	0	$-\frac{1}{2}$	1	$-\frac{1}{2}$	0	0	0	0
1	$\bar{\mathbf{a}}_7$	$\frac{1}{2}$	0	0	0	0	0	$-\frac{1}{2}$	1	$-\frac{1}{2}$	0	0	$\frac{1}{2}$	-1	$\frac{1}{2}$	0	0
0	$\bar{\mathbf{a}}_2$	$\frac{1}{4}$	0	1	0	$-\frac{1}{4}$	0	0	0	$\frac{1}{4}$	$\frac{1}{4}$	0	0	0	$-\frac{1}{4}$	0	0
1	$\bar{\mathbf{a}}_{14}$	1	0	0	0	-1	0	1	0	-1	1	0	-1	0	1	1	-1
	$\bar{C}_k - \bar{Z}_k$	$\bar{Z}=\frac{5}{2}$	0	0	0	$\frac{3}{2}$	2	0	0	$\frac{5}{2}$	$\frac{1}{2}$	0	2	2	$-\frac{1}{2}$	0	2

2.13 AN ALTERNATE FORMULATION OF THE MINMAD PROBLEM

The problem considered earlier, in Section 2.7, can be reformulated as follows:

$$\begin{aligned}
&\text{Minimize} && \mathbf{d}'\mathbf{e} \\
&\text{subject to} && -\mathbf{d} \leqslant \mathbf{Y} - \mathbf{X}\boldsymbol{\beta} \leqslant \mathbf{d} \\
& && \mathbf{d} \geqslant \mathbf{0} \\
& && \boldsymbol{\beta} \text{ unrestricted in sign.}
\end{aligned} \tag{2.13.1}$$

This formation can be written in matrix notation, writing out the two sets of constraints in (2.13.1) and introducing $\boldsymbol{\beta}=\boldsymbol{\beta}_1-\boldsymbol{\beta}_2$, where $\boldsymbol{\beta}_1$ and $\boldsymbol{\beta}_2$ are nonnegative vectors. Thus we

$$\begin{aligned}
&\text{minimize} \quad \mathbf{d}'\mathbf{e} \\
&\text{subject to} \quad \begin{pmatrix} \mathbf{X} & -\mathbf{X} & -\mathbf{I} \\ -\mathbf{X} & \mathbf{X} & -\mathbf{I} \end{pmatrix} \begin{bmatrix} \boldsymbol{\beta}_1 \\ \boldsymbol{\beta}_2 \\ \mathbf{d} \end{bmatrix} \leqslant \begin{pmatrix} \mathbf{Y} \\ -\mathbf{Y} \end{pmatrix} \\
&\qquad \boldsymbol{\beta}_1, \boldsymbol{\beta}_2, \mathbf{d} \geqslant \mathbf{0}.
\end{aligned} \tag{2.13.2}$$

There are $2n$ constraints and $2p+n$ variables in this problem. Yet we will be using this formulation for obtaining unbiased estimates of $\boldsymbol{\beta}$. In general a solution to the MINMAD problem need not yield an unbiased estimate of $\boldsymbol{\beta}$, even when the errors are symmetrically distributed and $E(\boldsymbol{\varepsilon})=\mathbf{0}$.

EXAMPLE 2.13.1 Consider $\mathbf{Y}'=[Y_1, Y_2]$, $\mathbf{X}'=[1,1]$, and $\boldsymbol{\varepsilon}'=(\varepsilon_1, \varepsilon_2)$. The linear model under consideration is $\mathbf{Y}=\mathbf{X}\boldsymbol{\beta}+\boldsymbol{\varepsilon}$. Suppose $\beta=1$ and ε_i is distributed as follows:

$$\varepsilon_i = \begin{cases} -1 & \text{with probability } \frac{1}{2} \\ +1 & \text{with probability } \frac{1}{2} \end{cases}$$

for $i=1,2$.

We wish to estimate β solving (2.13.1) corresponding to the different possible $[\varepsilon_1, \varepsilon_2]$, using the linear-programming formulation. We have, for instance when $\varepsilon_1=\varepsilon_2=-1$, the problem:

$$\begin{aligned} &\text{Minimize} \quad d_1+d_2 \\ &\text{subject to} \quad -\begin{pmatrix} d_1 \\ d_2 \end{pmatrix} \leqslant \begin{pmatrix} 0 \\ 0 \end{pmatrix} - \begin{pmatrix} 1 \\ 1 \end{pmatrix}\beta \leqslant \begin{pmatrix} d_1 \\ d_2 \end{pmatrix} \\ &\qquad\qquad\quad d_1, d_2 \geqslant 0. \end{aligned}$$

The minimum here is attained when $d_1=d_2=\hat{\beta}=0$, uniquely. Similarly, when $\varepsilon_1=\varepsilon_2=1$, the corresponding problem,

$$\begin{aligned} &\text{Minimize} \quad d_1+d_2 \\ &\text{subject to} \quad -\begin{pmatrix} d_1 \\ d_2 \end{pmatrix} \leqslant \begin{pmatrix} 2 \\ 2 \end{pmatrix} - \begin{pmatrix} 1 \\ 1 \end{pmatrix}\beta \leqslant \begin{pmatrix} d_1 \\ d_2 \end{pmatrix} \\ &\qquad\qquad\quad d_1, d_2 \geqslant 0. \end{aligned}$$

has a unique solution $\hat{\beta}=2$, $d_1=d_2=0$. But corresponding to $\varepsilon_1=-1$ and $\varepsilon_2=+1$ or $\varepsilon_1=+1$ and $\varepsilon_2=-1$ we have problems which do not have unique optimal solutions. In these two cases, $0\leqslant\hat{\beta}\leqslant 2$ is an optimal solution to the corresponding problem. For instance, when $\varepsilon_1=-1$ and $\varepsilon_2=+1$, we have

$$\begin{aligned} &\text{Minimize} \quad d_1+d_2 \\ &\text{subject to} \quad -\begin{pmatrix} d_1 \\ d_2 \end{pmatrix} \leqslant \begin{pmatrix} 0 \\ 2 \end{pmatrix} - \begin{pmatrix} 1 \\ 1 \end{pmatrix}\beta \leqslant \begin{pmatrix} d_1 \\ d_2 \end{pmatrix} \\ &\qquad\qquad\quad d_1, d_2 \geqslant 0. \end{aligned}$$

Suppose, we choose $\hat{\beta}=0$ in these two cases. We find

Case	ε_1	ε_2	$\hat{\beta}$
1	-1	-1	0
2	1	1	2
3	-1	1	0
4	1	-1	0

Hence for the estimator thus obtained has $E(\hat{\beta})=\frac{1}{2}$, which is biased as $\beta=1$. In fact, if we consider the alternate optimal solution $\hat{\beta}=1$, in Cases 3 and 4, we get $E(\hat{\beta})=1=\beta$. Thus when the solution to the linear-programming problem is not unique, we have the danger of getting biased estimates.

2.14 UNBIASED ESTIMATION USING MINMAD

To obtain unbiased estimates using MINMAD regression, let us assume $\boldsymbol{\varepsilon}$ is symmetric and $\mathbf{E}(\boldsymbol{\varepsilon})=\mathbf{0}$.

DEFINITION 2.14.1 We define an estimator of $\boldsymbol{\beta}$ denoted by $\boldsymbol{\beta}_0$ to be *antisymmetrical* if $\boldsymbol{\beta}-\boldsymbol{\beta}_0(\boldsymbol{\varepsilon})=-(\boldsymbol{\beta}-\boldsymbol{\beta}_0(-\boldsymbol{\varepsilon}))$ for any value $\boldsymbol{\varepsilon}$ taken by $\boldsymbol{\varepsilon}$.

EXAMPLE 2.14.1

$$\boldsymbol{\beta}_0=(\mathbf{X}'\mathbf{X})^{-1}\mathbf{X}'\mathbf{Y} \text{ is the least-squares estimate}$$

$$\begin{aligned}\boldsymbol{\beta}_0(\boldsymbol{\varepsilon})&=(\mathbf{X}'\mathbf{X})^{-1}\mathbf{X}'(\mathbf{X}\boldsymbol{\beta}+\boldsymbol{\varepsilon})\\&=\boldsymbol{\beta}+(\mathbf{X}'\mathbf{X})^{-1}\mathbf{X}'\boldsymbol{\varepsilon}\\\boldsymbol{\beta}_0(-\boldsymbol{\varepsilon})&=(\mathbf{X}'\mathbf{X})^{-1}\mathbf{X}'(\mathbf{X}\boldsymbol{\beta}-\boldsymbol{\varepsilon})\\&=\boldsymbol{\beta}-(\mathbf{X}'\mathbf{X})^{-1}\mathbf{X}'\boldsymbol{\varepsilon}.\end{aligned}$$

Thus $\boldsymbol{\beta}-\boldsymbol{\beta}_0(\boldsymbol{\varepsilon})=-(\boldsymbol{\beta}-\boldsymbol{\beta}_0(-\boldsymbol{\varepsilon}))$, so $\boldsymbol{\beta}_0$ is antisymmetrical.

EXAMPLE 2.14.2

$$\boldsymbol{\beta}_0=\mathbf{X}_A^{-1}\mathbf{Y}_A$$

where $\mathbf{X}_A$ is any $p\times p$ submatrix of $\mathbf{X}$ with rank p, and $\mathbf{Y}_A$ is the corresponding vector of p observations.

$$\boldsymbol{\beta}_0(\boldsymbol{\varepsilon}) = \mathbf{X}_A^{-1}(\mathbf{X}_A\boldsymbol{\beta} + \boldsymbol{\varepsilon}_A) = \boldsymbol{\beta} + \mathbf{X}_A^{-1}\boldsymbol{\varepsilon}_A$$

$$\begin{aligned}\boldsymbol{\beta}_0(-\boldsymbol{\varepsilon}) &= \mathbf{X}_A^{-1}(\mathbf{X}_A\boldsymbol{\beta} - \boldsymbol{\varepsilon}_A)\\ &= \boldsymbol{\beta} - \mathbf{X}_A^{-1}\boldsymbol{\varepsilon}_A.\end{aligned}$$

It can be seen that $\boldsymbol{\beta}_0$ is antisymmetrical.

Let us now consider Problem 2.13.2. The problem can be written in an equivalent form:

$$\text{Minimize} \quad \mathbf{d}'\mathbf{e}$$

$$\text{subject to} \quad \begin{pmatrix} -\mathbf{X} & \mathbf{X} & -\mathbf{I} \\ \mathbf{X} & -\mathbf{X} & -\mathbf{I} \end{pmatrix} \begin{bmatrix} \boldsymbol{\beta}_2 \\ \boldsymbol{\beta}_1 \\ \mathbf{d} \end{bmatrix} \leqslant \begin{pmatrix} \mathbf{Y} \\ -\mathbf{Y} \end{pmatrix}. \tag{2.14.1}$$

$$\boldsymbol{\beta}_1, \boldsymbol{\beta}_2, \mathbf{d} \geqslant \mathbf{0}$$

Now we can define new variables using any antisymmetrical estimator $\boldsymbol{\beta}_0$.

$$\bar{\boldsymbol{\beta}}_1 = \boldsymbol{\beta}_1 + \boldsymbol{\beta}_0^{(2)}, \qquad \bar{\boldsymbol{\beta}}_2 = \boldsymbol{\beta}_2 + \boldsymbol{\beta}_0^{(1)}$$

where $\boldsymbol{\beta}_0 = \boldsymbol{\beta}_0^{(1)} - \boldsymbol{\beta}_0^{(2)}$ such that $\boldsymbol{\beta}_0^{(1)}, \boldsymbol{\beta}_0^{(2)} \geqslant \mathbf{0}$. Problems 2.13.2 and 2.14.1 can be stated equivalently as given by (2.14.2) and (2.14.3).

$$\text{Minimize} \quad \mathbf{d}'\mathbf{e}$$

$$\text{subject to} \quad \begin{pmatrix} -\mathbf{X} & \mathbf{X} & -\mathbf{I} \\ \mathbf{X} & -\mathbf{X} & -\mathbf{I} \end{pmatrix} \begin{bmatrix} \bar{\boldsymbol{\beta}}_1 \\ \bar{\boldsymbol{\beta}}_2 \\ \mathbf{d} \end{bmatrix} \leqslant \begin{pmatrix} -\mathbf{Y} & +\mathbf{X}\boldsymbol{\beta}_0 \\ \mathbf{Y} & -\mathbf{X}\boldsymbol{\beta}_0 \end{pmatrix}$$

$$\bar{\boldsymbol{\beta}}_1, \bar{\boldsymbol{\beta}}_2, \mathbf{d} \geqslant \mathbf{0}. \tag{2.14.2}$$

$$\text{Minimize} \quad \mathbf{d}'\mathbf{e}$$

$$\text{subject to} \quad \begin{pmatrix} -\mathbf{X} & \mathbf{X} & -\mathbf{I} \\ \mathbf{X} & -\mathbf{X} & -\mathbf{I} \end{pmatrix} \begin{bmatrix} \bar{\boldsymbol{\beta}}_2 \\ \bar{\boldsymbol{\beta}}_1 \\ \mathbf{d} \end{bmatrix} \leqslant \begin{pmatrix} \mathbf{Y} & -\mathbf{X}\boldsymbol{\beta}_0 \\ -\mathbf{Y} & +\mathbf{X}\boldsymbol{\beta}_0 \end{pmatrix} \tag{2.14.3}$$

$$\bar{\boldsymbol{\beta}}_1, \bar{\boldsymbol{\beta}}_2, \mathbf{d} \geqslant \mathbf{0}.$$

We can obtain Problem 2.14.3 from (2.14.2) by rearranging the rows and columns.

We define a new estimator $\boldsymbol{\beta}_0^*$, obtained as follows: Problems 2.14.2 and 2.14.3 are selected with equal probabilities. Let $\mathbf{P}_1$ and $\mathbf{P}_2$ denote Problems 2.14.2 and 2.14.3, respectively. Suppose $\mathbf{P}_1$ and $\mathbf{P}_2$ have alternate basic optimal solutions. Any time these problems are chosen for solving, we choose one of the optimal solutions by giving probability to all the alternate optimal basic solutions, a finite number. Let such an optimal solution to the problem chosen be given by $\bar{\boldsymbol{\beta}}_1^*$ and $\bar{\boldsymbol{\beta}}_2^*$. Then, define

$$\boldsymbol{\beta}_0^* = \bar{\boldsymbol{\beta}}_1^* - \bar{\boldsymbol{\beta}}_2^* + \boldsymbol{\beta}_0$$

RESULT 2.14.1 $\boldsymbol{\beta}_0^*$ is an unbiased estimator of $\boldsymbol{\beta}$.

Proof

$$\begin{aligned}\mathbf{E}(\boldsymbol{\beta} - \boldsymbol{\beta}_0^*) &= \mathbf{E}\left[\boldsymbol{\beta} - \left(\bar{\boldsymbol{\beta}}_1^* - \bar{\boldsymbol{\beta}}_2^* + \boldsymbol{\beta}_0\right)\right] \\ &= \mathbf{E}[\boldsymbol{\beta} - \boldsymbol{\beta}_0] - \mathbf{E}\left[\bar{\boldsymbol{\beta}}_1^* - \bar{\boldsymbol{\beta}}_2^*\right].\end{aligned}$$

Now $\mathbf{E}(\boldsymbol{\beta}_0) = \boldsymbol{\beta}$ as $\boldsymbol{\beta}_0$ is antisymmetrical and $\boldsymbol{\varepsilon}$ is symmetrical. Therefore $\mathbf{E}(\boldsymbol{\beta} - \boldsymbol{\beta}_0^*) = -\mathbf{E}\{[\bar{\boldsymbol{\beta}}_1^* - \bar{\boldsymbol{\beta}}_2^* | \boldsymbol{\varepsilon}] + [\bar{\boldsymbol{\beta}}_1^* - \bar{\boldsymbol{\beta}}_2^* | -\boldsymbol{\varepsilon}]\}$. At this point we observe, for any value $\boldsymbol{\varepsilon}$ taken by $\boldsymbol{\varepsilon}$, the right-hand-side elements

$$\begin{aligned}\mathbf{Y}(-\boldsymbol{\varepsilon}) - \mathbf{X}\boldsymbol{\beta}_0(-\boldsymbol{\varepsilon}) &= \mathbf{X}\boldsymbol{\beta} - \boldsymbol{\varepsilon} - \mathbf{X}\boldsymbol{\beta}_0(-\boldsymbol{\varepsilon}) \\ &= -\boldsymbol{\varepsilon} + \mathbf{X}[\boldsymbol{\beta} - \boldsymbol{\beta}_0(-\boldsymbol{\varepsilon})] \\ &= -\boldsymbol{\varepsilon} - \mathbf{X}[\boldsymbol{\beta} - \boldsymbol{\beta}_0(\boldsymbol{\varepsilon})], \text{ as } \boldsymbol{\beta}_0 \text{ is antisymmetrical} \\ &= -\mathbf{Y}(\boldsymbol{\varepsilon}) + \mathbf{X}\boldsymbol{\beta}_0(\boldsymbol{\varepsilon}).\end{aligned}$$

Hence Problem 2.14.2 when $\boldsymbol{\varepsilon}$ takes the value $\boldsymbol{\varepsilon}$ and Problem 2.14.3 when $\boldsymbol{\varepsilon}$ takes the value $-\boldsymbol{\varepsilon}$ are the same, except that the constraints have been written in different order. Thus, $\mathbf{E}(\bar{\boldsymbol{\beta}}_1^* - \bar{\boldsymbol{\beta}}_2^* | \boldsymbol{\varepsilon}, \mathbf{P}_1)$ obtained from problem $\mathbf{P}_1$ when $\boldsymbol{\varepsilon}$ takes the value $\boldsymbol{\varepsilon}$ is the same as $\mathbf{E}(\bar{\boldsymbol{\beta}}_2^* - \bar{\boldsymbol{\beta}}_1^* | -\boldsymbol{\varepsilon}, \mathbf{P}_2)$ obtained from problem $\mathbf{P}_2$ when $\boldsymbol{\varepsilon}$ takes the value $-\boldsymbol{\varepsilon}$. Now,

$$\begin{aligned}\mathbf{E}(\boldsymbol{\beta} - \boldsymbol{\beta}_0^*) = -\mathbf{E}\Big\{&\left[\bar{\boldsymbol{\beta}}_1^* - \bar{\boldsymbol{\beta}}_2^* | \boldsymbol{\varepsilon}, \mathbf{P}_1\right]\tfrac{1}{2} + \left[\bar{\boldsymbol{\beta}}_1^* - \bar{\boldsymbol{\beta}}_2^* | \boldsymbol{\varepsilon}, \mathbf{P}_2\right]\tfrac{1}{2} \\ &+ \left[\bar{\boldsymbol{\beta}}_1^* - \bar{\boldsymbol{\beta}}_2^* | -\boldsymbol{\varepsilon}, \mathbf{P}_1\right]\tfrac{1}{2} + \left[\bar{\boldsymbol{\beta}}_1^* - \bar{\boldsymbol{\beta}}_2^* | -\boldsymbol{\varepsilon}, \mathbf{P}_2\right]\tfrac{1}{2}\Big\}\end{aligned}$$

Using the observation made above, we get

$$E(\boldsymbol{\beta}-\boldsymbol{\beta}_0^*)=E\{\mathbf{0}\}=\mathbf{0}.$$

Thus $\boldsymbol{\beta}_0^*$ is an unbiased estimate of $\boldsymbol{\beta}$.

Thus if we have an antisymmetric estimate $\boldsymbol{\beta}_0$, we can formulate two linear-programming problems and choose one of them with equal probability to obtain unbiased estimates.

2.15 A SIMULATION STUDY

To examine what would be the bias in the estimates if we select the optimal solution given by the linear programming package, and do not go through the method described earlier, a simulation study was conducted. This simulation study was also designed for evaluating the performances of different methods of estimation using three different distributions. The following model was used:

$$Y=16+4X_{i1}+X_{i2}+0.25X_{i3}+\varepsilon_i, \qquad i=1,\dots,20.$$

A set of three vectors X_1, X_2 and X_3 with no significant correlations ($r_{12}=0.03$, $r_{13}=-0.06$, and $r_{23}=-0.28$) was obtained using a set of random numbers within the range -10 and 10.

The errors are chosen to be independently and symmetrically distributed with mean 0 and variance 1. We chose uniform, normal, and Laplace distributions for consideration.

Five different methods of estimation were considered: least squares (L_2-Norm), MINMAD, MINMAXAD, MINSADBED, and MINSADBAD.

A CDC 6400 random-number-generator routine (RANF) was used for generating purposes. Means of 100 samples per parameter for five different methods along with their variances were computed. This result is presented in Table 2.15.1.

Notice that the bias is not appreciable, even when the estimates are obtained straight from the linear-programming optimal solution given by a computer package. Thus for all practical purposes it may be enough to solve the corresponding problem as a linear-programming problem and pick the estimates from the optimal solution corresponding to the final basis.

Table 2.15.1

	L_2-Norm		L_1-Norm(MINMAD)		MINMAXAD		MINSADBED		MINSADBAD	
Distrib.	Mean	Var.	Mean	Var.	Mean	Var.	Mean	Var.	Mean	Var.
	16.017	0.065	16.075	0.261	15.992	0.492	16.080	0.672	15.982	0.206
	3.999	0.002	4.004	0.003	3.995	0.002	3.001	0.002	4.007	0.005
Uniform	1.001	0.002	1.002	0.004	0.988	0.011	1.092	0.820	1.008	0.014
	0.250	0.002	0.264	0.003	0.252	0.001	0.250	0.002	0.241	0.008
	16.016	0.064	16.007	0.784	15.997	0.124	16.082	0.066	16.012	0.234
	4.006	0.002	4.003	0.002	4.003	0.003	4.005	0.002	4.007	0.007
Normal	1.002	0.002	1.005	0.003	1.002	0.009	1.002	0.001	1.003	0.012
	0.256	0.001	0.265	0.002	0.253	0.003	0.258	0.001	0.249	0.010
	16.010	0.096	16.026	0.101	16.005	0.306	16.075	0.095	15.930	0.622
	4.003	0.003	4.002	0.004	4.019	0.010	4.002	0.002	4.000	0.016
Laplace	1.009	0.003	1.012	0.003	1.028	0.015	1.003	0.006	1.010	0.022
	0.254	0.003	0.266	0.004	0.248	0.009	0.263	0.010	0.258	0.022

2.16 MAXIMUM-LIKELIHOOD ESTIMATES AND MINMAD ESTIMATES

Suppose the errors are independently and identically distributed as the double exponential distribution, with known σ. That is,

$$f(\varepsilon_i)=(2\sigma)^{-1}e^{-|\varepsilon_i|}/\sigma \quad -\infty<\varepsilon_i<\infty.$$

It is known that the application of the maximum-likelihood method for estimating β would imply minimizing

$$\sum_{i=1}^{n}|\varepsilon_i|.$$

Thus we have a justification for considering MINMAD estimation, and the application of linear programming thus comes handy in such situations. Several researchers find these estimates to be superior to least-squares estimates in certain cases when normality assumption on errors is violated. (A discussion on other properties of MINMAD estimates appears in the bibliographical notes.)

2.17 MINIMIZING MAXIMUM ABSOLUTE DEVIATIONS (MINMAXAD)

In this section we consider the minimization of the maximum deviation and formulate the seemingly nonlinear problem as a linear-programming problem. We also consider different ways of looking at the same problem, for a geometrical insight into the problem.

The problem is stated as follows:

$$\underset{\beta}{\text{Minimize}} \qquad \max_{1<i<n} |Y_i - \sum_{j=1}^{p} X_{ij}\beta_j| \tag{2.17.1}$$

$$\beta \text{ unrestricted in sign.}$$

Observe that (1) if $\mathbf{Y}$ is a linear combination of $\mathbf{X}_1,\ldots,\mathbf{X}_p, \mathbf{X}\boldsymbol{\beta}-\mathbf{Y}=\mathbf{0}$ has a solution $\boldsymbol{\beta}^*$, and corresponding to that solution,

$$\max_{1<i<n} |Y_i - \sum_{j=1}^{p} X_{ij}\beta_j^*|=0.$$

Hence $\boldsymbol{\beta}^*$ is an optimal solution to Problem 2.17.1. (2) If for any $j, \mathbf{X}_j = \mathbf{0}$, we can drop that column and corresponding β_j from the problem. Hence we assume that

(1) $\mathbf{Y}$ is not a linear combination of $\mathbf{X}_1,\ldots,\mathbf{X}_p$.

(2) $\mathbf{X}_j \neq \mathbf{0}$ for all $j=1,\ldots,p$.

Let

$$d= \max_{1<i<n} |Y_i - \sum_{j=1}^{p} X_{ij}\beta_j|.$$

Observe that d depends on $\boldsymbol{\beta}$. Under these assumptions, we can assume that $\mathbf{X}_j$'s are normalized, so that

$$\sum_{j=1}^{p} X_{ij}^2=1$$

for all $i=1,\ldots,n$. Then,

$$Y_i - \sum_{j=1}^{p} \beta_j X_{ij}=0, \qquad i=1,\ldots,n, \tag{2.17.2}$$

gives the equation of n hyperplanes in R^p consisting of the points $\boldsymbol{\beta}$. If $\boldsymbol{\beta}$ is an arbitrary point in $R^p, |Y_i - \sum_{j=1}^{p}\beta_j X_{ij}|$ is the distance of the point from the ith hyperplane. d is the maximum of these; the problem is to find a point $\boldsymbol{\beta}\in R^p$ for which this maximal distance is minimized. These observations are true for MINMAD regression as well.

We can restate the problem (2.17.1) as a linear-programming problem.

$$\begin{aligned} &\text{Minimize} \quad d \\ &\text{subject to} \quad d+\sum_{j=1}^{p} \beta_j X_{ij} \geqslant Y_i \\ &\qquad\qquad d-\sum_{j=1}^{p} \beta_j X_{ij} \geqslant -Y_i \\ &\qquad\qquad i=1,\ldots,n \\ &\qquad\qquad d, \beta_j \quad \text{unrestricted in sign.} \end{aligned} \tag{2.17.3}$$

In this problem d is not explicitly restricted in sign, but it will be nonnegative. Also d is bounded below. Hence we always have an optimal solution to this problem, by the assumption (2), $d>0$. Therefore, transform the variables as follows:

$$b_0=\frac{1}{d}, \qquad b_j=-\frac{\beta_j}{d}, \qquad j=1,\ldots,p.$$

We have an equivalent problem:

$$\begin{aligned} &\text{Maximize} \quad b_0 \\ &\text{subject to} \quad b_0 Y_i+\sum_{j=1}^{p} b_j X_{ij} \leqslant 1 \\ &\qquad\qquad -b_0 Y_i-\sum_{j=1}^{p} b_j X_{ij} \leqslant 1 \\ &\qquad\qquad i=1,\ldots,n \\ &\qquad\qquad b_0, b_j \quad \text{unrestricted in sign.} \end{aligned} \tag{2.17.4}$$

Now we can apply the simplex method to solve this problem, after adding slack variables to each of the constraints and minimizing $-b_0$.

Further modification is possible, which will reduce the size of the basis. We rewrite the last problem (Problem 2.17.4) as,

$$\begin{aligned} &\text{Maximize} \quad b_0 \\ &\text{subject to} \quad b_0 Y_i+\sum_{j=1}^{p} b_j X_{ij}+b_{s_i}=0 \\ &\qquad\qquad -1\leqslant b_{s_i} \leqslant 1 \qquad i=1,\ldots,n, \\ &\qquad\qquad b_0 \quad \text{unrestricted in sign.} \end{aligned} \tag{2.17.5}$$

Notice that this problem can be solved using the bounded-variables method after transforming once more the variables by $W_{s_i} = b_{s_i} + 1$, then $0 \leqslant W_{s_i} \leqslant 2$. Thus Problem 2.17.5 becomes:

$$\begin{aligned}
&\text{Maximize} \quad b_0 \\
&\text{subject to} \quad b_0 Y_i + \sum_{j=1}^{p} b_j X_{ij} + W_{s_i} = 1 \\
&\qquad\qquad 0 \leqslant W_{s_i} \leqslant 2 \\
&\qquad\qquad i = 1, \ldots, n.
\end{aligned} \tag{2.17.6}$$

Turning to the dual of Problem 2.17.3 we have, in matrix form,

$$\begin{aligned}
&\text{Maximize} \quad \mathbf{Y}'\mathbf{U}_1 - \mathbf{Y}'\mathbf{U}_2 \\
&\text{subject to} \quad [\mathbf{X}', -\mathbf{X}'] \begin{pmatrix} \mathbf{U}_1 \\ \mathbf{U}_2 \end{pmatrix} = \mathbf{0} \\
&\qquad\qquad \mathbf{U}_1'\mathbf{e} + \mathbf{U}_2'\mathbf{e} = 1 \\
&\qquad\qquad \mathbf{U}_1 \geqslant \mathbf{0}, \qquad \mathbf{U}_2 \geqslant \mathbf{0}.
\end{aligned} \tag{2.17.7}$$

The problem dual to Problem 2.17.4 is

$$\begin{aligned}
&\text{Minimize} \quad \mathbf{V}_1'\mathbf{e} + \mathbf{V}_2'\mathbf{e} \\
&\text{subject to} \quad \begin{pmatrix} \mathbf{Y}', -\mathbf{Y}' \\ \mathbf{X}', -\mathbf{X}' \end{pmatrix} \begin{pmatrix} \mathbf{V}_1 \\ \mathbf{V}_2 \end{pmatrix} = \begin{pmatrix} 1 \\ \mathbf{0} \end{pmatrix} \\
&\qquad\qquad \mathbf{V}_1 \geqslant \mathbf{0}, \qquad \mathbf{V}_2 \geqslant \mathbf{0}.
\end{aligned} \tag{2.17.8}$$

As Problem 2.17.3 has a finite optimal solution, so do Problems 2.17.7 and 2.17.8. In an optimal solution to Problem 2.17.8 either V_{1i}, or V_{2i} is zero. This is so because $\delta = \min(V_{1i}, V_{2i})$ is positive for some i implies that $V_{1i} - \delta, V_{2i} - \delta$ is also feasible for the problem and the objective function value is reduced by 2δ. Furthermore, if we are considering only basic feasible solutions, then both V_{1i} and V_{2i} cannot be positive as the columns are linearly dependent. Therefore we can write $V_i = V_{1i} - V_{2i}$ and $|V_i| = V_{1i} + V_{2i}$. Problem 2.17.8 then can be stated equivalently as,

$$\begin{aligned}
&\text{Minimize} \quad \sum_{i=1}^{n} |V_i| \\
&\text{subject to} \quad \mathbf{Y}'\mathbf{V} = 1 \\
&\qquad\qquad \mathbf{X}'\mathbf{V} = \mathbf{0} \\
&\qquad\qquad \mathbf{V} \text{ unrestricted in sign.}
\end{aligned} \tag{2.17.9}$$

Problem 2.17.9 can be transformed to an equivalent problem, Problem 2.17.10, which coincides with Problem 2.17.7.

Let

$$U_i = \frac{V_i}{\sum_{l=1}^{n} \|V_l\|}.$$

Then $\Sigma_{i=1}^{n} \|U_i\| = 1$. We have

$$\begin{aligned} &\text{maximize} \quad && \mathbf{Y}'\mathbf{U} \\ &\text{subject to} \quad && \mathbf{X}'\mathbf{U} = \mathbf{0} \\ & && \sum_{i=1}^{n} \|U_i\| = 1 \\ & && \mathbf{U} \text{ unrestricted in sign.} \end{aligned} \tag{2.17.10}$$

If we now replace U_i by $U_{1i} - U_{2i}$ and $|U_i|$ by $U_{1i} + U_{2i}$, we obtain Problem 2.17.7. In addition to Assumptions 1 and 2, make the following assumption: (3) Every p-rowed square submatrix of $\mathbf{X}$ is nonsingular. We can prove certain results on nondegeneracy of the extreme points of the set of feasible solutions to Problems 2.17.7 and 2.17.8.

RESULT 2.17.1 If Assumptions 1–3 are satisfied, then Problems 2.17.7 and 2.17.8 have no degenerate extreme points.

Proof Consider the constraints of Problem 2.17.7:

$$[\mathbf{X}', -\mathbf{X}'] \begin{pmatrix} \mathbf{U}_1 \\ \mathbf{U}_2 \end{pmatrix} = \mathbf{0}$$

$$\mathbf{U}_1'\mathbf{e} + \mathbf{U}_2'\mathbf{e} = 1$$

or

$$\sum_{j=1}^{n} X_{ij}(U_{1i} - U_{2i}) = 0, \qquad i = 1, \ldots, p$$

$$\sum_{j=1}^{n} (U_{1i} + U_{2i}) = 1,$$

where $U_{1i} \geqslant 0, U_{2i} \geqslant 0$; $U_{1i} - U_{2i}$ cannot be zero for all i because

$$\sum_i (U_{1i} + U_{2i}) = 1.$$

Since any p columns of matrix $\mathbf{X}'$ are linearly independent (Assumption 3), the constraints can be satisfied only if at least $p+1$ of the variables U_{1i}, U_{2i} are different from zero. Any basis for this problem will have size $p+1$. Suppose an extreme point $(\mathbf{U}_1, \mathbf{U}_2)$ of the set of feasible solutions to the problem is degenerate. Then the number of nonzero components in $(\mathbf{U}_1, \mathbf{U}_2)$ will be less than $p+1$, which will contradict the fact that such a solution is feasible. This proves the result for Problem 2.17.7. Similarly we can prove this result for Problem 2.17.8.

As a corollary to this result, using the complementary slackness property of duality theory of linear programming we obtain:

RESULT 2.17.2 If Assumptions 1–3 hold and an optimal solution to Problem 2.17.3 has been found, then at least $p+1$ of the constraints for this problem are satisfied with equality. This also implies that the maximal value, d, will be attained for at least $p+1$ of the observations.

From the foregoing results (2.17.1 and 2.17.2) under the assumptions made, none of the extreme points is degenerate, so exactly $p+1$ of the variables U_{1i}, U_{2i}, in Problem 2.17.7, are different from zero at each such extreme point. Also both U_{1i} and U_{2i} cannot be positive, for any i, as the corresponding columns are linearly dependent. Thus, letting $U_i = U_{1i} - U_{2i}$, there are exactly $p+1$ indices i for which U_i is nonzero, for any extreme point of the feasible region of Problem 2.17.7.

Let the set of such $p+1$ indices corresponding to an extreme point be denoted by S. On the other hand any such set S of indices defines a linear system of equations

$$\sum_{i \in S} X_{ij} U_i = 0, \qquad j = 1, \ldots, p \tag{2.17.11}$$

and

$$\sum_{i \in S} \|U_i\| = 1 \tag{2.17.12}$$

Thus, $U_i = 0, i \in S$ is not a solution to (2.17.11) and (2.17.12). We claim all the variables U_j are nonzero, in such a case. If $U_i = 0$ for some i then the rest of the columns can express $\mathbf{0}$, the right-hand side of (2.17.11), only if the corresponding U_j's are all zero, because of Assumption 3. Furthermore we have these U_i's uniquely determined up to a common factor. This fact can be easily verified, again using Assumption 3. Suppose the common factor is λ. Then, to satisfy (2.17.12),

$$|\lambda| \sum_{i\in S} \|U_i\| = 1 \qquad \text{or} \quad \lambda = \pm 1.$$

Thus we have exactly two solutions to the system (2.17.11) and (2.17.12).

We now consider the restricted regression problem, corresponding to a given set S of $p+1$ indices,

$$\underset{\beta}{\text{Minimize}} \max_{i\in S} \left| Y_i - \sum_{j=1}^{p} X_{ij}\beta_j \right|$$

The solution $\boldsymbol{\beta}_S$ to this restricted problem is the center of the ball inscribed in the simplex bounded by the $p+1$ hyperplanes in R^p,

$$Y_i - \sum_{j=1}^{p} X_{ij}\beta_j = 0, \qquad i\in S \tag{2.17.13}$$

The minimal value d_S is the radius of the inscribed ball. Therefore, Problem 2.17.7 is equivalent to finding from all subsets $\mathbf{S}$ of $p+1$ indices, that one for which the corresponding inscribed ball in the simplex given by (2.17.13) has the maximal radius.

Even though this geometrical interpretation of MINMAXAD problem under some assumptions, proves the above equivalence, it does not provide a simple method to solve the problem, unless C^n_{p+1} is small and n is 2 or 3 —in which case we can solve the problem graphically. Otherwise we resort to solving the problem by linear programming.

2.18 OBTAINING UNBIASED ESTIMATES USING MINMAXAD

In this case, as in Section 2.14, we can obtain unbiased estimates by identifying the two equivalent problems and using an estimate $\boldsymbol{\beta}_0$ which is antisymmetric. We have the two problems:

Minimize $\mathbf{d}$

$$\text{subject to} \quad \begin{pmatrix} -\mathbf{X} & \mathbf{X} & -\mathbf{e} \\ \mathbf{X} & -\mathbf{X} & -\mathbf{e} \end{pmatrix} \begin{bmatrix} \bar{\boldsymbol{\beta}}_1 \\ \bar{\boldsymbol{\beta}}_2 \\ \mathbf{d} \end{bmatrix} \leqslant \begin{pmatrix} -\mathbf{Y} & +\mathbf{X}\boldsymbol{\beta}_0 \\ \mathbf{Y} & -\mathbf{X}\boldsymbol{\beta}_0 \end{pmatrix} \tag{2.18.1}$$

$$\bar{\boldsymbol{\beta}}_1 \geqslant \mathbf{0}, \bar{\boldsymbol{\beta}}_2 \geqslant \mathbf{0}, \mathbf{d} \geqslant \mathbf{0}$$

and

$$\begin{aligned}&\text{Minimize} \quad \mathbf{d}\\ &\text{subject to} \quad \begin{pmatrix} -\mathbf{X} & \mathbf{X} & -\mathbf{e} \\ \mathbf{X} & -\mathbf{X} & -\mathbf{e} \end{pmatrix} \begin{bmatrix} \bar{\boldsymbol\beta}_2 \\ \bar{\boldsymbol\beta}_1 \\ \mathbf{d} \end{bmatrix} \leqslant \begin{pmatrix} \mathbf{Y} & -\mathbf{X}\boldsymbol\beta_0 \\ -\mathbf{Y} & +\mathbf{X}\boldsymbol\beta_0 \end{pmatrix} \qquad (2.18.2)\\ &\qquad \bar{\boldsymbol\beta}_1 \geqslant \mathbf{0}, \bar{\boldsymbol\beta}_2 \geqslant \mathbf{0}, \mathbf{d} \geqslant \mathbf{0},\end{aligned}$$

where $\bar{\boldsymbol\beta} = \bar{\boldsymbol\beta}_1 - \bar{\boldsymbol\beta}_2$ and $\bar{\boldsymbol\beta}_1 = \boldsymbol\beta_1 + \boldsymbol\beta_0^{(2)}, \bar{\boldsymbol\beta}_2 = \boldsymbol\beta_2 + \boldsymbol\beta_0^{(1)}$ as defined in Section 2.14. Let Problems 2.18.1 and 2.18.2 be denoted by $\mathbf{P}_1^*$ and $\mathbf{P}_2^*$, respectively. Then we obtain an unbiased estimated of $\boldsymbol\beta$ as follows:

1 Choose a $\boldsymbol\beta_0$ that is antisymmetric.

2 Select the problems $\mathbf{P}_1^*$ and $\mathbf{P}_2^*$ with equal probability.

3 Find an optimal solution as described in Selection 2.14 to the selected problem. Let $\bar{\boldsymbol\beta}_1^*, \bar{\boldsymbol\beta}_2^*$ denote such an optimal solution. Then estimate $\boldsymbol\beta^* = \bar{\boldsymbol\beta}_1^* - \bar{\boldsymbol\beta}_2^* + \boldsymbol\beta_0$.

The proof of the unbiasedness of this estimate is shown along the same lines as in Section 2.14.

2.19 CONVEX COMBINATION OF LEAST SQUARES AND MINMAD REGRESSION

Different researchers have suggested the use of combining least squares and MINMAD for obtaining estimates in regression problems. This section is devoted to formulation of the problem as a quadratic-programming problem, and the theory required for solving such problems. The least-squares regression problem can be stated equivalently as

$$\begin{aligned}&\text{Minimize} \quad \mathbf{d}'\mathbf{d}\\ &\text{subject to} \quad \mathbf{X}\boldsymbol\beta + \mathbf{d} = \mathbf{Y} \qquad (2.19.1)\\ &\qquad \boldsymbol\beta, \mathbf{d} \quad \text{unrestricted in sign}\end{aligned}$$

Problem 2.19.1 can be also formulated as

$$\begin{aligned}&\text{Minimize} \quad \mathbf{d}'\mathbf{d}\\ &\text{subject to} \quad -\mathbf{d} \leqslant \mathbf{Y} - \mathbf{X}\boldsymbol\beta \leqslant \mathbf{d} \qquad (2.19.2)\\ &\qquad \boldsymbol\beta \ \text{unrestricted in sign}, \mathbf{d} \geqslant \mathbf{0},\end{aligned}$$

or

$$\begin{aligned}
\text{Minimize} \quad & \mathbf{d}'\mathbf{d} \\
\text{subject to} \quad & \mathbf{X}\boldsymbol{\beta} - \mathbf{d} \leqslant \mathbf{Y} \\
& -\mathbf{X}\boldsymbol{\beta} - \mathbf{d} \leqslant -\mathbf{Y} \\
& \boldsymbol{\beta} \ \text{unrestricted in sign}, \ \mathbf{d} \geqslant 0.
\end{aligned}$$

Let $\boldsymbol{\beta} = \boldsymbol{\beta}_1 - \boldsymbol{\beta}_2, \boldsymbol{\beta}_1, \boldsymbol{\beta}_2 \geqslant 0$. Let

$$\mathbf{Q} = \begin{bmatrix} \mathbf{0}_{p\times p} & \mathbf{0}_{p\times p} & \mathbf{0}_{p\times n} \\ \mathbf{0}_{p\times p} & \mathbf{0}_{p\times p} & \mathbf{0}_{p\times n} \\ \mathbf{0}_{n\times p} & \mathbf{0}_{n\times p} & \mathbf{I}_{n\times n} \end{bmatrix}, \qquad \mathbf{A} = \begin{pmatrix} \mathbf{X} & -\mathbf{X} & -\mathbf{I} \\ -\mathbf{X} & \mathbf{X} & -\mathbf{I} \end{pmatrix},$$

$$\mathbf{V}' = (\boldsymbol{\beta}_1, \boldsymbol{\beta}_2, \mathbf{d}) \qquad \text{and} \quad \mathbf{b} = \begin{pmatrix} \mathbf{Y} \\ -\mathbf{Y} \end{pmatrix}.$$

Here $\mathbf{Q}$ is a $2p+n$ by $2p+n$ and $\mathbf{A}$ is a $2n$ by $2p+n$ matrix, respectively. So we have

$$\begin{aligned}
\text{Minimize} \quad & \mathbf{V}'\mathbf{Q}\mathbf{V} \\
\text{subject to} \quad & \mathbf{A}\mathbf{V} \leqslant \mathbf{b} \qquad (2.19.3) \\
& \mathbf{V} \geqslant \mathbf{0}
\end{aligned}$$

It is known that the performance of least-squares estimation is poor when normality assumption on errors is violated. In such cases we can estimate the parameters $\boldsymbol{\beta}$ by using a convex combination of the mean absolute deviation and mean square deviations, as an alternative.

Let $\gamma_1 + \gamma_2 = 1, \gamma_i \geqslant 0, i = 1,2$. Notice that with

$$\mathbf{C}' = \left(\underset{1\times 2p}{\mathbf{0}}, \underset{1\times n}{\mathbf{e}'} \right), \mathbf{C}'\mathbf{V} \text{ is equal to}$$

$$\sum_{i=1}^{n} |d_i| = \sum_{i=1}^{n} d_i$$

as d_i's are nonnegative. Thus we have the following problem:

$$\begin{aligned}
\text{Minimize} \quad & \gamma_1 \mathbf{d}'\mathbf{d} + \gamma_2 \sum_{i=1}^{n} d_i \\
& \qquad\qquad\qquad\qquad (2.19.4) \\
\text{subject to} \quad & -\mathbf{d} \leqslant \mathbf{Y} - \mathbf{X}\boldsymbol{\beta} \leqslant \mathbf{d} \\
& \boldsymbol{\beta} \quad \text{unrestricted in sign}, \qquad \mathbf{d} \geqslant \mathbf{0},
\end{aligned}$$

or

$$\begin{aligned} \text{Minimize} \quad & \gamma_1[\mathbf{V}'\mathbf{Q}\mathbf{V}] + \gamma_2\mathbf{C}'\mathbf{V} \\ \text{subject to} \quad & \mathbf{A}\mathbf{V} \leqslant \mathbf{b} \\ & \mathbf{V} \geqslant \mathbf{0}. \end{aligned} \tag{2.19.5}$$

Here, notice that $\mathbf{Q}$ is a positive semidefinite matrix, which implies that the objective function is a convex function as $\gamma_1 \geqslant 0$ and $\gamma_2\mathbf{C}'\mathbf{V}$ is linear. We prove certain results on the optimal solution to Problem 2.19.5 and develop a method which is a modification of the simplex procedure given in Section 2.7.

Let

$$\mathbf{f}(\mathbf{V}) = \gamma_1\mathbf{V}'\mathbf{Q}\mathbf{V} + \gamma_2\mathbf{C}'\mathbf{V} \qquad \text{and} \quad \mathbf{g}_i(\mathbf{V}) = \mathbf{A}_i\mathbf{V} - b_i, \qquad i = 1,\ldots,2n.$$

Then we wish to minimize $\mathbf{f}(\mathbf{V})$ subject to $\mathbf{g}_i(\mathbf{V}) \leqslant 0$, $i = 1,\ldots,2n$.

DEFINITION 2.19.1 The Lagrangian function $F(\mathbf{V}, \boldsymbol{\lambda})$ is defined as

$$F(\mathbf{V}, \boldsymbol{\lambda}) = \mathbf{f}(\mathbf{V}) + \boldsymbol{\lambda}G(\mathbf{V})$$

where

$$G(\mathbf{V}) = \begin{bmatrix} \mathbf{g}_1(\mathbf{V}) \\ \vdots \\ \mathbf{g}_{2n}(\mathbf{V}) \end{bmatrix}$$

and $\boldsymbol{\lambda}' = (\lambda_1,\ldots,\lambda_{2n})$ is a vector in $\mathbf{R}^{2n}$. The components of $\boldsymbol{\lambda}$ are called the *Lagrangian multipliers*.

DEFINITION 2.19.2 A point $(\mathbf{V}^0, \boldsymbol{\lambda}^0)'$ in $\mathbf{R}^{2p+3n}$ with $\mathbf{V}^0 \geqslant {}^0, \boldsymbol{\lambda}^0 \geqslant \mathbf{0}$ is called a *saddle point* of $F(\mathbf{V}, \boldsymbol{\lambda})$ if

$$F(\mathbf{V}^0, \boldsymbol{\lambda}) \leqslant F(\mathbf{V}^0, \boldsymbol{\lambda}^0) \leqslant F(\mathbf{V}, \boldsymbol{\lambda}^0) \tag{2.19.6}$$

for all $\mathbf{V} \geqslant \mathbf{0}, \boldsymbol{\lambda} \geqslant \mathbf{0}$.

RESULT 2.19.1 If $(\mathbf{V}^0, \boldsymbol{\lambda}^0)$ is a saddle point of $F(\mathbf{V}, \boldsymbol{\lambda})$, then $\mathbf{V}^0$ is an optimal solution to Problem 2.19.5.

Proof (2.19.6) implies that, for $\mathbf{V} \geqslant \mathbf{0}$ and $\boldsymbol{\lambda} \geqslant \mathbf{0}$,

$$\mathbf{f}(\mathbf{V}^0) + \boldsymbol{\lambda}'G(\mathbf{V}^0) \leqslant \mathbf{f}(\mathbf{V}^0) + \boldsymbol{\lambda}^{0\prime}G(\mathbf{V}^0) \leqslant \mathbf{f}(\mathbf{V}) + \boldsymbol{\lambda}^{0\prime}G(\mathbf{V})$$

$$\Rightarrow \boldsymbol{\lambda}'G(\mathbf{V}^0) \leqslant \boldsymbol{\lambda}^{0\prime}G(\mathbf{V}^0) \qquad \text{for all} \quad \boldsymbol{\lambda} \geqslant \mathbf{0}.$$

This in turn implies that $G(\mathbf{V}^0) \leqslant \mathbf{0}$. So $\boldsymbol{\lambda}^{0\prime}G(\mathbf{V}^0) \leqslant 0$. On the other hand for $\boldsymbol{\lambda}=\mathbf{0}$, $\boldsymbol{\lambda}'G(\mathbf{V}^0)=\mathbf{0} \leqslant \boldsymbol{\lambda}^{0\prime}G(\mathbf{V}^0)$. Thus $\mathbf{0} \leqslant \boldsymbol{\lambda}^{0\prime}G(\mathbf{V}^0) \leqslant \mathbf{0}$ or $\boldsymbol{\lambda}^{0\prime}G(\mathbf{V}^0)=\mathbf{0}$. Now if, $\mathbf{V}$ is feasible, i.e., $G(\mathbf{V}) \leqslant \mathbf{0}$,

$$\mathbf{f}(\mathbf{V}^0)+\boldsymbol{\lambda}^{0\prime}G(\mathbf{V}^0)=\mathbf{f}(\mathbf{V}^0) \leqslant \mathbf{f}(\mathbf{V})+\boldsymbol{\lambda}^{0\prime}G(\mathbf{V}) \leqslant \mathbf{f}(\mathbf{V}).$$

Or $\mathbf{V}^0$ is optimal for Problem 2.19.5. Hence the result.

The converse is shown in Result 2.19.2.

RESULT 2.19.2 Suppose $\mathbf{V}^0$ is optimal for Problem 2.19.5. Then there exists $\boldsymbol{\lambda}^0 \geqslant 0$ such that

$$F(\mathbf{V}^0, \boldsymbol{\lambda}) \leqslant F(\mathbf{V}^0, \boldsymbol{\lambda}^0) \leqslant F(\mathbf{V}, \boldsymbol{\lambda}^0)$$

for all $\boldsymbol{\lambda} \geqslant \mathbf{0}, \mathbf{V} \geqslant \mathbf{0}$.

Proof Let $\mathbf{V}^0$ be an optimal solution for Problem 2.19.5. Let $\mathbf{U}=(U_0, U_1, \ldots, U_{2n})'$ be vectors. Define sets D_1 and D_2, by

$$\mathbf{D}_1 = \{\mathbf{U} | U_0 \geqslant \mathbf{f}(\mathbf{V}), U_i \geqslant \mathbf{g}_i(\mathbf{V}), \qquad i=1,\ldots,2n$$
$$\text{for at least one } \mathbf{V} \geqslant \mathbf{0}\}$$

and

$$\mathbf{D}_2 = \{\mathbf{U} | U_0 < \mathbf{f}(\mathbf{V}^0), U_i < 0, \qquad i=1,\ldots,2n\}.$$

Since $\mathbf{f}(\mathbf{V})$ is convex and $\mathbf{g}_i(\mathbf{V})$ is linear, $\mathbf{D}_1$ and $\mathbf{D}_2$ are convex sets, and $\mathbf{D}_2$ is open. As $\mathbf{V}^0$ is optimal, $\mathbf{f}(\mathbf{V}^0) \leqslant \mathbf{f}(\mathbf{V})$ for all feasible $\mathbf{V}$. Therefore $\mathbf{D}_1 \cap \mathbf{D}_2 = \varnothing$. Both $\mathbf{D}_1$ and $\mathbf{D}_2$ are proper subsets of $\mathbf{R}^{2n+1}$. We now use the separation theorem for convex sets, which is stated as follows:

Let $\mathbf{D}_1, \mathbf{D}_2$ be two proper convex subsets of $\mathbf{R}^z$ which have no point in common. Let $\mathbf{D}_2$ be open. Then there exists a hyperplane $\mathbf{W}\mathbf{U}=\alpha$ which separates $\mathbf{D}_1$ and $\mathbf{D}_2$; that is, there is a vector $\mathbf{W} \neq \mathbf{0}$ and a real number α such that $\mathbf{W}\mathbf{U}^2 \leqslant \alpha < \mathbf{W}\mathbf{U}^1$ for all $\mathbf{U}^1 \in \mathbf{D}_1$ and $\mathbf{U}^2 \in \mathbf{D}_2$.

As the conditions stated in this theorem are satisfied, we have a vector $\mathbf{W}=(W_0, W_1, \ldots, W_{2n}), \mathbf{W} \neq \mathbf{0}$ such that

$$\mathbf{W}\mathbf{U}^1 > \mathbf{W}\mathbf{U}^2 \qquad \text{for} \quad \mathbf{U}^1 \in \mathbf{D}_1, \qquad \mathbf{U}^2 \in \mathbf{D}_2. \tag{2.19.7}$$

Since the components of $\mathbf{U}^2 \in \mathbf{D}_2$ can be arbitrarily large negatives, it follows that $\mathbf{W} \geqslant \mathbf{0}$. For a boundary point of $\mathbf{D}_2$ and a point $\mathbf{U}^1$ we still have (2.19.7), but the strict inequality is replaced by greater-than or equal-to inequality. Note that $\mathbf{U}^2=(\mathbf{f}(\mathbf{V}^0),0,\ldots,0,\ldots,0)'$ is in the boundary of $\mathbf{D}_2$ and $\mathbf{U}^1=(\mathbf{f}(\mathbf{V}),\mathbf{g}_1(\mathbf{V}),\ldots,\mathbf{g}_{2n}(\mathbf{V}))'$ belongs to $\mathbf{D}_1$ as long as $\mathbf{V}\geqslant\mathbf{0}$. Thus (2.19.7) becomes for these points $\mathbf{U}^2$ and $\mathbf{U}^1$

$$\mathbf{WU}^1\geqslant\mathbf{WU}^2.$$

Or

$$W_0\mathbf{f}(\mathbf{V})+\sum_{i=1}^{2n}W_i\mathbf{g}_i(\mathbf{V})\geqslant W_0\mathbf{f}(\mathbf{V}^0) \tag{2.19.8}$$

for all $\mathbf{V}\geqslant\mathbf{0}$. Now, either $W_0=0$ or $W_0>0$. We shall show that $W_0>0$. Suppose $W_0=0$. Then

$$\sum_{i=1}^{2n}W_i\mathbf{g}_i(\mathbf{V})\geqslant 0 \qquad \text{for all} \quad \mathbf{V}\geqslant\mathbf{0}. \tag{2.19.9}$$

As $\mathbf{W}\neq\mathbf{0}$ at least one $W_{i_0}>0, 1\leqslant i_0\leqslant 2n$. Consider the i_0th constraint $\mathbf{g}_{i_0}(\mathbf{V})$,

$$\mathbf{g}_{i_0}(\mathbf{V})=\begin{cases}(\mathbf{X}\boldsymbol{\beta}_1-\mathbf{X}\boldsymbol{\beta}_2-\mathbf{Id}-\mathbf{Y})_{i_0}\leqslant 0 & 1\leqslant i_0\leqslant n\\ (-\mathbf{X}\boldsymbol{\beta}_1+\mathbf{X}\boldsymbol{\beta}_2-\mathbf{Id}+\mathbf{Y})_{i_0}\leqslant 0 & n+1\leqslant \mathbf{i}_0\leqslant 2n.\end{cases}$$

As $\mathbf{d}_{i_0}>0$ can be chosen arbitrarily large still having $\mathbf{V}\geqslant\mathbf{0}$ and $G(\mathbf{V})\leqslant\mathbf{0}$, we have $\mathbf{g}_i(\mathbf{V})<0$ for such a choice. Hence $\mathbf{W}_{i_0}\mathbf{g}_{i_0}(\mathbf{V})<0$, and arbitrary small—that is, (2.19.9) is not true. So we conclude that $W_0>0$. Now let $\boldsymbol{\lambda}^0=(1/\mathbf{W}_0)\cdot(\mathbf{W}_1,\ldots,\mathbf{W}_{2n})'$, so $\boldsymbol{\lambda}^0\geqslant\mathbf{0}$ and (2.19.8) becomes

$$\mathbf{f}(\mathbf{V})+\sum_{i=1}^{2n}\boldsymbol{\lambda}_i^0\mathbf{g}_i(\mathbf{V})\geqslant\mathbf{f}(\mathbf{V}^0) \qquad \text{for all} \quad \mathbf{V}\geqslant\mathbf{0}. \tag{2.19.10}$$

For $\mathbf{V}=\mathbf{V}^0$, we have $\boldsymbol{\lambda}^{0\prime}G(\mathbf{V}^0)\geqslant 0$. Also, as $\mathbf{V}^0$ is feasible for Problem 2.19.5, $G(\mathbf{V}^0)\leqslant\mathbf{0}$. Since $\boldsymbol{\lambda}^0\geqslant\mathbf{0}$, we have $\boldsymbol{\lambda}^{0\prime}G(\mathbf{V}^0)\leqslant 0$, implying

$$\boldsymbol{\lambda}^{0\prime}G(\mathbf{V}^0)=0, \tag{2.19.11}$$

and

$$\boldsymbol{\lambda}'G(\mathbf{V}^0)\leqslant 0 \qquad \text{for} \quad \boldsymbol{\lambda}\geqslant\mathbf{0}. \tag{2.19.12}$$

Thus we have from (2.19.10), (2.19.11), and (2.19.12),

$$\mathbf{f}(\mathbf{V}^0)+\boldsymbol{\lambda}'G(\mathbf{V}^0)\leqslant \mathbf{f}(\mathbf{V}^0)+\boldsymbol{\lambda}^{0\prime}G(\mathbf{V}^0)\leqslant \mathbf{f}(\mathbf{V})+\boldsymbol{\lambda}^{0\prime}G(\mathbf{V})$$

for $\mathbf{V}\geqslant\mathbf{0}, \boldsymbol{\lambda}\geqslant\mathbf{0}$. Or $(\mathbf{V}^0, \boldsymbol{\lambda}^0)$ is a saddle point of $F(\mathbf{V}, \boldsymbol{\lambda})=\mathbf{f}(\mathbf{V})+\boldsymbol{\lambda}'G(\mathbf{V})$. Hence the result.

Result 2.19.2 is not in general true for any G.

EXAMPLE 2.19.1 Consider minimizing $-V$ subject to $V^2\leqslant 0, V\geqslant 0$. It is obvious that only $V=0$ satisfies the constraints and so is optimal. The Lagrangian function is $F(V,\lambda)=-V+\lambda V^2$. If $V^0=0$ and a corresponding $\lambda^0\geqslant 0$ such that (V^0,λ^0) is a saddle point of $F(V,\lambda)$, we have $0\leqslant -V+\lambda^0V^2$ for all $V\geqslant 0$, which is not true.

In the proof the crucial step is showing $W_0>0$. This was possible because of the nature of $G(\mathbf{V})$, namely, for each $i, \mathbf{g}_i(\mathbf{V})<0$ for at least one $\mathbf{V}\geqslant\mathbf{0}$.

In general this case can be stated as a condition to be satisfied by the constraints (known as Slater's *constraint qualification*) as follows:

The constraints $\mathbf{g}_i(\mathbf{V})\leqslant 0$ are such that for every i there exists a $\mathbf{V}^i\geqslant\mathbf{0}$ with $\mathbf{g}_i(\mathbf{V}^i)<0$. Or, equivalently, the constraints $\mathbf{g}_i(\mathbf{V})\leqslant 0$ are such that there exists a $\mathbf{V}\geqslant\mathbf{0}$ with $\mathbf{g}_i(\mathbf{V})<0$ for all i.

We now can state conveniently a summation of Results 2.19.1 and 2.19.2 as Result 2.19.3, which characterizes the optimality of a solution to Problem 2.19.5.

RESULT 2.19.3 $\mathbf{V}^0\geqslant\mathbf{0}$ is an optimal solution to Problem 2.19.5 if and only if there exists a $\boldsymbol{\lambda}^0\geqslant\mathbf{0}$ such that $(\mathbf{V}^0,\boldsymbol{\lambda}^0)$ is a saddle point of $F(\mathbf{V},\boldsymbol{\lambda})=\mathbf{f}(\mathbf{V})+\boldsymbol{\lambda}'G(\mathbf{V})$.

To declare a $(\mathbf{V}^0,\boldsymbol{\lambda}^0)$ to be a saddle point of $F(\mathbf{V},\boldsymbol{\lambda})$ we need to compare $\mathbf{F}(\mathbf{V}^0,\boldsymbol{\lambda}^0)$ with $\mathbf{F}(\mathbf{V},\boldsymbol{\lambda}^0)$ and $\mathbf{F}(\mathbf{V}^0,\boldsymbol{\lambda})$. We avoid this by noticing that the functions $\mathbf{F}(\mathbf{V})$ and $G(\mathbf{V})$ are differentiable; we try to find equivalent conditions in terms of partial derivatives or the gradients, which have been in use in classical optimization methods. But the optimality conditions in terms of gradients, only provide local optimality conditions. Fortunately, the type of problem that we are dealing with is such that any local optimum is also globally optimal. We can show this result in a general setup and specialize the required result from there.

DEFINITION 2.19.3 A function $\mathbf{f}(\mathbf{V})$ defined over a convex region $\mathcal{F}\subset\mathbf{R}^z$ is *quasiconvex* if for all $\mathbf{V}^1,\mathbf{V}^2\in\mathcal{F}, \mathbf{V}^0=\lambda\mathbf{V}^1+(1-\lambda)\mathbf{V}^2, 1>\lambda>0, \mathbf{f}(\mathbf{V}^1)\leqslant \mathbf{f}(\mathbf{V}^2)$ implies $\mathbf{f}(\mathbf{V}^0)\leqslant\mathbf{f}(\mathbf{V}^2)$ or equivalently

$$\mathbf{f}(\mathbf{V}^0)\leqslant\max\left[\mathbf{f}(\mathbf{V}^1),\mathbf{f}(\mathbf{V}^2)\right]. \tag{2.19.13}$$

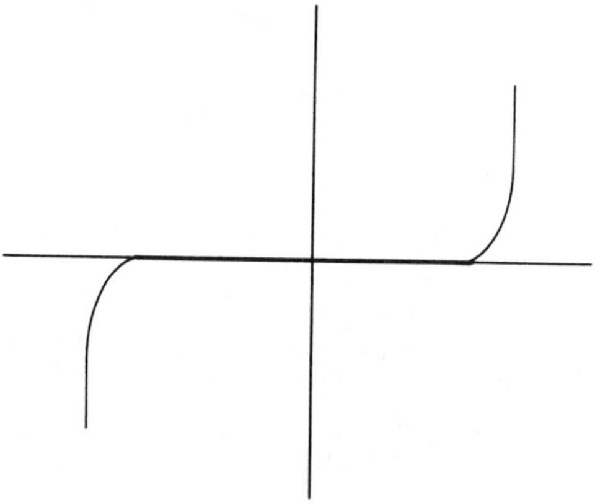

Figure 2.19.1 A function that is quasiconvex but not convex.

REMARK 2.19.1 Any convex function is *quasiconvex*. But the converse is not true. See Fig. 2.19.1, corresponding to:

$$\mathbf{f}(V)=\begin{cases}0 & \text{for} \quad |V|\leqslant 1\\ (V-1)^3 & \text{for} \quad V\geqslant 1\\ (V+1)^3 & \text{for} \quad V\leqslant -1\end{cases}$$

It is easy to verify that $\mathbf{f}(V)$ given above is quasiconvex but not convex.

DEFINITION 2.19.4 A function $\mathbf{f}(\mathbf{V})$ defined over a convex region $\mathcal{F}\subset\mathbf{R}^z$ is *explicitly quasiconvex* in $\mathcal{F}$ if it is quasiconvex in $\mathcal{F}$ and for all $\mathbf{V}^1,\mathbf{V}^2\in\mathcal{F},\mathbf{V}^0=\lambda\mathbf{V}^1+(1-\lambda)\mathbf{V}^2, 1>\lambda>0$,

$$\mathbf{f}(\mathbf{V}^1)<\mathbf{f}(\mathbf{V}^2) \qquad \text{implies} \quad \mathbf{f}(\mathbf{V}^0)<\mathbf{f}(\mathbf{V}^2).$$

Or

$$\mathbf{f}(\mathbf{V}^1)\neq\mathbf{f}(\mathbf{V}^2) \qquad \text{implies} \quad \mathbf{f}(\mathbf{V}^0)<\max\left[\mathbf{f}(\mathbf{V}^1),\mathbf{f}(\mathbf{V}^2)\right]. \quad (2.19.14)$$

REMARK 2.19.2 If $\mathbf{f}(\mathbf{V})$ is convex in the convex set $\mathcal{F}\subset\mathbf{R}^z$ it is explicitly quasiconvex in $\mathcal{F}$.

Proof Suppose $\mathbf{f}(\mathbf{V})$ is convex in $\mathfrak{F}$ and $\mathbf{f}(\mathbf{V}^1) \leq \mathbf{f}(\mathbf{V}^2)$. Then

$$\begin{aligned}\mathbf{f}(\mathbf{V}^0) &= \mathbf{f}(\lambda\mathbf{V}^1 + (1-\lambda)\mathbf{V}^2)\\ &\leq \mathbf{f}(\mathbf{V}^1) + (1-\lambda)\mathbf{f}(\mathbf{V}^2)\\ &\qquad \text{by definition of convexity of } \mathbf{f}\\ &= \mathbf{f}(\mathbf{V}^2) + \lambda\left[\mathbf{f}(\mathbf{V}^1) - \mathbf{f}(\mathbf{V}^2)\right]\\ &\leq \mathbf{f}(\mathbf{V}^2) \qquad \text{for all} \quad 1 > \lambda > 0.\end{aligned}$$

So, $\mathbf{f}(\mathbf{V})$ is quasiconvex in $\mathfrak{F}$. Furthermore,

$$\begin{aligned}\mathbf{f}(\mathbf{V}^0) &\leq \mathbf{f}(\mathbf{V}^2) + \lambda\left[\mathbf{f}(\mathbf{V}^1) - \mathbf{f}(\mathbf{V}^2)\right]\\ &< \mathbf{f}(\mathbf{V}^2) \qquad \text{if} \quad 1 > \lambda > 0 \qquad \text{and} \quad \mathbf{f}(\mathbf{V}^1) < \mathbf{f}(\mathbf{V}^2).\end{aligned}$$

Thus $\mathbf{f}(\mathbf{V})$ is explicitly quasiconvex in $\mathfrak{F}$.

REMARK 2.19.3 Notice that any quasiconvex function need not be explicitly quasiconvex. See Fig. 2.19.2, corresponding to:

$$\mathbf{f}(V) = \begin{cases} 1 & \text{if} \quad 0 \leq V \leq 1 \\ 0 & \text{if} \quad V > 1 \end{cases}$$

RESULT 2.19.4 If $\mathbf{f}(\mathbf{V})$ is explicitly quasiconvex in the convex set $\mathfrak{F} \subset \mathbf{R}^z$, then any local minimum point globally minimizes $\mathbf{f}(\mathbf{V})$ in $\mathfrak{F}$.

Proof Let $\mathbf{V}^1 \in \mathfrak{F}$ be a local minimum point that is not gobal. The implication is there exists $\mathbf{V}^2 \in \mathfrak{F}$ such that $\mathbf{f}(\mathbf{V}^2) < \mathbf{f}(\mathbf{V}^1)$. By explicitly

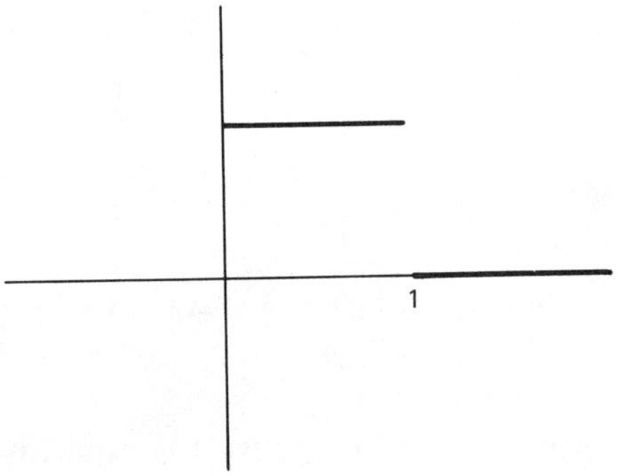

Figure 2.19.2 A function that is quasiconvex but not explicitly quasi convex.

quasiconvexity of $\mathbf{f}$,

$$\mathbf{f}(\mathbf{V}^0) < \mathbf{f}(\mathbf{V}^1) \qquad \text{for all} \quad \mathbf{V}^0 = \lambda \mathbf{V}^1 + (1-\lambda)\mathbf{V}^2$$

with $1 > \lambda > 0$. Now for every $\varepsilon > 0$, the open ball around $\mathbf{V}^1$ with radius ε, $\mathbf{B}(\mathbf{V}^1, \varepsilon)$ we have $\mathbf{B}(\mathbf{V}^1, \varepsilon) \cap (\mathbf{V}^1, \mathbf{V}^2) \neq \varnothing$. So $\mathbf{V}^1$ cannot be a local minimum point as it produces a contradiction. Hence the result.

Thus it is enough to find local optimality conditions in terms of gradients for Problem 2.19.5 as the objective function we have is convex, hence explicitly quasiconvex.

Let $\mathbf{F}_V$ and $\mathbf{F}_\lambda$ denote the gradients of $\mathbf{F}(\mathbf{V}, \boldsymbol{\lambda})$ with respect to $\mathbf{V}$ and $\boldsymbol{\lambda}$, respectively.

$$\mathbf{F}_V = \nabla_V \mathbf{F} = \left(\frac{\partial \mathbf{F}}{\partial \mathbf{V}_1}, \ldots, \frac{\partial \mathbf{F}}{\partial \mathbf{V}_{2p+n}} \right)',$$

$$\mathbf{F}_\lambda = \nabla_\lambda \mathbf{F} = \left(\frac{\partial \mathbf{F}}{\partial \lambda_1}, \ldots, \frac{\partial \mathbf{F}}{\partial \lambda_{2n}} \right)$$

Also $\mathbf{F}_V(\hat{\mathbf{V}}, \hat{\boldsymbol{\lambda}})$ means $\nabla_V \mathbf{F}$ evaluated at $(\hat{\mathbf{V}}, \hat{\boldsymbol{\lambda}})$; similarly $\mathbf{F}_\lambda(\hat{\mathbf{V}}, \hat{\boldsymbol{\lambda}})$ is understood.

RESULT 2.19.5 $\mathbf{V}^0 \geqslant \mathbf{0}$ is an optimal solution to Problem 2.19.5 if and only if there exists $\boldsymbol{\lambda}^0 \geqslant \mathbf{0}$ such that

$$\mathbf{F}_V(\mathbf{V}^0, \boldsymbol{\lambda}^0) \geqslant \mathbf{0}, \qquad \mathbf{V}^{0\prime}\mathbf{F}_V(\mathbf{V}^0, \boldsymbol{\lambda}^0) = 0, \tag{2.19.15}$$

$$\mathbf{F}_\lambda(\mathbf{V}^0, \boldsymbol{\lambda}^0) \leqslant \mathbf{0}, \qquad \boldsymbol{\lambda}^{0\prime}\mathbf{F}_\lambda(\mathbf{V}^0, \boldsymbol{\lambda}^0) = 0. \tag{2.19.16}$$

Proof We shall show that Conditions 2.19.15 and 2.19.16 are equivalent to saddle-point conditions, that is

$$\mathbf{F}(\mathbf{V}^0, \boldsymbol{\lambda}) \leqslant \mathbf{F}(\mathbf{V}^0, \boldsymbol{\lambda}^0) \leqslant \mathbf{F}(\mathbf{V}, \boldsymbol{\lambda}^0) \tag{2.19.17}$$

for $\mathbf{V} \geqslant \mathbf{0}$, $\boldsymbol{\lambda} \geqslant \mathbf{0}$. First we show that (2.19.17) implies (2.19.15) and (2.19.16).

Suppose for some k, $\partial F/\partial V_k$ is evaluated at $(\mathbf{V}^0, \boldsymbol{\lambda}^0)$ is negative violating (2.19.15). The implication is that we can find a vector $\mathbf{V} \geqslant \mathbf{0}$ with components $V_l = V_l^0$ for $l \neq k$ and $V_k > V_k^0$ such that $\mathbf{F}(\mathbf{V}, \boldsymbol{\lambda}^0) < \mathbf{F}(\mathbf{V}^0, \boldsymbol{\lambda}^0)$, contradicting (2.19.17). So (2.19.17) implies $\mathbf{F}_V(\mathbf{V}^0, \boldsymbol{\lambda}^0) \geqslant \mathbf{0}$.

Because $\mathbf{V}^0 \geqslant 0$, all the components in the inner product $\mathbf{V}^{0\prime}\mathbf{F}_V(\mathbf{V}^0, \boldsymbol{\lambda}^0)$ are nonnegative. Suppose for some k, $V_k^0 \cdot \partial \mathbf{F}(\mathbf{V}^0, \boldsymbol{\lambda}^0)/\partial V_k > 0$, then $V_k^0 > 0$, $\partial \mathbf{F}(\mathbf{V}^0, \boldsymbol{\lambda}^0)/\partial V_k > 0$. The implication is that there exists a vector $\mathbf{V}$ such that

$$V_l = \begin{cases} V_l^0, & l \neq k \\ V_l^0 > V_l \geqslant 0, & l = k. \end{cases}$$

with $\mathbf{F}(\mathbf{V}, \boldsymbol{\lambda}^0) < \mathbf{F}(\mathbf{V}^0, \boldsymbol{\lambda}^0)$, again contradicting (2.19.17). Hence $\mathbf{V}_k^{0\prime} \cdot \partial \mathbf{F}(\mathbf{V}^0, \boldsymbol{\lambda}^0)/\partial V_k$ is equal to zero, for all k or (2.19.17)⇒(2.19.15). Similarly we can show that (2.19.17)⇒(2.19.16). On the other hand, we shall show that (2.19.15) and (2.19.16)⇒(2.19.17). Notice that $\mathbf{F}(\mathbf{V}, \boldsymbol{\lambda}^0)$ is a convex function of $\mathbf{V}$. By definition of differentiable convex functions, the implication is that

$$\mathbf{F}(\mathbf{V}, \boldsymbol{\lambda}^0) \geqslant \mathbf{F}(\mathbf{V}^0, \boldsymbol{\lambda}^0) + (\mathbf{V} - \mathbf{V}^0)'\mathbf{F}_V(\mathbf{V}^0, \boldsymbol{\lambda}^0) \tag{2.19.18}$$

for $\mathbf{V} \geqslant \mathbf{0}$. Since $\mathbf{F}(\mathbf{V}^0, \boldsymbol{\lambda})$ is linear in $\boldsymbol{\lambda}$,

$$\mathbf{F}(\mathbf{V}^0, \boldsymbol{\lambda}) = \mathbf{F}(\mathbf{V}^0, \boldsymbol{\lambda}^0) + (\boldsymbol{\lambda} - \boldsymbol{\lambda}^0)'\mathbf{F}_\lambda(\mathbf{V}^0, \boldsymbol{\lambda}^0) \tag{2.19.19}$$

for $\boldsymbol{\lambda} \geqslant \mathbf{0}$. Now, $(\mathbf{V} - \mathbf{V}^0)'\mathbf{F}_V(\mathbf{V}^0, \boldsymbol{\lambda}^0) = \mathbf{V}'\mathbf{F}_V(\mathbf{V}^0, \boldsymbol{\lambda}^0) \geqslant \mathbf{0}$ [because of (2.19.15) and $\mathbf{V} \geqslant \mathbf{0}$]. Therefore (2.19.18)⇒

$$\mathbf{F}(\mathbf{V}, \boldsymbol{\lambda}^0) \geqslant \mathbf{F}(\mathbf{V}^0, \boldsymbol{\lambda}^0). \tag{2.19.20}$$

Also $(\boldsymbol{\lambda} - \boldsymbol{\lambda}^{0\prime})\mathbf{F}_\lambda(\mathbf{V}^0, \boldsymbol{\lambda}^0) = \boldsymbol{\lambda}'\mathbf{F}_\lambda(\mathbf{V}^0, \boldsymbol{\lambda}^0) \leqslant 0$ [because (2.19.16) and $\boldsymbol{\lambda} \geqslant \mathbf{0}$]. Therefore (2.19.19)⇒

$$\mathbf{F}(\mathbf{V}^0, \boldsymbol{\lambda}^0) \geqslant \mathbf{F}(\mathbf{V}^0, \boldsymbol{\lambda}). \tag{2.19.21}$$

Since (2.19.20) and (2.19.21) together yield (2.19.17), hence the result.

Observe that we have made use only of the convexity and differentiability of $\mathbf{f}(\mathbf{V}), \mathbf{g}_i(\mathbf{V})$ so far. Hence this result is in general true for a problem in which the objective function $\mathbf{f}(\mathbf{V})$ and the functions $\mathbf{g}_i(\mathbf{V})$ are convex and differentiable and the constraints satisfy the constraint qualification mentioned earlier. Going back to Problem 2.19.5, we have

$$\mathbf{f}(\mathbf{V}) = \gamma_1 \mathbf{V}'\mathbf{Q}\mathbf{V} + \gamma_2 \mathbf{C}'\mathbf{V}$$

and

$$\mathbf{g}_i(\mathbf{V}) = (\mathbf{A}\mathbf{V})_i - b_i.$$

Hence Conditions 2.19.5 and 2.19.6 read as follows:

$$\mathbf{F}_V(\mathbf{V}^0, \boldsymbol{\lambda}^0) = \gamma_2\mathbf{C} + 2\gamma_1\mathbf{Q}\mathbf{V}^0 + \mathbf{A}'\boldsymbol{\lambda}^0 \geqslant \mathbf{0}; \mathbf{V}^{0\prime}\mathbf{F}_V(\mathbf{V}^0, \boldsymbol{\lambda}^0) = 0,$$

and

$$\mathbf{F}_\lambda(\mathbf{V}^0, \boldsymbol{\lambda}^0) = \mathbf{A}\mathbf{V}^0 - \mathbf{b} \leqslant \mathbf{0}, \boldsymbol{\lambda}^{0\prime}\mathbf{F}_\lambda(\mathbf{V}^0, \boldsymbol{\lambda}^0) = 0.$$

Let $\boldsymbol{\mu}^0 = \mathbf{F}_V(\mathbf{V}^0, \boldsymbol{\lambda}^0)$ and $\boldsymbol{\eta}^0 = -\mathbf{F}_\lambda(\mathbf{V}^0, \boldsymbol{\lambda}^0)$. Then we have $\mathbf{V}^{0\prime}\boldsymbol{\mu}^0 = 0$ and $\boldsymbol{\lambda}^{0\prime}\boldsymbol{\eta}^0 = 0$. Or we can write $\mathbf{V}^{0\prime}\boldsymbol{\mu}^0 + \boldsymbol{\lambda}^{0\prime}\boldsymbol{\eta}^0 = 0$. Since $\mathbf{V}^0$, $\boldsymbol{\lambda}^0$, $\boldsymbol{\mu}^0$, and $\boldsymbol{\eta}^0$ are all nonnegative, Result 2.19.5 yields:

$\mathbf{V}^0 \geqslant \mathbf{0}$ is an optimal solution to Problem 2.19.5 if there exist $\boldsymbol{\lambda}^0$, $\boldsymbol{\mu}^0$, and $\boldsymbol{\eta}^0$, such that

$$\begin{aligned} &\mathbf{A}\mathbf{V}^0 + \boldsymbol{\eta}^0 = \mathbf{b} \\ &\boldsymbol{\mu}^0 - 2\gamma_1\mathbf{Q}\mathbf{V}^0 - \mathbf{A}'\boldsymbol{\lambda}^0 = \gamma_2\mathbf{C} \\ &\boldsymbol{\lambda}^0 \geqslant \mathbf{0}, \boldsymbol{\mu}^0 \geqslant \mathbf{0}, \boldsymbol{\eta} \geqslant \mathbf{0} \end{aligned} \tag{2.19.22}$$

and

$$\mathbf{V}^{0\prime}\boldsymbol{\mu}^0 + \boldsymbol{\lambda}^{0\prime}\boldsymbol{\eta}^0 = 0. \tag{2.19.23}$$

From the theory developed so far characterizing an optimal solution to Problem 2.19.5, we can devise a method to solve (2.19.5) which is a modification of the simplex procedure shown in Section 2.7. As the simplex procedure considers only basic feasible solutions, while searching for an optimal solution to a problem, here is a similar result that guarantees an optimal solution among the basic feasible solutions to the system of linear equalities of (2.19.22).

RESULT 2.19.6 Consider any solution $\mathbf{V}, \boldsymbol{\lambda}, \boldsymbol{\mu}, \boldsymbol{\eta}$ satisfying

$$\begin{aligned} &(1) \quad \mathbf{V}, \boldsymbol{\lambda}, \boldsymbol{\mu}, \boldsymbol{\eta} \geqslant \mathbf{0} \\ &(2) \quad \mathbf{A}\mathbf{V} + \boldsymbol{\eta} = \mathbf{b} \\ &(3) \quad \boldsymbol{\mu} - 2\gamma_1\mathbf{Q}\mathbf{V} - \mathbf{A}'\boldsymbol{\lambda} = \gamma_2\mathbf{C} \end{aligned} \tag{2.19.24}$$

and

$$(4) \quad \mathbf{V}'\boldsymbol{\mu} + \boldsymbol{\eta}'\boldsymbol{\lambda} = 0. \tag{2.19.25}$$

Then the $(\mathbf{V}, \boldsymbol{\lambda})$ part of such a solution is an extreme point of

$$\hat{\mathfrak{F}}=\{(\mathbf{V}, \boldsymbol{\lambda})|\mathbf{AV}\leqslant \mathbf{b}, -2\gamma_1\mathbf{QV}-\mathbf{A}'\boldsymbol{\lambda}\leqslant \gamma_2\mathbf{C}, \mathbf{V}\geqslant \mathbf{0}, \boldsymbol{\lambda}\geqslant \mathbf{0}\}.$$

Proof Let $(\mathbf{V}^*, \boldsymbol{\lambda}^*, \boldsymbol{\mu}^*, \boldsymbol{\eta}^*)$ satisfy (2.19.24) and (2.19.25). Then $(\mathbf{V}^*, \boldsymbol{\lambda}^*)\in\hat{\mathfrak{F}}$. Notice that there are $(2n+(2p+n)+(2p+n)+2n)=4p+6n$ constraints in all, including the nonnegative restrictions, which define $\hat{\mathfrak{F}}$. As $(\mathbf{V}^*, \boldsymbol{\lambda}^*, \boldsymbol{\mu}^*, \boldsymbol{\eta}^*)$ satisfy (2.19.25), at least $(2p+n)$ of the $2(2p+n)$ components of $(\mathbf{V}^*, \boldsymbol{\mu}^*)$ and at least $2n$ of the $4n$ components of $(\boldsymbol{\eta}^*, \boldsymbol{\lambda}^*)$ vanish. Thus $(\mathbf{V}^*, \boldsymbol{\lambda}^*)$ satisfies at least $2p+3n$ of the $4p+6n$ inequalities as equalities. Or equivalently, $(\mathbf{V}^*, \boldsymbol{\lambda}^*)$ lies on at least $2p+3n$ of the $4p+6n$ hyperplanes defining $\hat{\mathfrak{F}}$ or $(\mathbf{V}^*, \boldsymbol{\lambda}^*)$ constitutes a basic feasible solution. Hence the result.

So in spite of the additional nonlinear restrictions (2.19.25), it is enough to consider only the basic feasible solutions of the system given by (2) and (3), satisfying (4). Next we develop an algorithm for solving problem 2.19.5.

Algorithm 2.19.1

Step 1 We find a feasible solution $(\mathbf{A},\mathbf{I})\begin{bmatrix}\mathbf{V}\\ \boldsymbol{\eta}\end{bmatrix}=\mathbf{b}$, using the artifical variable technique described in Section 2.9. (If no such solution exists we stop. Such should not be the case for Problem 2.19.5.)

Let the solution be $\bar{\mathbf{V}}\geqslant \mathbf{0}$ $\bar{\boldsymbol{\eta}}\geqslant \mathbf{0}$ and the corresponding indices of the basic columns be denoted by $\mathbf{J}_{\bar{V}}$ and $\mathbf{J}_{\bar{\eta}}$, respectively.

Step 2 Calculate $\mathbf{h}=\gamma_2\mathbf{C}+2\gamma_1\mathbf{Q}\bar{\mathbf{V}}$. If $\mathbf{h}=\mathbf{0}$, $\mathbf{V}=\bar{\mathbf{V}}$, $\boldsymbol{\mu}=\mathbf{0}, \boldsymbol{\lambda}=\mathbf{0}$, and $\bar{\boldsymbol{\eta}}=\mathbf{b}-\mathbf{A}\bar{\mathbf{V}}\geqslant \mathbf{0}$, (2.19.24) and (2.19.25) are satisfied and we stop. Otherwise go to Step 3.

Step 3 Form the linear programming problem:

Minimize ζ

subject to
$$\begin{pmatrix}\mathbf{A} & \mathbf{I} & \mathbf{0} & \mathbf{0} & \mathbf{0}\\ -2\gamma_1\mathbf{Q} & \mathbf{0} & \mathbf{I} & -\mathbf{A}' & \mathbf{h}\end{pmatrix}\begin{bmatrix}\mathbf{V}\\ \boldsymbol{\eta}\\ \boldsymbol{\mu}\\ \boldsymbol{\lambda}\\ \zeta\end{bmatrix}=\begin{pmatrix}\mathbf{b}\\ \gamma_2\mathbf{C}\end{pmatrix}$$

$$\mathbf{V}, \boldsymbol{\eta}, \boldsymbol{\mu}, \boldsymbol{\lambda}\geqslant \mathbf{0}, \zeta\geqslant 0. \qquad (2.19.26)$$

We have an initial basis given by the $2p+3n$ columns as chosen below:

(1) The $2n$ columns $\begin{pmatrix} \mathbf{a}_j \\ -2\gamma_1 \mathbf{Q}_j \end{pmatrix}$ for $j \in \mathbf{J}_{\bar{V}}$ and $\begin{pmatrix} \mathbf{e}_j \\ \mathbf{0} \end{pmatrix}$ for $j \in \mathbf{J}_{\bar{\eta}}$

(2) The $2p+n$ columns from $\begin{bmatrix} \mathbf{0} \\ \mathbf{I} \end{bmatrix}$ corresponding to $j \notin \mathbf{J}_{\bar{V}}$ and from $\begin{pmatrix} \mathbf{0} \\ -\mathbf{A}' \end{pmatrix}$ corresponding to $j \notin \mathbf{J}_{\bar{\eta}}$.

In (1) and (2) we have $2p+3n$ linearly independent columns in all. (The proof is left as an exercise.)

(3) Now consider the column $\begin{bmatrix} \mathbf{0} \\ \mathbf{h} \end{bmatrix}$ corresponding to the variable ζ. In all, in (1), (2), and (3) we have $2p+3n+1$ columns. Remove one of the columns in (1) and (2) such that the remaining columns in (1) and (2) and the column $\begin{bmatrix} \mathbf{0} \\ \mathbf{h} \end{bmatrix}$ are lineary independent. Then we have a basis of size $2p+3n$ with $\mathbf{V}=\bar{\mathbf{V}}$, $\boldsymbol{\eta}=\bar{\boldsymbol{\eta}}$, $\boldsymbol{\mu}=\mathbf{0}$, $\boldsymbol{\lambda}=\mathbf{0}$ and $\zeta=1$. Notice that $\mathbf{V}'\boldsymbol{\mu}=0$ and $\boldsymbol{\eta}'\boldsymbol{\lambda}=0$ in this case. Thus we have an initial basis for Problem 2.19.26. Go to Step 4.

Step 4 Apply the simplex method to the problem, starting with the basis obtained in Step 3, with the additional restrictions that

$$\text{if} \quad \left.\begin{matrix} V_i \\ \eta_i \\ \mu_i \\ \lambda_i \end{matrix}\right\} \quad \text{is basic} \quad \left\{\begin{matrix} \mu_i \\ \lambda_i \\ V_i \\ \eta_i \end{matrix}\right. \quad \text{must not enter the basis.}$$

If we obtain a basis with no more exchange possible, we stop.

RESULT 2.19.7 The algorithm finds a solution to (2.19.26): $\hat{\mathbf{V}}, \hat{\boldsymbol{\mu}}, \hat{\boldsymbol{\lambda}}, \hat{\boldsymbol{\eta}} \geqslant \mathbf{0}$ and $\hat{\zeta}$ in which $\hat{\zeta}=0$. Or equivalently we have a solution to (2.19.24) and (2.19.25).

Proof Suppose $\zeta>0$. Then from $\hat{\mathbf{V}}$, $\hat{\boldsymbol{\mu}}$, $\hat{\boldsymbol{\lambda}}$, $\hat{\boldsymbol{\eta}}$, and $\hat{\zeta}$, we have a solution to:

Minimize ζ

$$\text{subject to} \quad \begin{pmatrix} \mathbf{A} & \mathbf{I} & \mathbf{0} & \mathbf{0} & \mathbf{0} \\ -2\gamma_1 \mathbf{Q} & \mathbf{0} & \mathbf{I} & -\mathbf{A}' & \mathbf{h} \end{pmatrix} \begin{bmatrix} \mathbf{V} \\ \boldsymbol{\eta} \\ \boldsymbol{\mu} \\ \boldsymbol{\lambda} \\ \zeta \end{bmatrix} = \begin{pmatrix} \mathbf{b} \\ \gamma_2 \mathbf{C} \end{pmatrix} \quad (2.19.27)$$

and $\hat{\boldsymbol{\mu}}'\mathbf{V}+\hat{\mathbf{V}}'\boldsymbol{\mu}=0, \mathbf{V} \geqslant \mathbf{0}, \boldsymbol{\mu} \geqslant \mathbf{0}, \boldsymbol{\eta} \geqslant \mathbf{0}, \boldsymbol{\lambda} \geqslant \mathbf{0}, \zeta \geqslant 0.$

But the dual of this problem is

$$\text{Maximize}\quad \mathbf{b}'\mathbf{W}^1+\gamma_2\mathbf{C}'\mathbf{W}^2$$

$$\text{subject to}\quad \begin{bmatrix} \mathbf{A}' & -2\gamma_1\mathbf{Q} & \hat{\boldsymbol{\mu}} \\ \mathbf{I} & \mathbf{0} & \mathbf{0} \\ \mathbf{0} & \mathbf{I} & \hat{\mathbf{V}} \\ \mathbf{0} & -\mathbf{A} & \mathbf{0} \\ \mathbf{0} & \mathbf{h}' & \mathbf{0} \end{bmatrix} \begin{bmatrix} \mathbf{W}^1 \\ \mathbf{W}^2 \\ \mathbf{W}^3 \end{bmatrix} \leqslant \begin{bmatrix} \mathbf{0} \\ \mathbf{0} \\ \mathbf{0} \\ \mathbf{0} \\ 1 \end{bmatrix} \tag{2.19.28}$$

$$W^i,\qquad i=1,2,3 \text{ unrestricted in sign.}$$

This problem also has a solution (from the duality theory developed in Section 2.10). Let this solution be $\hat{\mathbf{W}}^1$, $\hat{\mathbf{W}}^2$, and $\hat{\mathbf{W}}^3$, such that

$$\mathbf{b}'\hat{\mathbf{W}}^1+\gamma_2\mathbf{C}'\hat{\mathbf{W}}^2=\hat{\zeta}>0.$$

The solution also derives from the complementary slackness property, as $\hat{\zeta}>0$, the corresponding constraint in the dual problem, holds as an equality—i.e.,

$$\mathbf{h}'\hat{\mathbf{W}}^2=1.$$

Moreover, corresponding to the solution $\hat{\mathbf{V}}$, $\hat{\boldsymbol{\eta}}$, $\hat{\boldsymbol{\mu}}$, $\hat{\boldsymbol{\lambda}}$, we have exactly one of the following cases occurring:

(1) $\hat{V}_j>0, \hat{\mu}_j=0,$

(2) $\hat{V}_j=0, \hat{\mu}_j>0,$

(3) $\hat{V}_j=\hat{\mu}_j=0.$

CASE 1 $\Rightarrow(\mathbf{A}'\hat{\mathbf{W}}^1-2\gamma_1\mathbf{Q}\hat{\mathbf{W}}^2)_j=0$

CASE 2 $\Rightarrow(\hat{\mathbf{W}}^2)_j=0$

CASE 3 $\Rightarrow(\mathbf{A}'\hat{\mathbf{W}}^1-2\gamma_1\mathbf{Q}\hat{\mathbf{W}}^2)_j\leqslant 0,$ and $(\hat{\mathbf{W}}^2)_j\leqslant 0.$

(All three implications follow from the complementary slackness property.) Thus, we have

$$\hat{\mathbf{W}}^{2\prime}(\mathbf{A}'\hat{\mathbf{W}}^1-2\gamma_1\mathbf{Q}\hat{\mathbf{W}}^2)\geqslant 0.$$

Similarly, considering the three cases corresponding to $\hat{\eta}_i$ and $\hat{\lambda}_i$'s being positive or zero, we can show that

$$\hat{\mathbf{W}}^{1'}\mathbf{A}\hat{\mathbf{W}}^{2} \leqslant 0.$$

Hence,

$$\mathbf{W}^{2'}\mathbf{Q}\hat{\mathbf{W}}^{2} \leqslant 0.$$

Recall that

$$\mathbf{Q}=\begin{pmatrix} \mathbf{0} & \mathbf{0} \\ \mathbf{0} & \mathbf{I} \end{pmatrix} \quad \text{and} \quad \mathbf{C}=(\mathbf{0},\mathbf{e}).$$

Therefore for $\mathbf{V}$, and

$$\mathbf{h}=\gamma_2\mathbf{C}+2\gamma_1\mathbf{Q}\hat{\mathbf{V}}=\begin{pmatrix} \mathbf{0} \\ \gamma_2\mathbf{e}+2\gamma_1\hat{\mathbf{V}}_R \end{pmatrix}$$

where $\hat{\mathbf{V}}_R$ is the vector giving variables $(\hat{V}_{2p+1},\ldots,\hat{V}_{2p+n})'$. Hence $\mathbf{h}$ is such that $h_1,\ldots,h_{2p}$ are all zeroes. This implies that

$$\mathbf{h}'\hat{\mathbf{W}}^{2}=1=\sum_{j=2p+1}^{2p+n} h_j\hat{W}_j^2.$$

Now looking back at $\hat{\mathbf{W}}^{2'}\mathbf{Q}\hat{\mathbf{W}}^{2} \leqslant 0$, with the fact that $\mathbf{Q}$ is positive semidefinite, we have $\hat{\mathbf{W}}^{2'}\mathbf{Q}\hat{\mathbf{W}}^{2} \geqslant 0$. Therefore, $\mathbf{W}^{2'}\mathbf{Q}\hat{\mathbf{W}}^{2}=0$. Or

$$\sum_{j=2p+1}^{2p+n} \left(\hat{W}_j^2\right)^2=0 \qquad \text{(from the structure of } \mathbf{Q}\text{)}.$$

This is possible if and only if $\hat{W}_j^2=0$ for $j=2p+1,\ldots,2p+n$. This contradicts the fact that $\mathbf{h}'\hat{\mathbf{W}}^2=1$. Hence the result.

OBSERVATIONS AND DISCUSSIONS

(1) In Problem 2.19.2, $\mathbf{d}$ need not be explicitly restricted in sign. Also $\boldsymbol{\beta}$ could be kept as such, without replacing it by $\boldsymbol{\beta}_1-\boldsymbol{\beta}_2$. Then we require that Results 2.19.1 through 2.19.3 and 2.19.5 be proved for unrestricted variables. This is possible and we do not have the restriction $\mathbf{V}'\boldsymbol{\mu}=0$ as $\mathbf{V}$ is no longer nonnegative.

(2) Notice that if we are interested in solving the least-squares problem, we just need to let $\gamma_2 = 0$ in Problem 2.19.5. But it is interesting to consider Problem 2.19.1 in this case, and apply the following result.

RESULT 2.19.8 $\mathbf{V}^0$ is an optimal solution to the problem:

$$\begin{aligned} &\text{Minimize} && \mathbf{V}'\mathbf{Q}\mathbf{V} + \mathbf{C}'\mathbf{V} \\ &\text{subject to} && \mathbf{A}\mathbf{V} = \mathbf{b} \\ & && \mathbf{V} \text{ unrestricted in sign} \end{aligned} \tag{2.19.29}$$

if and only if $\mathbf{A}\mathbf{V}^0 = \mathbf{b}$ and there exists $\boldsymbol{\lambda}^0$ such that

$$-2\mathbf{Q}\mathbf{V}^0 - \mathbf{A}'\boldsymbol{\lambda}^0 = \mathbf{C}'.$$

In Problem 2.19.1, $\mathbf{C} = \mathbf{0}$, $\mathbf{A} = [\mathbf{X}, \mathbf{I}]$, $\mathbf{V} = [\boldsymbol{\beta}, \mathbf{d}]$, $\mathbf{Q} = \begin{bmatrix} \mathbf{0} & \mathbf{0} \\ \mathbf{0} & \mathbf{I} \end{bmatrix}$, and $\mathbf{b} = \mathbf{Y}$. Hence the result above implies

$$\mathbf{X}\boldsymbol{\beta}^0 + \mathbf{d}^0 = \mathbf{Y} \tag{2.19.30}$$

$$-\mathbf{X}'\boldsymbol{\lambda}^0 = \mathbf{0}, \tag{2.19.31}$$

and

$$-2\mathbf{d}^0 - \boldsymbol{\lambda}^0 = \mathbf{0}. \tag{2.19.32}$$

Or

$$\boldsymbol{\lambda}^0 = -2\mathbf{d}^0 \tag{2.19.33}$$

and

$$\mathbf{d}^0 = \mathbf{Y} - \mathbf{X}\boldsymbol{\beta}^0. \tag{2.19.34}$$

Using (2.19.33) and (2.19.34) in (2.19.31) and simplifying, we get

$$\mathbf{X}'\mathbf{X}\boldsymbol{\beta}^0 = \mathbf{X}'\mathbf{Y}. \tag{2.19.35}$$

If the rank of $\mathbf{X}'\mathbf{X}$ is p we can write

$$\boldsymbol{\beta}^0 = (\mathbf{X}'\mathbf{X})^{-1}\mathbf{X}'\mathbf{Y}, \tag{2.19.36}$$

which is the classical least-squares solution. Thus we observe that there is no need to go through the mathematical-programming approach if we are interested in only the usual least-squares regression. Of course, we do need this approach as soon as we have nonnegativity and inequality restrictions on $\boldsymbol{\beta}$ to be satisfied in addition.

(3) Algorithm 2.19.1 is applicable to general quadratic programming (QP) problems with $\mathbf{Q}$, positive definite. In this case, in the proof of Result 2.19.7 we have $\hat{\mathbf{W}}^{2\prime}\mathbf{Q}\hat{\mathbf{W}}^2=0 \Rightarrow \hat{\mathbf{W}}^2=\mathbf{0}$, but then we have a contradiction, as $\mathbf{h}'\hat{\mathbf{W}}^2=1$. This algorithm is also applicable if $\mathbf{Q}$ is positive-semidefinite. However, that proof is involved.

(4) There are other algorithms to solve QP based on the results involving the necessary and sufficient conditions for optimality.

(5) Computationally superior methods are obtainable with use of the results from the linear-complementary programming, as outlined in the next section.

(6) The basis size in Algorithm 2.19.1 is $2p+3n$, which is large even for small p and n. Hence we use a computer program implementing this algorithm, while solving Problem 2.19.5.

2.20 LINEAR COMPLIMENTARY PROGRAMMING AND ITS APPLICATION TO REGRESSION PROBLEMS

Given a square matrix $\mathbf{M}$ of order k and a vector $\mathbf{q}$ in R^k, the problem of finding nonnegative solutions in the variables W_i's, $i=1,\ldots,k$ and Z_i's, $i=1,\ldots,k$ to the system of linear equalities

$$\mathbf{W}-\mathbf{M}\mathbf{Z}=\mathbf{q}, \qquad \mathbf{W}\in\mathbf{R}^k, \qquad \mathbf{Z}\in\mathbf{R}^k,$$

and

$$\sum_{i=1}^{k} W_i Z_i = 0 \tag{2.20.1}$$

is known as the linear complementary-programming problem (LCP). Linear-programming (LP) and convex quadratic-programming problems (CQP) can be shown to be equivalent to a corresponding LCP problem. Hence a method available for solving LCP can be used to solve either of LP or CQP problems. Also computationally this approach turns out to be better than the simplex algorithm developed for LP and the algorithm discussed for CQP. With this in mind we go on to show the equivalence of the problems and state an algorithm for solving LCP.

DEFINITION 2.20.1 A nonnegative solution $(\mathbf{W},\mathbf{Z})$ to the system $\mathbf{W}=\mathbf{MZ}+\mathbf{q}$ is called a *feasible* solution to the LCP problem.

DEFINITION 2.20.2 A feasible solution to the LCP problem, $(\mathbf{W},\mathbf{Z})$, if in addition it satisfies $\mathbf{W}'\mathbf{Z}=0$, is called a *complementary* solution.

REMARK 2.20.1 As $\mathbf{W}\geqslant\mathbf{0}$ and $\mathbf{Z}\geqslant\mathbf{0}, \mathbf{W}'\mathbf{Z}=0$, implies $W_iZ_i=0$ for all i, or if W_i is positive Z_i has to be zero and vice versa. Of course, both W_i and Z_i could be zero.

REMARK 2.20.2 If $\mathbf{q}\geqslant\mathbf{0}$, then $\mathbf{W}=\mathbf{q}$ and $\mathbf{Z}=\mathbf{0}$ solves the LCP problem. Thus the problem is interesting only if some q_i's are negative. Given the linear-programming problem,

$$\begin{aligned} &\text{Minimize} && \mathbf{C}'\mathbf{X} \\ &\text{subject to} && \mathbf{AX}\geqslant\mathbf{b} \\ & && \mathbf{X}\geqslant\mathbf{0}, \end{aligned} \tag{2.20.2}$$

we have the dual of the problem,

$$\begin{aligned} &\text{Maximize} && \mathbf{b}'\mathbf{Y} \\ &\text{subject to} && \mathbf{A}'\mathbf{Y}\leqslant\mathbf{C} \\ & && \mathbf{Y}\geqslant\mathbf{0}. \end{aligned} \tag{2.20.3}$$

From the complementary slackness property of linear programming, if Problems 2.20.2 and 2.20.3 have feasible solutions satisfying complementary slackness conditions, then the feasible solutions are optimal for the respective problems. Hence we have the problem,

$$\begin{aligned} &\text{find } \mathbf{X},\mathbf{U},\mathbf{Y},\mathbf{V} \text{ such that} \\ &\mathbf{V}=-\mathbf{A}'\mathbf{Y}+\mathbf{C} \\ &\mathbf{U}=\mathbf{AX}-\mathbf{b} \\ &\mathbf{X},\mathbf{Y},\mathbf{U},\mathbf{V}\geqslant\mathbf{0} \\ &\mathbf{V}'\mathbf{X}+\mathbf{U}'\mathbf{Y}=0 \end{aligned}$$

where $\mathbf{U}$ and $\mathbf{V}$ are the surplus and slack variables introduced in problems 2.20.2 and 2.20.3.

Now defining

$$\mathbf{M}=\begin{pmatrix}\mathbf{0} & -\mathbf{A}'\\ \mathbf{A} & \mathbf{0}\end{pmatrix},\quad \mathbf{W}=\begin{pmatrix}\mathbf{V}\\ \mathbf{U}\end{pmatrix},\quad \mathbf{Z}=\begin{pmatrix}\mathbf{X}\\ \mathbf{Y}\end{pmatrix},\quad \text{and}\quad \mathbf{q}=\begin{pmatrix}\mathbf{C}\\ -\mathbf{b}\end{pmatrix}$$

we have a corresponding LCP problem. This establishes the fact that solving this LCP problem will provide optimal solutions, if they exist, to both the problems [primal and dual]. Here we notice that **M** is positive semidefinite.

Given the problem with a convex quadratic objective function and linear inequalities, we can similarly show its equivalence to an LCP problem. Consider a problem similar to Problem 2.19.5. We have

$$\begin{aligned}\text{Minimize}\quad & \mathbf{f(V)}=\mathbf{V}'\mathbf{QV}+\mathbf{C}'\mathbf{V}\\ \text{subject to}\quad & \mathbf{AV}\leqslant\mathbf{b}\\ & \mathbf{V}\geqslant\mathbf{0}\end{aligned}$$

where $\mathbf{Q},\mathbf{A},\mathbf{C},\mathbf{b}$ were as defined earlier (we can in general consider a **Q**, positive semidefinite matrix in the discussion).

From Result 2.19.6 we have (2.19.24) and (2.19.25), restated below:

(1) $\mathbf{V},\boldsymbol{\lambda},\boldsymbol{\mu},\boldsymbol{\eta}\geqslant\mathbf{0}$

(2) $\mathbf{AV}+\boldsymbol{\eta}=\mathbf{b}$

(3) $\boldsymbol{\mu}-2\mathbf{QV}-\mathbf{A}'\boldsymbol{\lambda}=\mathbf{C}$

and

(4) $\mathbf{V}'\boldsymbol{\mu}+\boldsymbol{\lambda}'\boldsymbol{\eta}=0.$

If we find any $\mathbf{V},\boldsymbol{\lambda},\boldsymbol{\mu},\boldsymbol{\eta}$ satisfying these four conditions we have an optimal solution to the problem given by **V**. So defining

$$\mathbf{M}=\begin{pmatrix}2\mathbf{Q} & \mathbf{A}'\\ -\mathbf{A} & \mathbf{0}\end{pmatrix},\quad \mathbf{q}=\begin{pmatrix}\mathbf{C}\\ \mathbf{b}\end{pmatrix},\quad \mathbf{W}=\begin{pmatrix}\boldsymbol{\mu}\\ \boldsymbol{\eta}\end{pmatrix},\quad \text{and}\quad \mathbf{Z}=\begin{pmatrix}\mathbf{V}\\ \boldsymbol{\lambda}\end{pmatrix},$$

we have an LCP problem giving the four conditions above. Hence it is possible to solve Problem 2.19.5 by the LCP approach. Notice that **M** is positive semidefinite as **Q** is positive semidefinite. Below we outline an algorithm for solving the LCP formulation of the problem.

Algorithm 2.20.1

Step 0 Consider the problem

$$\begin{aligned}\mathbf{W}-\mathbf{MZ}-\mathbf{e}Z_0&=\mathbf{q}\\ \mathbf{W},\mathbf{Z}\geqslant\mathbf{0},&\qquad Z_0\geqslant 0\\ \mathbf{W}'\mathbf{Z}&=0\end{aligned}$$

Table 2.20.1

Basic Variables	$\mathbf{q}$	$W_1 \cdots$	W_α	$\cdots$	W_k	$Z_1 \cdots$	$Z_\alpha \cdots$	Z_k	Z_0
W_1	q_1	1	0		0	$-m_{11}$	$-m_{1\alpha}$	$-m_{1k}$	-1
$\vdots$	$\vdots$	$\vdots$	$\vdots$		$\vdots$				$\vdots$
W_α	q_α	0	1	$\cdots$	0	$-m_{\alpha 1}$	$-m_{\alpha\alpha}$	$-m_{\alpha k}$	-1
$\vdots$	$\vdots$	$\vdots$	$\vdots$		$\vdots$				$\vdots$
W_k	q_k	0	0		1	$-m_{k1}$	$-m_{k\alpha}$	$-m_{kk}$	-1

where Z_0 is an artificial variable introduced to obtain an initial feasible solution that is complementary. (The idea is to obtain a solution in which $Z_0=0$ if possible.) The initial table corresponding to $\mathbf{W}=\mathbf{q}$, $\mathbf{Z}=\mathbf{0}$, and $Z_0=0$ is Table 2.20.1, where $\mathbf{M}=(m_{ij})$, $i, j=1,\ldots,k$. If the solution is infeasible, go to Step 1; otherwise stop.

Step 1 The variable Z_0 enters the basis, replacing W_α such that $q_\alpha = \min q_i < 0$. Transform the table using the ring-around-the-rosy method. At least one complementary pair of variables (W_α, Z_α) is nonbasic. The transformation yields Table 2.20.2. Go to Step 2.

Step 2 At each iteration the complement of the variable which was removed from the basis in the previous iteration is to be introduced into the basis. (For instance, initially, W_α is removed so Z_α enters in the next iteration.) Go to Step 3.

Step 3 If the variable Z_α entering as per Step 2 is such that the coefficients in the previous column corresponding to Z_α are less than or equal to zero, the implication is that Z_α can be increased arbitrarily. Then

Table 2.20.2

Basic Variables	$\mathbf{q}$	$W_1 \cdots$	$W_\alpha \cdots$	W_k	Z_1	$\cdots$	Z_α	$\cdots$	Z_k	Z_0
W_1	q_1-q_α	1	-1	0	$-m_{11}+m_{\alpha 1}$	$\cdots$	$-m_{1\alpha}+m_{\alpha\alpha}$	$\cdots$	$-m_{1\alpha}+m_{\alpha k}$	0
$\vdots$	$\vdots$	$\vdots$	$\vdots$	$\vdots$	$\vdots$		$\vdots$		$\vdots$	$\vdots$
Z_0	$-q_\alpha$	0	-1	0	$m_{\alpha 1}$	$\cdots$	$m_{\alpha\alpha}$	$\cdots$	$m_{\alpha k}$	1
$\vdots$	$\vdots$	$\vdots$	$\vdots$	$\vdots$	$\vdots$		$\vdots$		$\vdots$	$\vdots$
W_k	q_k-q_α	0	-1	1	$-m_{k1}+m_{\alpha 1}$	$\cdots$	$-m_{k\alpha}+m_{\alpha\alpha}$	$\cdots$	$-m_{kk}+m_{\alpha k}$	0

the procedure is said to terminate in a *secondary ray*. We are unable to determine whether the problem has a solution, unless certain assumptions on M are made. Stop.

Otherwise, determine as per simplex criterion given for nonnegative variables for the vector to leave the basis. If the variable W_j or Z_j leaves the basis, transform the table and go to step 2, If the variable to leave the basis is Z_0, we stop. We have a complementary feasible solution to the problem.

EXAMPLE 2.20.1 Consider the quadratic programming problem given below:

$$\begin{aligned} \text{Minimize} \quad & \mathbf{V'QV+C'V} \\ \text{subject to} \quad & \mathbf{AV \leqslant b} \\ & \mathbf{V \geqslant 0} \end{aligned}$$

where

$$\mathbf{Q}=\begin{bmatrix} 1 & -1 \\ -1 & 2 \end{bmatrix}, \qquad \mathbf{C'}=[-2,-6]$$

$$\mathbf{A}=\begin{bmatrix} 1 & 1 \\ -1 & 2 \end{bmatrix}, \qquad \text{and} \quad \mathbf{b}=\begin{bmatrix} 2 \\ 2 \end{bmatrix}.$$

We have the corresponding LCP problem as shown below:

$$\mathbf{M}=\begin{bmatrix} \mathbf{2Q} & \mathbf{A'} \\ -\mathbf{A} & \mathbf{0} \end{bmatrix}$$

$$=\begin{bmatrix} 2 & -2 & 1 & -1 \\ -2 & 4 & 1 & 2 \\ -1 & -1 & 0 & 0 \\ 1 & -2 & 0 & 0 \end{bmatrix}$$

and

$$\mathbf{q}=\begin{bmatrix} \mathbf{C} \\ \mathbf{b} \end{bmatrix}=\begin{bmatrix} -2 \\ -6 \\ 2 \\ 2 \end{bmatrix}$$

$$\mathbf{W}=\begin{bmatrix} \boldsymbol{\mu} \\ \boldsymbol{\eta} \end{bmatrix} \qquad \text{and} \quad \mathbf{Z}=\begin{bmatrix} \mathbf{V} \\ \boldsymbol{\lambda} \end{bmatrix}$$

Table 2.20.3

Basic Variables	q	W_1	W_2	W_3	W_4	Z_1	Z_2	Z_3	Z_4	Z_0
W_1	−2	1	0	0	0	−2	2	−1	1	−1
→W_2	−6	0	1	0	0	2	−4	−1	−2	−1
W_3	2	0	0	1	0	1	1	0	0	−1
W_4	2	0	0	0	1	−1	2	0	0	−1

Step 0 We have the equivalent problem

$$\mathbf{W}-\mathbf{M}\mathbf{Z}-\mathbf{e}Z_0=\mathbf{q}$$

$$\mathbf{W},\mathbf{Z}\geqslant\mathbf{0},\ Z_0\geqslant 0$$

$$\mathbf{W}'\mathbf{Z}=0$$

Let the initial solution be $\mathbf{W}=\mathbf{q}$, $\mathbf{Z}=\mathbf{0}$ and $Z_0=0$. Table 2.20.3 is the corresponding table. As this solution is infeasible, we go to Step 1.

Step 1 The variable Z_0 enters the basis replacing $\mathbf{W}_2$ as $\min_i\{q_i\}=q_2=-6$. Transforming the table we get Table 2.20.4.

Step 2 As W_2 was removed from the basis in the previous interchange, we introduce Z_2 into the basis. By the simplex exit criterion we find W_1 leaves the basis. We transform the table. The resultant table is given in Table 2.20.5. As W_1 left the basis in the previous iteration, we introduce Z_1 into the basis. And we transform the table to obtain Table 2.20.6. This time Z_3 enters. We find Z_0 leaving the basis. We just have to find the new basic solution the rest of the table is not necessary. Thus we find after

Table 2.20.4

Basic Variables	q	W_1	W_2	W_3	W_4	Z_1	Z_2	Z_3	Z_4	Z_0
W_1	4	1	−1	0	0	−4	6	0	3	0
Z_0	6	0	−1	0	0	−2	4	1	2	1
W_3	8	0	−1	1	0	−1	5	1	2	0
W_4	8	0	−1	0	1	−3	6	1	2	0

Table 2.20.5

Basic Variables	$\mathbf{q}$	W_1	W_2	W_3	W_4	Z_1	Z_2	Z_3	Z_4	Z_0
Z_2	$\frac{2}{3}$	$\frac{1}{6}$	$-\frac{1}{6}$	0	0	$-\frac{2}{3}$	1	0	$\frac{1}{2}$	0
Z_0	$\frac{10}{3}$	$-\frac{2}{3}$	$-\frac{1}{3}$	0	0	$\frac{2}{3}$	0	1	0	1
$\rightarrow W_3$	$\frac{14}{3}$	$-\frac{5}{6}$	$-\frac{1}{6}$	1	0	$\frac{7}{3}$	0	1	$-\frac{1}{2}$	0
W_4	4	-1	0	0	1	1	0	0	-1	0

transformation,

$$Z_2=\tfrac{6}{5}$$

$$Z_3=\tfrac{14}{5}$$

$$Z_1=\tfrac{4}{5}$$

$$W_4=\tfrac{2}{5}.$$

This solution is a complementary solution. As $\mathbf{Z}=\begin{bmatrix}\mathbf{V}\\ \boldsymbol{\lambda}\end{bmatrix}$, $V_1=Z_1=\frac{4}{5}$, $V_2=Z_2=\frac{6}{5}$, $\lambda_1=Z_3=\frac{14}{5}$ and $\lambda_2=Z_4=0$. Also $W_1=\mu_1=0$, $W_2=\mu_2=0$, $W_3=\eta_1=0$ and $W_4=\eta_2=\frac{2}{5}$. Therefore an optimal solution to the given quadratic programming problem is

$$V_1=\tfrac{4}{5}, \qquad V_2=\tfrac{6}{5}.$$

The objective function value is -7.2.

Table 2.20.6

Basic Variables	$\mathbf{q}$	W_1	W_2	W_3	W_4	Z_1	Z_2	Z_3	Z_4	Z_0
Z_2	2	$-\frac{1}{14}$	$-\frac{3}{14}$	$\frac{2}{7}$	0	0	1	$\frac{2}{7}$	$\frac{5}{14}$	0
$\rightarrow Z_0$	2	$-\frac{3}{7}$	$-\frac{2}{7}$	$-\frac{2}{7}$	0	0	0	$\frac{5}{7}$	$\frac{1}{7}$	0
Z_1	2	$-\frac{5}{14}$	$-\frac{1}{14}$	$\frac{3}{7}$	0	1	0	$\frac{3}{7}$	$-\frac{3}{14}$	0
W_4	2	$-\frac{9}{14}$	$\frac{1}{14}$	$-\frac{3}{7}$	1	0	0	$\frac{4}{7}$	$-\frac{11}{14}$	1

Application of Algorithm

It should be noted that in general when Algorithm 2.20.1 terminates in a secondary ray no conclusion can be reached about the existence of a solution to the LCP problem. We then say that Algorithm 2.20.1 is not applicable to that LCP problem with given **M** and **q**. There are many sufficient conditions on **M**, so the fact that the algorithm either computes a solution to the LCP or terminates in a secondary ray implies that the LCP has no solution. We state the following: Let **M** be a square matrix.

DEFINITION 2.20.3 A matrix M is said to be *copositive plus* if (1) for all $\mathbf{X} \geqslant \mathbf{0}$, $\mathbf{X}'\mathbf{M}\mathbf{X} \geqslant 0$, and (2) $\mathbf{X}'\mathbf{M}\mathbf{X} = 0, \mathbf{X} \geqslant \mathbf{0} \Rightarrow (\mathbf{M}' + \mathbf{M})\mathbf{X} = \mathbf{0}$.

DEFINITION 2.20.4 A matrix **M** is said to be *strictly copositive* if $\mathbf{X}'\mathbf{M}\mathbf{X} > 0$ for all $\mathbf{X} \geqslant \mathbf{0}$.

Let $CP^+ = \{\mathbf{M} | \mathbf{M}$ is copositive plus$\}$, and $SCP = \{\mathbf{M} | \mathbf{M}$ is strictly copositive$\}$.

We state the following result without proof.

RESULT 2.20.1 Consider solving a LCP problem using Algorithm 2.20.1 when the corresponding matrix $\mathbf{M} \in CP^+$. If Algorithm 2.20.1 terminates in a secondary ray the problem does not have a solution. If $\mathbf{M} \in SCP$, then the algorithm never terminates in a secondary ray.

REMARK 2.20.3 The linear-programming problem and quadratic-programming problem with **Q** positive semidefinite can be solved using Algorithm 2.20.1, as the corresponding **M** matrices are positive semidefinite.

DEFINITION 2.20.5 A square matrix **M** is called a **Z**-matrix if $m_{ij} \leqslant 0$ for all $i \neq j$.

RESULT 2.20.2 Consider the LCP problem with the matrix **M** being a **Z**-matrix. Then Algorithm 2.20.1 is applicable to this problem.

RESULT 2.20.3 If **M** is a **Z**-matrix, then solving the following linear-programming problem by the simplex method finds a solution to the LCP problem or shows that none exists.

$$
\begin{aligned}
\text{Minimize} \quad & Z_0 \\
\text{subject to} \quad & \mathbf{W} - \mathbf{M}\mathbf{Z} - \mathbf{d}Z_0 = \mathbf{q} \\
& \mathbf{W}, \mathbf{Z} \geqslant \mathbf{0}, \; Z_0 \geqslant 0
\end{aligned}
$$

where **d** is any positive vector.

REMARK 2.20.4 Nonexistence of a solution to the LCP in the Result 2.20.3 is shown by obtaining $Z_0 > 0$ in the optimal table, at termination of the simplex procedure.

REMARK 2.20.5 If the LCP has a solution in fact we require only k iterations if $\mathbf{M}$ is of size k.

Introducing nonnegativity restriction is not a problem if we use the quadratic program (2.14.1) as we have the corresponding problem,

$$\begin{aligned} &\text{Minimize} && \mathbf{d}'\mathbf{d} \\ &\text{subject to} && \mathbf{X}\boldsymbol{\beta} - \mathbf{d} = \mathbf{Y} \\ & && \boldsymbol{\beta} \geqslant \mathbf{0}, \mathbf{d} \text{ unrestricted in sign.} \end{aligned}$$

An interesting point to note is that the corresponding LCP in this case is of a reduced dimension, as the matrix $\mathbf{Q}$ is

$$\mathbf{Q} = \begin{bmatrix} \mathbf{0} & \mathbf{0} & \mathbf{0} \\ \mathbf{0} & \mathbf{I} & \mathbf{0} \\ \mathbf{0} & \mathbf{0} & \mathbf{I} \end{bmatrix}$$

and is of dimension $3n \times 3n$ and the matrix

$$\mathbf{A} = \begin{pmatrix} \mathbf{X} & \mathbf{I} & -\mathbf{I} \\ -\mathbf{X} & -\mathbf{I} & \mathbf{I} \end{pmatrix}, \qquad \mathbf{V} = [\boldsymbol{\beta}', \mathbf{d}_1', \mathbf{d}_2'].$$

Thus we can see that the $\mathbf{M}$ matrix in the LCP is only of dimension $5n \times 5n$ in this case.

If we have to impose linear consistent constraints on $\boldsymbol{\beta}$ and wish to estimate $\boldsymbol{\beta}$ using least-squares method, only matrix $\mathbf{A}$ and $\mathbf{b}$ in Formulation 2.11.5 need to be changed, and they are given by

$$\mathbf{A} = \begin{bmatrix} \mathbf{X} & -\mathbf{X} & \mathbf{I} & -\mathbf{I} \\ -\mathbf{X} & \mathbf{X} & -\mathbf{I} & \mathbf{I} \\ \mathbf{D} & -\mathbf{D} & \mathbf{0} & \mathbf{0} \end{bmatrix}; \qquad \mathbf{b} = \begin{bmatrix} \mathbf{Y} \\ -\mathbf{Y} \\ \mathbf{R} \end{bmatrix}$$

where $\mathbf{D}$ corresponds to the constraints on $\boldsymbol{\beta}$—namely, we require $\mathbf{D}\boldsymbol{\beta} = \mathbf{R}$.

It is to be noted that in all the cases above we need not make any assumption on the rank of $\mathbf{X}$. Also we are able to solve the problem using a simplex-like method. For other approaches to these problems see the bibliographical notes.

2.21 GENERALIZED LEAST-SQUARES AND LINEAR COMPLEMENTARY PROGRAMMING

Suppose ε_i's are correlated random variables, in particular if we assume that the variance-covariance matrix of $\boldsymbol{\varepsilon}$ is given by $\sigma^2\mathbf{V}$, where $\mathbf{V}$ is a positive definite matrix, and we have the generalized least-squares problem.

From the theory on positive definite matrices we know there exists a nonsingular matrix $\mathbf{K}$ such that $\mathbf{V}=\mathbf{K}\mathbf{K}'$. Let $\mathbf{Z}=\mathbf{K}^{-1}\mathbf{Y}$, $\mathbf{B}=\mathbf{K}^{-1}\mathbf{X}$, and $\boldsymbol{\eta}=\mathbf{K}^{-1}\boldsymbol{\varepsilon}$. We now have $\mathbf{Z}=\mathbf{B}\boldsymbol{\beta}+\boldsymbol{\eta}$, where $\mathbf{B}$ has the same rank as $\mathbf{X}$—namely, P if we assume $\mathbf{X}$ is of full rank. $\mathbf{E}(\boldsymbol{\eta})=\mathbf{0}$, and the variance-covariance matrix of $\boldsymbol{\eta}$ is given by $\sigma^2\mathbf{I}$. Therefore, minimizing $\boldsymbol{\eta}'\boldsymbol{\eta}$ with respect to $\boldsymbol{\beta}$ and using the results obtained in Section 2.4, we find

$$\begin{aligned}\boldsymbol{\beta}^* &= (\mathbf{B}'\mathbf{B})^{-1}\mathbf{B}'\mathbf{Z}\\ &= (\mathbf{X}'\mathbf{V}^{-1}\mathbf{X})^{-1}\mathbf{X}'\mathbf{V}^{-1}\mathbf{Y}\end{aligned} \tag{2.21.1}$$

with $\mathbf{E}(\boldsymbol{\beta}^*)=(\mathbf{X}'\mathbf{V}^{-1}\mathbf{X})^{-1}\mathbf{X}'\mathbf{V}^{-1}\mathbf{X}\boldsymbol{\beta}=\boldsymbol{\beta}$, and the variance-covariance matrix of $\boldsymbol{\beta}^*$ is given by

$$\begin{aligned}\mathbf{D}(\boldsymbol{\beta}^*) &= \sigma^2(\mathbf{B}'\mathbf{B})^{-1}\\ &= \sigma^2(\mathbf{X}'\mathbf{V}^{-1}\mathbf{X})^{-1}.\end{aligned} \tag{2.21.2}$$

As we noted earlier we wish to minimize $\boldsymbol{\eta}'\boldsymbol{\eta}$ with respect to $\boldsymbol{\beta}$. Let us now look at $\boldsymbol{\eta}'\boldsymbol{\eta}$.

$$\begin{aligned}\boldsymbol{\eta}'\boldsymbol{\eta} &= (\mathbf{K}^{-1}\boldsymbol{\varepsilon})'(\mathbf{K}^{-1}\boldsymbol{\varepsilon})\\ &= \boldsymbol{\varepsilon}(\mathbf{K}\mathbf{K}')^{-1}\boldsymbol{\varepsilon}\\ &= \boldsymbol{\varepsilon}'\mathbf{V}^{-1}\boldsymbol{\varepsilon}.\end{aligned} \tag{2.21.3}$$

Thus we see that the objective function in this case is more general than in the simple least-squares approach, in which it is $\boldsymbol{\varepsilon}'\mathbf{I}\boldsymbol{\varepsilon}$. Therefore we would like to consider this problem and formulate it as an LCP.

Let $\mathbf{W}$ be any positive definite matrix. Consider the problem

$$\begin{aligned}&\text{Minimize} && \mathbf{d}'\mathbf{W}\mathbf{d}\\ &\text{subject to} && \mathbf{X}\boldsymbol{\beta}-\mathbf{d}=\mathbf{Y}\\ & && \boldsymbol{\beta},\mathbf{d}\ \text{unrestricted in sign.}\end{aligned} \tag{2.21.4}$$

It is easy to notice that this introduces changes only in the matrix **Q** of the LCP. We have

$$\mathbf{Q}=\begin{bmatrix} \mathbf{0} & \mathbf{0} & \mathbf{0} & \mathbf{0} \\ \mathbf{0} & \mathbf{0} & \mathbf{0} & \mathbf{0} \\ \mathbf{0} & \mathbf{0} & \mathbf{W} & -\mathbf{W} \\ \mathbf{0} & \mathbf{0} & -\mathbf{W} & \mathbf{W} \end{bmatrix}. \tag{2.21.5}$$

Thus, as in Problem 2.11.1, we have an LCP with **Q** given by (2.21.5).

It is useful to note that when **X** is not of full rank, $\mathbf{X}'\mathbf{V}^{-1}\mathbf{X}$ does not have the inverse and we have to use a g-inverse of $(\mathbf{X}'\mathbf{V}^{-1}\mathbf{X})$ to obtain $\boldsymbol{\beta}$. But while we are using the LCP, **X** need not be of full rank.

2.22 MINIMIZING SUM OF ABSOLUTE DIFFERENCES BETWEEN DEVIATIONS (MINSADBED Regression)

Let d_i denote the deviation from the expected and the observed values of Y_i, as before; that is,

$$d_i = \mathbf{Y}_i - (\mathbf{X}\boldsymbol{\beta})_i, \qquad i=1,\dots,n \tag{2.22.1}$$

where **X** is an $n\times p$ matrix, $\boldsymbol{\beta}$ is a $p\times 1$ vector, and $(\mathbf{X}\boldsymbol{\beta})_i$ denotes the ith row of $\mathbf{X}\boldsymbol{\beta}$.

Consider the problem of finding $\boldsymbol{\beta}$ so as to

$$\text{Minimize} \sum_{i<j} |d_i - d_j|. \tag{2.22.2}$$

Expression 2.22.2 is the minimization of sum absolute differences between deviations. We now proceed to describe this problem as a linear programming problem.

Let $d_{ij} = d_i - d_j$. Now d_{ij} is unrestricted in sign. So as usual we can write $d_{ij} = d_{ij}^1 - d_{ij}^2$ where $d_{ik}^k \geqslant 0$ for $k=1,2$. Therefore $|d_i - d_j|$, the absolute difference between the deviations, can be given by

$$|d_i - d_j| = d_{ij}^1 + d_{ij}^2 \tag{2.22.3}$$

with

$$d_i - d_j - d_{ij}^1 + d_{ij}^2 = 0$$

$$d_{ij}^k \geqslant 0, \qquad k=1, \qquad 2, 1 \leqslant i < j \leqslant n. \tag{2.22.4}$$

Thus our objective can be written as,

$$\text{Minimize } \sum_{i<j} d_{ij}^1 + \sum_{i<j} d_{ij}^2.$$

We can put this now in matrix notation, using

$$\mathbf{D}^k = \left[d_{12}^k, d_{13}^k, \cdots, d_{n-1n}^k \right]' \text{ for } k=1,2. \text{ We have:}$$

$$\begin{array}{ll} \text{Minimize} & \mathbf{D}^{1}\mathbf{e} + \mathbf{D}^2\mathbf{e} \\ \text{subject to} & \mathbf{X}\boldsymbol{\beta} + \mathbf{I}\mathbf{d} = \mathbf{Y} \\ & \mathbf{d}, \mathbf{D}^1, \mathbf{D}^2 \quad \text{satisfy (2.22.4)} \\ & \boldsymbol{\beta} \text{ unrestricted in sign.} \end{array} \tag{2.22.5}$$

The matrix corresponding to (2.22.4) can now be given. Observe that we have one equation for each choice of the pair (i, j), $i<j$. Therefore there are $r=n(n-1)/2$ constraints. The variable d_i appears in $n-i$ constraints where d_j appears with coefficient -1 for all $i+1 \leqslant j \leqslant n$ and d_{ij}^k appears only once in these r constraints—namely, $d_{ij}^1 - d_{ij}^2$ appears in the constraint corresponding to (i, j). Thus, we can express these constraints by $[\mathbf{0}, \mathbf{H}, \mathbf{I}_r, -\mathbf{I}_r]\mathbf{V} = \mathbf{0}$, where $\mathbf{V}' = (\boldsymbol{\beta}', \mathbf{d}', \mathbf{D}^{1'}, \mathbf{D}^{2'})$ and $\mathbf{D}^1 \geqslant \mathbf{0}$, $\mathbf{D}^2 \geqslant \mathbf{0}$ where

$$\mathbf{H} = \begin{bmatrix} \mathbf{e}_{n-1}, & -\mathbf{I}_{n-1} & \\ \mathbf{0}_{n-2\times 1}, & \mathbf{e}_{n-2}, & -\mathbf{I}_{n-2} \\ \vdots & & \\ \mathbf{0}_{1\times n-2}, & \mathbf{e}_1, & -\mathbf{I}_1 \end{bmatrix}.$$

Now we have Problem 2.22.5 stated as follows:

$$\begin{array}{ll} \text{Minimize} & \mathbf{D}^{1'}\mathbf{e} + \mathbf{D}^{2'}\mathbf{e} \\ \text{subject to} & \begin{pmatrix} \mathbf{X}_{n\times p} & \mathbf{I}_n & \mathbf{0}_{n\times r} & \mathbf{0}_{n\times r} \\ \mathbf{0}_{r\times p} & \mathbf{H}_{r\times n} & -\mathbf{I}_r & \mathbf{I}_r \end{pmatrix} \mathbf{V} = \begin{pmatrix} \mathbf{Y} \\ \mathbf{0} \end{pmatrix} \\ & \boldsymbol{\beta}, \mathbf{d} \text{ unrestricted in sign} \\ & \mathbf{D}^1 \geqslant \mathbf{0},\ \mathbf{D}^2 \geqslant \mathbf{0}, \\ & \mathbf{V}' = (\boldsymbol{\beta}', \mathbf{d}', \mathbf{D}^{1'}, \mathbf{D}^{2'}). \end{array} \tag{2.22.6}$$

This problem has $n+n(n-1)/2$ constraints and $(n-1)n+n+p$ variables.

But by considering the dual of (2.22.6) we obtain essentially a reduced-size problem with some variables having upper bounds, similar to that of the MINMAD regression problem.

We have,

$$\text{Maximize} \quad \mathbf{Y}'\mathbf{U}^1$$

$$\text{subject to} \quad \begin{bmatrix} \mathbf{X}' & \mathbf{0} \\ \mathbf{I} & \mathbf{H}' \\ \mathbf{0} & -\mathbf{I} \\ \mathbf{0} & \mathbf{I} \end{bmatrix} \begin{pmatrix} \mathbf{U}^1 \\ \mathbf{U}^2 \end{pmatrix} \begin{Bmatrix} = \\ \leqslant \\ \leqslant \end{Bmatrix} \begin{Bmatrix} \begin{bmatrix} \mathbf{0} \\ \mathbf{e}_r \\ \mathbf{e}_r \end{bmatrix} \end{Bmatrix}$$

$$\mathbf{U}^1, \mathbf{U}^2 \text{ unrestricted in sign.}$$

Now the problem is equivalent to,

$$\text{Maximize} \quad \mathbf{Y}'\mathbf{U}'$$

$$\text{subject to} \quad \begin{pmatrix} \mathbf{X}' & 0 \\ \mathbf{I} & \mathbf{H}' \end{pmatrix} \begin{pmatrix} \mathbf{U}^1 \\ \mathbf{U}^2 \end{pmatrix} = 0$$

$$-\mathbf{U}^2 \leqslant \mathbf{e}_r$$

$$\mathbf{U}^2 \leqslant \mathbf{e}_r$$

$$\mathbf{U}^1, \mathbf{U}^2 \text{ unrestricted in sign.} \tag{2.22.7}$$

Let $\mathbf{W} = \mathbf{U}^2 + \mathbf{e}$. Then $2\mathbf{e} \geqslant \mathbf{W} \geqslant \mathbf{0}$ and $\mathbf{H}'\mathbf{U}^2 = \mathbf{H}'(\mathbf{W} - \mathbf{e}) = \mathbf{H}'\mathbf{W}' - \mathbf{H}'\mathbf{e}$. Now $(\mathbf{H}'\mathbf{e})' = \mathbf{e}'\mathbf{H} = [(n-1), (n-3), \ldots, -(n-1)]$. Let $\mathbf{b}' = \mathbf{e}'\mathbf{H}$. Thus, finally, we obtain

$$\text{Maximize} \quad \mathbf{Y}'\mathbf{U}^1$$

$$\text{subject to} \quad \begin{pmatrix} \mathbf{X}' & \mathbf{0} \\ \mathbf{I} & \mathbf{H}' \end{pmatrix} \begin{pmatrix} \mathbf{U}^1 \\ \mathbf{W} \end{pmatrix} = \begin{pmatrix} \mathbf{0} \\ \mathbf{b} \end{pmatrix}$$

$$\mathbf{U}^1 \quad \text{unrestricted in sign}$$

$$\mathbf{0} \leqslant \mathbf{W} \leqslant 2\mathbf{e}. \tag{2.22.8}$$

Now notice that the matrix in (2.22.8) can be rewritten by interchanging rows, as

$$\begin{pmatrix} \mathbf{I} & \mathbf{H}' \\ \mathbf{X}' & \mathbf{0} \end{pmatrix} \begin{pmatrix} \mathbf{U}^1 \\ \mathbf{W} \end{pmatrix} = \begin{pmatrix} \mathbf{b} \\ \mathbf{0} \end{pmatrix}.$$

This matrix is in decomposable form, as the variables $\mathbf{U}^1$ appear only in $\mathbf{X}'\mathbf{U}^1=0$. Moreover, from $[\mathbf{I},\mathbf{H}']\begin{bmatrix}\mathbf{U}^1\\ \mathbf{W}\end{bmatrix}=\mathbf{b}$ we obtain the following bounds for U_i^1's:

$$-(n-1)\leqslant U_i^1\leqslant n-1 \qquad i=1,\dots,n. \tag{2.22.9}$$

That is, $-(n-1)\mathbf{e}\leqslant\mathbf{U}^1\leqslant(n-1)\mathbf{e}$. Let $\mathbf{W}^1=\mathbf{U}^1+(n-1)\mathbf{e}$. Then $\mathbf{0}\leqslant\mathbf{W}^1\leqslant 2(n-1)\mathbf{e}$. Using the variables $\mathbf{W}^1$ and $\mathbf{W}$, the problem now can be written as:

$$\text{Maximize} \quad \mathbf{Y}'\mathbf{W}^1-(n-1)\mathbf{Y}'\mathbf{e}$$

or

$$\text{Maximize} \quad \mathbf{Y}'\mathbf{W}^1$$

$$\text{subject to} \quad \begin{pmatrix}\mathbf{I} & \mathbf{H}'\\ \mathbf{X}' & 0\end{pmatrix}\begin{pmatrix}\mathbf{W}^1\\ \mathbf{W}\end{pmatrix}=(n-1)\begin{pmatrix}\mathbf{e}\\ \mathbf{X}'\mathbf{e}\end{pmatrix}.$$

$$2(n-1)\mathbf{e}\geqslant\mathbf{W}^1\geqslant\mathbf{0};2\mathbf{e}\geqslant\mathbf{W}\geqslant\mathbf{0}. \tag{2.22.10}$$

Now Problem 2.22.10 is a bounded-variable linear-programming problem, and as $\mathbf{H}'$ has a special structure the usual bounded-variable method can be modified to exploit it.

2.23 PROPERTIES OF THE MINSADBED REGRESSION ESTIMATOR

Properties of this new estimator are discussed next. We consider the following example, shown in Fig. 2.23.1. The deviation from the line and

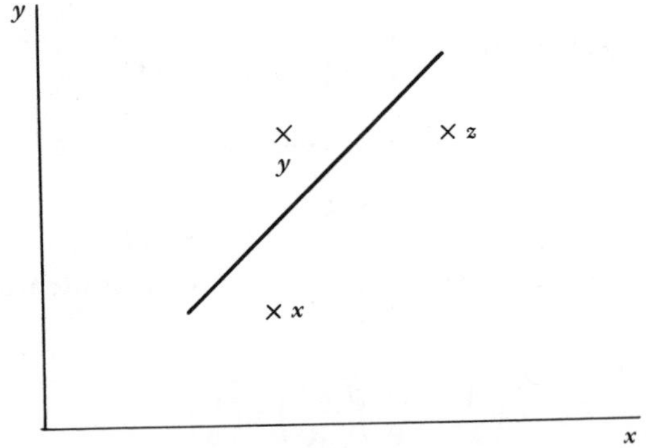

Figure 2.23.1 Using MINSADBED β_0 cannot be estimated.

the points x, y, and z are -1, $+1$, and -1, respectively. If we move the line in a parallel manner and make it pass through x and z we have the deviations 0, 2, and 0, respectively. Now if we consider the sum of the absolute differences between the deviations we get $|-1-1|+|1+1|+|-1+1|=2+2+0=4$ while the deviations are -1, 1, and -1, and it is $|0-2|+|2-0|+|0-0|=4$ again when the deviations are 0, 2, and 0. In fact, for all lines parallel to the given line we will get the same value of 4 for the sum of differences between the deviations. Such is not the case if we consider the sum of absolute deviations or the sum of squared deviations. For this reason while using MINSADBED we cannot estimate β_0. For the general case we have the following argument.

Consider

$$d_i = Y_i - \sum_{k=0}^{p} \beta_k X_{ki}.$$

Therefore

$$|d_i - d_j| = \left| Y_i - Y_j - \sum_{k=1}^{p} \beta_k (X_{ki} - X_{kj}) \right|,$$

which does not involve β_0.

For this reason, as in least squares regression, we make the line pass through $(\overline{\mathbf{X}}, \overline{\mathbf{Y}})$ by adding the constraint $\overline{\mathbf{X}}\boldsymbol{\beta} = \overline{\mathbf{Y}}$ where

$$\overline{\mathbf{X}} = \left(\overline{X}_0, \ldots, \overline{X}_{p-1}\right), \overline{\mathbf{Y}} = \frac{1}{n} \sum_{j=1}^{n} Y_j \text{ and } \overline{X}_k = \frac{1}{n} \sum_{j=1}^{n} X_{kj}.$$

Unbiasedness can be proved in the same way as for MINMAD regression estimates. Let $\boldsymbol{\beta}_0$ be any antisymmetric estimator. Define $\overline{\boldsymbol{\beta}}_1 = \boldsymbol{\beta}_1 + \boldsymbol{\beta}_0^{(2)}, \overline{\boldsymbol{\beta}}_2 = \boldsymbol{\beta}_2 + \boldsymbol{\beta}_0^{(1)}$ where $\boldsymbol{\beta} = \boldsymbol{\beta}_1 - \boldsymbol{\beta}_2$, $\boldsymbol{\beta}_0 = \boldsymbol{\beta}_0^{(1)} - \boldsymbol{\beta}_0^{(2)}$, $\boldsymbol{\beta}_k \geqslant \mathbf{0}$, $\boldsymbol{\beta}_0^{(k)} \geqslant \mathbf{0}$, $k = 1, 2$. Notice that problem 2.22.6 can be equivalently written as,

$\mathbf{P}_1$: Minimize $\mathbf{D}^{1'}\mathbf{e} + \mathbf{D}^{2'}\mathbf{e}$

$$\text{subject to} \quad \begin{bmatrix} -\mathbf{X} & \mathbf{X} & -\mathbf{I} & \mathbf{0} & \mathbf{0} \\ \mathbf{X} & -\mathbf{X} & \mathbf{I} & \mathbf{0} & \mathbf{0} \\ \mathbf{0} & \mathbf{0} & \mathbf{H} & \mathbf{I} & -\mathbf{I} \end{bmatrix} \mathbf{V} \left\{ \begin{matrix} \leqslant \\ \leqslant \\ = \end{matrix} \right\} \begin{bmatrix} -\mathbf{Y} + \mathbf{X}\boldsymbol{\beta}_0 \\ \mathbf{Y} - \mathbf{X}\boldsymbol{\beta}_0 \\ \mathbf{0} \end{bmatrix} \tag{2.23.1}$$

where $\mathbf{V}' = [\overline{\boldsymbol{\beta}}_1', \overline{\boldsymbol{\beta}}_2', \mathbf{d}', \mathbf{D}^{1'}, \mathbf{D}^{2'}]$, $\overline{\boldsymbol{\beta}}_1, \overline{\boldsymbol{\beta}}_2, \mathbf{D}^1, \mathbf{D}^2 \geqslant \mathbf{0}$, and $\mathbf{d}$ unrestricted in sign.

Interchanging variables $\bar{\beta}_1$ and $\bar{\beta}_2$ of (2.23.1), we can write (2.23.1) equivalently as follows:

$$\mathbf{P}_2: \quad \text{Minimize} \quad \mathbf{D}^{1'}\mathbf{e}+\mathbf{D}^{2'}\mathbf{e}$$

$$\text{subject to} \quad \begin{bmatrix} -\mathbf{X} & \mathbf{X} & -\mathbf{I} & \mathbf{0} & \mathbf{0} \\ \mathbf{X} & -\mathbf{X} & \mathbf{I} & \mathbf{0} & \mathbf{0} \\ \mathbf{0} & \mathbf{0} & \mathbf{H} & \mathbf{I} & -\mathbf{I} \end{bmatrix} \boldsymbol{\lambda} \begin{matrix} \leqslant \\ \leqslant \\ = \end{matrix} \begin{bmatrix} \mathbf{Y}-\mathbf{X}\boldsymbol{\beta}_0 \\ -\mathbf{Y}+\mathbf{X}\boldsymbol{\beta}_0 \\ \mathbf{0} \end{bmatrix}$$

$$\boldsymbol{\lambda}' = \left[\bar{\boldsymbol{\beta}}_2', \bar{\boldsymbol{\beta}}_1', \mathbf{d}', \mathbf{D}^{1'}, \mathbf{D}^{2'}\right]$$

$$\bar{\boldsymbol{\beta}}_1, \bar{\boldsymbol{\beta}}_2, \mathbf{D}^1, \mathbf{D}^2 \geqslant \mathbf{0}; \qquad \mathbf{d} \text{ unrestricted in sign.} \tag{2.23.2}$$

Observe that $\boldsymbol{\lambda}'$ is obtained from $\mathbf{V}'$ by interchanging $\bar{\boldsymbol{\beta}}_1'$ and $\bar{\boldsymbol{\beta}}_2'$.

Now estimate $\boldsymbol{\beta}$ as follows:

1 Select one of the problems $\mathbf{P}_1$ or $\mathbf{P}_2$, with probability of selection $\frac{1}{2}$.

2 Determine an optimal solution to the problem selected, as discussed in Section 2.14. Consider the corresponding $\bar{\boldsymbol{\beta}}_1^*$ and $\bar{\boldsymbol{\beta}}_2^*$.

3 $\boldsymbol{\beta} = \bar{\boldsymbol{\beta}}_1^* - \bar{\boldsymbol{\beta}}_2^* + \boldsymbol{\beta}_0$ is an unbiased estimate of $\boldsymbol{\beta}$. The proof of this follows from:

(a) $\mathbf{Y}(-\boldsymbol{\varepsilon}) - \mathbf{X}\boldsymbol{\beta}_0(-\boldsymbol{\varepsilon}) = -\mathbf{Y}(\boldsymbol{\varepsilon}) + \mathbf{X}\boldsymbol{\beta}_0(\boldsymbol{\varepsilon})$

(b) $\mathbf{P}_1(\boldsymbol{\varepsilon})$ and $\mathbf{P}_2(-\boldsymbol{\varepsilon})$ are identical; so are $\mathbf{P}_2(\boldsymbol{\varepsilon})$ and $\mathbf{P}_1(-\boldsymbol{\varepsilon})$.

The rest of the proof is exactly similar to that in Section 2.14.

2.24 MINIMIZING SUM OF ABSOLUTE DIFFERENCES BETWEEN ABSOLUTE DEVIATIONS (MINSADBAD Regression)

We next consider the estimation of $\boldsymbol{\beta}$ using the objective of minimizing the sum of absolute differences between absolute deviations.

We have:

$$\text{Minimize} \quad \sum_{i<j} |\underline{d}_i - \underline{d}_j|$$

$$\text{subject to} \quad \underline{\mathbf{d}} \leqslant \mathbf{Y} - \mathbf{X}\boldsymbol{\beta} \leqslant \underline{\mathbf{d}}$$

$$\underline{\mathbf{d}} \geqslant \mathbf{0}$$

$$\boldsymbol{\beta} \text{ unrestricted in sign.} \tag{2.24.1}$$

Here $\underline{\mathbf{d}}' = (\underline{d}_1, \ldots, \underline{d}_n)$ the absolute deviations in the observations.

This problem can be formulated as an LP problem:

$$\text{Minimize} \quad \mathbf{D}^{1\prime}\mathbf{e}+\mathbf{D}^{2\prime}\mathbf{e}$$

$$\text{subject to} \quad \begin{bmatrix} \mathbf{X}_{n\times p} & \mathbf{I}_n & \mathbf{0} & \mathbf{0} \\ -\mathbf{X} & \mathbf{I}_n & \mathbf{0} & \mathbf{0} \\ \mathbf{0}_{r\times p} & \mathbf{H}_{r\times n} & \mathbf{I}_r & -\mathbf{I}_r \end{bmatrix} \begin{bmatrix} \boldsymbol{\beta} \\ \underline{\mathbf{d}} \\ \mathbf{D}_1 \\ \mathbf{D}_2 \end{bmatrix} \begin{matrix} \geqslant \\ \geqslant \\ = \end{matrix} \begin{bmatrix} \mathbf{Y} \\ -\mathbf{Y} \\ \mathbf{0} \end{bmatrix}$$

$$\underline{\mathbf{d}},\mathbf{D}^1,\mathbf{D}^2 \geqslant \mathbf{0}, \boldsymbol{\beta} \text{ unrestricted in sign.} \tag{2.24.2}$$

where $\mathbf{H},\mathbf{D}^1,\mathbf{D}^2$ are as defined in Section 2.22. We can identify the dual of this problem as:

$$\text{Maximize} \quad \mathbf{Y}'\mathbf{U}^1-\mathbf{Y}'\mathbf{U}^2$$

$$\text{subject to} \quad \begin{bmatrix} \mathbf{X}' & -\mathbf{X}' & \mathbf{0} \\ \mathbf{I}_n & \mathbf{I}_n & \mathbf{H}' \\ \mathbf{0} & \mathbf{0} & \mathbf{I}_r \\ \mathbf{0} & \mathbf{0} & -\mathbf{I}_r \end{bmatrix} \begin{bmatrix} \mathbf{U}^1 \\ \mathbf{U}^2 \\ \mathbf{U}^3 \end{bmatrix} \begin{matrix} = \\ \leqslant \\ \leqslant \\ \leqslant \end{matrix} \begin{bmatrix} \mathbf{0} \\ \mathbf{0} \\ \mathbf{e} \\ \mathbf{e} \end{bmatrix}$$

$$\mathbf{U}^1,\mathbf{U}^2 \geqslant \mathbf{0}$$

$$\mathbf{U}^3 \qquad \text{unrestricted in sign.}$$

As in the case of MINSADBED, we can rewrite this dual problem, introducing new variables $\mathbf{W}=\mathbf{U}^3+\mathbf{e}$, $0 \leqslant \mathbf{W} \leqslant 2\mathbf{e}$.

$$\text{Maximize} \quad \mathbf{Y}'\mathbf{U}^1-\mathbf{Y}'\mathbf{U}^2$$

$$\text{subject to} \quad \begin{pmatrix} \mathbf{X}' & -\mathbf{X}' & \mathbf{0} \\ \mathbf{I}_n & \mathbf{I}_n & \mathbf{H}' \end{pmatrix} \begin{bmatrix} \mathbf{U}^1 \\ \mathbf{U}^2 \\ W \end{bmatrix} \begin{matrix} = \\ \leqslant \end{matrix} \begin{bmatrix} \mathbf{0} \\ \mathbf{H}'\mathbf{e} \end{bmatrix}$$

$$\mathbf{U}^1 \geqslant \mathbf{0}, \mathbf{U}^2 \geqslant \mathbf{0}, \mathbf{W} \geqslant \mathbf{0}, \mathbf{W} \leqslant 2\mathbf{e}.$$

Here again the structure can be studied to develop modified methods. Also unbiased estimation is possible as in the case of MINSADBED regression.

2.25 CHOOSING THE "BEST" REGRESSION USING THE MINMAD CRITERION

Suppose that $x_1, x_2, \ldots, x_p$ is the complete set of all possible independent variables including any functions such as squares, cross product, and so on, from which a linear regression is to be chosen for the purpose of prediction. The problem is to decide which independent variables should be included in the model.

In choosing such a subset, it is appropriate to include as many independent variables in the model as possible for reliable predictions. On the other hand, because of costs involved in obtaining data on a large number of independent variables, we like our prediction equation to involve as few independent variables as possible. A compromise between these two extremes is called "*choosing the best regression equation*."

There is no unique statistical procedure for selecting the best subset of independent variables to be included in the model, and in many situations personal judgment is required. Moreover, the available methods do not necessarily lead to the same solution when they are applied to the same problem.

Determining the "best" q independent variables from among a set of p independent variables is the problem under consideration for $q=p, p-1,\ldots,k$; $1 \leqslant k \leqslant p$. We are considering the MINMAD regression in this section instead of the least squares regression to distinguish between the different choices of independent variables. One way is to completely enumerate all the possible subsets of size q from the set of p independent variables and compare them to find the best subset of independent variables. First we show that the problem can be posed as a mixed-integer programming problem, in which some variables can take only values 0 or 1. Then we explain a branch-and-bound approach.

Let L_j and U_j be the lower and upper bounds on $\beta_j, j=1,\ldots,p$. Here β_1 corresponds to the constant term. Let $V_j=0$ or 1, depending on whether β_j is included in the subset of size q or not.

The problem can be formulated as a mixed-integer programming problem, as follows:

$$
\begin{aligned}
&\text{Minimize} \quad && Z=\sum_{i=1}^{n}(d_{i1}+d_{i2}) \\
&\text{subject to} \quad && \sum_{j=1}^{p} X_{ij}\beta_j + d_{i1} - d_{i2} = Y_i \qquad i=1,\ldots,n \\
& && L_jV_j \leqslant \beta_j \leqslant U_jV_j \qquad j=1,\ldots,p \\
& && V_j = 0 \quad \text{or} \quad 1 \\
& && \sum_{j=1}^{p} V_j = q \\
& && d_{i1}, d_{i2} \geqslant 0, \qquad i=1,\ldots,n \\
& && \beta_j \quad \text{unrestricted in sign} \qquad j=1,\ldots,p. \qquad (2.25.1)
\end{aligned}
$$

If we have to determine the best sets of independent variables according to the MINMAD criterion, for $q=p,\ldots,k$, for any specified k, then we need to solve $p-k+1$ such problems. Instead we use a branch-and-bound procedure to solve those problems.

Branch and Bound Scheme

Any branch-and-bound approach for solving discrete, combinational, or integer-restricted programming problems is composed of a well-defined branching strategy that partitions the original set of feasible solutions into two or more sets at each branching and a bounding strategy that involves one or more tests to declare that certain branches cannot yield any optimal solution. The efficiency of a branch-and-bound scheme lies in its ability to enumerate most of the branches implicitly. This in turn depends on the bounding efficiency and the pruning tests.

In this approach, which is essentially a search procedure, the partitioning and pruning continues until the problem is solved.

We require certain definitions and explanations before we describe the algorithm.

DEFINITION 2.25.1 (Candidate List) The candidate list will be the list of candidate problems, each representing a disjoint subset of solutions to the original problem.

DEFINITION 2.25.2 (Branching) If a set of feasible solutions, in general a subset of the set of all feasible solutions to the problem, has been partitioned, we say that the current candidate problem has been branched.

DEFINITION 2.25.3 (Incumbent) The incumbent solution is the best solution obtained so far at any given point of the search. When the algorithm starts we can assume a large objective function value if no incumbent is known.

DEFINITION 2.25.4 (Fathoming) A candidate problem is fathomed if its optimal solution is known; hence, it can be removed from the candidate list and the incumbent can be updated if possible.

DEFINITION 2.25.5 (Pruning) A candidate problem is pruned if a lower bound on its set of solutions is greater than or equal to the objective function of the incumbent at a given point in the algorithm. Hence, when a candidate problem is pruned it can be removed from the candidate list.

Bounding strategy

If at any point of the search process a new candidate problem is generated, we calculate a lower bound on the objective function value in the set of feasible solutions defined by the candidate problem. Usually, the structure of the original problem is exploited to find good lower bounds. Also general approaches like Lagrangian relaxation can be used towards this end.

With this general understanding of the branch-and-bound scheme, we can approach an algorithm of the stepwise MINMAD regression problem.

The search procedure for the stepwise problem

We are using a branch-and-bound approach for solving the optimal stepwise MINMAD regression. We define certain terms before proceeding to describe the search procedure.

DEFINITION 2.25.6 (Assignment) Any specification of values for all p of the V_j will be termed an assignment. V_j's are fixed at either zero or one. Let $V_j = 1$, when the jth independent variable is *in the regression*. When $V_j = 0$, the jth independent variable is *out* of the regression.

DEFINITION 2.25.7 (Partial assignment) A partial assignment S is a specification of values, zeroes and ones, for a subset of V_j's. $V_j \in S$ is called *fixed* and the rest of the $V_j \notin S$ are termed free variables with respect to S. Let S_1 and S_0 be the subsets of S such that S_1 is the set of V_j's in S fixed as 1 and $S_0 = S - S_1$.

DEFINITION 2.25.8 (Completion of a partial assignment) Given any S, if we fix the $V_j \notin S$ at the zero or one level we have a completion of the partial assignment S, denoted by C^S. There are 2^{F^S} such completions where F^S is the cardinality of $M - S$, where $M = \{V_j \mid j = 1, 2, \ldots, p\}$.

The following results provide the necessary tests for pruning and lower bounds.

Let $Z(S)$ be the lower bound on the objective for any completion of S. Let C_*^S be the completion of S in which $V_j = 1$ for all free V_j's.

RESULT 2.25.1 $Z(C_*^S) \leqslant Z(C^S)$ for any completion of S.

Proof Proof follows from the fact that the set of V_j's that are fixed as 1 in any C^S is a subset of the set of V_j's that are fixed as 1 in C_*^S. Hence the result.

$Z(C_*^S)$ is the objective function values (corresponding to C_*^S) of the MINMAD problem:

$$\text{Minimize} \quad \sum_{i=1}^{n} (d_{i1} + d_{i2})$$

$$\text{subject to} \quad \sum_{j \in S_1 \cup (M-S)} X_{ij}\beta_j + d_{i1} - d_{i2} = Y_i$$

$$i = 1, \ldots, n$$

$$\beta \text{ unrestricted in sign, } d_{i1}, d_{i2} \geqslant 0.$$

Given any partial assignment we have two offsprings, namely $S' = \{S \cup V_j \text{ with } V_j = 1\}$ and $S'' = \{S \cup V_j \text{ with } V_j = 0\}$ for some $V_j \in M - S$. From the nature of the problem it is sufficient to consider S'', though the lower bound for $S', Z(C_*^{S'})$ will always be less than or equal to $Z(C_*^{S''})$. The reason is that the set of V_j's that are fixed as 1 in $C_*^{S'}$ is the same as the set of V_j's that are fixed as 1 in C_*^S. But S'' distinguishes between the different choices. So we consider S' after S'' is pruned or fathomed. Thus we have the following result.

RESULT 2.25.2 It is sufficient to consider while branching from any S the offspring $S'' = \{S \cup V_j\}$ with $V_j = 0$ for some $j \in M - S$.

Next we observe that certain completions of S need not be considered, as the number of independent variables in the regression will be less than k in those completions.

Let I^S be the cardinality of the set of V_j's fixed in the regression by S, that is, $V_j = 1$ in S. We have $I^S + F^S \leqslant p$.

Let $I^S \leqslant k$, those completions that fix $I^S, I^S + 1, \ldots, k - 1$ independent variables in the regression need not be considered.

If $I^S \geqslant k$, every completion of S needs to be considered implicitly or explicitly.

Let $Z_i, i = k, \ldots, p$ be the objective function value, for the incumbent corresponding to a problem with i independent variables.

We have $Z_r \geqslant Z_{r+1}$ for $p - 1 \geqslant r \geqslant k$.

Now we have Result 2.25.3, which is a pruning test.

Let $t^S = \max(k, I^S)$.

RESULT 2.25.3 If $Z_{t^S} < Z(C_*^S)$, S can be pruned.

Proof $Z_{t^S} \geqslant Z_i$ for $p \geqslant i > t^S$, and as mentioned earlier it is sufficient to consider the completions of S that fix t^S or more independent variables in the regression. Therefore if $Z_{t^S} < Z(C_*^S)$, no completion of S can yield a better solution than the best known solutions—namely, the incumbents for each i, $t^S \leqslant i \leqslant I^S + F^S$.

RESULT 2.25.4 If $Z_{t^S} < \pi_r$ then it is not necessary to consider the completion of S in which V_r is fixed as zero, where π_r is the objective function value of the assignment in which the rth independent variable alone is out of the regression (i.e., we are using all the remaining $p-1$ independent variables in the regression).

Proof π_r by definition is a lower bound for any completion in which V_r is fixed as zero. Therefore the result follows.

This discussion yields the following algorithm.

Algorithm 2.25.1

Corresponding to the set S of fixed variables, we define an ordered index set $R(S)$ of elements in S, in the order in which they are included in the set S. Let us also mark the indices in $R(S)$ as follows. If $j \in R(S)$ is fixed as 0 then we mark j with a "*" as a suffix; that is, we write j_*. Otherwise we do not mark j.

In the beginning all independent variables are free. That is $S = \varnothing$, $R(S) = \varnothing$, $I^S = 0$, $F^S = p$, $t^S = k$. Let Z_i be a very large number for all $i = k, \ldots, p$. Compute π_i, the objective function value of the assignment in which the ith independent variable alone is out of the regression. That is,

$$V_l = \begin{cases} 1, & l \neq i, \quad l = 1, \ldots, p, \\ 0, & l = i \end{cases}.$$

Go to Step 1.

Step 1 Find $Z(C_*^S)$ for the current partial assignment. (The corresponding MINMAD problem is the current candidate problem.) Go to Step 2.

Step 2 If $Z(C_*^S) \leqslant Z_{t^S}$ then go to Step 3. Otherwise go to Step 5.

Step 3 If $Z(C_*^S) \geqslant Z_{I^S+F^S}$ then go to Step 4. Otherwise, $Z(C_*^S) < Z_{I^S+F^S}$; then we have the best solution found so far for the problem with $I^S + F^S$ independent variables in the regression. Store $Z_{I^S+F^S} = Z(C_*^S)$ and store the corresponding solution as the best incumbent for the problem with $I^S + F^S$ independent variables. Go to Step 4.

Step 4 If $t^S \geqslant I^S + F^S$, go to Step 5. Otherwise go to Step 7.

Step 5 If $R(S)$ has no marked element, stop. Otherwise, locate the last marked element j_* in the ordered index set $R(S)$. Drop all the elements that have been generated after j from the set S also remove the mark on j. Consider the modified S and $R(S)$. Compute I^S, F^S, and t^S. (Removing the mark on j is equivalent to fixing $V_j = 1$ in the new partial assignment.) Go to Step 6.

Step 6 If $Z_{t^s} \geqslant \pi_r$ for every free variable, then go to Step 7. Otherwise consider the partial assignment obtained by fixing every free variable l such that $Z_{t^s} < \pi_l$, in the regression. That is, the new partial assignment obtained is

$$S = S_0 \cup [S_1 \cup \{l \mid Z_{t^s} < \pi_l\}].$$

Recompute $R(S)$, I^S, F^S, and t^S, go to Step 4.

Step 7 Select a free variable $V_j \in M - S$. Consider $S_0 = S_0 \cup \{V_j\}$. Recompute $R(S)$, I^S, F^S, and t^S. Go to Step 1.

EXAMPLE 2.25.1 This example illustrates Algorithm 2.25.1. Consider the data given in Example 2.4.1. Suppose we wish to find the "best two" independent variables, using MINMAD criterion and keeping the β_0 term always in the model. We use Algorithm 2.25.1 for this purpose. This algorithm can find the best q independent variables for all q such that $q = k, k+1, \ldots, p$.

Initially we have $S = \varnothing$, $I^S = 0$, $F^S = 4$, and $t^S = \max(I^S, k) = 2$. Also by solving the MINMAD problem without the ith independent variable in the model, $i = 1, 2, 3, 4$, we get $\pi_1 = 22.54$, $\pi_2 = 20.20$, $\pi_3 = 19.66$, and $\pi_4 = 19.06$. We go to Step 1 with $S = \varnothing$. We store $Z_i = 9999$.

Step 1 We calculate $Z(C_*^S)$ by solving the corresponding problem, namely, the problem with all the independent variables in the model. We find $Z(C_*^S) = 18.83$.

Step 2 $Z(C_*^S) \leqslant Z_2 = 9999$. So we go to Step 3.

Step 3 $Z(C_*^S) < Z_4 = 9999$, as $I^S + F^S = 4$. So set $Z_4 = 18.83$ and the corresponding incumbent as the optimal solution corresponding to S. We have $\beta_0 = -13.34$, $\beta_1 = 2.35$, $\beta_3 = 1.01$, and $\beta_4 = 0.60$. We go to Step 4.

Step 4 $t^S = 2 < I^S + F^S = 4$. We go to Step 7.

Step 7 We select the free variable V_1. $S = \{V_1\}$. Now $R(S) = \{1_*\}$, $I^S = 0$, $F^S = 3$, and $t^S = 2$. We go to Step 1.

Step 1 $Z(C_*^S)=22.54$. $Z(C_*^S)$ in this case corresponds to π_1 as the problems are the same, that is, fixing the variable V_1 out of the regression and finding the optimal MINMAD regression estimates. We go to Step 2.

Step 2 $Z(C_*^S) \leqslant 9999 = Z_{t^S} = Z_2$. We go to Step 3.

Step 3 $Z(C_*^S) < Z_{I^S+F^S} = Z_3 = 9999$. We set $Z_3 = 22.54$ and store the corresponding optimal estimates for the incumbent problem. We have $\hat{\beta}_0 = 214.56$, $\hat{\beta}_2 = -1.05$, $\hat{\beta}_3 = -1.59$, and $\hat{\beta}_4 = -1.65$. We go to Step 4.

Step 4 $t^S = 2 < 3 = I^S + F^S$. We go to Step 7.

Step 7 We select the free variable V_2. $S = \{V_1, V_2\}$. Now $R = \{1_*, 2_*\}$, $I^S = 0$, $F^S = 2$, and $t^S = 2$. We go to Step 1.

Step 1 We find $Z(C_*^S) = 35.31$. We go to Step 2.

Step 2 $Z(C_*^S) \leqslant 9999 = Z_{t^S} = Z_2$. We go to Step 3.

Step 3 $Z(C_*^S) < Z_{I^S+F^S} = Z_2$.
We get $Z_2 = 35.31$ and store the corresponding incumbent solution. We have $\hat{\beta}_0 = 129.23$, $\hat{\beta}_3 = -1.11$, and $\hat{\beta}_4 = -0.73$. We go to Step 4.

Step 4 $t_S = 2 = I^S + F^S$. We go to Step 5.

Step 5 $R(S)$ has a marked element, and the last marked element is 2. We have the corresponding $S = \{V_1, V_2\}$. Now $R(S) = \{1_*, 2\}$, $I^S = 1$, $F^S = 2$, and $t^S = 2$. We go to Step 6.

Step 6 $Z_2 = 35.31 > \pi_3$ and π_4, corresponding to the free variables. So we go to Step 7.

And this process is repeated until we revisit Step 5 when $R(S)$ has no marked element; we then stop the process. The tree corresponding to the problem is given in Fig. 2.25.1. The "best" independent variables obtained are the independent variables 1 and 2, with the corresponding Z_2 equal to 23.13, and the estimates given by $\hat{\beta}_0 = 49.96$, $\hat{\beta}_1 = 1.56$, and $\hat{\beta}_2 = 0.68$. Incidentally, the best three independent variables corresponding to the present value of $Z_3 = 19.06$ are the independent variables 1, 2 and 3. The estimates are $\hat{\beta}_0 = 46.85$, $\hat{\beta}_1 = 1.72$, $\hat{\beta}_2 = 0.67$, and $\hat{\beta}_3 = 0.37$.

This example is only to explain the steps in the Algorithm 2.25.1. No conclusion about computational burden should be drawn from it.

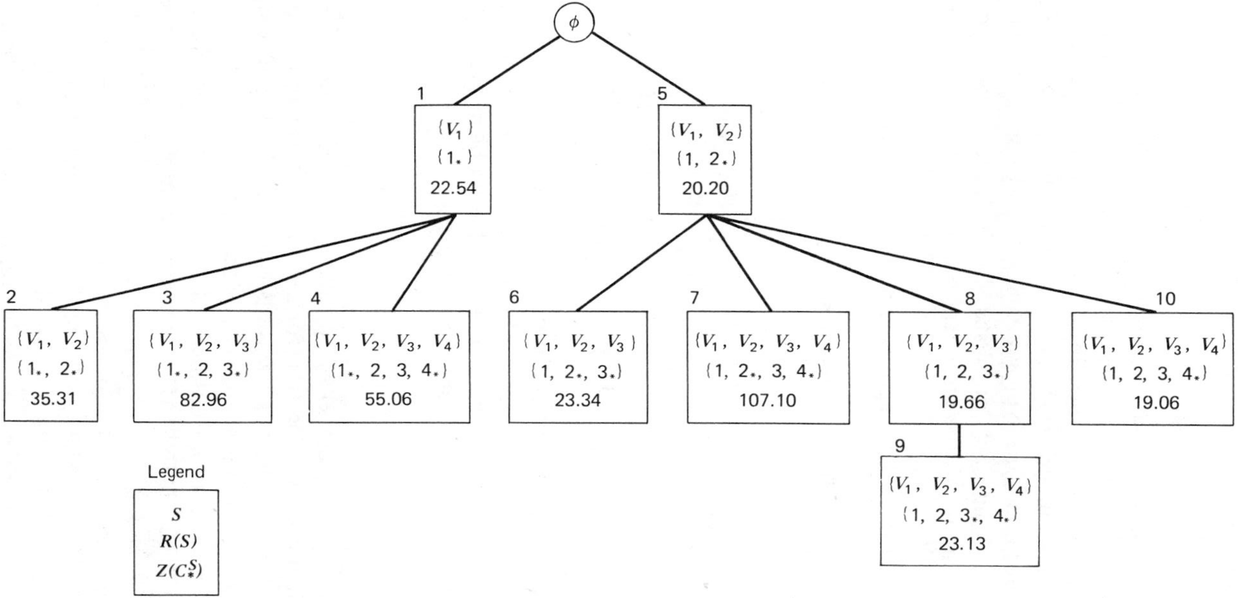

Figure 2.25.1 The tree corresponding to the branch and bound algorithm.

2.26 FINDING "k-BEST" INDEPENDENT VARIABLES IN THE MINMAD REGRESSION PROBLEM—ALTERNATE APPROACH

In Section 2.25 we discussed the optimal stepwise MINMAD regression problem. There we used a branch-and-bound algorithm to find an optimal set of k independent variables from among the p variables, with the MINMAD regression criterion. Now we pose the same problem as a cardinality constrained linear-programming (CCLP) problem and describe a method for solving such problems. In the process we also develop algorithms for finding all the basic feasible solutions of a linear programming problem.

Consider the MINMAD regression problem. We have

$$\text{Minimize} \quad \mathbf{d}_1'\mathbf{e}+\mathbf{d}_2'\mathbf{e}$$

$$\text{subject to} \quad (\mathbf{X} \; -\mathbf{X} \quad \mathbf{I} \; -\mathbf{I})\begin{pmatrix} \boldsymbol{\beta}_1 \\ \boldsymbol{\beta}_2 \\ \mathbf{d}_1 \\ \mathbf{d}_2 \end{pmatrix} = \mathbf{Y}$$

$$\boldsymbol{\beta}_1, \boldsymbol{\beta}_2, \mathbf{d}_1, \mathbf{d}_2 \geqslant \mathbf{0}$$

In an optimal solution to this problem we may have at most p among the β_{1i} and β_{2i} will be positive. Now if we require at most k among the β_{1i} and β_{2i} to be positive in any optimal solution to the problem, we can add this additional restriction to the problem as follows:

$$|\boldsymbol{\beta}_1|^+ + |\boldsymbol{\beta}_2|^+ \leqslant k \tag{2.26.1}$$

where $|\mathbf{x}|^+$ denotes the number of positive x_i's in vector $\mathbf{x}$.

This restriction cannot be transformed into a linear restriction. However, if we can find all the basic feasible solutions to the problem, we can find among those solutions a subset of solutions that satisfy (2.26.1), and if that subset is nonempty an optimal solution among them is an optimal solution to the k-best MINMAD regression problem.

However, it is not always necessary to find all the basic feasible solutions to the MINMAD problem in order to find a solution to the k-best MINMAD regression problem. That is achieved by modifying an algorithm developed for finding all the basic feasible solutions to a linear-programming problem, so as to omit some of the basic feasible solutions, which may not yield an optimal solution or may violate the constraint of (2.26.1). Some theory is necessary for applying this approach.

2.27 FINDING ALL THE BASIC FEASIBLE SOLUTIONS TO A SYSTEM OF LINEAR EQUATIONS

Consider the system

$$\mathbf{AW}=\mathbf{b}, \mathbf{W} \geqslant \mathbf{0}. \tag{2.27.1}$$

Let $\mathcal{F}=\{\mathbf{W}|\mathbf{AW}=\mathbf{b}, \mathbf{W}\geqslant \mathbf{0}\}$, where $\mathbf{A}$ is an $m\times n$ matrix. Let $\mathcal{F}^*=\{\mathbf{W}|\mathbf{W}$ is a basic feasible solution to 2.27.1$\}$.

RESULT 2.27.1 There is a one-to-one correspondence between the solutions $\mathbf{W}\in\mathcal{F}$ and the solutions $\mathbf{Z}=\begin{pmatrix}\mathbf{W}\\ \zeta\end{pmatrix}$ of the system

$$\begin{aligned} \mathbf{Z} &\geqslant \mathbf{0} \\ \mathbf{DZ} &= \mathbf{0} \\ \mathbf{e}'_{n+1}\mathbf{Z} &= 1 \end{aligned} \tag{2.27.2}$$

where

$$\mathbf{D}=(\mathbf{A},-\mathbf{b}) \text{ and } \mathbf{e}'_{n+1}=(0,\ldots,0,1).$$

Proof Any solution $\mathbf{W}$ satisfies $\mathbf{AW}=\mathbf{b}, \mathbf{W}\geqslant \mathbf{0}$. So $\mathbf{Z}=\begin{pmatrix}\mathbf{W}\\ 1\end{pmatrix}$ satisfies System 2.27.2, and vice versa.

RESULT 2.27.2 $\mathbf{W}^*\in\mathcal{F}^*$ if and only if $\begin{pmatrix}\mathbf{W}^*\\ 1\end{pmatrix}$ is a basic feasible solution to (2.27.2).

Proof If $\mathbf{W}^*$ is a basic feasible solution to $\mathcal{F}$, then the columns corresponding to positive W_j^*'s are linearly independent and they are less than or equal to m in number. So $\mathbf{Z}=\begin{pmatrix}\mathbf{W}^*\\ 1\end{pmatrix}$ has at most $m+1$ positive Z_j's, and thus is a basic feasible solution to (2.27.2). The converse follows similarly.

Let $\mathbf{D}^i$ denote the ith row of $\mathbf{D}$. The following result shows that System 2.27.2 is equivalent to a system with one less equality constraint. Repeated application of this result can produce an equivalent system in which we have only one equality constraint, in which case finding the basic feasible solutions to the system will be no problem. Then these feasible solutions can be related to the basic feasible solution of System 2.27.2. This in fact provides an outline of an algorithm for finding all the basic feasible solutions of $\mathcal{F}$, using Result 2.27.2.

RESULT 2.27.3 There exists a matrix $\mathbf{Q}^1$ with $n+1$ rows and q_1 columns, for some q_1 satisfying the conditions

(1) $\mathbf{Q}^1 \geqslant \mathbf{0}$

(2) $\mathbf{D}^1\mathbf{Q}^1 = \mathbf{0}$

such that given any solution $\mathbf{Z}$ to System 2.27.2 there is a solution $\mathbf{U}$ to the following system (2.27.3):

$$(\mathbf{D}^i\mathbf{Q}^1)\mathbf{U} = 0 \qquad i=2,\dots,m$$

$$(\mathbf{e}'_{n+1}\mathbf{Q}^1)\mathbf{U} = 1 \tag{2.27.3}$$

$$\mathbf{U} \geqslant \mathbf{0}$$

satisfying $\mathbf{Q}^1\mathbf{U} = \mathbf{Z}$. On the other hand, if $\mathbf{U}$ is a solution to System 2.27.3 then $\mathbf{Z} = \mathbf{Q}^1\mathbf{U}$ is a solution to System 2.27.2.

Proof In fact we construct a $\mathbf{Q}^1$ which satisfies the stipulated conditions, while proving this result. Construct $\mathbf{Q}^1$ as follows:

(1) If $D_{1g} = 0$, that is, the gth component of $\mathbf{D}^1$ is zero, we include the column $\mathbf{e}_g$ in $\mathbf{Q}^1$ [that is, $\mathbf{e}_g = (0,\dots,1,\dots,0)'$ is a $n+1$-column vector with 1 in the gth coordinate].

(2) For any pair of components of $\mathbf{D}^1$ such that $D_{1s} > 0$ and $D_{1t} < 0$, include the column $\mathbf{e}_{st} = D_{1s}\mathbf{e}_t - D_{1t}\mathbf{e}_s$ in $\mathbf{Q}^1$. The number of columns in $\mathbf{Q}^1$ is equivalent to q_1 equaling the number of zeroes in $\mathbf{D}^1$ plus the number of pairs of components of $\mathbf{D}^1$ with opposite signs. Now, notice that $\mathbf{Q}^1$ satisfies both $\mathbf{Q}^1 \geqslant \mathbf{0}$ and $\mathbf{D}^1\mathbf{Q}^1 = \mathbf{0}$. Also, if there is no zero in $\mathbf{D}^1$ and no pair (s,t) such that $D_{1s} > 0$ and $D_{1t} < 0$, we have no solution $\mathbf{Z}$ to the System 2.27.2 and in a vaccous sense, the result is true.

For any solution $\mathbf{U}$ to System 2.27.3, $\mathbf{Z} = \mathbf{Q}^1\mathbf{U}$ is a solution to System 2.27.2, because $\mathbf{Q}^1 \geqslant \mathbf{0}$ and $\mathbf{U} \geqslant \mathbf{0}$. Therefore, $\mathbf{Z} = \mathbf{Q}^1\mathbf{U} \geqslant \mathbf{0}$. Next, $\mathbf{D}^1\mathbf{Z} = \mathbf{D}^1\mathbf{Q}^1\mathbf{U} = 0$. $\mathbf{D}^i\mathbf{Z} = \mathbf{D}^i\mathbf{Q}^1\mathbf{U} = 0$, $i=2,\dots,m$ and $\mathbf{e}'_{n+1}\mathbf{Z} = \mathbf{e}'_{n+1}\mathbf{Q}^1\mathbf{U} = 1$ as $\mathbf{U}$ is a solution to (2.27.3). Thus we have shown one part of the result: If $\mathbf{U}$ is a solution to (2.27.3), $\mathbf{Q}^1\mathbf{U} = \mathbf{Z}$ is a solution to (2.27.2). To prove the converse, suppose $\mathbf{Z}$ is a solution to (2.27.2). We have $\mathbf{D}^1\mathbf{Z} = \Sigma_+ D_{1s}Z_s + \Sigma_- D_{1t}Z_t = 0$, where Σ_+ is the summation over all s such that $D_{1s} > 0$, and similarly, Σ_- is the summation over all t such that $D_{1t} < 0$. As $\mathbf{Z} \geqslant \mathbf{0}$, both $\Sigma_+ D_{1s}Z_s$ and $-\Sigma_- D_{1t}Z_t$ are $\geqslant 0$. Let $\rho = \Sigma_+ D_{1s}Z_s = -\Sigma_- D_{1t}Z_t$. Denote the components of any solution $\mathbf{U}$ to (2.27.3) as follows:

U_g corresponding to Column $\mathbf{e}_g$,

U_{st} corresponding to Column $\mathbf{e}_{st}$, in $\mathbf{Q}^1$.

Define the solution $\mathbf{U}$ such that

$$U_g = Z_g \text{ for } g \text{ such that } D_{1g} = 0$$

$$U_{st} = \begin{cases} Z_s Z_t/\rho & \text{for } s \text{ and } t \text{ such that } D_{1s} > 0,\ D_{1t} < 0, \quad \rho \neq 0 \\ 0 & \text{if } \rho = 0. \end{cases}$$

We shall show that $\mathbf{U}$ satisfies System 2.27.3. Notice that $\mathbf{U} \geqslant \mathbf{0}$. To show that $\mathbf{U}$ satisfies

$$\mathbf{D}^i\mathbf{Q}^1\mathbf{U} = 0, \qquad i = 2, \ldots, m \qquad \text{and}$$

$$\mathbf{e}'_{n+1}\mathbf{Q}^1\mathbf{U} = 1.$$

It is enough to show $\mathbf{Q}^1\mathbf{U} = \mathbf{Z}$. Notice that in the construction of $\mathbf{Q}^1$ we have one row for each element of $\mathbf{D}^1$. Let the rows of $\mathbf{Q}^1$ be partitioned such that $\mathbf{R}_0 = \{g \mid D_{1g} = 0\}$; $\mathbf{R}_+ = \{s \mid D_{1s} > 0\}$ and $\mathbf{R}_- = \{t \mid D_{1t} < 0\}$. Also let us stick to the subscripts g, s, and t for the elements belonging to $\mathbf{R}_0$, $\mathbf{R}_+$ and $\mathbf{R}_-$, respectively. Let $\mathbf{Q}_i^1$ denote the ith row of $\mathbf{Q}^1$. From the construction of $\mathbf{Q}^1$, we have

$$\mathbf{Q}_g^1\mathbf{U} = U_g = Z_g, \qquad \text{for } g \in \mathbf{R}_0;$$

$$\mathbf{Q}_s^1\mathbf{U} = -\Sigma_- D_{1t} U_{st}, \qquad \text{for } s \in \mathbf{R}_+$$

$$= Z_s \frac{(-\Sigma_- D_{1t} Z_t)}{\rho}, \qquad \text{substituting for } U_{st}$$

$$= Z_s, \qquad \text{for } s \in \mathbf{R}_+$$

and

$$\mathbf{Q}_t^1\mathbf{U} = \Sigma_+ D_{1s} U_{st}, \qquad \text{for } t \in \mathbf{R}_-$$

$$= \frac{Z_t \Sigma_+ D_{1s} Z_s}{\rho} = Z_t, \qquad \text{for } t \in \mathbf{R}_-.$$

Thus $\mathbf{Q}^1\mathbf{U} = \mathbf{Z}$. Hence $\mathbf{U}$ is a solution to (2.27.3), and hence the result.

EXAMPLE 2.27.1 Suppose we have the system

$$\begin{bmatrix} 3 & -1 & 0 \\ 5 & 2 & -1 \\ -2 & -1 & 4 \end{bmatrix} \begin{bmatrix} W_1 \\ W_2 \\ W_3 \end{bmatrix} = \begin{bmatrix} 5 \\ 4 \\ 2 \end{bmatrix}$$

$$W_1, W_2, W_3 \geqslant 0$$

Then

$$\mathbf{D}=\begin{bmatrix} 3 & -1 & 0 & -5 \\ 5 & 2 & -1 & -4 \\ -2 & -1 & 4 & 2 \end{bmatrix}$$

and we have

$$\mathbf{DZ}=\mathbf{0},$$

$$\mathbf{e}_4'\mathbf{Z}=1.$$

Therefore, $\mathbf{D}^1=(3,-1,0,-5)$, $\mathbf{Q}^1$ is given by

$$\mathbf{Q}^1 = \begin{array}{c} \begin{matrix} \mathbf{e}_3 & \mathbf{e}_{12} & \mathbf{e}_{14} \end{matrix} \\ \begin{bmatrix} 0 & 1 & 5 \\ 0 & 3 & 0 \\ 1 & 0 & 0 \\ 0 & 0 & 3 \end{bmatrix} \end{array}$$

Here $R_0=\{3\}$, $R_+=\{1\}$, and $R_-=\{2,4\}$

REMARK 2.27.1 Notice that given a $\mathbf{Z}$, in the proof of Result 2.27.3, we have defined $\mathbf{U}$ in one fashion. There can be other $\mathbf{U}$'s as well. Therefore, given a $\mathbf{Q}^1$, satisfying the conditions in Result 2.27.3 any solution $\mathbf{U}$ to System 2.27.3 defines a unique solution $\mathbf{Z}$ to System 2.27.2, whereas given a $\mathbf{Z}$ to System 2.27.2, $\mathbf{Z}$ need not define a $\mathbf{U}$ for system 2.27.3 uniquely.

REMARK 2.27.2 The result goes through even if we have chosen any $\mathbf{D}^i$ instead of $\mathbf{D}^1$—i.e., we could have rearranged the rows of $\mathbf{D}$ to name $\mathbf{D}^i$ as $\mathbf{D}^1$. So choosing $\mathbf{D}^1=\mathbf{D}^i$ for the smallest number of columns in $\mathbf{Q}^1$, will reduce the computational burden.

REMARK 2.27.3 Now it is possible to repeat the procedure m times so as to find $\mathbf{Q}^1$ corresponding to $\mathbf{D}^1$ and $\mathbf{Q}^2$ corresponding to $\mathbf{D}^2\mathbf{Q}^1$, and so on, until all the m rows of $\mathbf{D}$ are eliminated. In general we find $\mathbf{Q}^i$ corresponding to $\mathbf{D}^i\mathbf{Q}^1\mathbf{Q}^2\cdots\mathbf{Q}^{i-1}$. At the mth stage we have the system

$$\mathbf{e}_{n+1}'\mathbf{Q}^1\mathbf{Q}^2\cdots\mathbf{Q}^m\mathbf{U}=1$$

$$\mathbf{U}\geqslant\mathbf{0} \tag{2.27.4}$$

and

$$\mathbf{Z}=\mathbf{Q}^1\mathbf{Q}^2\cdots\mathbf{Q}^m\mathbf{U}.$$

Let $\mathbf{Q}^* = \mathbf{Q}^1\mathbf{Q}^2 \cdots \mathbf{Q}^m$. Thus the set of basic feasible solutions to (2.27.4) is given by

$$\mathcal{U} = \left\{ \mathbf{U} \middle| \mathbf{U} = \frac{1}{Q^*_{n+1j}} \mathbf{e}_j \text{ for } Q^*_{n+1j} > 0 \right\}$$

where $\mathbf{e}_j$ is a q-vector and where q is the number of columns in $\mathbf{Q}^*$.

EXAMPLE 2.27.2 Now consider the data given in Example 2.27.1. We have

$$\mathbf{Q}^1 = \begin{bmatrix} 0 & 1 & 5 \\ 0 & 3 & 0 \\ 1 & 0 & 0 \\ 0 & 0 & 3 \end{bmatrix}$$

$\mathbf{D}^2\mathbf{Q}^1 = (-1, 11, 13)$; therefore

$$\mathbf{Q}^2 = \begin{matrix} \mathbf{e}_{12} & \mathbf{e}_{13} \\ \end{matrix} \begin{bmatrix} 11 & 13 \\ 1 & 0 \\ 0 & 1 \end{bmatrix} \qquad \text{and so,} \quad \mathbf{Q}^1\mathbf{Q}^2 = \begin{bmatrix} 1 & 5 \\ 3 & 0 \\ 11 & 13 \\ 0 & 3 \end{bmatrix}$$

$\mathbf{D}^3\mathbf{Q}^1\mathbf{Q}^2 = (39, 48)$. As there are no zeroes in $\mathbf{D}^3\mathbf{Q}^1\mathbf{Q}^2$ and no sign change, we get

$$\mathbf{Q}^3 = \begin{bmatrix} 0 \\ 0 \\ 0 \\ 0 \end{bmatrix}.$$

The implication is that there is no solution to (2.27.2) with $\mathbf{D}$ as given in this example.

REMARK 2.27.4 Notice that $\mathbf{Q}^r$ is such that in each column of $\mathbf{Q}^r$ there are at most two nonzero elements. Therefore, $\mathbf{Q}^r$ can be stored in a compact way. Notice that $\mathbf{Q}^1\mathbf{Q}^2, \ldots,$ and so on, are only multiplying the matrices $\mathbf{D}^{(-1)}, \mathbf{D}^{(-2)}, \ldots,$ respectively, where $\mathbf{D}^{(-1)}$ is the matrix obtained from $\mathbf{D}$ omitting Row 1; $\mathbf{D}^{(-2)}$ is the matrix obtained from $\mathbf{D}$ omitting Rows 1 and 2, and so on, in a step-by-step manner. So we can start the computation as follows: Let $\mathbf{Q} = \begin{pmatrix} \mathbf{D} \\ \mathbf{I} \end{pmatrix}$ where $\mathbf{I}$ is an $n+1 \times n+1$ matrix. Let $\bar{\mathbf{Q}} = \begin{pmatrix} \mathbf{D} \\ \mathbf{I} \end{pmatrix}\mathbf{Q}^1 = \begin{pmatrix} \mathbf{DQ}^1 \\ \mathbf{Q}^1 \end{pmatrix}$. We can perform the rest of the computations of finding $\mathbf{Q}^2, \mathbf{Q}^3, \ldots, \mathbf{Q}^m$ by taking $\mathbf{Q}=\overline{\mathbf{Q}}$. In the last stage, we have

$$\overline{\mathbf{Q}}=\begin{pmatrix} \mathbf{D}\mathbf{Q}^1\mathbf{Q}^2\cdots\mathbf{Q}^m \\ \mathbf{Q}^1\mathbf{Q}^2\cdots\mathbf{Q}^m \end{pmatrix}.$$

That is, we can get $\mathbf{Q}^*$ at the end from the last $n+1$ rows of $\overline{\mathbf{Q}}$. Also, the basic feasible solutions of (2.27.2) are obtained through dividing by the element in the last row of $\mathbf{Q}^*$, in each column of $\mathbf{Q}^*$ if the element is $\neq 0$.

In this setup, we can reduce the number of columns in $\mathbf{Q}^r$ by observing the following:

OBSERVATION 1 Let us consider the rth stage.

Let $S=\{(s,t)|$ the set of all s and t such that the elements $q_{rs}>0$ and $q_{rt}<0\}$, where $\mathbf{q}_i$ is the ith row of $\begin{pmatrix} \mathbf{D}\mathbf{Q}^1\mathbf{Q}^2\cdots\mathbf{Q}^{r-1} \\ \mathbf{Q}^1\mathbf{Q}^2\cdots\mathbf{Q}^{r-1} \end{pmatrix}$. Let $I(s,t)=\{i\,|\,i \geqslant m+1, q_{is}=q_{it}=0\}$, for $(s,t)\in S$.

If $I(s,t)=\varnothing$, then we need not add the column $\mathbf{e}_{st}=q_{rs}\mathbf{e}_t-q_{rt}\mathbf{e}_s$ to $\mathbf{Q}^r$. $\mathbf{Q}^1\mathbf{Q}^2\cdots\mathbf{Q}^{r-1}\mathbf{e}_{st}$ will have $n+1$ positive elements, which implies that such a column cannot yield a basic feasible solution to (2.27.2).

OBSERVATION 2 If at any stage r the $\mathbf{D}\mathbf{Q}^1\mathbf{Q}^2\cdots\mathbf{Q}^{r-1}$ is a zero matrix, then $\mathbf{Q}^*=\mathbf{Q}^1\mathbf{Q}^2\cdots\mathbf{Q}^{r-1}$, as $\mathbf{Q}^r, \mathbf{Q}^{r+1}, \ldots, \mathbf{Q}^m$ will all be identity matrices.

OBSERVATION 3 If a row of $\mathbf{D}\mathbf{Q}^1\mathbf{Q}^2\cdots\mathbf{Q}^{r-1}$ contains no zero element and no pair (s,t) such that $q_{rs}>0$ and $q_{rt}<0$, then there is no solution to System 2.27.2. Hence there is no solution to the original system $\mathbf{AW}=\mathbf{b}, \mathbf{W} \geqslant \mathbf{0}$, as we have already observed in the proof of Result 2.27.3.

Consider any $1 \leqslant \eta \leqslant m$. The following result further reduces the number of columns in $\mathbf{Q}^r$.

RESULT 2.27.4 At the rth stage of the computation, suppose a column of $\mathbf{Q}^1\mathbf{Q}^2\cdots\mathbf{Q}^{r-1}$ has more than η positive elements. Then any column obtained subsequently for $\mathbf{Q}^r$, as a linear combination of this column and some other column, will also have more than η positive elements.

Proof Let $\mathbf{q}_j$ denote any column of $\begin{pmatrix} \mathbf{D}_r\mathbf{Q}^1\ldots\mathbf{Q}^{r-1} \\ \mathbf{Q}^1\cdots\mathbf{Q}^{r-1} \end{pmatrix}$. Let $\bar{\mathbf{q}}_j$ denote the portion of $\mathbf{q}_j$ corresponding to the last $n+1$ rows.

Suppose $q_{rj}=0$. Then we have $\mathbf{e}_j$ in $\mathbf{Q}^r$ by construction of $\mathbf{Q}^r$, and so we will have $\mathbf{q}_j$ also in $\mathbf{Q}^1\mathbf{Q}^2\cdots\mathbf{Q}^r$. Suppose $q_{rj}\neq 0$. If q_{rj} and all nonzero q_{rs} are of the same sign, then we do not combine $\mathbf{q}_j$ with any other columns and so these columns do not appear subsequently in $\mathbf{Q}^i, i \geqslant r$.

So we could possibly combine $\mathbf{q}_j$ with another column of $\mathbf{Q}^1 \cdots \mathbf{Q}^{r-1}$ if there is a $\mathbf{q}_u$ such that q_{rj} and q_{ru} are of opposite signs. In such case the new column formed is

$$\mathbf{e}_{ju} = |q_{ru}|\mathbf{e}_j + |q_{rj}|\mathbf{e}_u.$$

Therefore $\mathbf{Q}^1\mathbf{Q}^2 \cdots \mathbf{Q}^{r-1}\mathbf{e}_{ju}$ will be $|q_{ru}|\mathbf{q}_j + |q_{rj}|\mathbf{q}_u$, which will have more than η positive components as $\mathbf{q}_j$ has more than η positive components. Hence the result.

RESULT 2.27.5 Suppose in the rth stage we have $q_{rs} > 0$, $q_{rt} < 0$ and there exists a u such that $q_{iu} = 0$ for all $i \in I(s, t)$ and $\bar{\mathbf{q}}_u$ has more than η positive components. For any α_1, α_2 positive, let $\mathbf{q} = \alpha_1\mathbf{q}_s + \alpha_2\mathbf{q}_t; \bar{\mathbf{q}}^\alpha$, denoting the last $n+1$ rows of $\mathbf{q}^\alpha$, has more than η positive elements. Notice that if the conditions stated in the result are true we need not add column $\mathbf{e}_{st} = q_{rs}\mathbf{e}_t - q_{rt}\mathbf{e}_s$ to $\mathbf{Q}^r$ (see Observation 2). However, this result shows that such columns, that are not added to $\mathbf{Q}^r$, do not subsequently produce a column with positive components less than or equal to η among the last $(n+1)$ rows.

Proof Suppose q_{vu} is a positive component of $\bar{\mathbf{q}}_u$. Then $v \notin I(s, t)$ as $q_{iu} = 0$ for all $i \in I(s, t)$. Hence at least one of q_{vs} or q_{vt} will be strictly positive; and since α_1, α_2 are positive vth component of $\mathbf{q}^\alpha$ is strictly positive. Therefore, $\bar{\mathbf{q}}^\alpha$ has at least as many positive components as $\bar{\mathbf{q}}_u$. Hence the result. Thus combining Results 2.27.4 and 2.27.5 we get Corollary 2.27.1.

COROLLARY 2.27.1 If in any stage of computation a column $\bar{\mathbf{q}}_j$ of $\mathbf{Q}^1 \cdots \mathbf{Q}^{r-1}$ has more than $m+1$ positive elements, then the column $\mathbf{q}_j$ can be discarded from further computation.

Proof From Results 2.27.4 and 2.27.5 we know that $\mathbf{q}_j$ can only produce in the subsequent stages a column which has more than $\eta = m+1$ positive components in the last $m+1$ components.

But any basic feasible solution to (2.27.3) can have at most $m+1$ positive components. Therefore in $\mathbf{Q}^*$ the corresponding column will have more than $m+1$ positive components, and cannot yield a basic feasible solution to (2.27.3). So we can discard $\mathbf{q}_j$ from subsequent computation. Hence the result.

Thus we are in a position to state an algorithm for finding all the basic feasible solutions to a system like (2.27.3). We consider the problem of finding all the extreme points of

$$\mathbf{DW}=\mathbf{0}$$
$$\mathbf{W}\geqslant\mathbf{0}.$$

Let $\mathcal{C}=\{\mathbf{W}|\mathbf{D}=\mathbf{0},\mathbf{W}\geqslant\mathbf{0}\}$. Consider the matrix $(\mathbf{D}',\mathbf{I}')'$. The algorithm produces a series of transformations of this matrix that generates the required extreme points of the cone of nonnegative solutions of the given system of equations. Let $\mathbf{Q}$ denote the matrix that is transformed at any stage of the computation and $\overline{\mathbf{Q}}$ the resulting matrix. $\mathbf{Q}$ is partitioned into $(\mathbf{U}',\mathbf{L}')$; $\mathbf{Q}=(\mathbf{U}',\mathbf{L}')'$. Initially $\mathbf{U}'=\mathbf{D}'$ and $\mathbf{L}'=\mathbf{I}$. Notice that $\mathbf{U},\mathbf{L}$ do not refer to upper and lower triangular matrices.

Algorithm 2.27.1

Step 0 If all the rows of the matrix $\mathbf{U}$ are zeroes then the matrix $\mathbf{L}$ gives the extreme vectors of the cone of nonnegative solutions of the system of equations. Otherwise go to Step 1.

Step 1 If a row of $\mathbf{U}$ has neither zeroes nor any sign change then $\mathbf{W}=\mathbf{0}$ is the only solution to the system. Otherwise go to Step 2.

Step 2 Consider row r for processing. Let $R=\{j|q_{rj}=0\}$. Let v be the number of elements of R. Then the first v columns of the new matrix $\overline{\mathbf{Q}}$ are the $\mathbf{q}_j$ for $j\in R$, where $\mathbf{q}_j$ denotes the jth column of $\mathbf{Q}$. Go to Step 2′.

Step 2′ If $\mathbf{Q}$ has only two columns and $q_{r1}q_{r2}<0$, adjoin the column $|q_{r2}|\mathbf{q}_1+|q_{rq}|\mathbf{q}_2$ to the matrix $\overline{\mathbf{Q}}$. Go to Step 4. Otherwise, go to Step 3.
Let $S=\{(s,t)|q_{rs}q_{rt}<0, s<t\}$.

Step 3 Let $I(s,t)=\{i|i\in L, q_{is}=q_{it}=0\}$, for an $(s,t)\in S$, where L also denote the set of indices of the rows in L namely $\{m+1,\ldots,m+n+1\}$.

1. If $I(s,t)=\varnothing$ then $\mathbf{q}_s$ and $\mathbf{q}_t$ do not contribute a column to the matrix $\mathbf{Q}$. Go to Step 4.
2. If $I(s,t)\neq\varnothing$, check for a u not equal to s or t such that $q_{iu}=0$ for all $i\in I(s,t)$. If such a u exists, the $\mathbf{q}_s,\mathbf{q}_t$ do not contribute a column to $\overline{\mathbf{Q}}$. Go to Step 4. Otherwise, choose $\alpha_1=|q_{rs}|, \alpha_2=|q_{rt}|$. Adjoin the column $\alpha_1\mathbf{q}_s+\alpha_2\mathbf{q}_t$ to the matrix $\overline{\mathbf{Q}}$.

Step 4 When all the pairs in S have been examined, and the additional columns (if any) have been added, we say that Row r has been processed. Now if $\overline{\mathbf{Q}}=\varnothing$, stop; $\mathbf{W}=\mathbf{0}$ is the only solution to the system. Otherwise, let $\mathbf{Q}=\overline{\mathbf{Q}}$ go to Step 0. Now this algorithm can be modified to solve a problem with cardinality restriction using Results 2.27.4 and 2.27.5. We illustrate this with an example.

EXAMPLE 2.27.3 Consider the following problem of finding all feasible solutions to the given system of equations with cardinality constraint:

Given

$$\mathbf{D}=\begin{bmatrix} 0 & 1 & 0 & 1 & -4 \\ 1 & -1 & 1 & -1 & -2 \\ 0 & 0 & 1 & 1 & -6 \end{bmatrix}$$

find $\mathbf{DW}=\mathbf{0}, \mathbf{W} \geqslant \mathbf{0}$, with $|\mathbf{W}|^{+} \leqslant 3$.

The initial matrix is

$$\mathbf{Q}=[\mathbf{D}',\mathbf{I}']',$$

that is

$$\mathbf{Q}=\left[\begin{array}{ccccc} 0 & 1 & 0 & 1 & -4 \\ 1 & -1 & 1 & -1 & -2 \\ 0 & 0 & 1 & 1 & -6 \\ \hline 1 & 0 & 0 & 0 & 0 \\ 0 & 1 & 0 & 0 & 0 \\ 0 & 0 & 1 & 0 & 0 \\ 0 & 0 & 0 & 1 & 0 \\ 0 & 0 & 0 & 0 & 1 \end{array}\right].$$

We apply Algorithm 2.27.1 using if possible Results 2.27.4 and 2.27.5. Step 0 and Step 1 are not applicable. We go to Step 2. We consider row 1 for processing.

Step 2 $R=\{1,3\}$. Therefore columns 1 and 3 are included in the new matrix $\overline{\mathbf{Q}}$. We go to Step 2′. But 2′ is not applicable. We go on to Step 3. Now

$$\begin{aligned} S &= \{(s,t) | q_{rs} \cdot q_{rt} < 0, s < t\} \\ &= \{(2,4),(4,5)\} \end{aligned}$$

Step 3 $I(2,5)=\{i | i \in L, q_{is}=q_{it}=0\}=4,6,7\}$. But there is no u as stipulated in Step 3(2). So we adjoin $|q_{15}|\mathbf{q}_2+|q_{12}|\mathbf{q}_5$ to $\overline{\mathbf{Q}}$. That is we multiply the column 2 by 4 and column 5 by 1 and add them together, to obtain a new column.

Similarly (4, 5) yields another column. The new matrix then obtained is $\overline{\mathbf{Q}}$. Below each column in $\overline{\mathbf{Q}}$ we indicate the column (s) of the previous matrix with which we produced that column. Let $\mathbf{Q}=\overline{\mathbf{Q}}$. We go to Step 1.

Row 1 has been processed.

$$\overline{\mathbf{Q}}=\left[\begin{array}{cccc} 0 & 0 & 0 & 0 \\ 1 & 1 & -6 & -6 \\ 0 & 1 & -6 & -2 \\ \hline 1 & 0 & 0 & 0 \\ 0 & 0 & 4 & 0 \\ 0 & 1 & 0 & 0 \\ 0 & 0 & 0 & 4 \\ 0 & 0 & 1 & 1 \end{array}\right].$$
$$\begin{array}{cccc} 1 & 3 & 2,5 & 4,5 \end{array}$$

We take Row 2 for processing.

Step 2 $R=\phi$.

$$S=\{(1,3),(1,4),(3,4),(2,4)\}$$

Step 3 $I(s,t)$ is not empty, however, no u as per Step 3(2) exists so we obtain four new columns as follows:

$$\overline{\mathbf{Q}}=\left[\begin{array}{cccc} 0 & 0 & 0 & 0 \\ 0 & 0 & 0 & 0 \\ -6 & -2 & 0 & 4 \\ \hline 6 & 6 & 0 & 0 \\ 4 & 0 & 4 & 0 \\ 0 & 0 & 6 & 6 \\ 0 & 4 & 0 & 4 \\ 1 & 1 & 1 & 1 \end{array}\right].$$
$$\begin{array}{cccc} 1,3 & 1,4 & 2,3 & 2,4 \end{array}$$

Now Row 2 is processed. Let $\mathbf{Q}=\overline{\mathbf{Q}}$. We go to Step 1. We process Row 3.

Step 2 $R=\{3\}$. So Column 3 is included in $\overline{\mathbf{Q}}$.

$$S=\{(1,4),(2,4)\}.$$

Step 3 $I(1,4)=\phi$. So we do not get any new column for $\overline{\mathbf{Q}}$ from $(1,4)$ as per Step 3(1). $I(2,4)=\{2\}$ but no u exists as per Step 3(2).

We combine Columns 2 and 4. Thus we get $\overline{\mathbf{Q}}$ given by,

$$\overline{\mathbf{Q}}=\begin{bmatrix} 0 & 0 \\ 0 & 0 \\ 0 & 0 \\ \hline 0 & 24 \\ 4 & 0 \\ 6 & 12 \\ 0 & 24 \\ 1 & 6 \end{bmatrix}.$$
$$\begin{matrix} 3 & 2,4 \end{matrix}$$

As there are no more rows to be processed, we stop. We have the following solutions:

$$x_1=0, \quad x_2=4, \quad x_3=6, \quad \text{and} \quad x_4=0$$

and

$$x_1=\tfrac{24}{6}=4, \quad x_2=0, \quad x_3=\tfrac{12}{6}=2, \quad \text{and} \quad x_4=\tfrac{24}{6}=4.$$

Both these solutions satisfy the cardinality constraint.

REMARK 2.27.5 From the computational experience derived, we find for large matrices, unless the programming is done efficiently by the methods of sparce matrices, it may not be a feasible approach for solving the k-best regression problem. However, further research on the computational and storage aspects is required.

REMARK 2.27.6 If the problem is to minimize certain objective functions in the k-best MINMAD regression problem we can eliminate certain basic feasible solutions on the basis of the objective function value. Also in the k-best MINMAD problem we only require $|\beta_1|^+ + |\beta_2|^+ \leqslant k$. So we only consider the corresponding components for cardinality restriction. This modification can be easily incorporated in the method described.

BIBLIOGRAPHICAL NOTES

2.1 Seber (1977), Draper and Smith (1966), Daniel and Wood (1971), Williams (1959), Albert (1972) and Ezekiel and Fox (1959) have written books exclusively on least-squares linear-regression models. Harter in a series of six articles (1974a, 1974b, 1975a, 1975b, 1975c,

1976) has accomplished the enormous job of collecting the available literature on regression methods, and supplies a chronological development of L_1-norm estimation. Gentle (1977) also provides a good bibliography on L_1-norm estimation. Chebyshev norm estimation is discussed in Collatz and Wetterling (1975).

2.2 The result presented in Section 2.2 and Example 2.3.1 are from Karst (1958). In Stiefel (1960) we find a geometrical interpretation of MINMAXAD estimation.

2.3 Jaeckel (1972) considers estimators similar to the MINSADBED estimator.

2.4 Example 2.4.1 is from Hald (1952). Presentation in the section follows Seber (1977).

2.6 Hoerl and Kennard (1970a, b) introduced the class of estimation known as Ridge estimators. Stein (1960) used shrinkage estimators. An extensive simulation study was conducted by Dempster, Schatzoff, and Wermuth (1977) to compare 56 alternatives arising from Ridge and Stein methods, against the least-squares multiple regression.

2.7 MINMAD estimator was studied as early as 1757 by Boscovich (see Eisenhart 1961, 1962). Laplace (1793) also suggested and studied this estimator (see Stigler 1973). Edgeworth (1887) presented a method for the simple regression, with MINMAD estimator. However, Turner (1887) questioned Edgeworth's claim of his method's computational superiority over the least-squares method, and also pointed out the nonuniqueness of the MINMAD estimator. It was only after the work of Charnes, Cooper, and Ferguson (1955), that a renewed interest in using the MINMAD estimator for regression problems was created. They showed the equivalence between a MINMAD problem and a linear-programming problem. Wagner (1959) suggested solving the problem through the dual approach. More recently, because of the efficient modification of the simplex method for solving the MINMAD problem by Barrodale and Roberts (1973), the possibility has further increased of using MINMAD as an alternative to least squares.

2.8 and 2.9 The simplex method for linear-programming problems was developed by Dantzig. There are several books on linear programming, written at different levels of difficulty. An authentic chronological development of the subject can be found in Dantzig (1963). In linear-programming literature the problem is usually formulated with all nonnegative variables. This warrants transforming unrestricted variables to nonnegative variables. The presentation here differs from the usual in that we have developed the method to deal with the unrestricted variables as they are. Collatz and Wetterling (1975) contains a brief discussion of this modification. For further reading in linear programming see Dantzig (1963), Hadley (1962), Gale (1960), Gass (1964), Simmonard (1966), Llewelyn (1964), Grass (1975), Leunberger (1973), and Murty (1976). The observations made at the end of Algorithm 2.8.1 for the MINMAD problem were used by Barrodale and Roberts (1973) to modify the simplex method so as to bypass some iterations. The MINMAD problem with additional linear restrictions (restricted MINMAD problem) is considered along the same lines in Barrodale and Roberts (1978). Special purpose algorithms for the MINMAD problem have also been given by Armstrong and Hultz (1977), and Bartels and Conn (1977). Computer comparisons have established the Barrodale and Roberts algorithm as an efficient method for solving the MINMAD problem. A recent revised simplex version of this algorithm by Armstrong, Frome, and Kung (1979) is claimed to be even more efficient than the Barrodale and Roberts algorithm.

2.10 and 2.11 Wagner (1959) suggested that the MINMAD problem can be solved by solving the dual of the MINMAD problem. He also observed that the dual problem can be reduced to a problem with a smaller basis but that the dual variables have upper-bound restrictions. Material in this section presents a specialization of duality results in linear programming.

2.12 See Dantzig (1963) and Hadley (1962).

2.13 Example 2.13.1 is from Sielken and Hartley (1973).

2.14 Material in this section closely follows Sielken and Hartley. Unbiased estimation when the error distribution is symmetric with mean zero in a general setup is considered in Seely and Hogg (1977).

2.15 The simulation study appears in Arthanari et al. (1977).

2.16 The behavior of MINMAD estimators has been considered under different conditions. Ashar and Wallace (1963), Rice and White (1964), Glahe and Hunt (1970), Kiountouzis (1973), Bourdon (1974), Rosenberg and Carlson (1977), and Pfaffenberger and Dinkel (1978), among others, have tried to find the distribution of MINMAD estimators using the Monte Carlo method.

Taylor (1973) suggested the combination of MINMAD and LS in which MINMAD should be applied first as a means of identifying outliers to be terminated, and then the least squares applied after the trimming has been done. He also gives excellent arguments for the use of MINMAD in econometric analysis. Wilson (1978) via Monte Carlo sampling investigated the cases in which the disturbances are normally distributed with constant variance except for one or more outliers whose disturbances are generated from normal distribution with larger variance. Among other results he found that MINMAD estimation retains its advantage over least squares under different conditions such as variations in outlier variance, number of independent variables, number of observations, and number of outliers.

In brief the MINMAD estimator will tend to be more efficient than the least-squares (L_2) estimator in the presence of a few large disturbances. Pfaffenberger and Dinkel (1978) in their Monte Carlo study found that MINMAD (L_1) performs much better than least squares when the error distribution is cauchy. They used the ratio MSE $(L_1)/\text{MSE}(L_2)$ with sample sizes of 20, 40, and 100 for the purpose of comparison, and found that L_1 is more efficient than L_2 when errors are distributed cauchy, using the model $Y=b_0+b_1x_1+b_2x_2+e$. Rosenberg and Carlson (1977) attempted to find the distribution of $\hat{\beta}$ on the basis of extensive Monte Carlo methodology. Using symmetrical disturbance distributions of the error they found that the distribution of $\hat{\beta}$ is approximately multivariate normal, with mean β and covariance matrix $\lambda(X'X)^{-1}$, where λ/T is the variance of the median of a sample of size T drawn from the disturbance distribution. For low kurtosis such approximation seems to be good, but in high kurtosis they find the multivariate normal approximation much less adequate. The shape of the distribution of MINMAD estimates yet remains an open question. Thus, in spite of the evidence provided various researchers for the superiority of MINMAD estimator, until a distribution theory is developed for MINMAD for testing statistical hypotheses about the underlying population of regression coefficients, wide applications of such an estimator may be delayed. Taylor (1973) provides a possible approach for the problem of obtaining the distribution of the estimators.

2.17 Wagner (1959) gives a linear-programming formulation of the MINMAXAD problem. Also he suggests the dual approach for solving this problem. Steifel (1960) considers this problem and also brings out the connections between linear programming and Jordan elimination. Additionally, he gives examples to bring out the geometrical aspects of the problem. Collatz (1965) also discusses this problem in the context of Chebyshev approximation theory. The material in this section closely follows that of Collatz and Wetterling (1975). See Barrodale and Young (1966) for computational simplifications.

2.18 See Sielken and Hartley (1973).

2.19 Combining least squares in MINMAD regression in various ways has been suggested by Gentle, Kennedy and Sposito (1977a, 1977b), Taylor (1973), and McCormick and Sposito

(1976). The theory of convex programming developed in this section stems from the celebrated work of Kuhn and Tucker (1951) on nonlinear programming. For books on nonlinear programming see Arrow, Hurwiticz and Uzawa (1958), Bazaraa and Shetty (1979), Hadley (1964), Mangasarian (1969), Martos (1975), Zangwill (1969), and Zoutendijk (1976). Result 2.19.1 is from to Kuhn and Tucker (1951). Result 2.19.2 is proved in a more general setup in Karlin (1959).

Different types of constraint qualifications and their implications are discussed in Martos (1975) and Mangasarian (1968). Also quasiconvexity and other weak versions of convexity are discussed in Mangasarian (1968) and Martos (1975).

The algorithm described is from Wolfe (1959). For other methods refer to Bazaraa and Shetty (1979). Certain of the proofs in this section are specialized, exploring the structure of the problem. It is possible to bring in computational improvements in the method by studying the special structure of the $\mathbf{q}$ and $\mathbf{A}$ matrices in Problem 2.19.5.

2.20 Lemke and Howson (1964), in connection with bimatrix games, first developed an algorithm similar to the one given in this section. The algorithm in this section, known as "complementary pivot algorithm", is from Lemke (1965). For examples see Murty (1976). Also a ghost-story analogy of the complementary pivot algorithm, given by Eavas, appears there. For a computer code of the algorithm implemented in the revised simplex fashion see Ravindran (1972).

Waterman (1974) gives a complete search procedure for the problem with nonnegativity restrictions on β that uses least-squares regression. His approach requires solving 2^p unrestricted least-squares problems. Armstrong and Frome (1976) give a branch-and-bound scheme for the same problem. Numerical stability of restricted least-squares problem is studied by Stoer (1971). Judge and Takayama (1966) also consider the restricted least-squares problem and suggest a quadratic-programming formulation of the problem. Example 2.20.1 is from Bazaraa and Shetty (1979).

2.21 See Seber (1977) for the unrestricted problem.

2.22–2.24 The material presented in these sections is from Arthanari et al. (1977). A computer simulation study mentioned in Section 2.15 also includes these estimators, for comparison.

2.25 For methods available for choosing the best regression using the least-squares criterion see Seber (1977). The material in this section is based on Roodman (1974). Narula and Wellington (1977b) consider the problem with the minimum sum of weighted absolute errors.

2.27 See Chernikova (1965), Rubin (1975), and Martos (1975).

REFERENCES

Abdelmalek, N. N. (1976). *L_1 Solution of Overdetermined system of Linear Equations by a Dual Simplex Method and LU Decomposition*. Technical Report, Division of Electrical Engineering, National Research Council, Ottawa.

Albert, A. (1972). *Regression and the Moore-Penrose Pseudo Inverse*. Academic Press, New York.

Appa, G., and Smith, C. (1973). "On L_1 and Chebyshev Estimation." *Math. Program.* **5**, 73.

Armstrong, R. D., and Frome, E. L. (1976a). "A Comparison of Two Algorithms for Absolute Deviation Curve Fitting." *J. Am. Stat. Assoc.* **71**, 328.

Armstrong, R. D., and Frome, E. L. (1976b). "A Branch and Bound Solution of a Restricted Least Squares Problem." *Technometrics* **18**, 447.

Armstrong, R. D., Frome, E. L., and Kung, D. S. (1979). "A Revised Simplex Algorithm for the Absolute Deviation Curve Fitting Problem." *Commun. Stat.* **B 8**, 175.

Armstrong, R. D., and Frome, E. L. (1977). "A Special Purpose Linear Programming Algorithm for Obtaining Least Absolute Value Estimates in a Linear Model with Dummy Variables." *Commun. Stat.* **B 6**, 383.

Armstrong, R. D., and Hultz, J. W. (1977). "An Algorithm for a Restricted Discrete Approximation Problem in the L_1 Norm." *SIAM J. Numer. Anal.* **14**, 555.

Arrow, K. J., Hurwiticz, L., and Uzawa, H. (1958). *Studies in Linear and Nonlinear Programming*. Stanford Univ. Press, Palo Alto, Calif.

Arthanari, T. S., Dodge, Y., Nazemi, I., and Panosian, E. (1977). *On Performance of Five Methods of Estimation: A Simulation Study*. School of Planning and Computer Applications, Tehran.

Ashar, V. G., and Wallace, T. D. (1963). "A Sampling Study of Minimum Absolute Deviations Estimators." *Oper. Res.* **11**, 747.

Barrodale, I. (1968). "L_1 Approximation and the Analysis of Data." *Appl. Stat.* **17**, 51.

Barrodale, I., and Roberts, F. D. K. (1970). "Applications of Mathematical Programming to l_p Approximation." In *Nonlinear Programming*. J. B. Rosen, O. L. Mangasarian, and K. Ritter, Eds. Academic Press, New York, pp. 447–64.

Barrodale, I., and Roberts, F. D. K. (1973). "An Improved Algorithm for Discrete l_1 Linear Approximation." *SIAM J. Numer. Anal.* **10**, 839.

Barrodale, I., and Roberts, F. D. K. (1974). "Algorithm 478: Solution of an Overdetermined System of Equations in the l_1 Norm." *Commun. Assoc. Comput. Mach.* **17**, 319.

Barrodale, I., and Roberts, F. D. K. (1977). "Algorithms for Restricted Estimation." *Commun. Stat.* **B 6**, 353.

Barrodale, I., and Roberts, F. D. K. (1978). "An Efficient Algorithm for Discrete l_1 Linear Approximation with Linear Constraints." *SIAM J. Numer. Anal.* **15**, 603.

Barrodale, I., Roberts, F. D. K., and Hunt, D. R. (1970). "Computing Best l_p Approximations by Functions Nonlinear in One Parameter." *Computer J.* **13**, 382.

Barrodale, I., and Young, A. (1966). "Algorithms for Best L_1 and L_∞ Linear Approximations on a Discrete Set." *Numer. Math.* **8**, 295.

Bartels, R. H., and Conn, A. R. (1977). "LAV Regression: A Special Case of Piecewise Linear Minimization." *Commun. Stat.* **B 6**, 329.

Bartels, R. H., Conn, A. R., and Sinclair, J. W. (1978). "Minimization Techniques for Piecewise Differentiable Functions: The L_1 Solution to an Overdetermined Linear System. *SIAM J. Numer. Anal.* **15**, 224.

Bazaraa, M. S., and Shetty, C. M. (1979). *Nonlinear Programming*. John Wiley. New York.

Ben-Israel, A., and Charnes, A. (1968). "An Explicit Solution of a Special Class of Linear Programming Problems." *Oper. Res.* **16**, 1166.

Bowley, A. L. (1928). *F. Y. Edgeworth's Contributions to Mathematical Statistics*. Royal Statistical Society, London.

Charnes, A., Cooper, W. W., and Ferguson, R. O. (1955). "Optimal Estimation of Executive Compensation by Linear Programming." *Manage. Sci.* **1**, 138.

Chernikova, N. V. (1965). "Algorithm for Finding a General Formula for the Non-Negative Solutions of a System of Linear Equalities." *U.S.S.R. Comput. Math. & Math. Phys.*, **5**, 228.

Claerbout, J. F., and Muir, F. (1973). "Robust Modeling with Erratic Data." *Geophysics* **38**, 826.

Collatz, L., and Wetterling, W. (1975). *Optimization Problems*. Springer-Verlag, New York.

Crocker, D. C. (1969). "Linear Programming Techniques in Regression Analysis: The Hidden Danger." *AIEE Trans.* **1**, 112.

Daniel, C., and Wood, F. S. (1971). *Fitting Equations to Data*. Wiley-Interscience, New York.

Dantzig, G. B. (1963). *Linear Programming and Extensions*. Princeton Univ. Press, Princeton, N.J.

Davies, M. M. (1967). "Linear Approximation Using the Criterion of Least Total Deviations." *J. R. Stat. Soc.* **B 29**, 101.

Dempster, A. P., Schatzoff, M., and Wermuth, N. (1977). "A Simulation Study of Alternatives to Ordinary Least Squares." *J. Am. Stat. Assoc.* **72**, 77.

Draper, N. R., and Smith, H. (1966). *Applied Regression Analysis*. Wiley, New York.

Duris, C. S., and Sreedharan, V. P. (1968). "Chebyshev and l_1 Solutions of Linear Equations Using Least Squares Solutions." *SIAM J. Numer. Anal.* **5**, 491.

Dyer, M. E., and Proll, L. G. (1977). "An Algorithm for Determining All Extreme Points of a Convex Polytope." *Math. Program.* **12**, 81.

Edgeworth, F. Y. (1887). "On observations Relating to Several Quantities. *Phil. Mag.* (Ser. 5) **24**, 222.

Edgeworth, F. Y. (1888). "On a New Method of Reducing Observations Relating to Several Quantities." *Phil. Mag.* (Ser. 5) **25**, 184.

Eisenhart, C. (1961). *Boscovich and the Combination of Observations. Roger Joseph Boscovich, S. J., F. R. S., 1711–1787: Studies of His Life and Work on the 250th Anniversary of His Birth.* L. L. White, Ed. Allen and Unwin, London; 200–12.

Eisenhart, C. (1962). "Roger Joseph Boscovich and the Combination of Observations." *Actes Symp. Internat. R. J. Boskovic 1961,* Belgrade-Zagreb-Lublin, pp. 19–25.

Ekblom, H., and Henriksson, S. (1969). "L_p-Criteria for the Estimation of Location Parameters." *SIAM J. Appl. Math.* **17**, 1130.

Ezekiel, M., and Fox, K. A. (1959). *Methods of Correlation and Regression Analysis* (ed. 3). Wiley, New York.

Fisher, W. D. (1961). "A Note on Curve Fitting with Minimum Deviations by Linear Programming." *J. Am. Stat. Assoc.* **56**, 359.

Forsythe, A. B. (1972). "Robust Estimation of Straight Line Regression Coefficients by Minimizing pth Power Deviations." *Technometrics* **14**, 159.

Gale, D. (1960). *The Theory of Linear Economic Models*. McGraw-Hill, New York.

Gass, S. I. (1964). *Linear Programming* (ed. 2). McGraw-Hill, New York.

Gentle, J. E. (1977). "Least Absolute Values Estimation: An Introduction." *Commun. Stat.* **B 6** 313.

Gentle, J. E., Kennedy, W. J., and Sposito, V. A. (1977a). "On Least Absolute Deviations Estimations." *Commun. Stat.* **A 6**, 839.

Gentle, J. E., Kennedy, W. J., and Sposito, V. A. (1977b). "On Properties of L_1 Estimators." *Math. Program.* **12**, 139.

Gilsinn, J., et al. (1977). "Methodology and Analysis for Comparing Discrete Linear L_1 Approximation Codes." *Commun. Stat.* **B 6**, 399.

Glahe, F. R., and Hunt, J. G. (1970). "The Small Sample Properties of Simultaneous Equation Least Absolute Estimators Vis-a-Vis Least Squares Estimators." *Econometrica* **38**, 742.

Grass, S. I. (1975). *Linear Programming: Methods and Applications* (ed. 4). McGraw-Hill, New York.

Graves, J. S. (1974). "On the Storage and Handling of Binary Data Using FORTRAN with Applications to Integer Programming." *Oper. Res.* **27**, 534.

Hadley, G. (1962). *Linear Programming*. Addison-Wesley, Reading, Mass.

Hadley, G. (1964). *Nonlinear and Dynamic Programming*. Addison-Wesley, Reading, Mass.

Hald, A. (1952). *Statistical Theory with Engineering Applications*. Wiley, New York.

Hanson, R. J., and Lawson, C. L. (1974). *Solving Least Squares Problems*. Prentice-Hall, Englewood Cliffs, N.J.

Harris, T. E. (1950). "Regression Using Minimum Absolute Deviations." *Am. Stat.* **4**, 14.

Harter, H. L. (1974a). "The Method of Least Squares and Some Alternatives. I." *Int. Stat. Rev.* **42**, 147.

Harter, H. L. (1974b). "The Method of Least Squares and Some Alternatives. II." *Int. Stat. Rev.* **42**, 235.

Harter, H. L. (1975a). "The Method of Least Squares and Some Alternatives. III. *Int. Stat. Rev.* **43**, 1.

Harter, H. L. (1975b). "The Method of Least Squares and Some Alternatives. IV." *Int. Stat. Rev.* **43**, 125–90 and 273–8.

Harter, H. L. (1975c). "The Method of Least Squares and Some Alternatives. V." *Int. Stat. Rev.* **43**, 269.

Harter, H. L. (1976). "The Method of Least Squares and Some Alternatives. VI." *Int. Stat.* **44**, 113.

Harter, H. L. (1977). "Nonuniqueness of Least Absolute Values Regression." *Comm. Stat.* **A 6**, 829.

Hill, R. W., and Holland, P. W. (1977). "Two Robust Alternatives to Least-Squares Regression." *J. Am. Stat. Assoc.* **72**, 828.

Hoerl, A. E., and Kennard, R. W. (1970a). "Ridge Regression. Biased Estimation for Non Orthogonal Problems." *Technometrics* **12**, 55.

Hoerl, A. E., and Kennard, R. W. (1970b). "Ridge Regression. Applications to Non Orthogonal Problems." *Technometrics* **12**, 69.

Hogg, R. V. (1974). "Adaptive Robust Procedures: A Partial Review and Some Suggestions for Future Applications and Theory." *J. Am. Stat. Assoc.* **69**, 909.

Jaeckel, A. L. (1972). "Estimating Regression Coefficients by Minimizing the Dispersion of the Residuals." *Ann. Math. Stat.* **43**, 1449.

Judge, G. G., and Takayama, T. (1966). "Inequality Restrictions in Regression Analysis." *J. Am. Stat. Assoc.* **61**, 166.

Karlin, S. (1959). *Mathematical Methods and Theory in Games, Programming and Economics* (Vols. 1, 2). Pergamon Press, London.

Karst, O. J. (1958). "Linear Curve Fitting Using Least Deviations." *J. Am. Stat. Assoc.* **53**, 118.

Kelley, J. E. (1958). "An Application of Linear Programming to Curve Fitting." *SIAM J. Appl. Math.* **6**, 15.

Kennedy, W. J., and Gentle, J. E. (1977). "Examining Rounding Error in LAV Regression Computations." *Commun. Stat.* **B 6**, 415.

Kennedy, W. J., Gentle, J. E., and Sposito, V. A. (1977). "A Computer Oriented Method for Generating Test Problems for L_1 Regression." *Commun. Stat.* **B 6**, 21.

Kiountouzis, E. A. (1973). "Linear Programming Techniques in Regression Analysis." *Appl. Stat.* **22**, 69.

Kough, P. F. (1979). "The Indefinite Quadratic Programming Problem." *Oper. Res.* **27**, 516.

Kuhn, H. W., and Tucker, A. W. (1951). "Nonlinear Programming." In *Proceedings of the Second Berkeley Symposium on Mathematical Statistics and Probability*, J. Neyman, Ed. Univ. California Press, Berkeley, Calif. pp. 481–92.

Lemke, C. E. (1965). "Bimatrix Equilibrium Points and Mathematical Programming." *Manage. Sci.* **11**, 681.

Lemke, C. E. and Howson, J. T. (1964). "Equilibrium Points of Bimatrix Games." *SIAM J.* **12**.

Llewellyn, R. W. (1964). *Linear Programming*. Holt, Rinehart, and Winston, New York.

Luenberger, D. G. (1973). *Introduction to Linear and Nonlinear Programming*. Addison-Wesley, Reading, Mass.

McCormick, G. F. and Sposito, V. A. (1976). Using the L_2-estimator in L_1-estimation. SIAM J. Numer. Anal. **13**, 337.

Mangasarian, O. L. (1969). *Nonlinear Programming*. Wiley, New York.

Martos, B. (1975). *Non Linear Programming Theory and Methods*. North-Holland Publishing, Amsterdam.

Murty, K. G. (1976). *Linear and Combinatorial Programming*. Wiley, New York.

Narula, S. C., and Wellington, J. F. (1977a). "Prediction, Linear Regression and Minimum Sum of Relative Errors." *Technometrics* **19**, 185.

Narula, S. C., and Wellington, J. F. (1977b). "An Algorithm for the Minimum Sum of Weighted Absolute Error Regression." *Commun. Stat.* **B 6**, 341.

Orchard-Hays, W. (1968). *Advanced Linear Programming Computing Techniques*. McGraw-Hill, New York.

Osborne, M. R., and Watson, G. A. (1971). "On an Algorithm for Discrete Nonlinear L_1 Approximation." *Comp. J.* **14**, 184.

Pfaffenberger, R. C., and Dinkel, J. J. (1978). "Absolute Deviations Curve Fitting: An Alternative to Least Squares." In *Contributions to Survey Sampling and Applied Statistics*. H. A. David (ed.) Academic Press, New York.

Ravindran, A. (1970). "Computational Aspects of Lamke's Complementary Algorithm Applied to Linear Programs." *Oper. Res.* **7**, 241.

Rhodes, E. C. (1930). "Reducing Observations by the Method of Minimum Deviations." *Phil. Mag.* (Ser. 7), **9**, 974.

Rice, J. R., and White, J. S. (1964). "Norms for Smoothing and Estimation." *SIAM Rev.* **6**, 243.

Roodman, G. (1974). "A procedure for Optimal Stepwise MSAE Regression Analysis." *Oper. Res.* **22**, 393.

Rosenberg, B., and Carlson, D. (1977). "A Simple Approximation of the Sampling Distribution of Least Absolute Residuals Regression Estimates." *Commun. Stat.* **B 6**, 421.

Rubin, D. S. (1975). "Vertex Generation and Cardinality Constrained Linear Programs." *Oper. Res.* **23**, 555.

Sadovski, A. N. (1974). "Algorithm AS74. L_1-Norm Fit of a Straight Line." *Appl. Stat.* **23**, 244.

Seber, G. A. F. (1977). *Linear Regression Analysis.* Wiley-Interscience, New York.

Seely, J., and Hogg, R. V. (1977). *Unbiased Estimation in Linear Models.* Technical Report 57, Department of Statistics, University of Iowa, Iowa City, Iowa.

Sharpe, W. F. (1971). "Mean-Absolute-Deviation Characteristic Lines for Securities and Portfolios." *Manage. Sci.* **18**, B1.

Sielken, R. L., and Hartley, H. O. (1973). "Two Linear Programming Algorithms for Unbiased Estimation of Linear Models." *J. Am. Stat. Assoc.* **68**, 639.

Simmonnard, M. (1966). *Linear Programming.* Transl. from French by Jewell, W. S. Prentice-Hall, Englewood Cliffs, N.J.

Singleton, R. R. (1940). "A Method of Minimizing the Sum of Absolute Values of Deviations." *Ann. Math. Stat.* **11**, 301.

Sposito, V. A., Kennedy, W. J., and Gentle, J. E. (1977). "Algorithm AS110: L_p Norm Fit of a Straight Line." *Appl. Stat.* **26**. 114.

Sposito, V. A., and Smith, W. C. (1976). "On a Sufficient and a Necessary Condition for L_1 Estimation." *Appl. Stat.* **25**, 154.

Spyropoulos, K., Kiountouzis, E., and Young, A. (1973). "Discrete Approximation in the L_1 Norm." *Comput. J.* **16**, 18.

Stein, C. (1960). "Multiple Regression." In *Contributions to Probability and Statistics: Essays in Honor of Harold Hotelling.* Stanford Univ. Press. Palo Alto, Calif., pp. 424–43.

Stiefel, E. (1960). "Note on Jordan Elimination, Linear Programming and Tchebyscheff Approximation." *Nemer. Math.* **2**, 1.

Stigler, S. M. (1973). "Studies in the History of Probability and Statistics. XXXII: Laplace, Fisher, and the Discovery of Sufficiency." *Biometrika* **60**, 439.

Stoer, J. (1971). "On the Numerical Solution of Constrainted Least Squares Problems." *SIAM J. Numer. Anal.* **8**, 382.

Taylor, L. D. (1973). "Estimation by Minimizing the Sum of Absolute Errors." In *Frontiers in Econometrics*, P. Zarembka, Ed., Academic Press, New York, pp. 169–90.

Turner, H. H. (1887). "On Mr. Edgeworth's Method of Reducing Observations Relating to Several Quantities." *Phil. Mag.* (Ser. 5) **24**, 466.

Wagner, H. M. (1959). "Linear Programming Techniques for Regression Analysis." *J. Am. Stat. Assoc.* **54**, 206.

Wagner, H. M. (1962). "Nonlinear Regression with Minimal Assumptions." *J. Am. Stat. Assoc.* **57**, 572.

Waterman, M. S. (1974). "A Restricted Least Squares Problem." *Technometrics* **16**, 135.

Wilson, H. G. (1978). "Least Squares Versus Minimum Absolute Deviations Estimation in Linear Models." *Decis. Sci.* **9**, 322.

Williams, E. J. (1959). *Regression Analysis.* Wiley, New York.

Wolfe, Ph. (1959). "The Simplex Method for Quadratic Programming." *Econometrica* **27**, 382.

Zangwill, W. I. (1969). *Nonlinear Programming.* Prentice-Hall, Englewood Cliffs, N.J.

Zoutendijk, G. (1976). *Mathematical Programming Methods.* North Holland, Amsterdam.

CHAPTER 3

Generalized Inverses in Linear Statistical Models

3.1 INTRODUCTION

The generalized inverses of matrices (g-inverses) are used in many situations in linear statistical inference. A brief review of some of the interesting results available on g-inverses of a matrix will help to show how the linear-programming artificial-variable technique can be used to compute the g-inverse of a matrix.

Thus, the situations where g-inverses are to be computed in linear models emphasize the use of mathematical programming in these areas.

DEFINITION 3.1.1 Let $\mathbf{A}$ be an $m\times n$ matrix of any rank. Let $\mathbf{R(A)}$ be the rank of $\mathbf{A}$. A g-inverse of $\mathbf{A}$ is an $n\times m$ matrix denoted by $\mathbf{A}^-$ such that $\mathbf{AA^-A}=\mathbf{A}$. $\mathbf{A}^-$ is not unique in general; given $\mathbf{A}^-$ since $\mathbf{AA^-A}=\mathbf{A}$, then $\mathbf{A^-AA^-A}=\mathbf{A^-A}$ or $\mathbf{A^-A}$ is idempotent. Also $\mathbf{R(A^-A)}=\mathbf{R(A)}$ as $\mathbf{R(A)}\geqslant \mathbf{R(A^-A)}\geqslant \mathbf{R(AA^-A)}=\mathbf{R(A)}$. Suppose $\mathbf{A}^-$ is any $n\times m$ matrix such that $\mathbf{A^-A}$ is idempotent and $\mathbf{R(A^-A)}=\mathbf{R(A)}$. Then

$$\mathfrak{M}((\mathbf{A^-A})^*)=\mathfrak{M}(\mathbf{A}^*)\Rightarrow\mathcal{O}((\mathbf{A^-A}^*))=\mathcal{O}(\mathbf{A}^*)$$

where symbols $\mathfrak{M}$ and $\mathcal{O}$ denote the column space (linear space spanned by the columns), and orthogonal complement of $\mathfrak{M}(\cdot)$ in $\mathbf{R}^m$, respectively, and * represents the conjugate transpose.

Hence $\mathbf{A^-A(I-A^-A)}=\mathbf{0}\Rightarrow\mathbf{A(I-A^-A)}=\mathbf{0}$, or $\mathbf{A}=\mathbf{AA^-A}$. Thus we have an equivalent definition of a g-inverse of $\mathbf{A}$.

DEFINITION 3.1.2 A g-inverse of $\mathbf{A}$ of order $m\times n$ is a matrix $\mathbf{A}^-$ of order $n\times m$ such that $\mathbf{A^-A}$ is idempotent and $\mathbf{R(A^-A)}=\mathbf{R(A)}$.

We next consider solution of the homogeneous equation $\mathbf{AX}=\mathbf{0}$ in terms of a g-inverse of $\mathbf{A}$. We have:

RESULT 3.1.1 A general solution of the homogeneous equation $\mathbf{AX}=\mathbf{0}$ is $\mathbf{X}=(\mathbf{I}-\mathbf{A}^-\mathbf{A})\mathbf{Z}$ where $\mathbf{Z}$ is an arbitrary vector. This result follows from the fact that $\mathbf{A}(\mathbf{I}-\mathbf{A}^-\mathbf{A})=\mathbf{0}$ and $\mathbf{R}(\mathbf{I}-\mathbf{A}^-\mathbf{A})=n-\mathbf{R}(\mathbf{A})$.

RESULT 3.1.2 A general solution of a consistent nonhomogeneous equation $\mathbf{AX}=\mathbf{Y}$ is $\mathbf{X}=\mathbf{A}^-\mathbf{Y}+(\mathbf{I}-\mathbf{A}^-\mathbf{A})\mathbf{Z}$ where $\mathbf{Z}$ is an arbitrary vector. Result 3.1.2 follows from Result 3.1.1 and the fact that a general solution of $\mathbf{AX}=\mathbf{Y}$ is the sum of a particular solution of $\mathbf{AX}=\mathbf{Y}$ and a general solution of $\mathbf{AX}=\mathbf{0}$.

DEFINITION 3.1.3 $\mathbf{X}_b$ is said to be a basic solution of the equation $\mathbf{AX}=\mathbf{Y}$ if (1) $\mathbf{AX}_b=\mathbf{Y}$ and (2) $\mathbf{X}_b$ has utmost r nonzero components, where $r=\mathbf{R}(\mathbf{A})$. We are interested in finding $\mathbf{G}$, a g-inverse which provides $\mathbf{X}_b$.

RESULT 3.1.3 Let $\mathbf{A}$ be an $m\times n$ matrix with $\mathbf{R}(\mathbf{A})=r$. $\mathbf{G}$ is a g-inverse providing a basic solution of a consistent equation $\mathbf{AX}=\mathbf{Y}$ if there exists a permutation matrix

$$\mathbf{E}=\begin{pmatrix}\mathbf{E}_1\\ \mathbf{E}_2\end{pmatrix}$$

such that $\mathbf{E}_1\mathbf{G}=(\mathbf{AE}_1')_L^{-1}, \mathbf{E}_2\mathbf{GA}=\mathbf{0}$ where $\mathbf{E}_1$ is of order $r\times n$ and $\mathbf{E}_2$ of order $(n-r)\times n$, and $(\mathbf{AE}_1')_L^{-1}(\mathbf{AE}_1)\doteq\mathbf{I}$, that is subscript L denote the left inverse.

Proof Let $\mathbf{G}$ be a g-inverse that provides $\mathbf{X}_b$. Then $\mathbf{GY}$ has $(n-r)$ components zero whenever $\mathbf{Y}\in\mathfrak{M}(\mathbf{A})$. We obtain $\mathbf{G}$ so that $\mathbf{GA}$ has $(n-r)$ null rows. Suppose rows $i_1, i_2,\ldots, i_r$ of $\mathbf{GA}$ are nonnull. Let $\mathbf{E}$ denote the matrix obtained from the identity matrix by bringing its i_1th, i_2th,..., i_rth rows to the first r positions. Then if $\mathbf{EG}$ is partitioned as

$$\mathbf{EG}=\begin{pmatrix}\mathbf{G}_1\\ \mathbf{G}_2\end{pmatrix},$$

we have

$$\begin{pmatrix}\mathbf{G}_1\\ \mathbf{G}_2\end{pmatrix}\mathbf{A}=\begin{pmatrix}\mathbf{G}_1\mathbf{A}\\ \mathbf{0}\end{pmatrix}$$

where $\mathbf{G}_1$ is of order $r\times m$ and $\mathbf{G}_2$ of order $(n-r)\times m$. Also, since $\mathbf{EE}'=\mathbf{I}$, if $\mathbf{G}=\mathbf{A}^-$, then $\mathbf{EG}=(\mathbf{AE}')^-$. Let $\mathbf{AE}'$ be partitioned as

$$\mathbf{AE}'=(\mathbf{A}_1:\mathbf{A}_2)$$

where $\mathbf{A}_1$ has r columns and $\mathbf{A}_2$ has $n-r$ columns. Now

$$\mathbf{AE}'(\mathbf{EG})\mathbf{AE}'=\mathbf{AE}'$$

$$\Rightarrow(\mathbf{A}_1:\mathbf{A}_2)\begin{pmatrix}\mathbf{G}_1\\ \mathbf{G}_2\end{pmatrix}\mathbf{AE}'=\mathbf{AE}',$$

$$\Rightarrow(\mathbf{A}_1:\mathbf{A}_2)\begin{pmatrix}\mathbf{G}_1\\ \mathbf{0}\end{pmatrix}(\mathbf{A}_1:\mathbf{A}_2)=(\mathbf{A}_1:\mathbf{A}_2)$$

$$\Rightarrow\mathbf{A}_1\mathbf{G}_1\mathbf{A}_1=\mathbf{A}_1;\qquad \mathbf{A}_1\mathbf{G}_1\mathbf{A}_2=\mathbf{A}_2.$$

Now since $\mathbf{E}'$ is nonsingular, $\mathbf{R}(\mathbf{A}_1:\mathbf{A}_2)=R(\mathbf{A})=r$. Also

$$\mathbf{A}_1\mathbf{G}_1(\mathbf{A}_1:\mathbf{A}_2)=(\mathbf{A}_1:\mathbf{A}_2)\Rightarrow\mathbf{R}(\mathbf{A}_1\mathbf{G}_1)\geqslant\mathbf{R}(\mathbf{A}_1:\mathbf{A}_2)=r$$

$$\Rightarrow\mathbf{R}(\mathbf{A}_1)=\mathbf{R}(\mathbf{G}_1)=r,$$

since $\mathbf{A}_1$ has only r columns and $\mathbf{G}_1$ has r rows. Thus $\mathfrak{M}(\mathbf{A}_2)\subset\mathfrak{M}(\mathbf{A}_1)$, so $\mathbf{A}_1\mathbf{G}\mathbf{A}_2=\mathbf{A}_2$ is automatically satisfied and so

$$\mathbf{G}_1=\mathbf{A}_1^-.$$

Since $\mathbf{A}_1$ is a matrix with full rank, clearly $\mathbf{G}_1=(\mathbf{A}_1)_L^{-1}$. Hence we have shown the existence of

$$\mathbf{E}=\begin{pmatrix}\mathbf{E}_1\\ \mathbf{E}_2\end{pmatrix}\quad\text{with}\quad \mathbf{E}_1\mathbf{G}=\mathbf{G}_1=(\mathbf{AE}_1')_L^{-1},\mathbf{E}_2\mathbf{GA}=\mathbf{0}.$$

Hence the result.

The general solution of $\mathbf{AXA}=\mathbf{A}$ can be expressed in two alternative forms:

$$\mathbf{X}=\mathbf{A}^-+\mathbf{U}-\mathbf{A}^-\mathbf{AUAA}^- \tag{3.1.1}$$

or

$$\mathbf{X}=\mathbf{A}^-+\mathbf{V}(\mathbf{I}-\mathbf{AA}^-)+(\mathbf{I}-\mathbf{A}^-\mathbf{A})\mathbf{W} \tag{3.1.2}$$

where $\mathbf{A}^-$ is any particular g-inverse, and $\mathbf{U},\mathbf{V},\mathbf{W}$ are arbitrary matrices. It can be easily verified that $\mathbf{X}$ satisfies $\mathbf{AXA}=\mathbf{A}$. Any given $\mathbf{X}$ can be expressed in one of the two forms by choosing

(1) $\mathbf{U}=\mathbf{X}-\mathbf{A}^-$, or

(2) $\mathbf{V}=\mathbf{G}-\mathbf{A}^-$ and $\mathbf{W}=\mathbf{GAA}^-$, respectively.

DEFINITION 3.1.4 $\mathbf{A}^-$, a g-inverse of $\mathbf{A}$, is said to be reflexive if $\mathbf{A}^-\mathbf{A}\mathbf{A}^-=\mathbf{A}^-$.

3.2 SOME COMPUTATIONS OF A g-INVERSE

Let $\mathbf{A}$ be an $m\times n$ matrix of rank r. Then there exists a nonsingular submatrix $\mathbf{B}$ of order r. If $\mathbf{A}$ can be partitioned as

$$\mathbf{A}=\begin{pmatrix}\mathbf{B} & \mathbf{C}\\ \mathbf{F} & \mathbf{E}\end{pmatrix},\qquad \text{then}\quad \mathbf{A}^-=\begin{pmatrix}\mathbf{B}^{-1} & \mathbf{0}\\ \mathbf{0} & \mathbf{0}\end{pmatrix},\qquad \text{or}\quad \begin{pmatrix}\mathbf{B}^{-1} & -\mathbf{B}^{-1}\mathbf{C}\\ \mathbf{0} & \mathbf{I}\end{pmatrix}.$$

Here $\mathbf{A}^-\mathbf{A}\mathbf{A}^-=\mathbf{A}^-$ when we choose $\mathbf{A}^-=\begin{pmatrix}\mathbf{B}^{-1} & \mathbf{0}\\ \mathbf{0} & \mathbf{0}\end{pmatrix}$; that is, $\mathbf{A}^-$ is a reflexive g-inverse of $\mathbf{A}$. When

$$\mathbf{A}^-=\begin{pmatrix}\mathbf{B}^{-1} & -\mathbf{B}^{-1}\mathbf{C}\\ \mathbf{0} & \mathbf{I}\end{pmatrix},$$

$\mathbf{A}^-$ has maximum rank, namely, $\min(n,m)$.

We now focus our attention to obtain g-inverses of these types. In the literature, we can obtain $\mathbf{A}$ in the partitioned form $\begin{pmatrix}\mathbf{B} & \mathbf{C}\\ \mathbf{F} & \mathbf{E}\end{pmatrix}$ with $\mathbf{B}^{-1}$ of rank r by suitable rearrangement of rows and columns. A systematic approach that produces this rearrangement and also provides a g-inverse of $\mathbf{A}$ is attempted here.

Consider $\mathbf{AX}=\mathbf{0}$. Add one artificial variable to each row of $\mathbf{A}$. We obtain $[\mathbf{A},\mathbf{I}]\begin{bmatrix}\mathbf{X}\\ \mathbf{X}_a\end{bmatrix}=\mathbf{0}$. A readily available basis for this system is $\mathbf{I}$, corresponding to the artificial variables. Now to find a basis containing a maximum number of columns from $\mathbf{A}$, we proceed as follows:

$$\text{Maximize}\quad -\mathbf{X}_a'\mathbf{e}$$

$$\text{subject to}\quad [\mathbf{A},\mathbf{I}]\begin{bmatrix}\mathbf{X}\\ \mathbf{X}_a\end{bmatrix}=\mathbf{0}$$

$$\mathbf{X},\mathbf{X}_a\geqslant\mathbf{0}.$$

This is a linear-programming problem, and the objective function has a maximum equal to zero, corresponding to $\mathbf{X}_a=\mathbf{0}$. In terms of linear-programming terminology, we are at the end of Phase I and only artificial variables at zero level are in the basis. Even if optimality conditions, $Z_j - C_j \geqslant 0$, do not hold, we go on to Phase II. In Phase II, we replace the original costs for the columns of **A** and zero cost for artificial variables, and proceed with simplex iterations. But as we do not have any original objective function, we simply attempt to remove as many artificial variables as possible from the basis. Until no more artificial variables can be removed from the basis we continue the iterations, removing one artificial variable each time. This approach corresponds to giving a very high cost for legitimate variables and zero cost for artificial variables, and continuing until optimality conditions hold. If artificial variables still in the basis can be replaced by legitimate variables, we do so. Applying the simplex method, now we introduce any $\mathbf{a}_j$ from **A** into the basis such that one of the artificial variables can be removed from the basis. We repeat this process until no more artificial variables can be removed from the basis. At this point in the LP table, the rows corresponding to the artificial variables in the basis are such that if X_{a_i} is in the basis, we have a 1 in the column corresponding to the variables X_{a_i} and zeroes corresponding to the X_j's. This is true because if there is a nonzero element corresponding to any column $\mathbf{a}_j$ in the basic row corresponding to X_{a_i}, we can introduce $\mathbf{a}_j$ into the basis and remove X_{a_i} from the basis. So all the elements corresponding to the $\mathbf{a}_j$'s are zero.

Now if we could remove all the artificial variables from the basis, rank (**A**) would equal m. Otherwise, we have rank $(\mathbf{A}) < m$ and some artificial variables still in the basis: the rank (**A**) is m minus the number of artificial variables in the basis corresponding to the final table.

A submatrix of **A**, **B** with rank equal to that of **A** is now available by considering the columns (in order) that are in the basis and dropping the rows corresponding to the artificial variables in the basis. These rows are redundant.

The inverse of this matrix **B** is readily available in the last table. If **B** is obtained from the last basis and $(\mathbf{B}_{i1},\ldots,\mathbf{B}_{ir})$ give the columns in **B**, then $\mathbf{B}^{-1}$ is available in the columns corresponding to the artificial variables $X_{a_{i1}},\ldots,X_{a_{ir}}$.

So far we have shown how the linear-programming artificial-variable technique could be used to find a submatrix **B** of **A** and its inverse, where **B** has the same rank as **A**.

EXAMPLE 3.2.1 Consider the matrix

$$\mathbf{A}=\begin{bmatrix} 4 & 8 & 12 \\ 8 & 16 & 24 \\ 12 & 24 & 48 \end{bmatrix}.$$

Table 3.2.1 is the initial table corresponding to this matrix after adding artificial variables. Notice that we do not require the $\mathbf{X}_B$ column and $Z_j - C_j$ row for our computations as all the variables enter the basis at zero level, so $\mathbf{X}_B = \mathbf{0}$ for all the bases we are considering. The $\mathbf{C}_B$ column is also not required. Now $\mathbf{a}_1$ can replace $\mathbf{a}_4$, as we have a nonzero element, 4, in Row 1, Column 1, of the table. Using the usual LP transformation, we get (Table 3.2.2). Now we find in Table 3.2.2 that $\mathbf{a}_6$ can be replaced by $\mathbf{a}_3$. We get Table 3.2.3. Thus we find that no more artificial variables can be removed from the basis. **B** here is given by

$$\mathbf{B} = \begin{pmatrix} 4 & 12 \\ 12 & 48 \end{pmatrix}$$

Table 3.2.1

$\mathbf{C}_B$	Vectors in the basis	$\mathbf{X}_B$	$\mathbf{Y}_1$	$\mathbf{Y}_2$	$\mathbf{Y}_3$	$\mathbf{Y}_4$	$\mathbf{Y}_5$	$\mathbf{Y}_6$
-1	$\mathbf{a}_4$		4	8	12	1	0	0
-1	$\mathbf{a}_5$		8	16	24	0	1	0
-1	$\mathbf{a}_6$		12	24	48	0	0	1

Table 3.2.2

Vectors in the Basis	$\mathbf{Y}_1$	$\mathbf{Y}_2$	$\mathbf{Y}_3$	$\mathbf{Y}_4$	$\mathbf{Y}_5$	$\mathbf{Y}_6$
$\mathbf{a}_1$	1	2	3	$\frac{1}{4}$	0	0
$\mathbf{a}_5$	0	0	0	-2	1	0
$\mathbf{a}_6$	0	0	12	-3	0	1

Table 3.2.3

Vectors in the Basis	$\mathbf{Y}_1$	$\mathbf{Y}_2$	$\mathbf{Y}_3$	$\mathbf{Y}_4$	$\mathbf{Y}_5$	$\mathbf{Y}_6$
$\mathbf{a}_1$	1	2	0	1	0	$-\frac{1}{4}$
$\mathbf{a}_5$	0	0	0	-2	1	0
$\mathbf{a}_3$	0	0	1	$-\frac{1}{4}$	0	$\frac{1}{12}$

as Columns 1 and 3 are linearly independent and Row 2 is redundant. $\mathbf{B}^{-1}$ is available from the table as

$$\begin{bmatrix} 1 & -\frac{1}{4} \\ -\frac{1}{4} & \frac{1}{12} \end{bmatrix}.$$

Therefore a g-inverse of $\mathbf{D}$ a rearrangement of $\mathbf{A}$ is given by

$$\mathbf{D}^{-}=\begin{bmatrix} 1 & -\frac{1}{4} & 0 \\ -\frac{1}{4} & \frac{1}{12} & 0 \\ 0 & 0 & 0 \end{bmatrix}.$$

The rearrangements that are done in the columns and rows of $\mathbf{A}$ to obtain $\mathbf{D}$ are done on $\mathbf{D}^{-}$, interchanging the roles of columns and rows to obtain $\mathbf{A}^{-}$. Also we can get the g-inverse $\mathbf{G}$ with maximum rank from the table as

$$\mathbf{G}=\begin{bmatrix} 1 & -2 & -\frac{1}{4} \\ 0 & 1 & 0 \\ -\frac{1}{4} & 0 & \frac{1}{12} \end{bmatrix},$$

$$\mathbf{A}^{-}=\begin{bmatrix} 1 & 0 & -\frac{1}{4} \\ 0 & 0 & 0 \\ -\frac{1}{4} & 0 & \frac{1}{12} \end{bmatrix}.$$

After having developed the method using simplex tables we can now reduce the amount of computation required by resorting to the revised simplex table. In that case we do not need to transform all the $\mathbf{a}_j$'s at each stage; we just keep the inverse of the present basis.

Now for an explanation of the revised simplex method for finding a g-inverse. We keep the inverse of the corresponding matrix of each basis $\mathbf{B}$ and calculate $\mathbf{Y}_j=\mathbf{B}^{-1}\mathbf{a}_j$ for an $\mathbf{a}_j$ outside the basis. If we find an rk such that $Y_{rk}\neq 0$ corresponding to an artificial vector, we introduce the vector $\mathbf{a}_k$ into the basis and remove the vector $\mathbf{B}_r$ from the basis. The inverse of the new basis $\mathbf{B}$ is obtained by the usual transformation that is used in the simplex procedure. If at any stage we find that corresponding to every artificial vector $\mathbf{B}_i$ in the basis, each vector $\mathbf{a}_k$ outside the basis has $Y_{ik}=0$, we stop. We have obtained the maximum number of linearly independent columns from $\mathbf{A}$ in the current basis. If we denote by $\mathbf{A}_1$ the submatrix

Table 3.2.4

Vectors in the Basis	$\mathbf{C}_1$	$\mathbf{C}_2$	$\cdots$	$\mathbf{C}_m$	$\mathbf{Y}_k$
$\mathbf{B}_1$	C_{11}	C_{12}		C_{1m}	Y_{1k}
$\vdots$	$\vdots$	$\vdots$	$\cdots$	$\vdots$	$\vdots$
$\mathbf{B}_m$	C_{m1}	C_{m2}		C_{mm}	Y_{mk}

obtained by deleting the rows corresponding to the artificial variables from the basic columns of $\mathbf{A}$, $\mathbf{A}_1^{-1}$ is readily available in the last table. We now have Table 3.2.4 where $[\mathbf{C}_1, \mathbf{C}_2, \ldots, \mathbf{C}_m]$ is the inverse of the basis $[\mathbf{B}_1, \mathbf{B}_2, \ldots, \mathbf{B}_m]$.

We illustrate the method using the same example discussed earlier in this chapter.

EXAMPLE 3.2.2 Consider Example 3.2.1. We have the initial table corresponding to artificial variables; as the basis is an identity matrix, the inverse is the same. See Table 3.2.5 and note also that

$$\mathbf{A} = \begin{bmatrix} 4 & 8 & 12 \\ 8 & 16 & 24 \\ 12 & 24 & 48 \end{bmatrix}.$$

Now, $\mathbf{Y}_j = \mathbf{B}^{-1}\mathbf{a}_j = \mathbf{a}_j$, and we find $Y_{11} \neq 0$, so we introduce $\mathbf{a}_1$ into the basis. So $k=1$ and $r=1$. We get, by using the transformation

$$\hat{C}_{ij} = C_{ij} - \frac{Y_{ik}}{Y_{rk}} C_{rj} \qquad i \neq r, \qquad i = 1, \ldots, n$$

Table 3.2.5

Vectors in the Basis	$\mathbf{C}_1$	$\mathbf{C}_2$	$\mathbf{C}_3$	$\mathbf{Y}_k$
$\mathbf{a}_4$	1	0	0	4
$\mathbf{a}_5$	0	1	0	8
$\mathbf{a}_6$	0	0	1	12

Table 3.2.6

Vectors in the Basis	C_1	C_2	C_3	Y_k
$\mathbf{a}_1$	$\frac{1}{4}$	0	0	3
$\mathbf{a}_5$	-2	1	0	0
$\mathbf{a}_6$	-3	0	1	12

and

$$\hat{C}_{rj} = \frac{C_{rj}}{Y_{rk}} \qquad j=1,\ldots,m$$

and also Table 3.2.6.

Since $(\mathbf{B}^{-1}\mathbf{a}_2)'=(2,0,0)$, $\mathbf{a}_2$ cannot be introduced into the basis. Also notice that we need not calculate Y_{ik} for an $\mathbf{a}_k$ for i corresponding to the legitimate vector in the basis, as we are seeking only to remove artificial vectors. We next consider $(\mathbf{B}^{-1}\mathbf{a}_3)'=(3,0,12)$. Therefore, $\mathbf{a}_3$ is introduced and $\mathbf{a}_6$ is removed; $r=3$, $k=3$. Then we obtain Table 3.2.7. Now $(\mathbf{B}^{-1}\mathbf{a}_2)' =(-,0,-)$. Therefore we find for all $\mathbf{a}_k$ outside the basis and all corresponding artificial vectors $\mathbf{B}_i$ in the basis $Y_{ik}=0$. Here we stop.

$\mathbf{A}_1^{-1}$ is obtained by omitting the second row and second column of the final table. It is the same as the one obtained earlier.

Notice that if $Y_{ik}=0$ corresponding to an artificial vector $\mathbf{B}_i$ in the basis, the corresponding row in the table is unaltered by introduction of vector $\mathbf{a}_k$ and removal of any other vector. The implication is that we have found $(\mathbf{B}^{-1}\mathbf{a}_j)_i$ in the previous step; it remains the same in this step.

Table 3.2.7

Vectors in the Basis	C_1	C_2	C_3
$\mathbf{a}_1$	1	0	$-\frac{1}{4}$
$\mathbf{a}_5$	-2	1	0
$\mathbf{a}_3$	$-\frac{1}{4}$	0	$\frac{1}{12}$

3.3 LEAST-SQUARES SOLUTION AND g-INVERSE COMPUTATION

In Chapter 2 we discussed least-squares regression and the method of finding estimates of the regression coefficients solving the normal equations. Now we would like to use the linear-programming artificial-variable method with unrestricted legitimate variables.

We have $\mathbf{X}\boldsymbol{\beta}=\mathbf{Y}$ as earlier. The normal equations are given by $\mathbf{X}'\mathbf{X}\boldsymbol{\beta}=\mathbf{X}'\mathbf{Y}$. Let $\mathbf{X}_a$ denote the vector of artificial variables. Then we can restate the problem of solving the normal equations as follows:

$$\text{Maximize} \quad -\mathbf{X}_a'\mathbf{e}$$

$$\text{subject to} \quad [\mathbf{X}'\mathbf{X}:\mathbf{I}]\begin{bmatrix}\boldsymbol{\beta}\\ \mathbf{X}_a\end{bmatrix}=\mathbf{X}'\mathbf{Y}$$

$$\mathbf{X}_a \geqslant \mathbf{0}, \qquad \boldsymbol{\beta} \ \text{unrestricted in sign.}$$

For this LP problem, if we have to use the ordinary simplex method $\boldsymbol{\beta}$ must be expressed as the difference of two nonnegative vectors, which will increase the size of the problem. This increase may be avoided by the modified simplex method for unrestricted variables, discussed earlier.

As before, we proceed with Phase I with the modified method. At the end of Phase I we have a least-squares estimate of the parameters. Also we can try to remove as many artificial variables from the basis as possible, to introduce legitimate variables in place of artificial variables.

The advantage of this approach is that we can also get a g-inverse of $\mathbf{X}'\mathbf{X}$ as a byproduct, which in turn can be used to compute the residual sum of squares and standard error.

EXAMPLE 3.3.1 Consider

$$\mathbf{X}'\mathbf{X}=\begin{bmatrix}4 & 8 & 12\\ 8 & 16 & 24\\ 12 & 24 & 48\end{bmatrix}$$

and

$$\mathbf{X}'\mathbf{Y}=\begin{bmatrix}16\\ 32\\ 72\end{bmatrix}.$$

The first table is Table 3.3.1. After introducing Vector $\mathbf{a}_3$ in place of $\mathbf{a}_6$ and

Table 3.3.1

C_B	Vectors in the Basis	X_B	Y_1	Y_2	Y_3	Y_4	Y_5	Y_6
−1	a_4	16	4	8	12	1	0	0
−1	a_5	32	8	16	24	0	1	0
−1	a_6	72	12	24	48	0	0	1
	$Z_j - C_j$	$Z = -120$	−24	−48	−84	0	0	0

Table 3.3.2

C_B	Vectors in the Basis	X_B	Y_1	Y_2	Y_3	Y_4	Y_5	Y_6
0	a_3	1.33	.333	.667	1	.083	0	0
−1	a_5	0	0	0	0	−2.0	1	0
−1	a_6	8.0	−4.0	−8.0	0	−4.0	0	1
	$Z_j - C_j$	$Z = -8.0$	4.0	8.0	0	4.5	0	0

Table 3.3.3

C_B	Vectors in the Basis	X_B	Y_1	Y_2	Y_3	Y_4	Y_5	Y_6
0	a_3	2	0	0	1	−.25	0	.083
−1	a_5	0	0	0	0	−2.0	1	0
0	a_1	−2	1	2	0	1	0	−.25
	$Z_j - C_j$	$Z = 0$	0	0	0	2.5	0	1

transforming Table 3.3.1 we obtain Table 3.3.2. As per the modified criterion for choosing the vector to enter the basis we choose $\mathbf{a}_1$ to enter the basis and as $Z_j - C_j = 4 > 0$. We allow negative Y_{ij}'s also for consideration in forming the ratio X_{B_i}/Y_{ij}. Thus $\mathbf{a}_6$ is removed from the basis. The resultant table is Table 3.3.3.

Now we find that the optimality conditions hold, namely, $Z_j - C_j = 0$ for all unrestricted variables. So we stop. Observe that $\mathbf{a}_5$ cannot be removed from the basis. From the $\mathbf{X}_B$ column we read out

$$\hat{\beta}_1 = -2.0, \qquad \hat{\beta}_2 = 0, \qquad \hat{\beta}_3 = 2.$$

By rearranging the elements under the columns corresponding to the artificial variables, and replacing by zeroes in columns that are not in the basis and rows that are redundant, we obtain a g-inverse of $\mathbf{X}'\mathbf{X}$, as in Example 3.2.1.

3.4 GENERALIZED INVERSE IN LINEAR ESTIMATION

The Gauss-Markov setup in linear estimation is given by

$$\mathbf{Y}=\mathbf{X}\boldsymbol{\beta}+\boldsymbol{\varepsilon}, \qquad \mathbf{E}(\boldsymbol{\varepsilon})=\mathbf{0}, \qquad \mathbf{D}(\boldsymbol{\varepsilon})=\sigma^2\mathbf{I} \tag{3.4.1}$$

where $\mathbf{D}$ stands for variances and covariances and $\boldsymbol{\beta}$ and $\boldsymbol{\sigma}^2$ are unknown. The problem is that of estimating the parameters $\boldsymbol{\beta}$ on the basis of the observations Y_i. A more general setup is obtained if we have $\mathbf{E}(\mathbf{Y})=\mathbf{X}\boldsymbol{\beta}$ and $\mathbf{D}(\mathbf{Y})=\boldsymbol{\sigma}^2\mathbf{G}$ where $\mathbf{G}$ is supposed to be known and nonsingular. $\boldsymbol{\beta}$ and $\boldsymbol{\sigma}^2$ are unknown. By transforming $\mathbf{Y}$ to $\mathbf{Z}$ by $\mathbf{Z}=\mathbf{G}^{-1/2}\mathbf{Y}$, we again have the same setup as that of (3.4.1). A special case of the general model is to have $\mathbf{D}(\mathbf{Y})=\Sigma$, Σ nonsingular and unknown. This model also can be transformed into (3.4.1). The problem of estimation of $\boldsymbol{\beta}$ in the general model is solved using the theory of least squares. If $\mathbf{G}$ is singular, we need to use results obtained in generalized inverses of matrices. Also in the general model, with $\mathbf{G}$ nonsingular, if $\mathbf{X}$ is not of full rank we need to use g-inverses of matrices.

These problems can be viewed as linear-programming problems. In certain cases, we need to compute a g-inverse of a matrix, use this g-inverse in another expression, and find a g-inverse of the resultant matrix. In such cases, we need to solve more than one linear-programming problem.

We briefly discuss the unified theory of least squares below. The linear-programming technique could be used to compute the g-inverses appearing therein.

3.5 UNIFIED THEORY OF LEAST SQUARES

The problem of estimating $\boldsymbol{\beta}$ in the generalized regression model was considered in Chapter 2 under the assumption that the matrix of variances and covariances of $\mathbf{Y}$ is given by $\sigma^2\mathbf{V}$, which is nonsingular (in fact, it is positive definite). Now consider a more general setup and show how the estimation of $\boldsymbol{\beta}$ and related problems can be reduced to that of computing a g-inverse of a matrix.

By a general Gauss-Markoff (GGM) model, we refer to

$$\mathbf{Y}=\mathbf{X}\boldsymbol{\beta}+\boldsymbol{\varepsilon}$$

$$E(\mathbf{Y})=\mathbf{X}\boldsymbol{\beta}$$

$$D(\mathbf{Y})=\sigma^2\mathbf{G}$$

where $\mathbf{X}$ may not be of full rank and $\mathbf{G}$ is possibly singular. This general model includes the case of having constraints on $\boldsymbol{\beta}$, namely, $\mathbf{R}\boldsymbol{\beta}=\mathbf{C}$.

Such can be observed by considering $\mathbf{Y}_e=\begin{bmatrix}\mathbf{Y}\\ \mathbf{C}\end{bmatrix}$, $\mathbf{X}_e=\begin{bmatrix}\mathbf{X}\\ \mathbf{R}\end{bmatrix}$ and $\mathbf{G}_e=\begin{bmatrix}\mathbf{G} & \mathbf{0}\\ \mathbf{0} & \mathbf{0}\end{bmatrix}$. Thus we have the same form as (3.4.1). We now prove the following results to obtain the best linear unbiased estimates (BLUE), and their variances and covariances. Let a g-inverse of $\begin{bmatrix}\mathbf{G} & \mathbf{X}\\ \mathbf{X}' & \mathbf{0}\end{bmatrix}$ be given by

$$\begin{pmatrix}\mathbf{C}_1 & \mathbf{C}_2\\ \mathbf{C}_3 & -\mathbf{C}_4\end{pmatrix}.$$

RESULT 3.5.1 The BLUE of an estimable parametric function $\mathbf{p}'\boldsymbol{\beta}$ is $\mathbf{p}'\hat{\boldsymbol{\beta}}$ where $\hat{\boldsymbol{\beta}}=\mathbf{C}_3\mathbf{Y}$ or $\hat{\boldsymbol{\beta}}=\mathbf{C}_2'\mathbf{Y}$.

Proof If $\mathbf{L}'\mathbf{Y}$ is such that $E(\mathbf{L}'\mathbf{Y})=\mathbf{p}'\boldsymbol{\beta}$, then $\mathbf{X}'\mathbf{L}=\mathbf{p}$. Therefore we have the problem

$$\begin{aligned}&\text{Minimize} && \mathbf{L}'\mathbf{G}\mathbf{L}\\ &\text{subject to} && \mathbf{X}'\mathbf{L}=\mathbf{p},\end{aligned} \tag{3.5.1}$$

as we wish to minimize $\text{Var}(\mathbf{L}'\mathbf{Y})=\sigma^2\mathbf{L}'\mathbf{G}\mathbf{L}$. Let $\mathbf{L}^*$ be an optimum choice and $\mathbf{L}$ any feasible solution to $\mathbf{X}'\mathbf{L}=\mathbf{p}$. Then

$$\begin{aligned}\mathbf{L}'\mathbf{G}\mathbf{L}&=(\mathbf{L}-\mathbf{L}^*+\mathbf{L}^*)'\mathbf{G}(\mathbf{L}-\mathbf{L}^*+\mathbf{L}^*)\\ &=(\mathbf{L}-\mathbf{L}^*)'\mathbf{G}(\mathbf{L}-\mathbf{L}^*)+\mathbf{L}^{*\prime}\mathbf{G}\mathbf{L}^*+2\mathbf{L}^{*\prime}\mathbf{G}(\mathbf{L}-\mathbf{L}^*)\\ &\geqslant \mathbf{L}^{*\prime}\mathbf{G}\mathbf{L}\text{ if and only if }\mathbf{L}^{*\prime}\mathbf{G}(\mathbf{L}-\mathbf{L}^*)=\mathbf{0}\end{aligned}$$

whenever $\mathbf{X}'(\mathbf{L}-\mathbf{L}^*)=\mathbf{0}$, that is, $\mathbf{G}\mathbf{L}^*=-\mathbf{X}\mathbf{K}^*$ for a suitable $\mathbf{K}^*$. Then there exist $\mathbf{L}^*$ and $\mathbf{K}^*$ to satisfy the equations

$$\begin{pmatrix}\mathbf{G} & \mathbf{X}\\ \mathbf{X}' & \mathbf{0}\end{pmatrix}\begin{pmatrix}\mathbf{L}^*\\ \mathbf{K}^*\end{pmatrix}=\begin{pmatrix}\mathbf{0}\\ \mathbf{p}\end{pmatrix}. \tag{3.5.2}$$

As a g-inverse of

$$\begin{pmatrix} \mathbf{G} & \mathbf{X} \\ \mathbf{X}' & \mathbf{0} \end{pmatrix} \quad \text{is} \quad \begin{pmatrix} \mathbf{C}_1 & \mathbf{C}_2 \\ \mathbf{C}_3 & -\mathbf{C}_4 \end{pmatrix}.$$

We have $\mathbf{L}^* = \mathbf{C}_2\mathbf{p}$ or $\mathbf{C}_3'\mathbf{p}$ and $\mathbf{K}^* = \mathbf{C}_4\mathbf{p}$ as a solution. The BLUE of $\mathbf{p}'\boldsymbol{\beta}$ is

$$\mathbf{L}^{*\prime}\mathbf{Y} = \mathbf{p}'\mathbf{C}_2' \cdot \mathbf{Y} = \mathbf{p}'\mathbf{C}_3\mathbf{Y}.$$

Thus if we can compute a g-inverse of $\begin{pmatrix} \mathbf{G} & \mathbf{X} \\ \mathbf{X}' & \mathbf{0} \end{pmatrix}$, we have the BLUE of $\mathbf{p}'\boldsymbol{\beta}$.

Notice that 3.5.1 is a quadratic programming problem, with the variables unrestricted in sign and the constraints are all equalities. Result 2.19.8 will imply 3.5.2. However, 3.5.2 can be solved doing the simplex algorithm, after adding artificial variables.

RESULT 3.5.2 The dispersion matrix of $\hat{\boldsymbol{\beta}}$ is $\sigma^2\mathbf{C}_4$ in the sense $\mathbf{Var}(\mathbf{p}'\hat{\boldsymbol{\beta}}) = \sigma^2\mathbf{p}'\mathbf{C}_4\mathbf{p}$, $\text{cov}(\mathbf{p}'\hat{\boldsymbol{\beta}}, \mathbf{q}'\hat{\boldsymbol{\beta}}) = \sigma^2\mathbf{p}'\mathbf{C}_4\mathbf{q} = \sigma^2\mathbf{q}'\mathbf{C}_4\mathbf{p}$ whenever $\mathbf{p}'\boldsymbol{\beta}$ and $\mathbf{q}'\boldsymbol{\beta}$ are estimable.

Proof We have $\mathbf{p} = \mathbf{X}'\mathbf{M}$ for some $\mathbf{M}$ (which is a necessary and sufficient condition that $\mathbf{L}'\mathbf{Y}$ is unbiased for $\mathbf{p}'\boldsymbol{\beta}$). We observe that

$$\mathbf{X}'\mathbf{C}_2\mathbf{X}' = \mathbf{X}' \quad \text{as} \quad \begin{pmatrix} \mathbf{G} & \mathbf{X} \\ \mathbf{X}' & \mathbf{0} \end{pmatrix}\begin{pmatrix} \mathbf{a} \\ \mathbf{b} \end{pmatrix} = \begin{pmatrix} \mathbf{0} \\ \mathbf{X}'\mathbf{d} \end{pmatrix}$$

are solvable for any $\mathbf{d}$ and a g-inverse of

$$\begin{pmatrix} \mathbf{G} & \mathbf{X} \\ \mathbf{X}' & \mathbf{0} \end{pmatrix}, \quad \text{namely} \quad \begin{pmatrix} \mathbf{C}_1 & \mathbf{C}_2 \\ \mathbf{C}_3 & -\mathbf{C}_4 \end{pmatrix};$$

we get $\mathbf{a} = \mathbf{C}_2\mathbf{X}'\mathbf{d}, \mathbf{b} = -\mathbf{C}_4\mathbf{X}'\mathbf{d}$ as a solution. Substituting these in the equations, we get:

$$\mathbf{G}\mathbf{C}_2\mathbf{X}'\mathbf{d} - \mathbf{X}\mathbf{C}_4\mathbf{X}'\mathbf{d} = \mathbf{0}$$

$$\mathbf{X}'\mathbf{C}_2\mathbf{X}'\mathbf{d} = \mathbf{X}'\mathbf{d} \quad \text{for all} \quad \mathbf{d}.$$

The implication is that $GC_2X' = XC_4X'$ and $X'C_2X' = X'$. Now

$$\begin{aligned}\mathrm{Var}(p'C_2'Y) &= \sigma^2 M'(XC_2'G)C_2X'M \\ &= \sigma^2 M'XC_4(X'C_2X')M \quad \text{as} \quad XC_2'G = XC_4X'. \\ &= \sigma^2 M'XC_4X'M \quad \text{as} \quad X'C_2X' = X' \\ &= \sigma^2 p'C_4p.\end{aligned}$$

On the same lines,

$$\mathrm{cov}(p'C_2'Y, q'C_2'Y) = \sigma^2 p'C_4q = \sigma^2 q'C_4p.$$

It can be also shown that an unbiased estimator of σ^2 is:

$$\hat{\sigma}^2 = f^{-1}Y'C_1Y \quad \text{where} \quad f \quad \text{is of rank equal to} \quad R(G:X) - R(X).$$

These results bring out the use of computing a g-inverse of $A = \begin{pmatrix} G & X \\ X' & 0 \end{pmatrix}$. Now we can add artificial columns to **A** and do the simplex-like iterations to get a g-inverse of **A**.

BIBLIOGRAPHICAL NOTES

3.1 Rao (1962) introduced the concept of generalized inverse into the area of statistics. Ben-Israel and Greville (1974), Bjerhammar (1973), Rao and Mitra (1971), and Boullion and Odell (1971) have written books exclusively on generalized inverses.

3.2 There are various ways of computing generalized inverses. Nobe (1976) gives different methods of computing the Moore-Penrose g-inverse. Dodge and Majumdar (1979) give an algorithm which finds the least-squares generalized inverses for classification models without any rounding-off errors. Example 3.2.1 is from Rao and Mitra (1971).

3.3–3.4 Materials presented here may not be found elsewhere.

3.5 Nashed and Rall (1976) give a list of 1776 references on generalized inverses. It starts with Adam and Eve and ends with Thomas Jefferson (Declaration of Independence). An enormous and notable job.

REFERENCES

Ben-Israel, A., and Greville, T. N. E. (1974). *Generalized Inverses: Theory and Applications*. Wiley, New York.

Bjerhammar, A. (1973). *Theory of Errors and Generalized Matrix Inverses*. Elsevier, New York.

Bjerhammer, A. (1977). *Hyper Estimates*. Royal Institute of Technology, Division of Geodesy, Stockholm.

Boullion, T. L., and Odell, P. L. (1971). *Generalized Inverse Matrices*. Wiley, New York.

Dodge, Y., and Majumdar, D. (1979). "An Algorithm for Finding Least Square Generalized Inverses for Classification Models with Arbitrary Patterns." *J. Stat. Comput. Simul.* **9**, 1.

Nashed, M. Z., and Rall, L. B. (1976). "Annotated Bibliography on Generalized Inverses and Applications." In *Proceedings of an Advanced Seminar*, M. Z. Nashed, Ed. Academic Press, New York, 771.

Nobe, B. (1976). "Methods for Computing the Moore-Penrose Generalized Inverse and Related Matters." In *Proceedings of an Advanced Seminar*, M. Z. Nashed, Ed. Academic Press, New York, 245.

Rao, C. R. (1962). "A Note on Generalized Inverse of a Matrix with Applications to Problems in Mathematical Statistics." *J. R. Stat. Soc.* **B 24**, 152.

Rao, C. R. (1967). "Calculus of Generalized Inverse of Matrices. Part I: General Theory." *Sankhya* **A 29**, 317.

Rao, C. R., and Mitra, S. K. (1971). *Generalized Inverse of Matrices and Its Applications*. Wiley, New York.

CHAPTER 4

Theory of Testing Statistical Hypotheses

4.1 INTRODUCTION

In this chapter we introduce the basic ideas behind testing statistical hypotheses, and describe the fundamental lemma of testing statistical hypotheses due to Neyman and Pearson. The main aim of this chapter is to bring out the duality results related to the Neyman-Pearson problem. In fact, we consider a certain generalization of the Neyman-Pearson problem and prove the duality results. Before approaching the generalized problem, we consider the finite-sample-space case in great detail.

4.2 THE PROBLEM OF TESTING STATISTICAL HYPOTHESES

Let S denote the sample space of all outcomes of an experiment, and let x denote an arbitrary element of S. Treat x as a realization of a random variable X. Let H_0 be a *hypothesis* (a conjecture) to be called *null hypothesis*, which specifies partly or completely the probability distribution of X.

The problem of testing H_0 is to decide, on the basis of an observed x, whether H_0 is true or not.

EXAMPLE 4.2.1 H_0 might be the hypothesis that a 6-faced die is unbiased. We have to test this hypothesis on the basis of r_i, $i=1,\ldots,6$, where r_i is the number of times i appeared on the top face in n independent trials of throwing the die, where $n=\Sigma_{i=1}^{6} r_i$. The null hypothesis here completely specifies the probability distribution of $\mathbf{r}=(r_1,\ldots,r_6)$. We call such a hypothesis *simple*.

EXAMPLE 4.2.2 H_0 might be the hypothesis that the diameters of a certain component produced by a machine are distributed normally with unknown mean and variance. In this case the null hypothesis specifies only the type of probability distribution and not the two parameters involved. That is, the null hypothesis simply says that the probability distribution of the diameters belongs to a certain class of probability distributions. We call such a hypothesis a *composite hypothesis*. We wish to test this hypothesis on the basis of a sample of size n drawn from the production of the machine.

From these examples we can state in general:

A *statistical hypothesis* is an assertion about the distribution of one or more random variables. If the statistical hypothesis completely specifies the distribution it is called a *simple* hypothesis; otherwise it is called a *composite* hypothesis.

A *test* of a statistical hypothesis is a procedure that, based on the observed data, leads to a decision to *reject* H_0, the hypothesis under consideration, or not to reject H_0. There are two types of errors involved here. Rejecting H_0 when it is true is called Type I error; not rejecting H_0 when it is not true is called Type II error.

Any test partitions the sample space S into two regions, W and $S-W$, such that whenever $x \in W$ we reject the null hypothesis H_0 and whenever $x \in S-W$ we do not reject H_0. The region W associated with a test is called the *critical region* of the test.

Generally, we are interested in tests which simply specify the probability of rejecting H_0, on the basis of the observed data. Such tests are called *randomized* tests. When S is divided into two regions, W and $S-W$, we have the probabilities 1 and 0, respectively, for rejecting or not rejecting the hypothesis H_0. In the randomized case in general we have three regions, $W1$, $W2$, and $S-W1-W2$, with $W1 \cap W2 = \varnothing$. We reject H_0 if $x \in W1$; we do not reject H_0 if $x \in W2$; and we reject H_0 with probability $0 \leqslant \phi(x) \leqslant 1$ when $x \in S-W1-W2$.

For a simple hypothesis H_0, we denote the probability of committing Type I error by $P(W|H_0)$, called the *level of significance*. The level of significance for a composite hypothesis H_0, that is, H_0 is a class of simple hypotheses, is given by

$$\alpha = \operatorname*{Sup}_{h \in H_0} P(W|h).$$

The probability of committing Type II error for a particular hypothesis $h \neq H_0$ ($h \notin H_0$ in the case of a composite H_0) is given by

$$P(S-W|h) = \beta(h).$$

We call the function $\nu(h)=[1-\beta(h)]$, defined over all $h\notin H_0$, *the power function* of the test. When $h\notin H_0$, h is called an *alternate* hypothesis.

EXAMPLE 4.2.3 Consider the random variable X which has a binomial distribution, with $n=5$ and $p=\theta$. That is, $P(X=r)=C_r^n\theta^r(1-\theta)^{n-r}$, $r=0,1,\ldots,5$, $0\leqslant\theta\leqslant 1$.

Let H_0: $\theta=\frac{1}{2}$. The sample space $S=\{0,1,2,3,4,5\}$.

Suppose we consider the test: Reject H_0 if the observed value r is greater than 3; otherwise do not reject H_0—that is, $W=\{4,5\}$, the critical region of the test. The level of significance of this test is

$$\begin{aligned}P(W\mid H_0)&=P\left(\{4,5\}\mid\theta=\tfrac{1}{2}\right)\\&=C_4^5\cdot\left(\tfrac{1}{2}\right)^4\left(1-\tfrac{1}{2}\right)^1+C_5^5\cdot\left(\tfrac{1}{2}\right)^5\left(1-\tfrac{1}{2}\right)^0\\&=\tfrac{6}{32}.\end{aligned}$$

Consider H_1: $\theta=\frac{3}{4}$, which is an alternate hypothesis. Corresponding to this hypothesis,

$$\begin{aligned}P(S-W\mid H_1)&=P\left[\{0,1,2,3\}\mid\theta=\tfrac{3}{4}\right]\\&=\sum_{r=0}^{3}C_r^5\cdot\left(\tfrac{3}{4}\right)^r\cdot\left(\tfrac{1}{4}\right)^{5-r}=\tfrac{376}{1024}.\end{aligned}$$

Therefore $\beta(H_1)=\frac{376}{1024}$, $\nu(H_1)=1-\frac{376}{1024}=\frac{648}{1024}$.

Suppose instead we consider the test: Reject H_0 if the observed value r is greater than 4; otherwise do not reject H_0—that is, $W=\{5\}$. Then we have

$$P\left(\{5\}\mid\theta=\tfrac{1}{2}\right)=\tfrac{1}{32},\quad \nu(H_1)=1-\tfrac{781}{1024}=\tfrac{243}{1024},$$

and

$$P\left(\{0,1,2,3,4\}\mid\theta=\tfrac{3}{4}\right)=\tfrac{781}{1024}.$$

Notice that even though the probability of rejecting H_0 when it is true is reduced in the later test, the probability of committing Type II error, that of assuming H_1 is true, has increased. Therefore, in general, it may not be possible to minimize both the errors simultaneously, while testing. So naturally we try for tests that keep the probability of Type I error at a certain level, and minimize the Type II error, or maximize the power function. This brings us to the *Neyman-Pearson problem*.

The problem of determining a critical region such that for a given level of significance, the second kind of error is minimized or equivalently the power function is maximized is known as the Neyman-Pearson problem. We restrict our attention to a simple H_0 and simple H_1 alternate hypothesis in the beginning. Neyman and Pearson also established a lemma which is of fundamental importance in the theory of testing statistical hypotheses. The Neyman-Pearson problem can be stated as:

$$\underset{W \subset S}{\text{Maximize}} \qquad 1 - P(S - W \mid H_1)$$

$$\text{subject to} \qquad P(W \mid H_0) = \alpha(\text{a given value}). \tag{4.2.1}$$

RESULT 4.2.1 W^* is optimal for Problem 4.2.1 if there exists a real number k such that

$$W^* = \{x \mid P(x \mid H_1) \geqslant kP(x \mid H_0)\} \tag{4.2.2}$$

and

$$P(W^* \mid H_0) = \alpha.$$

Before we give a proof of this result, which is the celebrated Neyman-Pearson fundamental lemma, we consider the Neyman-Pearson problem when S is finite and H_0, H_1 are simple.

We approach this problem from the mathematical-programming point of view. We go on to consider a generalized version of the Neyman-Pearson problem, through duality theory, and obtain results that include a generalized version of the Neyman-Pearson fundamental lemma.

4.3 SIMPLE H_0 AGAINST SIMPLE H_1 WITH FINITE SAMPLE SPACE

Let $S = \{x_1, x_2, \ldots, x_s\}$. Let $I^* = \{1, 2, \ldots, s\}$ be the set of all indices of $x \in S$.

Let ϕ_i be the indicator of any $W \subset S$, that is,

$$\phi_i = \begin{cases} 1 & \text{if } i \in W \\ 0 & \text{otherwise.} \end{cases}$$

Let $\boldsymbol{\phi} = (\phi_1, \ldots, \phi_s)'$.

Let α be given, $0 \leqslant \alpha \leqslant 1$. Let P_{0i} denote the probability that the realization is x_i, given that H_0 is true. Similarly P_{1i} is defined. Now the Neyman-Pearson problem is equivalent to Problem 4.3.1.

Problem 4.3.1

$$\text{Maximize} \quad 1 - \sum P_{1i}(1-\phi_i)$$

$$\text{subject to} \quad \sum P_{0i}\phi_i = \alpha$$

$$\phi_i = 0 \text{ or } 1 \qquad \text{for all} \quad i \in I^*$$

where the summation is over all $i \in I^*$.

Problem 4.3.1 can readily be seen as a linear-programming problem with the additional restriction that ϕ_i is 0 or 1. Such problems are known as 0–1 integer programming problems. Especially when we have only one constraint, as we have in Problem 4.3.1, we call such problems 0–1 knapsack problems.

REMARK 4.3.1 Notice that the objective function is equivalent to $\Sigma P_{1i}\phi_i$. Hereafter we consider Problem 4.3.1 with $\Sigma P_{1i}\phi_i$ as the objective function. In general, there may not exist any $\boldsymbol{\phi}$ satisfying $\Sigma P_{0i}\phi_i = \alpha$, for a given α.

Consider the Problem 4.3.1 without the integer restriction on $\boldsymbol{\phi}$—that is, $0 \leqslant \phi_i \leqslant 1$; $i \in I^*$. We can state such a problem, after including the upper bound on ϕ_i and nonnegativity restrictions on ϕ_i as constraints, in the matrix notation as follows:

Problem 4.3.2

$$\text{Maximize} \qquad \mathbf{P}_1'\boldsymbol{\phi}$$

$$\text{subject to} \begin{pmatrix} \mathbf{P}_0 \\ \mathbf{I} \\ -\mathbf{I} \end{pmatrix} \boldsymbol{\phi} \begin{matrix} = \alpha \\ \leqslant \mathbf{e} \\ \leqslant \mathbf{0} \end{matrix}$$

$$\boldsymbol{\phi} \quad \text{unrestricted in sign.}$$

We have $\boldsymbol{\phi}$ unrestricted in sign in Problem 4.3.2 as there is no explicit restriction on $\boldsymbol{\phi}$ to be nonnegative, other than what the constraints implicitly impose on $\boldsymbol{\phi}$. Now, the dual of this problem is Problem 4.3.3.

Problem 4.3.3

$$\text{Minimize} \qquad \alpha k + \mathbf{e}'\boldsymbol{\nu}_1$$

$$\text{subject to} \qquad [\mathbf{P}_0, \mathbf{I}, -\mathbf{I}]\begin{bmatrix} k \\ \boldsymbol{\nu}_1 \\ \boldsymbol{\nu}_2 \end{bmatrix} = \mathbf{P}_1$$

$$k \text{ unrestricted in sign}, \qquad \boldsymbol{\nu}_1, \boldsymbol{\nu}_2 \geqslant \mathbf{0}.$$

From the duality theory discussed in Section 2.11, we have two possibilities for the problems above:

(1) if both the problems have feasible solutions, then both the problems have optimal solutions.

(2) If $\boldsymbol{\phi}$ is feasible for Problem 4.3.2 and $(k, \boldsymbol{\nu}_1, \boldsymbol{\nu}_2)'$ feasible for Problem 4.3.3, and we have the complementary slackness conditions satisfied—that is, whenever

$$0 < \phi_i < 1 \Rightarrow \nu_{1i} = \nu_{2i} = 0$$

$$\phi_i = 1 \Rightarrow \nu_{2i} = 0$$

$$\phi_i = 0 \Rightarrow \nu_{1i} = 0$$

and whenever

$$\nu_{1i} > 0 \Rightarrow \phi_i = 1$$

$$\nu_{2i} > 0 \Rightarrow \phi_i = 0$$

then $\boldsymbol{\phi}$ is optimal for Problem 4.3.2, and $(k, \boldsymbol{\nu}_1, \boldsymbol{\nu}_2)'$ is optimal for Problem 4.3.3.

Conversely, complementary slackness conditions hold whenever $\boldsymbol{\phi}$ is optimal for Problem 4.3.2 and $(k, \boldsymbol{\nu}_1, \boldsymbol{\nu}_2)'$ is optimal for Problem 4.3.3.

REMARK 4.3.2 Problem 4.3.2 corresponds to the Neyman-Pearson problem, when the test is a randomized test. ϕ_i gives the probability of rejecting H_0 when x_i is observed.

So we prove the following results for a randomized test and consider the nonrandomized problem (4.3.1) later.

RESULT 4.3.1 There exists a real k such that, defining

$$\phi_i = \begin{cases} 1 & \text{when} \quad p_{1i} > kp_{0i} \\ 0 & \text{when} \quad p_{1i} < kp_{0i} \\ 0 \leqslant \phi_i \leqslant 1 & \text{when} \quad p_{1i} = kp_{0i} \end{cases}$$

yields an optimal $\boldsymbol{\phi}$ to Problem 4.3.2.

Proof Notice that Problems 4.3.2 and 4.3.3 have feasible solutions. $\phi_i = \alpha$, for all i, is feasible for Problem 4.3.2, with the further implication that $\sum p_{1i}\phi_i \geqslant \alpha$ for any feasible $\boldsymbol{\phi}$. For any real k defining

$$\nu_{1i} = \begin{cases} p_{1i} - kp_{0i} & \text{if} \quad p_{1i} - kp_{0i} > 0 \\ 0 & \text{otherwise} \end{cases}$$

and

$$\nu_{2i} = \begin{cases} kp_{0i} - p_{1i} & \text{if} \quad p_{1i} - kp_{0i} < 0 \\ 0 & \text{otherwise} \end{cases}$$

provides $(k, \boldsymbol{\nu}_1, \boldsymbol{\nu}_2)$ feasible for Problem 4.3.3. Hence both the problems have optimal solutions. Consider an optimal $(k, \boldsymbol{\nu}_1, \boldsymbol{\nu}_2)$ for Problem 4.3.3. In any optimal solution to Problem 4.3.2,

$$\nu_{1i} > 0 \Rightarrow \phi_i = 1$$

$$\nu_{2i} > 0 \Rightarrow \phi_i = 0$$

and

$$\nu_{1i} = \nu_{2i} = 0 \Rightarrow 0 \leqslant \phi_i \leqslant 1,$$

and these ϕ_i's are chosen so as to achieve

$$\sum p_{0i}\phi_i = \alpha.$$

Thus the existence and sufficiency of such a k is established. We prove the necessity of such a k in the next result.

RESULT 4.3.2 If $\boldsymbol{\phi}$ is optimal for Problem 4.3.2, then there exists a k such that

$$\phi_i = 1 \Rightarrow p_{1i} \geqslant kp_{0i}$$

$$\phi_i = 0 \Rightarrow p_{1i} \leqslant kp_{0i}$$

$$0 < \phi_i < 1 \Rightarrow p_{1i} = kp_{0i}.$$

Proof We use the complementary slackness property again here. In any optimal solution to Problem 4.3.3,

$$\phi_i = 1 \Rightarrow \nu_{2i} = 0$$

$$\Rightarrow kp_{0i} + \nu_{1i} = p_{1i}$$

$$\Rightarrow kp_{0i} \leqslant p_{1i} \qquad \text{as} \quad \nu_{1i} \geqslant 0.$$

Similarly,

$$\phi_i = 0 \Rightarrow kp_{0i} \geqslant p_{i1}, \qquad 0 < \phi_i < 1 \Rightarrow \nu_{1i} = \nu_{2i} = 0$$

or $kp_{0i} = p_{1i}$ as desired.

REMARK 4.3.3 If we use a strict complementary slackness property we can also show that there exist a $\boldsymbol{\phi}$, k such that

$$\phi_i = 1 \Leftrightarrow p_{1i} > kp_{0i}$$

$$\phi_i = 0 \Leftrightarrow p_{1i} < kp_{0i}$$

and

$$0 < \phi_i < 1 \Leftrightarrow p_{1i} = kp_{0i}$$

$\boldsymbol{\phi}$ is optimal for Problem 4.3.2 and k is optimal for (4.3.3).

Here we have seen the necessity, sufficiency, and existence of such a k, for randomized tests, through linear-programming duality theory.

Now for the nonrandomized case. We have the following result, for Problem 4.3.1.

RESULT 4.3.3 If there exists a k such that k divides S into two regions,

$$W = \{x_i \mid p_{1i} \geqslant kp_{0i}\}$$

$$S - W = \{x_i \mid p_{1i} \leqslant kp_{0i}\}$$

and $\boldsymbol{\phi}$, the indicator of W, satisfies, $\sum P_{0i}\phi_i = \alpha$, then $\boldsymbol{\phi}$ is optimal for Problem 4.3.1.

Proof Corresponding to a k as given in the result, we have a feasible $(k, \boldsymbol{\nu}_1, \boldsymbol{\nu}_2)$ for Problem 4.3.3.

$$\phi_i = 1 \quad \text{when} \quad p_{1i} \geqslant kp_{0i} \Rightarrow \nu_{2i} = 0, \qquad \nu_{1i} \geqslant 0$$

$$\phi_i = 0 \quad \text{when} \quad p_{1i} \leqslant kp_{0i} \Rightarrow \nu_{1i} = 0, \qquad \nu_{2i} \geqslant 0.$$

Therefore $\boldsymbol{\phi}$ and $(k, \boldsymbol{\nu}_1, \boldsymbol{\nu}_2)'$ satisfy complementary slackness conditions; also $\Sigma\, p_{0i}\phi_i = \alpha$. Therefore $\boldsymbol{\phi}$ is feasible for Problem 4.3.2. Hence $\boldsymbol{\phi}$ is optimal for Problem 4.3.2. Since ϕ_i is 0 or 1 for all i, $\boldsymbol{\phi}$ is optimal for Problem 4.3.1 as well. Hence the result.

REMARK 4.3.4 This result can be useful only when the objective function value for an optimal solution to Problem 4.3.2 is the same as the optimal objective function value of Problem 4.3.1.

Because of the 0–1 integer restriction, if the objective function corresponding to an optimal solution to Problem 4.3.1 is strictly less than that of Problem 4.3.2, we cannot find a k as stipulated in Result 4.3.3.

Such can be established by contradiction, if there exists such a k.

The following example also helps us understand this fact.

EXAMPLE 4.3.1 Let $I^* = \{1,2,3\}$. There are three possible outcomes. Let p_{0i} and p_{1i} be as given in the following table:

i	1	2	3
p_{0i}	$\frac{2}{8}$	$\frac{5}{8}$	$\frac{1}{8}$
p_{1i}	$\frac{3}{8}$	$\frac{1}{8}$	$\frac{4}{8}$
p_{1i}/p_{0i}	$\frac{3}{2}$	$\frac{1}{5}$	4

Let $\alpha = \frac{1}{4}$. Then $\phi_1 = 1$, $\phi_2 = 0$, $\phi_3 = 0$ is the only solution to $\Sigma\, p_{0i}\phi_i = \frac{1}{4}$. Hence it is also optimal. That is, $\Sigma\, p_{1i}\phi_i = \frac{3}{8}$ is the optimal objective function value for Problem 4.3.2. We reject the hypothesis H_0 if we observe $x = 1$, and do not reject otherwise.

But there is no k such that $p_{1i} \geqslant kp_{0i}$ for $i = 1$, and $p_{1i} \leqslant kp_{0i}$ for $i = 2, 3$, for these inequalities imply $k \leqslant \frac{3}{2} = p_{11}/p_{01}$ as well as $k \geqslant 4 = p_{13}/p_{03}$.

In this example the optimal value corresponding to Problem 4.3.2 is $\frac{11}{16}$ for $\phi_1 = \frac{1}{2}$, $\phi_2 = 0$, $\phi_3 = 1$, and $k = p_{11}/p_{01} = \frac{3}{2}$ satisfies the requirements of Result 4.3.1.

A Graphical Solution Procedure for the Problem

Even though the problem as stated is in the s-dimension—that is, finding ϕ_i, $i=1,2,\ldots,s$—we can solve this problem by plotting the points (p_{0i}, p_{1i}) in a two-dimensional plane, picking up those ϕ_i that are to be one and those that are to be zero and a ϕ_j such that $0<\phi_j<1$, if any such j exists, by the following simple procedure.

Rotate clockwise a ray with the origin as pivot point and p_1 axis as starting position. Points that are swept out by the ray are selected in turn until the sum $\Sigma_W p_{0i} > \alpha$; here the summation is over those points swept out by the ray. If upon selection of point (p_{0j}, p_{1j}), we have $\Sigma_W p_{0i} > \alpha$, the value of ϕ_j is chosen as $1/p_{0j}[\alpha - \Sigma_W p_{0i} + p_{0j}]$. With the exception of this j, all the points swept out have corresponding $\phi_i = 1$, and those that are not swept out have $\phi_i = 0$. This solution almost produces the 0–1 solution desired in Problem 4.3.2, with the exception of ϕ_j.

EXAMPLE 4.3.2 Let $S=\{1,2,3,4,5,6,7,8\}$. Let p_{0i}, p_{1i} be given as in the following table below. Let $\alpha = \frac{9}{32}$.

i	1	2	3	4	5	6	7	8
p_{0i}	$\frac{2}{16}$	$\frac{7}{32}$	$\frac{1}{16}$	$\frac{1}{16}$	$\frac{4}{16}$	$\frac{3}{32}$	$\frac{2}{16}$	$\frac{1}{16}$
p_{1i}	$\frac{3}{16}$	$\frac{1}{16}$	$\frac{2}{16}$	$\frac{4}{16}$	$\frac{2}{16}$	$\frac{1}{16}$	$\frac{2}{16}$	$\frac{1}{16}$
p_{1i}/p_{0i}	$\frac{3}{2}$	$\frac{2}{7}$	$\frac{2}{1}$	$\frac{4}{1}$	$\frac{2}{4}$	$\frac{2}{3}$	$\frac{2}{2}$	$\frac{1}{1}$
Rank	3	8	2	1	7	6	5	4

Figure 4.3.1 shows that as we rotate the ray, first we include Point 4, $p_{04} = \frac{1}{16}$. We proceed further and include Point 3. Now $p_{04} + p_{03} = \frac{1}{16} + \frac{1}{16} = \frac{2}{16}$. We proceed further and include Point 1. Now $p_{04} + p_{03} + p_{01} = \frac{4}{16}$. As we proceed further, we find the ray passing through Points 7 and 8. Assume that we include Point 8, in the swept-out set, as $p_{08} < p_{07}$. Then

$$p_{04} + p_{03} + p_{01} + p_{08} = \frac{4}{16} + \frac{1}{16} = \frac{5}{16} > \frac{9}{32}.$$

Therefore

$$\frac{1}{16} \cdot \phi_8 = \left(\frac{9}{32} - \frac{4}{16}\right) = \frac{1}{32} \qquad \text{or} \quad \phi_8 = \frac{1}{2}.$$

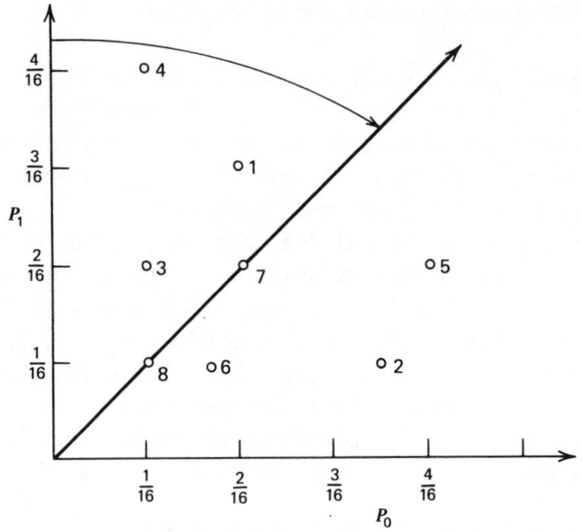

Figure 4.3.1 A graphic solution for Problem 4.3.2.

We stop. We have

$$\phi_4=\phi_3=\phi_1=1,\ \phi_8=\tfrac{1}{2} \qquad \text{and} \quad \phi_7=\phi_6=\phi_2=\phi_5=0.$$

This solution is optimal to Problem 4.3.2, with only one $0<\phi_i<1$.

REMARK 4.3.4 The optimality of a solution obtained in the procedure above can be seen by taking $k=p_{1j}/p_{0j}$, for the j which terminates the procedure. Then from Result 4.3.1, ϕ so obtained is optimal.

4.4 FINITE COMPOSITE H_0 AGAINST SIMPLE H_1 WITH FINITE SAMPLE SPACE

So far we have looked at the simple case of testing a simple H_0 against a simple H_1 when the sample space is finite, from the linear-programming point of view.

Next we consider H_0 to be a finite set of hypotheses. Let $H_0=\{h_1, h_2,\ldots, h_m\}$. Let S be finite, as earlier. Let $p_{0j}(i)$ be the probability of observing x_i, given that $h_j \in H_0$ is true. Let $p_1(i)$ be the probability of observing x_i, given that H_1 is true. Let ϕ be the indicator of any subset W of S.

Now the probability of committing Type I error when h_j is true is

$$P(W|h_j)=\sum_{i=1}^{s} p_{0j}(i)\phi_i.$$

The level of significance α is given by

$$\max_{1\leqslant j\leqslant m} \sum p_{0j}(i)\phi_i=\alpha, \qquad \text{as defined earlier.}$$

If for some ϕ, $\sum p_{0j}(i)\phi_i=\alpha$ for all $h_j\in H_0$, we call the corresponding $W\subset S$, a *similar critical region*. And ϕ is called a similar critical region test. Here, finding a most powerful test can be formulated as:

$$\begin{aligned} &\text{Maximize} && \sum_{i=1}^{s} p_1(i)\phi_i \\ &\text{subject to} && \sum_{i=1}^{s} p_{0j}(i)\phi_i=\alpha, \qquad j=1,\dots,m \\ &\phi_i \quad 0 \text{ or } 1 && \text{for all} \quad i=1,\dots,s. \end{aligned} \tag{4.4.1}$$

Now if we relax the condition that ϕ_i has to be 0 or 1, we have $0\leqslant\phi_i\leqslant 1$ and we have a *randomized* test. Let us call such a problem, Problem 4.4.2. Problem 4.4.1 can be readily identified as a bounded-variable linear-programming problem with the additional restriction that the variable must be either 0 or 1.

When we relax this 0–1 restriction, Problem 4.4.2, thus obtained, is a bounded-variable linear-programming problem. We have studied such problems in Section 2.12. So we can use the duality theory to obtain results similar to (4.3.2), (4.3.3), and so on.

Notice that in general there may not exist any ϕ such that similar critical-region restrictions are satisfied. If there exists such a ϕ we are trying to find the most powerful test in these problems.

Let $0<\alpha<1$. Suppose we have

$$\sum_{i=1}^{s} p_{0j}(i)\phi_i\leqslant\alpha, \qquad j=1,\dots,m$$

and

$$\sum_{i=1}^{s} p_1(i)\phi_i>\alpha. \tag{4.4.2}$$

We call any such ϕ an *unbiased* test. Further, if it maximizes $\Sigma p_{1i}\phi_i$, we that ϕ is a *most powerful unbiased test*.

In addition, if the alternative hypothesis is not simple, that is, we have several alternate hypotheses, we define the concept of uniformly most powerful test.

If for every $h \in H_1$, ϕ is such that $\Sigma_{i=1}^{s} p_{1h}(i)\phi_i$ is maximized and ϕ satisfies

$$\sum_{i=1}^{s} p_{0j}(i)\phi_i \leqslant \alpha, \qquad j=1,\ldots,m \tag{4.4.3}$$

$$0 \leqslant \phi_i \leqslant 1, \qquad i=1,\ldots,s.$$

Then we say ϕ is a *uniformly most powerful* test. In other words, ϕ remains most powerful for each $h \in H_1$.

Suppose H_1 contains only a finite number of alternatives. Let

$$H_1 = \{h_{m+1},\ldots,h_{m+q}\}.$$

Let $p_{1r}(i)$ be the probability of observing x_i given that h_{m+r} is true.

Then we have the objective of finding ϕ^* such that ϕ^* satisfies (4.4.3) and

$$\sum_{i=1}^{s} p_{1r}(i)\phi_i^* \geqslant \sum_{i=1}^{s} p_{1r}(i)\phi_i \qquad \text{for} \quad r=1,2,\ldots,q \tag{4.4.4}$$

and for all ϕ satisfying (4.4.3).

In general there may not exist a uniformly most powerful test or we may not find a ϕ^* feasible such that the above inequalities (4.4.4) are satisfied for all feasible ϕ. Hence, we search for something weaker. We try to maximize over the feasible region the minimum of the power functions corresponding to the alternative hypothesis in H_1. Such a test will be called a *maximin* test.

Problem 4.4.3

$$\text{Maximize} \qquad \left[\min_{1 \leqslant r \leqslant q} \sum_{i=1}^{s} p_{1r}(i)\phi_i\right]$$

$$\text{subject to} \qquad \sum_{i=1}^{s} p_{0j}(i)\phi_i \leqslant \alpha, \qquad j=1,\ldots,m$$

$$0 \leqslant \phi_i \leqslant 1 \qquad i=1,\ldots,s.$$

Now, the minimum of a set of linear functions is a concave function, which is piecewise linear. So the objective function is nonlinear. However, we introduce a new variable $\xi \geq 0$ and q new constraints to Problem 4.4.3, and consider Problem 4.4.4, a linear-programming problem.

Problem 4.4.4

$$\begin{aligned} &\text{Maximize} && \xi \\ &\text{subject to} && \sum_{i=1}^{s} p_{0j}(i)\phi_i \leq \alpha, && j=1,\dots,m \\ & && -\sum_{i=1}^{s} p_{1r}(i)\phi_i + \xi \leq 0, && r=1,\dots,q \\ & && \xi \geq 0, \quad 0 \leq \phi_i \leq 1, && i=1,\dots,s. \end{aligned}$$

RESULT 4.4.1 If there exists a (ξ^*, ϕ^*) optimal for Problem 4.4.4, then ϕ^* is optimal for Problem 4.4.3, in which case there exists a ξ^* such that (ξ^*, ϕ^*) is optimal for Problem 4.4.4.

Proof Suppose (ξ^*, ϕ^*) is optimal for Problem 4.4.4 and there exists a $\hat{\phi}$ feasible for Problem 4.4.3 such that

$$\min_{1 \leq r \leq q} \sum p_{1r}(i)\hat{\phi}_i > \min_{1 \leq r \leq q} \sum p_{1r}(i)\phi_i^*.$$

Then consider

$$\hat{\xi} = \min_{1 \leq r \leq q} \sum p_{1r}(i)\hat{\phi}_i \geq 0.$$

Now $(\hat{\xi}, \hat{\phi})$ is feasible for Problem 4.4.4 and $\hat{\xi} > \xi^*$ which leads to a contradiction. Hence ϕ^* must be optimal for Problem 4.4.3. Similarly if ϕ^* is optimal for Problem 4.4.3, consider (ξ^*, ϕ^*) such that $\xi^* = \min_{1 \leq r \leq q} \sum p_{1r}(i)\phi_i^*$, (ξ^*, ϕ^*), can be shown to be optimal for Problem 4.4.4.

Thus the problem of finding a maximin test is also a linear-programming problem with $m+k$ linear inequalities and $s+1$ nonnegative variables s of these bounded above by 1.

If ξ^* for Problem 4.4.4 is in addition greater than α we have an unbiased test that is also a maximin test.

We have deliberately avoided discussing nonrandomized tests for the time being.

Now consider the dual of Problem 4.4.4.

Problem 4.4.5

$$\text{Minimize} \quad \alpha \sum_{j=1}^{m} k_j + \sum_{i=1}^{s} \nu_{1i}$$

$$\text{subject to} \quad \sum_{j=1}^{m} p_{0j}(i)k_j - \sum_{r=1}^{q} p_{1r}(i)k_{m+r} + \nu_{1i} - \nu_{2i} = 0 \qquad i=1,\ldots,s.$$

$$k_{m+r} + \cdots + k_{m+q} = 1$$

$$k_j \geq 0, j=1,\ldots,m, m+1,\ldots,m+q$$

$$\nu_{1i}, \nu_{2i} \geq 0, i=1,\ldots,s.$$

REMARK 4.4.2 Notice that Problem 4.4.4 has a feasible solution. Also the objective function ξ is bounded above by 1. Hence we have an optimal solution to Problem 4.4.4. From duality theory, Problem (4.4.5) also has optimal solution. Also notice that given $k_j, j=1,\ldots,m+q$ nonnegative and $\sum_{r=1}^{q} k_{m+r} = 1$, we can find ν_{1i} and $\nu_{2i} \geq 0$, as follows:

$$\nu_{1i} = \sum_{r=1}^{q} p_{1r}(i)k_{m+r} - \sum_{j=1}^{m} p_{0j}(i)k_j$$

and

$$\nu_{2i} = 0, \text{ if } \sum_{r=1}^{q} p_{1r}(i)k_{m+r} - \sum_{j=1}^{m} p_{0j}(i)k_j \geq 0.$$

Otherwise

$$\nu_{1i} = 0$$

$$\nu_{2i} = \sum_{j=1}^{m} p_{0j}(i)k_j - \sum_{r=1}^{q} p_{1r}(i)k_{m+r}.$$

Also because of the complementary slackness property, if $(\mathbf{k}, \boldsymbol{\nu}_1, \boldsymbol{\nu}_2)$ is feasible for Problem 4.4.5 then $\mathbf{k}$ divides the sample space into three regions,

$$W1=\left\{x_i \,\middle|\, \sum_{j=1}^{m} p_{0j}(i)k_j < \sum_{r=1}^{q} p_{1r}(i)k_{m+r}\right\}$$

$$W2=\left\{x_i \,\middle|\, \sum_{j=1}^{n} p_{0j}(i)k_j > \sum_{r=1}^{q} p_{1r}(i)k_{m+r}\right\}$$

and

$$S-W1-W2=\left\{x_i \,\middle|\, \sum_{j=1}^{m} p_{0j}(i)k_j = \sum_{r=1}^{q} p_{1r}(i)k_{m+r}\right\}.$$

Suppose ϕ is defined as given below,

$$\phi_i=\begin{cases} 1 & \text{if } x_i \in W1 \\ 0 & \text{if } x_i \in W2 \\ 0 \leqslant \phi_i \leqslant 1 & \text{if } x_i \in S-W1-W2. \end{cases}$$

If ϕ is feasible for Problem 4.4.5, with $\xi \geqslant 0$; then (ξ, ϕ) is optimal for the Problem 4.4.4 and hence ϕ optimal for Problem 4.4.3. Further, if $k_1,\ldots, k_m$ are positive then we have a maximin-similar-region test. In addition if $\xi > \alpha$, we have a maximin-similar-region test that is unbiased. Remark 4.4.2 is summarized in the following results.

RESULT 4.4.2 Given H_0 consisting of m hypotheses, H_1 consisting of q hypotheses, and S consisting of s elements, if there exists a nonnegative vector $\mathbf{k}$ of $m+q$ components such that

(1) $$\sum_{r=1}^{q} k_{m+r} = 1$$

(2) $$W1=\left\{x_i \,\middle|\, \sum_{j=1}^{m} p_{0j}(i)k_j < \sum_{r=1}^{q} p_{1r}(i)k_{m+r}\right\}$$

$$W2=\left\{x_i \,\middle|\, \sum_{j=1}^{m} p_{0j}(i)k_j > \sum_{r=1}^{q} p_{1r}(i)k_{m+r}\right\}$$

and

(3)
$$\phi_i = \begin{cases} 1, & x_i \in W1 \\ 0 & x_i \in W2 \\ 0 \leqslant \phi_i \leqslant 1 & \text{if } x_i \in S - W1 - W2, \end{cases}$$

such that ϕ satisfy $\Sigma_{i=1}^{r} p_{0j}(i)\phi_i \leqslant \alpha$, $j=1,\ldots,m$, then ϕ is optimal for Problem 4.4.3.

RESULT 4.4.3 If there exists a nonnegative **k** such that

(1)
$$\sum_{r=1}^{q} k_{m+r} = 1$$

(2)
$$k_1,\ldots,k_m > 0$$

(3) ϕ_i is as defined in Result 4.4.2

and ϕ_i satisfies $\Sigma_{i=1}^{s} p_{0j}(i)\phi_i \leqslant \alpha, j=1,\ldots,m$, then, in fact, $\Sigma_{i=1}^{s} p_{0j}(i)\phi_i = \alpha$, $j=1,\ldots,m$, and ϕ is optimal for Problem 4.4.3. That is, ϕ is a maximin similar-region test.

This result follows from the fact when **k** provides a feasible solution to Problem 4.4.5 and complementary slackness property holds, for some ϕ such that (ξ,ϕ) is feasible for Problem 4.4.4, then the inequalities corresponding to $k_1,\ldots,k_m$ will hold as equalities if $k_1,\ldots,k_m$ are all positive. And (ξ,ϕ) is optimal for Problem 4.4.4 or ϕ is optimal for Problem 4.4.3.

So far we have considered the case when S, H_0, and H_1 are finite. The discussions provide a basis for an understanding of the duality results related to the Neyman-Pearson problem. We have considered the realization of a single random variable. The results can be extended when **X** is a random vector.

In the next section we relax the finiteness restriction on S and prove the Neyman-Pearson lemma. We also give a generalized version of the lemma, with a view to linking the subsequent discussion of the problem in Section 4.6 from the mathematical-programming point of view.

4.5 PROOF OF THE NEYMAN-PEARSON LEMMA AND THE GENERALIZED VERSION OF THE LEMMA

Here is a proof of the Neyman-Pearson lemma (Result 4.2.1), when H_0 and H_1 are simple. Let S be the set of all realizations of a random variable X. S

is not restricted to be finite in this proof. Let $P(x|H_0)$ and $P(x|H_1)$ be the densities at x under H_0 and H_1, respectively, relative to a measure ν. In Problem 4.2.1 we have

$$P(W|H_0)=\int_W P(x|H_0)\,d\nu$$

and

$$P(W|H_1)=\int_W P(x|H_1)\,d\nu$$

We wish to prove that if there exists a real number k such that

$$W^*=\{x|P(x|H_1)\geqslant kP(x|H_0)\}$$

further W^* is such that $\int_{W^*}P(x|H_0)\,d\nu=\alpha$, then W^* is optimal for Problem 4.2.1.

Proof Suppose there exists a real k such that for the region W^*

$$P(x|H_1)\geqslant kP(x|H_0)$$

and outside W^*, $P(x|H_1)\leqslant kP(x|H_0)$ and $\int_{W^*}P(x|H_0)\,d\nu=\alpha$. For any other region $W\subset S$ such that

$$\int_W P(x|H_0)\,d\nu=\alpha$$

we shall show that $\int_{W^*}P(x|H_1)\,d\nu\geqslant\int_W P(x|H_1)\,d\nu$. Let

$$W\cap W^*=\hat{W}.$$

Then,

$$\int_{W-\hat{W}}P(x|H_0)\,d\nu=\int_{W^*-\hat{W}}P(x|H_0)\,d\nu. \tag{4.5.1}$$

Consider

$$\int_{W^*} P(x|H_1)\,dv - \int_{W} P(x|H_1)\,dv$$

$$= \int_{W^*-\hat{W}} P(x|H_1)\,dv - \int_{W-\hat{W}} P(x|H_1)\,dv$$

$$\geqslant \int_{W^*-\hat{W}} P(x|H_0)\,dv - \int_{W-\hat{W}} kP(x|H_0)\,dv$$

$$=0, \qquad \text{from}(4.5.1).$$

Hence the result.

REMARK 4.5.1 Here we have shown the sufficient condition for a test to be most powerful. The existence and necessity of such a test also can be proved. In fact for a generalization of the Neyman-Pearson lemma, we have such a result. We state this result without proof, as the problem will be discussed later in Section 4.6.

RESULT 4.5.1 Let $f_0, f_1, \ldots, f_m$, be given bounded, Lebesgue-measurable functions, with domain R^n. Let $c_1, \ldots, c_m$ be given real numbers. Suppose ϕ is a Lebesgue-measurable function such that $0 \leqslant \phi(z) \leqslant 1$ for all $z \in R^n$ satisfying

$$\int_{R^n} f_i(z)\phi(z)\,dz = c_i, \qquad i=1,\ldots,m \tag{4.5.2}$$

Let Φ be the class of all such ϕ for which (4.5.2) holds.

(1) There exists a $\phi \in \Phi$ which maximizes

$$\int_{R^n} f_0(z)\phi(z)\,dz \tag{4.5.3}$$

(2) A sufficient condition for a member of Φ to maximize (4.5.3) is the existence of real numbers $k_1, \ldots, k_m$ such that

$$\phi(z) = \begin{cases} 1 & \text{when } f_0(z) > \sum_{i=1}^{m} k_i f_i(z) \\ 0 & \text{otherwise} \end{cases} \tag{4.5.4}$$

(3) If a member of Φ satisfies (4.5.4) with $k_1,\ldots,k_m \geqslant 0$ then it maximizes (4.5.3) among all Lebesgue-measurable functions, satisfying

$$\int_{R^n} f_i(z)\phi(z)\,dz \leqslant c_i \qquad i=1,\ldots,m \tag{4.5.5}$$

and $0 \leqslant \phi(z) \leqslant 1$.

(4) Let $\mathfrak{B}=\{\mathbf{d}|\mathbf{d}\in R^m,\ d_i=\int_{R^n} f_i(z)\phi(z)\,dz$ for some Lebesgue-measurable function ϕ such that $0\leqslant \phi(z)\leqslant 1\}$. If $\mathbf{C}=(c_1,\ldots,c_m)$ is an interior point of $\mathfrak{B}$, then a necessary condition for a member of Φ to maximize (4.5.3) is that (4.5.4) holds almost everywhere on R^n.

REMARK 4.5.2 The integrals involved are Lebesgue integrals. The necessity part of the result requires that $\mathbf{C}$ be an interior point of $\mathfrak{B}$. However, this requirement could be relaxed. Also this result could be proved in a more general measure theoretic setup. However, we restrict our attention to this generalization of the Neyman-Pearson lemma, so that elaborate measure theoretical knowledge is not required for understanding what follows. (Knowledge of Lebesque integrals is assumed.)

In the following section, we consider the corresponding Generalized Neyman-Pearson Problem and prove duality results through which a proof of Result 4.5.2 can be obtained.

4.6 DUALITY IN GENERALIZED NEYMAN-PEARSON PROBLEM

A generalized version of the Neyman-Pearson problem can be stated as follows:

Problem 4.6.1

$$\text{Maximize} \quad g(\phi)=\int_{R^n} f_0(z)\phi(z)\,dz$$

$$\text{subject to} \quad \int_{R^n} f_i(z)\phi(z)\,dz \leqslant c_i, \qquad i=1,\ldots,m \tag{4.6.1}$$

$$l(z)\leqslant \phi(z)\leqslant u(z) \qquad \text{for all} \quad z\in R^n. \tag{4.6.2}$$

Let this problem be called the *primal* problem in the ensuing discussion. The functions $f_0, f_1, \ldots, f_m, l, u$ are given bounded, Lebesgue-measurable functions, with domain R^n; c_i's are given real numbers. The integrals involved are Lebesque integrals. Assume that $l(z) \leqq u(z)$ for all $z \in R^n$. Any measurable function ϕ satisfying (4.6.1) and (4.6.2) will be a *feasible solution* to the primal problem. Any feasible solution that maximizes g will be called an *optimal solution* to the primal problem. Here we have $l(z)$ and $u(z)$ in place of 0 and 1 in Result 4.5.2. Also the equalities in Result 4.5.2 are now replaced by inequalities. Later we shall consider this problem with equality constraints.

Given the primal problem, consider this problem:

Problem 4.6.2

$$\text{Minimize} \quad D(\mathbf{k}, v_1, v_2) = \sum_{i=1}^{m} c_i k_i - \int_{R^n} l(z) v_1(z)\, dz + \int_{R^n} u(z) v_2(z)\, dz \tag{4.6.3}$$

$$\text{subject to} \quad \sum_{i=1}^{m} k_i f_i(z) - v_1(z) + v_2(z) = f_0(z), \qquad \text{for all} \quad z \in R^n \tag{4.6.4}$$

$$\mathbf{k} = (k_1, \ldots, k_m) \geqslant \mathbf{0}. \tag{4.6.5}$$

$$v_1(z) \geqslant 0, \qquad \text{for all} \quad z \in R^n \tag{4.6.6}$$

$$v_2(z) \geqslant 0, \qquad \text{for all} \quad z \in R^n. \tag{4.6.7}$$

Let this problem be called the *dual* problem. A feasible solution to this problem will be denoted by $(\mathbf{k}, v_1, v_2)$ and will satisfy (4.6.4) through (4.6.7), where v_1, v_2 are bounded measurable functions with domain R^n. An optimal solution to this problem will be a feasible solution that minimizes the objective function of Problem 4.6.2.

Now notice the similarity between the primal and dual problems, and the dual linear-programming problems (Problems 4.3.2 and 4.3.3) we considered earlier. However, now we have infinitely many constraints in the dual problem. Also we are now dealing with feasible solutions to the dual problem which involve a vector and two measurable functions, and we have m sets of inequalities in (4.6.1) instead of the one equality. However, we can state a third problem in which we are required to find an optimal vector to solve the problem, and establish the equivalence between these two problems.

Let for any $\mathbf{k} \in R^m$, $\mathbf{k} \geqslant \mathbf{0}$.

$$S^-(\mathbf{k}) = \left\{ z \in R^n \,|\, f_0(z) - \sum_{i=1}^{m} k_i f_i(z) \leqslant 0 \right\} \tag{4.6.8}$$

$$S^+(\mathbf{k}) = \left\{ z \in R^n \,|\, f_0(z) - \sum_{i=1}^{m} k_i f_i(z) > 0 \right\}$$

$$= R^n - S^-(\mathbf{k}). \tag{4.6.9}$$

Problem 4.6.3

$$\text{Minimize} \quad L(\mathbf{k}) = \left[\sum_{i=1}^{m} k_i c_i + \int_{S^-(\mathbf{k})} l(z) \left[f_0(z) - \sum_{i=1}^{m} k_i f_i(z) \right] dz \right.$$

$$\left. + \int_{S^+(\mathbf{k})} u(z) \left[f_0(z) - \sum_{i=1}^{m} k_i f_i(z) \right] dz \right] \quad \text{over} \quad \mathbf{k} \geqslant \mathbf{0}.$$

This problem will be referred to as the *Lagrangian* problem. To establish the connections between Problem 4.6.3 and the primal problem, we require a certain restriction, called Slater's constraints qualification introduced in Section 2.19.

DEFINITION 4.6.1 The primal problem is said to satisfy *Slater's constraint qualification* if there exists a feasible solution to the primal problem, such that all the $\leqslant$ inequalities in (4.6.1), hold as strict inequalities.

We assume this constraints qualification is satisfied, for the primal problem.

Equivalence of the Dual Problem and the Lagrangian Problem

We prove the following results to establish the equivalence.

RESULT 4.6.1 For any $\mathbf{k} \geqslant \mathbf{0}$, $\mathbf{k} \in R^m$, there exist functions v_1 and v_2 such that $(\mathbf{k}, v_1, v_2)$ is a feasible solution to the dual problem and $L(\mathbf{k}) = D(\mathbf{k}, v_1, v_2)$.

Proof Let $\mathbf{k} \geqslant \mathbf{0}$. Define v_1 and v_2 as follows:

$$v_1(z) = \begin{cases} \sum_{i=1}^{m} k_i f_i(z) - f_0(z), & \text{for all } z \in S^-(\mathbf{k}) \\ 0 & \text{otherwise.} \end{cases}$$

$$v_2(z) = \begin{cases} -\sum_{i=1}^{m} k_i f_i(z) + f_0(z) & \text{for all } z \in S^+(\mathbf{k}) \\ 0 & \text{otherwise.} \end{cases}$$

Then v_1, v_2 are nonnegative for all z and they satisfy (4.6.4). By definition, v_1, v_2 are bounded and measurable, and so $(\mathbf{k}, v_1, v_2)$ is a feasible solution to the dual problem.

Moreover,

$$L(\mathbf{k}) = \sum_{i=1}^{m} c_i k_i - \int_{S^-(\mathbf{k})} l(z) v_1(z)\, dz + \int_{S^+(\mathbf{k})} u(z) v_2(z)\, dz$$

$$- \int_{S^+(\mathbf{k})} l(z) v_1(z)\, dz + \int_{S^-(\mathbf{k})} u(z) v_2(z)\, dz$$

by definition of v_1, v_2. Therefore,

$$L(\mathbf{k}) = \sum_{i=1}^{m} c_i k_i - \int_{R^n} l(z) v_1(z)\, dz + \int_{R^n} u(z) v_2(z)\, dz.$$

$$= D(\mathbf{k}, v_1, v_2).$$

Hence the result.

RESULT 4.6.2 If $(\mathbf{k}, v_1, v_2)$ is a feasible solution to the dual problem such that

$$v_1(z) \cdot v_2(z) = 0 \qquad \text{almost everywhere (a.e.) on } R^n \tag{4.6.10}$$

then

$$L(\mathbf{k}) = D(\mathbf{k}, v_1, v_2).$$

Proof Given any feasible solution $(\mathbf{k}, v_1, v_2)$ satisfying the conditions stated in the result, let

$$W^-(\mathbf{k}) = \left\{ z \in R^n : f_0(z) - \sum_{i=1}^{m} k_i f_i(z) < 0 \right\}$$

$$W^0(\mathbf{k}) = \left\{ z \in R^n : f_0(z) = \sum_{i=1}^{m} k_i f_i(z) \right\}$$

$$= S^-(k) - W^-(\mathbf{k}).$$

Then it follows from Condition 4.6.10 that,

$$v_1(z) = \begin{cases} \sum_{i=1}^{m} k_i f_i(z) - f_0(z), & \text{a.e. on} \quad W^-(\mathbf{k}) \\ 0 & \text{a.e. on} \quad S^+(\mathbf{k}) \end{cases}$$

and

$$v_2(z) = \begin{cases} -\sum_{i=1}^{m} k_i f_i(z) + f_0(z), & \text{a.e. on} \quad S^+(\mathbf{k}) \\ 0 & \text{a.e. on} \quad W^-(\mathbf{k}). \end{cases}$$

Thus

$$D(\mathbf{k}, v_1, v_2) = \sum_{i=1}^{m} c_i k_i - \int_{W^-(\mathbf{k})} l(z) \left[\sum_{i=1}^{m} k_i f_i(z) - f_0(z) \right] dz$$

$$- \int_{W^0(\mathbf{k})} l(z) v_1(z)\, dz + \int_{S^+(\mathbf{k})} u(z) \left[f_0(z) - \sum_{i=1}^{m} k_i f_i(z) \right] dz$$

$$+ \int_{W^0(\mathbf{k})} u(z) v_2(z)\, dz.$$

Now we have, as $(\mathbf{k}, v_1, v_2)$ is feasible and $v_1(z) \cdot v_2(z) = 0$ a.e. on R^n,

$$v_1(z) = 0, \qquad \text{a.e. on} \quad W^0(\mathbf{k})$$

$$v_2(z) = 0, \qquad \text{a.e. on} \quad W^0(\mathbf{k}).$$

So that

$$\int_{W^0(\mathbf{k})} u(z)v_2(z)\,dz=0$$

and

$$\int_{W^0(\mathbf{k})} l(z)v_1(z)\,dz=0$$

$$=-\int_{W^0(\mathbf{k})} l(z)\left[\sum_{i=1}^{m} k_i f_i(z)-f_0(z)\right]dz.$$

Thus,

$$D(\mathbf{k},v_1,v_2)=\sum_{i=1}^{m} c_i k_i+\int_{S^-(\mathbf{k})} l(z)\left[f_0(z)-\sum_{i=1}^{m} k_i f_i(z)\right]dz$$

$$+\int_{S^+(\mathbf{k})} u(z)\left[f_0(z)-\sum_{i=1}^{m} k_i f_i(z)\right]dz$$

$$=L(\mathbf{k}).$$

Hence the result.

RESULT 4.6.3 Let $(\mathbf{k},v_1,v_2)$ be any feasible solution to the dual problem, and define the function $m(z)$ on R^n by

$$m(z)=\min[v_1(z),v_2(z)].$$

If T is any measurable subset of R^n and functions $\hat{v}_1$ and $\hat{v}_2$ are defined on R^n by

$$\hat{v}_1(z)=\begin{cases} v_1(z), & \text{for all } z\in R^n-T \\ v_1(z)-m(z), & \text{for all } z\in T \end{cases} \tag{4.6.11}$$

and

$$\hat{v}_2(z)=\begin{cases} v_2(z), & \text{for all } z\in R^n-T \\ v_2(z)-m(z), & \text{for all } z\in T \end{cases} \tag{4.6.12}$$

then $(\mathbf{k}, \hat{v}_1, \hat{v}_2)$ is a feasible solution to the dual problem and

$$D(\mathbf{k}, v_1, v_2) - D(\mathbf{k}, \hat{v}_1, \hat{v}_2) = \int_T [u(z) - l(z)] m(z)\, dz.$$

Proof We can verify easily that $(\mathbf{k}, \hat{v}_1, \hat{v}_2)$ is feasible for the dual problem. Further,

$$D(\mathbf{k}, v_1, v_2) - D(\mathbf{k}, \hat{v}_1, \hat{v}_2)$$

$$= -\int_T l(z) v_1(z)\, dz + \int_T u(z) v_2(z)\, dz + \int_T l(z) \hat{v}_1(z)\, dz - \int_T u(z) \hat{v}_2(z)\, dz$$

$$= -\int_T l(z) m(z)\, dz + \int_T u(z) m(z)\, dz$$

$$= \int_T [u(z) - l(z)] m(z)\, dz.$$

Hence the result.

This result is used in the proof of the following result.

RESULT 4.6.4 If there exists an optimal solution to the dual problem, then there exists an optimal solution $(\mathbf{k}, v_1, v_2)$ to the dual problem such that

$$v_1(z) \cdot v_2(z) = 0, \qquad \text{a.e. on} \quad R^n.$$

Proof Let $(\hat{\mathbf{k}}, \hat{v}_1, \hat{v}_2)$ be an optimal solution to the dual problem. Define the following sets:

$$T_1 = \{z \in R^n \mid \hat{v}_1(z) \cdot \hat{v}_2(z) > 0, u(z) - l(z) > 0\} \qquad (4.6.13)$$

$$T_2 = \{z \in R^n \mid \hat{v}_1(z) \cdot \hat{v}_2(z) > 0, u(z) - l(z) = 0\} \qquad (4.6.14)$$

$$T_3 = \{z \in R^n \mid \hat{v}_1(z) \cdot \hat{v}_2(z) = 0, u(z) - l(z) \geqslant 0\}. \qquad (4.6.15)$$

First, we consider T_1 and show that T_1 has the measure of zero. Suppose the measure on T_1 is positive. Let

$$m(z) = \min[\hat{v}_1(z), \hat{v}_2(z)] \text{ for } z \in R^n.$$

Define v_1^*, v_2^* obtained from $\hat{v}_1$ and $\hat{v}_2$, and $m(z)$ as follows:

$$\begin{aligned} v_1^*(z)&=\hat{v}_1(z), && \text{for all} \quad z\in R^n-T_1\\ v_2^*(z)&=\hat{v}_2(z), && \text{for all} \quad z\in R^n-T_1\\ v_1^*(z)&=\hat{v}_1(z)-m(z), && \text{for all} \quad z\in T_1\\ v_2^*(z)&=\hat{v}_2(z)-m(z), && \text{for all} \quad z\in T_1. \end{aligned}$$

By Result 4.6.3, $(\hat{\mathbf{k}}, v_1^*, v_2^*)$ is a feasible solution to the dual problem and

$$D(\hat{\mathbf{k}}, \hat{v}_1, \hat{v}_2)-D(\hat{\mathbf{k}}, v_1^*, v_2^*)=\int_{T_1}[u(z)-l(z)]m(z)\,dz>0.$$

So $D(\hat{\mathbf{k}}, \hat{v}_1, \hat{v}_2)$ is greater than $D(\hat{\mathbf{k}}, v_1^*, v_2^*)$, which leads to a contradiction as $(\hat{\mathbf{k}}, \hat{v}_1, \hat{v}_2)$ is optimal for the dual problem. Thus the measure of T_1 is zero. Next, define functions v_1 and v_2 by

$$\begin{aligned} v_1(z)&=\hat{v}_1(z), && \text{for all} \quad z\in R^n-T_2\\ v_2(z)&=\hat{v}_2(z), && \text{for all} \quad z\in R^n-T_2\\ v_1(z)&=\hat{v}_1(z)-m(z), && \text{for all} \quad z\in T_2\\ v_2(z)&=\hat{v}_2(z)-m(z), && \text{for all} \quad z\in T_2. \end{aligned}$$

Again $(\hat{\mathbf{k}}, v_1, v_2)$ is a feasible solution to the dual problem. Consider

$$D(\hat{\mathbf{k}}, \hat{v}_1, \hat{v}_2)-D(\hat{\mathbf{k}}, v_1, v_2)=\int_{T_2}[u(z)-l(z)]m(z)\,dz=0$$

by definition of T_2. Thus $(\hat{\mathbf{k}}, v_1, v_2)$ is also an optimal solution to the dual problem. The way we have defined v_1, v_2,

$$v_1(z)\cdot v_2(z)=0, \qquad \text{for all} \quad z\in T_2.$$

Anyway, for all $z\in T_3$, $\hat{v}_1(z)\cdot\hat{v}_2(z)=0$, and $\hat{v}_1(z)=v_1(z)$, $\hat{v}_2(z)=v_2(z)$ by definition, for $z\in T_3$. The implication is that

$$v_1(z)\cdot v_2(z)=0, \qquad \text{for all} \quad z\in T_3.$$

Since the measure of T_1 is zero, it follows that v_1 and v_2 as defined satisfy $v_1(z)\cdot v_2(z)=0$ almost everywhere on R^n. Hence the result.

This result is analogous to the fact that v_{1i} and v_{2i} cannot both be positive in an optimal solution to Problem 4.3.3—that is, $v_{1i}\cdot v_{2i}=0$. Here we have, except on a set of measure zero, $v_1(z)\cdot v_2(z)=0$.

The next result establishes the relationship between optimal solutions the dual and the Lagrangian problems.

RESULT 4.6.5 If $(\mathbf{k}, v_1, v_2)$ is an optimal solution to the dual problem, then $\mathbf{k}$ is optimal for the Lagrangian problem and $L(\mathbf{k})=D(\mathbf{k}, v_1, v_2)$.

Proof Let $(\mathbf{k}, v_1, v_2)$ be an optimal solution to the dual problem. From Result 4.6.4 there exists $(\mathbf{k}, v_1^*, v_2^*)$ an optimal solution to the dual problem such that $v_1^*(z)\cdot v_2^*(z)=0$, a.e. on R^n. From Result 4.6.2, $L(\mathbf{k})=D(\mathbf{k}, v_1^*, v_2^*)=D(\mathbf{k}, v_1, v_2)$. Let $\hat{\mathbf{k}}$ be any feasible solution to the Lagrangian problem. From Result 4.6.1 there exist functions $\hat{v}_1, \hat{v}_2$ such that $(\hat{\mathbf{k}}, \hat{v}_1, \hat{v}_2)$ is a feasible solution to the dual problem and $L(\hat{\mathbf{k}})=D(\hat{\mathbf{k}}, \hat{v}_1, \hat{v}_2)$. Since $L(\mathbf{k})=D(\mathbf{k}, v_1, v_2)\leqslant D(\hat{\mathbf{k}}, \hat{v}_1, \hat{v}_2)$ by optimality of $(\mathbf{k}, v_1, v_2)$ to the dual problem. The implication is that $L(\mathbf{k})\leqslant L(\hat{\mathbf{k}})$ for any feasible solution to the Lagrangian problem, or that $\hat{\mathbf{k}}$ is optimal for the Lagrangian problem. Hence the result.

The following result is needed to establish the existence of an optimal solution to the dual problem, when an optimal solution to the Lagrangian problem exists.

RESULT 4.6.6 If $(\mathbf{k}, v_1, v_2)$ is any feasible solution to the dual problem, then $L(\mathbf{k})\leqslant D(\mathbf{k}, v_1, v_2)$.

Proof Consider

$$D(\mathbf{k}, v_1, v_2)-L(\mathbf{k})=\int_{S^-(\mathbf{k})} l(z)\left[\sum_{i=1}^{m} k_i f_i(z)-v_1(z)+v_2(z)-f_0(z)\right]dz$$

$$+\int_{S^+(\mathbf{k})} u(z)\left[\sum_{i=1}^{m} k_i f_i(z)-v_1(z)+v_2(z)-f_0(z)\right]dz$$

$$+\int_{S^-(\mathbf{k})}[u(z)-l(z)]v_2(z)\,dz$$

$$+\int_{S^+(\mathbf{k})}[u(z)-l(z)]v_1(z)\,dz. \tag{4.6.16}$$

Substituting for $D(\mathbf{k}, v_1, v_2)$ and $L(\mathbf{k})$ and grouping yields the right-hand side of the expression above. But the first two integrals in the right-hand side are zeroes, as $\sum_{i=1}^{m} k_i f_i(z) - v_1(z) + v_2(z) = f_0(z)$, for $z \in R^n$ as $(\mathbf{k}, v_1, v_2)$ is feasible for the dual problem. The last two integrals are nonnegative as $u(z) \geqslant l(z)$, $v_1(z)$, $v_2(z) \geqslant 0$ for all $z \in R^n$. Hence,

$$D(\mathbf{k}, v_1, v_2) \geqslant L(\mathbf{k}) \qquad \text{as desired.}$$

The next result establishes the relationship between optimal solutions to the Lagrangian and the dual problems.

RESULT 4.6.7 If $\mathbf{k}$ is an optimal solution to the Lagrangian problem then there exist functions v_1, v_2 such that $(\mathbf{k}, v_1, v_2)$ is optimal for the dual problem and

$$L(\mathbf{k}) = D(\mathbf{k}, v_1, v_2).$$

Proof From Result 4.6.6, for any feasible solution $(\hat{\mathbf{k}}, \hat{v}_1, \hat{v}_2)$ to the dual problem

$$L(\hat{\mathbf{k}}) \leqslant D(\hat{\mathbf{k}}, \hat{v}_1, \hat{v}_2).$$

But

$$L(\mathbf{k}) \leqslant L(\hat{\mathbf{k}}), \text{ as } \mathbf{k} \text{ is optimal for the Lagrangian problem.}$$

$$\Rightarrow L(\mathbf{k}) \leqslant D(\hat{\mathbf{k}}, \hat{v}_1, \hat{v}_2).$$

However, from Result 4.5.1, there exist v_1, v_2 such that $(\mathbf{k}, v_1, v_2)$ is feasible for the dual problem and $L(\mathbf{k}) = D(\mathbf{k}, v_1, v_2)$. Therefore $D(\mathbf{k}, v_1, v_2) \leqslant D(\hat{\mathbf{k}}, \hat{v}_1, \hat{v}_2)$ for any feasible solution, $(\hat{\mathbf{k}}, \hat{v}_1, \hat{v}_2)$. Hence the result.

Duality Theorem and Related Results

Now, to prove the duality results between the primal problem and the dual problem, we first prove the connection between the Lagrangian and the primal problem through a saddle-point theorem similar to the Kuhn and Tucker saddle-point theorem (Result 2.19.2). From the already-established equivalence between the dual and the Lagrangian problems we obtain the required results.

Let us for the time being assume that we have to minimize $-g_0(\phi)$ instead of maximize $g_0(\phi)$ in the primal problem, where

$$g_0(\phi)=\int_{R^n} f_0(z)\phi(z)\,dz.$$

Also let

$$g_i(\phi)=\int_{R^n} f_i(z)\phi(z)dz-c_i, \qquad i=1,\dots,m.$$

Let

$$\mathcal{C}=\{\phi \mid l(z)\leqslant\phi(z)\leqslant u(z), \qquad \text{for all} \quad z\in R^n\}.$$

So the Lagrangian function is defined as follows:

$$F(\phi,\mathbf{k})=-g_0(\phi)+\sum_{i=1}^{m} k_i g_i(\phi), \qquad \text{for} \quad \phi\in\mathcal{C}, \mathbf{k}\geqslant\mathbf{0}, \mathbf{k}\in R^m.$$

The implication is that for $\mathbf{k}\geqslant\mathbf{0}$

$$-L(\mathbf{k})=\inf_{\phi\in\mathcal{C}} F(\phi,\mathbf{k}).$$

We can maximize $-L(\mathbf{k})$ over $\mathbf{k}\geqslant\mathbf{0}$ instead of minimizing $L(\mathbf{k})$ over $\mathbf{k}\geqslant\mathbf{0}$.

REMARK 4.6.1 Let $\mathbf{a}\in R^{m+1}$. Let $\mathcal{C}_1=\{\mathbf{a}\mid a_0\geqslant -g_0(\phi), a_i\geqslant g_i(\phi),\ i=1,\dots,n$, for at least one $\phi\in\mathcal{C}\}$. Let

$$\mathcal{C}_2=\{\mathbf{a}\mid a_0<-g_0(\phi^0), a_i<0, \qquad i=1,\dots,m\}$$

where ϕ^0 is an optimal solution to the primal problem. Then it can be verified that both $\mathcal{C}_1$ and $\mathcal{C}_2$ are convex and $\mathcal{C}_1\cap\mathcal{C}_2=\varnothing$; as ϕ^0 is optimal for the primal, no a_0 can be less than $-g_0(\phi^0)$ nor can it be greater than or equal to $-g_0(\phi)$ for some $\phi\in\mathcal{C}$, $g_i(\phi)\leqslant 0$, $i=1,\dots,m$. Also notice that $\mathcal{C}_2$ is open.

Now the separation theorem stated in Result 2.19.2 can be applied, between $\mathcal{C}_1$ and $\mathcal{C}_2$. From the separation theorem, we have a vector $\boldsymbol{\omega}\neq\mathbf{0}$ and we have

$$\boldsymbol{\omega}\mathbf{a}^1>\boldsymbol{\omega}\mathbf{a}^2, \qquad \mathbf{a}^1\in\mathcal{C}_1, \qquad \mathbf{a}^2\in\mathcal{C}_2.$$

RESULT 4.6.8 Suppose Slater's constraint qualification holds for the primal problem. That is, there exists a ϕ such that $g_i(\phi)<0$ for all $i=1,\ldots,m$, $\phi\in\mathcal{C}$. If ϕ^0 is optimal for the primal problem, then there exists a $\mathbf{k}^0 \geqslant \mathbf{0}$.

$$F(\phi^0,\mathbf{k}) \leqslant F(\phi^0,\mathbf{k}^0) \leqslant F(\phi,\mathbf{k}^0) \qquad \text{for all } \mathbf{k}\geqslant\mathbf{0}, \qquad \phi\in\mathcal{C}.$$

Proof From Remark 4.6.1, there exists a $\boldsymbol{\omega}\neq\mathbf{0}$ such that

$$\boldsymbol{\omega}\mathbf{a}^1 > \boldsymbol{\omega}\mathbf{a}^2, \qquad \text{for all} \quad \mathbf{a}^1\in\mathcal{C}_1, \mathbf{a}^2\in\mathcal{C}_2.$$

But the components of $\mathbf{a}^2$ can be made arbitrarily large negative, implying $\boldsymbol{\omega}\geqslant\mathbf{0}$. For a boundary point $\mathbf{a}^2$ of $\mathcal{C}_2$ we have $\boldsymbol{\omega}\mathbf{a}^1 \geqslant \boldsymbol{\omega}\mathbf{a}^2$.

Therefore, in particular for $\mathbf{a}^2=(-g_0(\phi^0),0,\ldots,0)$ that is on the boundary of $\mathcal{C}_2$, and $\mathbf{a}^1=(-g_0(\phi), g_1(\phi),\ldots, g_m(\phi))$ that belongs to $\mathcal{C}_1$ for any $\phi\in\mathcal{C}$, we have $\boldsymbol{\omega}\mathbf{a}^1 \geqslant \boldsymbol{\omega}\mathbf{a}^2$. Thus

$$-\omega_0 g_0(\phi) + \sum_{i=1}^{m} \omega_i g_i(\phi) \geqslant -\omega_0 g_0(\phi^0).$$

We show that $\omega_0>0$ by contradiction. Suppose $\omega_0=0$. Then

$$\sum_{i=1}^{m} \omega_i g_i(\phi) \geqslant 0 \qquad \text{for all} \quad \phi\in\mathcal{C}.$$

Moreover $\omega_i>0$ for at least one i as $\boldsymbol{\omega}\neq\mathbf{0}$. This contradicts the assumption that the constraint qualification holds. Therefore $\omega_0>0$. Now define $k_i^0=\omega_i/\omega_0$, $i=1,\ldots,m$. k^0 so defined is nonnegative, as $\omega_i\geqslant 0$.

Thus we have,

$$-g_0(\phi) + \sum_{i=1}^{m} k_i^0 g_i(\phi) \geqslant -g_0(\phi^0), \quad \text{for all } \phi\in\mathcal{C}. \tag{4.6.17}$$

Since $\phi^0\in\mathcal{C}$, we have from 4.6.17 for $\phi=\phi^0$ $\Sigma_{i=1}^m k_i^0 g_i(\phi^0)\geqslant 0$. Also, as ϕ^0 is feasible, that is $g_i(\phi^0)\leqslant 0$ for all $i=1,2,\ldots,m$. Since $\mathbf{k}^0\geqslant\mathbf{0}$, we have $\Sigma_{i=1}^m k_i^0 g_i(\phi^0)\leqslant 0$, implying

$$\sum_{i=1}^{m} k_i^0 g_i(\phi^0)=0 \tag{4.6.18}$$

for any

$$\mathbf{k}\geqslant\mathbf{0}, \ \sum_{i=1}^{m} k_i g_i(\phi^0)\leqslant 0. \tag{4.6.19}$$

Thus, we have from (4.6.17) through (4.6.19),

$$-g_0(\phi) + \sum_{i=1}^{m} k_i^0 g_i(\phi) \geqslant -g_0(\phi^0) + \sum_{i=1}^{m} k_i^0 g_i(\phi^0) \geqslant -g_0(\phi^0) + \sum_{i=1}^{m} k_i g_i(\phi^0)$$

$$\text{for all } \phi\in\mathcal{C}, \mathbf{k}\geqslant\mathbf{0}$$

as desired.

This result implies that whenever we have an optimal solution ϕ^0 to the primal problem, we have an optimal solution $\mathbf{k}^0$ to the Lagrangian problem. Also we have $L(\mathbf{k}^0)=g_0(\phi^0)$.

REMARK 4.6.2 We can also show that if $(\phi^0,\mathbf{k}^0)$ with $\phi\in\mathcal{C}$, $\mathbf{k}^0\geqslant\mathbf{0}$, is a saddle point of $F(\phi,\mathbf{k})$, that is,

$$F(\phi^0,\mathbf{k})\leqslant F(\phi^0,\mathbf{k}^0)\leqslant F(\phi,\mathbf{k}^0) \qquad \text{for all} \quad \phi\in\mathcal{C},\ \mathbf{k}\geqslant\mathbf{0},$$

then ϕ^0 is optimal for the primal and $\mathbf{k}^0$ is optimal for the Lagrangian problem. The proof is similar to that for Result 2.19.1.

REMARK 4.6.3 Suppose $\mathbf{k}^0$ is optimal for the Lagrangian problem. Then consider a ϕ^0 such that

$$\phi^0(z)=\begin{cases} l(z),\ z\in S^-(\mathbf{k}^0),\\ u(z),\ z\in S^+(\mathbf{k}^0).\end{cases}$$

Now consider

$$F(\phi^0,\mathbf{k}^0)=-g_0(\phi^0)+\sum_{i=1}^{m}k_i^0 g_i(\phi^0)$$

$$=-\int_{R^n}f_0(z)\phi^0(z)\,dz+\sum_{i=0}^{m}k_i^0\left[\int_{R^n}f_i(z)\phi^0(z)\,dz-c_i\right]$$

$$=-\int_{S^-(\mathbf{k})}f_0(z)l(z)\,dz-\int_{S^+(\mathbf{k})}f_0(z)u(z)\,dz$$

$$+\sum_{i=1}^{m}k_i^0\left[\int_{S^-(\mathbf{k}^0)}f_i(z)l(z)\,dz+\int_{S^+(\mathbf{k}^0)}f_i(z)u(z)\,dz-c_i\right]$$

$$=\int_{S^-(\mathbf{k}^0)}\left[\sum_{i=1}^{m}k_i^0f_i(z)-f_0(z)\right]l(z)\,dz$$

$$+\int_{S^+(\mathbf{k}^0)}\left[\sum_{i=1}^{m}k_i^0f_i(z)-f_0(z)\right]u(z)\,dz-\sum_{i=1}^{n}k_i^0c_i.$$

$$\leqslant\int_{S^-(\mathbf{k}^0)}\left[\sum_{i=1}^{m}k_i^0f_i(z)-f_0(z)\right]\phi(z)\,dz$$

$$+\int_{S^+(\mathbf{k}^0)}\left[\sum_{i=1}^{m}k_i^0f_i(z)-f_0(z)\right]\phi(z)\,dz-\sum_{i=1}^{n}k_i^0c_i$$

for any $\phi\in\mathcal{C}$.

$$\leqslant \sum_{i=1}^{m} k_i^0\left[\int_{R^n} f_i(z)\phi(z)\,dz - c_i\right] - \int_{R^n} f_0(z)\phi(z)\,dz = F(\phi,\mathbf{k}^0)$$

from the definition of $g_0(\phi)$ and $g_i(\phi)$. The implication is that

$$F(\phi^0,\mathbf{k}^0)\leqslant F(\phi,\mathbf{k}^0).$$

Notice that $-L(\mathbf{k})=\inf_{\phi\in\mathcal{C}} F(\phi,\mathbf{k})$ is concave over R^m. This can be seen as follows:

$$-L(\mathbf{k})=\inf_{\phi\in\mathcal{C}}\left\{-g_0(\phi)+\sum_{i=1}^{m} k_i g_i(\phi)\right\}$$

since g_0, g_i are continuous and $\mathcal{C}$ is closed and bounded, $L(\mathbf{k})$ is finite everywhere on R^m.

Let $1\geqslant\lambda\geqslant 0$, $\mathbf{k}^1$, $\mathbf{k}^2\in R^m$, then

$$-L\left[\lambda\mathbf{k}^1+(1-\lambda)\mathbf{k}^2\right]=\inf_{\phi\in\mathcal{C}}\left\{-g_0(\phi)+\sum_{i=1}^{m}\left[\lambda k_i^1+(1-\lambda)k_i^2\right]g_i(\phi)\right\}$$

$$=\inf_{\phi\in\mathcal{C}}\left\{\lambda\left[-g_0(\phi)+\sum_{i=1}^{m}k_i^1 g_i(\phi)\right]+(1-\lambda)\left[-g_0(\phi)\right.\right.$$

$$\left.\left.+\sum_{i=1}^{m}k_i^2 g_i(\phi)\right]\right\}\geqslant\lambda\inf_{\phi\in\mathcal{C}}\left\{-g_0(\phi)+\sum_{i=1}^{m}k_i^1 g_i(\phi)\right\}$$

$$+(1-\lambda)\inf_{\phi\in\mathcal{C}}\left\{-g_0(\phi)+\sum_{i=1}^{m}k_i^2 g_i(\phi)\right\}$$

$$=\lambda\left[-L(\mathbf{k}^1)\right]+(1-\lambda)\left[-L(\mathbf{k}^2)\right]$$

Hence $-L(\mathbf{k})$ is concave as desired.

REMARK 4.6.4 If $\mathbf{k}^0$ is optimal for the Lagrangian problem and $L(\mathbf{k})$ is differentiable at $\mathbf{k}^0$, then

$$F(\phi^0,\mathbf{k})\leqslant F(\phi^0,\mathbf{k}^0).$$

This can be proved along the same lines as in Section 2.19, noticing the fact $-L(\mathbf{k})$ is concave and we are maximizing $-L(\mathbf{k})$, $\mathbf{k}\geqslant\mathbf{0}$. Remarks 4.6.2–4 imply Result 4.6.9.

RESULT 4.6.9 If $\mathbf{k}^0$ is optimal for the Lagrangian problem, and $L(\mathbf{k})$ is differentiable at $\mathbf{k}^0$, then $(\phi^0,\mathbf{k}^0)$ is a saddle point of $F(\phi,\mathbf{k})$, where

$$\phi^0(z)=\begin{cases} l(z), & z\in S^-(\mathbf{k}^0),\\ u(z), & z\in S^+(\mathbf{k}^0).\end{cases}$$

Hence ϕ^0 is optimal for the primal problem and $g_0(\phi^0)=L(\mathbf{k}^0)$. Results 4.6.5 and 4.6.7, which establish the relationship between the optimal solutions to the dual and the Lagrangian problem, together with Results 4.6.8 and 4.6.9, which establish the relationship between the optimal solutions to the primal and the Lagrangian problems, imply the following main result on duality.

RESULT 4.6.10 Suppose the primal problem satisfies Slater's constraints qualification. If ϕ is an optimal solution to the primal, then there exists an optimal solution $(\mathbf{k}, v_1, v_2)$ to the dual problem and $g_0(\phi)=D(\mathbf{k}, v_1, v_2)$. On the other hand, suppose L is differentiable at the point $\mathbf{k}$. If $(\mathbf{k}, v_1, v_2)$ is an optimal solution to the dual problem then there exists an optimal solution ϕ to the primal problem and $g_0(\phi)=D(\mathbf{k}, v_1, v_2)$.

We also have the complementary slackness property as in linear programming. Given any feasible solution ϕ to the primal problem and any feasible solution $(\mathbf{k}, v_1, v_2)$ to the dual problem, then the *complementary slackness conditions* are stated as:

$$k_i\left[c_i-\int_{R^n} f_i(z)\phi(z)\,dz\right]=0, \qquad i=1,\ldots,m \tag{4.6.20}$$

$$v_1(z)[\phi(z)-l(z)]=0 \quad \text{a.e. on} \quad R^n \tag{4.6.21}$$

$$v_2(z)[u(z)-\phi(z)]=0 \quad \text{a.e. on} \quad R^n \tag{4.6.22}$$

Result 4.6.11 establishes the necessity and sufficiency of the complementary slackness property under the assumption that the Slater's constraints qualification holds or that L is differentiable at $\mathbf{k}$.

RESULT 4.6.11 Suppose ϕ is feasible for the primal problem and $(\mathbf{k}, v_1, v_2)$ is feasible for the dual problem. If the primal satisfies Slater's constraints qualification or if L is differentiable at $\mathbf{k}$, then ϕ is optimal for the primal and $(\mathbf{k}, v_1, v_2)$ is optimal for the dual problem, if and only if ϕ, and $(\mathbf{k}, v_1, v_2)$ satisfy the complementary slackness conditions, (4.6.20–22).

Before proving Result 4.6.11 we prove Result 4.6.12, which says that for any pair of feasible solutions to the primal and the dual problems, we have the objective function value of the primal problem is less than or equal to that of the dual problem. We have seen a similar result in Section 2.11.

RESULT 4.6.12 If ϕ is any feasible solution for the primal problem and $(\mathbf{k}, v_1, v_2)$ is any feasible solution to the dual problem, then $g_0(\phi) \leqslant D(\mathbf{k}, v_1, v_2)$.

Proof From the feasibility of ϕ and $(\mathbf{k}, v_1, v_2)$ we have

$$k_i\left[c_i - \int_{R^n} f_i(z)\phi(z)\,dz\right] \geqslant 0, \qquad i=1,\ldots,m \tag{4.6.23}$$

The implication is that

$$\sum_{i=1}^{m} c_i k_i - \int_{R^n} \Sigma k_i f_i(z)\phi(z)\,dz \geqslant 0. \tag{4.6.24}$$

Adding $g_0(\phi)$ to both sides of (4.6.24). Now

$$g_0(\phi) = \int_{R^n} f_0(z)\phi(z)\,dz \leqslant \int_{R^n} [\,f_0(z) - \Sigma k_i f_i(z)]\phi(z)\,dz + \sum_{i=1}^{m} c_i k_i.$$

However,

$$\int_{R^n} [\,f_0(z) - \Sigma k_i f_i(z)]\phi(z)\,dz = -\int_{R^n} v_1(z)\phi(z)\,dz + \int_{R^n} v_2(z)\phi(z)\,dz,$$

which leads us from (4.6.21) and the feasibility of $v_1(z)$ and $v_2(z)$ to

$$g_0(\phi) \leqslant \sum_{i=1}^{m} c_i k_i - \int_{R^n} v_1(z)\phi(z)\,dz + \int_{R^n} v_2(z)\phi(z)\,dz. \tag{4.6.25}$$

But

$$-v_1(z)\phi(z) \leqslant -v_1(z)l(z) \qquad \text{a.e. on} \quad R^n \tag{4.6.26}$$

$$v_2(z)\phi(z) \leqslant v_2(z)u(z) \qquad \text{a.e. on} \quad R^n \tag{4.6.27}$$

Using (4.6.25) through (4.6.27) we establish the required result. We use Result 4.6.12 in the proof of Result 4.6.11 that follows.

Proof (of Result 4.6.11) When complementary slackness conditions hold, the inequalities (4.6.23), (4.6.26), and (4.6.27) in the proof of Result 4.6.12 hold as equalities. Hence, $g_0(\phi)=D(\mathbf{k}, v_1, v_2)$.

On the other hand, if ϕ, $(\mathbf{k}, v_1, v_2)$ are optimal solutions to the primal and the dual problems, respectively, Result 4.6.8 or (4.6.9) implies that $g_0(\phi)=D(\mathbf{k}, v_1, v_2)$. Suppose complementary slackness conditions are not satisfied; then there exists an inequality among the inequalities (4.6.23), (4.6.26), and (4.6.27) which does not hold as an equality. This inequality will imply that $g_0(\phi)<D(\mathbf{k}, v_1, v_2)$, leading to a contradiction. Hence the result.

In this section, we have developed the duality theory related to a generalized Neyman-Pearson problem (Problem 4.6.1).

Looking at Result 4.5.2, we notice that we must consider Problem 4.6.1 with the equality restrictions

$$\int_{R^n} f_i(z)\phi(z)\,dz=c_i, \qquad i=1,\ldots,m$$

to establish duality results for the corresponding problem.

Such consideration requires certain modifications to the assumptions and the results proved for Problem 4.6.1. Notice that Slater's constraints qualification can no longer hold when we have equalities instead of inequalities. Other types of constraint qualifications are needed, such as the interior-point restriction mentioned in Result 4.5.2 (4). With such a constraints qualification these results can be proved for the equality case, with the obvious modifications.

We have concluded that studying the problem of testing statistical hypotheses from the mathematical-programming point of view, apart from reestablishing known results, can also awaken fresh interest in interpreting the duality properties of the problem.

BIBLIOGRAPHICAL NOTES

4.1 For an exclusive treatise on testing of statistical hypotheses see Lehmann (1959). For an introductory approach to the topic see Hogg and Craig (1978), Brunk (1975), Lindgren (1968), and Rao (1973). For decision theoretic approach see Ferguson (1967).

4.2 For a historical development of the theory of testing statistical hypotheses see page 120 of Lehmann (1959). The fundamental work by Neyman and Pearson appears in Neyman and Pearson (1936).

4.3 The treatment in this section, which is a special case of the problem, does not appear elsewhere. Barankin (1951) was the first to observe that linear programming might be used in this area. The graphical solution given is due to Dantzig (1957). A good treatment of knapsack problems with exhaustive references can be found in Chapter 10 of Salkin (1975). Schaafsma (1970) considers maximin tests and suggests the use of linear programming.

4.5 A discussion of different versions of the generalized Neyman-Pearson problem appears in Francis and Wright (1969); see Dantzig and Wald (1951), and Chernoff and Scheffe (1952). Also see Chapter 3, Section 6, of Lehmann (1959) for a proof of Result 4.5.2.

4.6 The treatment of the problem in this section clearly follows that of Francis (1971). The mathematical-programming approach to the generalized Neyman-Pearson problem has been considered by Francis and Wright (1969), Meeks and Francis (1973), Pukelsheim (1978), and Krafft (1970). The result on the duality relationship between the Lagrangian problem and the primal problem is due to Francis and Wright (1969). Wagner (1969), considers nonlinear functional versions of the Neyman-Pearson lemma, and discusses a number of applications of the nonlinear version. Duality relationships for a nonlinear version of the generalized Neyman-Pearson problem appear in Meeks and Francis (1971). Pukelsheim (1978) provides an introduction to abstract programming and duality results and applies them to the Neyman-Pearson problem. Krafft (1970) develops duality results for the problem in a general-measure theoretical setup.

Related problems in decision theory have also received the mathematical-programming treatment. Weiss (1961) shows the use of a simplex method for solving minimax decision functions. Similar duality results appear in Witting (1966), Krafft and Witting (1967), Schaafsma (1970), Baumann (1968), and Krafft and Schmitz (1970), in the German literature. For the theory of abstract programming see Gol'shtein (1972), Rockafellar (1970), and Luenberger (1969).

REFERENCES

Barankin, E. W. (1951). *On the System of Linear Equations, with Applications to Linear Programming and the Theory of Statistical Hypothesis*. Publ. Stat. **1**, Univ. Calif. Press, Berkeley, pp 161–214.

Baumann, V. (1968). Eine Parameterfreie Theorie der Ungunstingsten Verteilungen fur das Testen von Hypothesen. *Z. Wahrscheinlichkeitstheorie Verw. Geb.* **11**, 41.

Brunk, H. D. (1975). *An Introduction to Mathematical Statistics* (ed. 3). Wiley, New York.

Chernoff, H., and Scheffe, H. (1952). "A Generalization of the Neyman-Pearson Lemma." *Ann. Math. Stat.* **23**, 213.

Dantzig, G. B. (1957). "Discrete-Variable Extremum Problems." *Oper. Res.* **5**, 266.

Dantzig, G. B. and Wald, A. (1951). "On the Fundamental Lemma of Neyman-Pearson." *Ann. Math. Stat.* **22**, 87.

Duffin, R. J. (1956). "Infinite Programs." In *Linear Inequalities and Related Systems*. H. W. Kuhn and A. W. Tucker, Eds. *Ann. Math. Stud.* **38**, Princeton Univ. Press, Princeton, N.J., pp. 159–70.

Ferguson, T. S. (1967). *Mathematical Statistics*. Academic Press, New York.

Francis, R. L. (1971). "On Relationships Between the Neyman-Pearsons Problem and Linear Programming." In *Optimization Methods in Statistics*. J. S. Rustagi, Ed., Academic Press, New York, pp. 259–280.

Francis, R. L., and Wright, G. (1969). "Some Duality Relationships for the Generalised Neyman-Pearson Problems." *J. Optimiz. Theory Appl.* **4**, 394.

Gol'shtein, E. G. (1972). *Theory of Convex Programming*. American Mathematical Society, Providence, R.I.

Hogg, R. V., and Craig, A. T. (1978). *Introduction to Mathematical Statistics* (ed. 4). Macmillan, London.

Krafft, O. (1970). "Programming Methods in Statistics and Probability Theory." In *Nonlinear Programming*. J. B. Rosen, O. L. Mangasarian, and K. Ritter, Eds. Academic Press, New York, pp. 425–446.

Krafft, O., and Schmitz, N. (1970). "A Systematical Multiple Decision Problem and Linear Programming." *Oper. Res. Verfahren* **7**, 126.

Krafft, O., and Witting, H. (1967). "Optimale Tests and ungunstigste Verteilungen." *Z. Wahrscheinlichkeitstheorie Verw. Geb.* **7**, 289.

Lehmann, E. L. (1959). *Testing Statistical Hypotheses*. Wiley, New York.

Lindgren, B. W. (1968). *Statistical Theory* (ed. 2). Macmillan, New York.

Luenberger, D. G. (1969). *Optimization by Vector Space Methods*. Wiley, New York.

Meeks, H. D., and Francis, R. L. (1973). "Duality Relationships for Nonlinear Version of the Generalized Neyman-Pearson Problem." *J. Optimiz. Theory Appl.* **11**, 360.

Pukelsheim, F. (1978). *A Quick Introduction to Mathematical Programming with Applications to Most Powerful Tests, Nonnegative Variance Estimation and Optimal Design Theory*. Technical Report **128**, Stanford Univ. Press, Stanford, Calif.

Rao, C. R. (1973). *Linear Statistical Inference and Its Applications*. Wiley, New York.

Rockafellar, R. T. (1970). *Convex Analysis*. Princeton Univ. Press, Princeton, N.J.

Salkin, H. (1975). *Integer Programming*. Addison-Wesley, Reading, Mass.

Schaafsma, W. (1970). "Most Stringent and Maximin Tests as Solutions of Linear Programming Problems." *Z. Wahrscheinlichkeitstheorie Verw. Geb.* **14**, 290.

Schmetterer, L. (1974). *Introduction to Mathematical Statistics*. Springer-Verlag, New York.

Wagner, D. H. (1969). "Nonlinear Functional Versions of the Neyman-Pearson Lemma." *SIAM Rev.* **11**, 52.

Weiss, L. (1961). Statistical Decision Theory. McGraw-Hill, New York.

Witting, H. (1965). *Mathematische Statistik*. Teubner, Stuttgart.

CHAPTER 5

Sampling

5.1 INTRODUCTION

For proper planning in micro as well as macro economics, we require information on many factors of interest. Data are either obtained through design and control of statistical experiments, or collected and recorded by observation or inquiry. Such surveys can be complete enumerations or sample surveys.

Sampling theory deals with problems associated with the selection of samples from a population according to certain probability mechanisms. For example, the simplest procedure is to give equal chance to every unit in the population to be included in the sample. We call this *simple random sampling*, denoting by SRS (v, k) where v is the cardinality of the population and k is the sample size. We can consider the possibilities of either allowing or not allowing a unit to occur more than once in the sample: SRS with and without replacement, as the case may be. Another sampling scheme attaches probabilities according to some size measure and the units are included in the sample with these probabilities: πPS or PPS (probability proportional to size) sampling.

In general a *sampling design* is a probability measure on the set of all possible samples from a given population, for a given size of the sample. The error arising because of inferring about a population characteristic on the basis of the sample is known as *sampling error*. The errors otherwise arising at stages of processing and compilation of data are termed *nonsampling errors*. The sampling error in general decreases with increase in the size of the sample drawn. Nonsampling errors are larger in larger samples or in complete enumeration, but in small samples, by better organization in the field and tabulation stages, we can reduce the nonsampling error.

The problem of deriving statistical information on population characteristics, based on sample data, can be formulated as an optimization problem in which we wish to minimize the cost of the survey, which is a function of the sample size, size of the sampling unit, the sampling scheme, and the scope of the survey, subject to the restriction that the loss in precision arising out of making decisions on the basis of the survey results is within a certain prescribed limit. Or alternatively, we may minimize the loss in precision, subject to the restriction that the cost of the survey is within the given budget. Thus we are interested in finding the optimal sample size and the optimal sampling scheme which will enable us to obtain estimates of the population characteristics with prescribed properties.

In this chapter we illustrate the use of mathematical programming to choose optimal sample size in different situations. We also consider certain estimation problems related to sampling and apply mathematical-programming methods to solve them. These applications can easily be extended to other areas in sampling.

5.2 OPTIMUM ALLOCATION OF SAMPLE SIZES IN STRATIFIED RANDOM SAMPLING

In the theory of sampling, stratified sampling occupies an important place. In stratified sampling the total population $U=\{U_1,\ldots,U_N\}$ is first partitioned into several subpopulations (called *strata*). Population characteristics can be inferred with samples from each stratum, exploiting the gain in precision in the estimates, administrative convenience, and the flexibility of using different sampling procedures in the different subpopulations.

Let N_i be the number of units in the ith stratum and $\Sigma_{i=1}^{L} N_i = N$, where L is the number of strata into which the N units are divided. Let n_i be the size of the sample drawn from the ith stratum. Assume that the samples are drawn independently in different strata.

The problem of optimally choosing the n_i's is known as the "optimal allocation problem." The objective in this problem might be minimization of the variance of the estimate of the population characteristic under study, with restriction on the total number of samples drawn or on the total budget available. Also the objective might be minimization of the total cost of sampling for a desired precision.

First we consider an unbiased estimate of the population mean, $\overline{Y}$, where Y is the characteristic under study. Let $\bar{y}_i$ be an unbiased estimate of the stratum mean $\overline{Y}_i$—that is,

$$\bar{y}_i = \frac{1}{n_i} \sum_{h=1}^{n_i} y_{ih}.$$

Then $\bar{y}_{st}$, given by

$$\bar{y}_{st}=\frac{1}{N}\sum_{i=1}^{L} N_i\bar{y}_i, \tag{5.2.1}$$

is an unbiased estimate of the population mean $\bar{Y}$, seen as follows:

$$\begin{aligned} E(\bar{y}_{st}) &= E\left[\frac{1}{N}\sum_{i=1}^{L} N_i\bar{y}_i\right] \\ &= \frac{1}{N}\sum_{i=1}^{L} E(\bar{y}_i)N_i \\ &= \frac{1}{N}\sum_{i=1}^{L} \bar{Y}_i N_i \\ &= \frac{1}{N}\sum_{i=1}^{L}\sum_{h=1}^{N_i} y_{ih}=\bar{Y} \end{aligned}$$

as desired, where y_{ih} is the value of y for the hth unit in the ith stratum. As the precision of this estimate is measured by the variance of the sample estimate, we consider next the variance of $\bar{y}_{st}$, denoted by $V(\bar{y}_{st})$. $V(\bar{y}_{st})$, by definition, is the average of $(\bar{y}_{st}-\bar{Y})^2$ over all possible samples.

Now

$$\begin{aligned} (\bar{y}_{st}-\bar{Y})^2 &= \left[\sum_{i=1}^{L}\frac{N_i(\bar{y}_i-\bar{Y}_i)}{N}\right]^2 \\ &= \frac{1}{N^2}\sum_{i=1}^{L} N_i^2(\bar{y}_i-\bar{Y}_i)^2 \\ &\quad + \frac{2}{N^2}\sum_{i>i'} N_iN_{i'}(\bar{y}_i-\bar{Y}_i)(\bar{y}_{i'}-\bar{Y}_{i'}). \end{aligned}$$

Averaging over all samples, and noticing the fact that the cross-product terms vanish, we get

$$\begin{aligned} V(\bar{y}_{st}) &= \frac{1}{N^2}\sum_{i=1}^{L} N_i^2E(\bar{y}_i-\bar{Y}_i)^2 \\ &= \frac{1}{N^2}\sum_{i=1}^{L} N_i^2V(\bar{y}_i). \end{aligned}$$

However, $V(\bar{y}_i)$ has the expression

$$V(\bar{y}_i)=S_i^2\frac{(N_i-n_i)}{N_i n_i}=S_i^2\left(\frac{1}{n_i}-\frac{1}{N_i}\right),$$

where

$$S_i^2=\frac{1}{N_i-1}\sum_{h=1}^{N_i}(y_{ih}-\bar{Y}_i)^2.$$

Let $N_i/N=W_i$, and $x_i=1/n_i-1/N_i$. Then we have

$$V(\bar{y}_{st})=\sum_{i=1}^{L}W_i^2S_i^2x_i. \tag{5.2.2}$$

Problem A

Here we consider the problem of choosing n_i, $i=1,\dots,L$, such that the sum of these n_i equals n, a fixed total sample size, and the $V(\bar{y}_{st})$ is a minimum. This problem can be formulated as

$$\begin{aligned}&\text{Minimize} \quad \sum_{i=1}^{L}W_i^2S_i^2x_i\\ &\text{subject to} \quad \sum_{i=1}^{L}n_i=n\\ &\qquad N_i\geqslant n_i\geqslant 1,\ n_i \text{ integer}, \qquad i=1,\dots,L.\end{aligned} \tag{5.2.3}$$

Let $a_i=W_i^2S_i^2$, $i=1,\dots,L$. Then the objective function

$$\sum_{i=1}^{L}W_i^2S_i^2x_i=\sum_{i=1}^{L}a_ix_i$$

$$=\sum_{i=1}^{L}a_i\left(\frac{1}{n_i}-\frac{1}{N_i}\right)=\sum_{i=1}^{L}\frac{a_i}{n_i}-\sum_{i=1}^{L}\frac{a_i}{N_i}.$$

But $\Sigma_{i=1}^{L} a_i/N_i$ is a constant. Therefore, it is sufficient to consider minimizing $\Sigma_{i=1}^{L} a_i/n_i$. Thus Problem A becomes:

$$\begin{aligned} &\text{Minimize} \quad \sum_{i=1}^{L} \frac{a_i}{n_i} \\ &\text{subject to} \quad \sum_{i=1}^{L} n_i = n \\ &N_i \geqslant n_i \geqslant 1,\ n_i \text{ integer}, \qquad i=1,\ldots, L. \end{aligned} \tag{5.2.4}$$

If the restrictions that n_i must be a positive integer and bounded above by N_i for all i are relaxed, then the classical Lagrangian multiplier method can be used to find optimal n_i. We have

$$n_i = n\frac{\sqrt{a_i}}{\sum_{i=1}^{L} \sqrt{a_i}} \tag{5.2.5}$$

However, there are three eventualities: (1) $n_i > N_i$ for some i, or (2) n_i may not be an integer for every i, or (3) $n_i < 1$ for some i. In that case we do not have a solution to Problem 5.2.4.

In the sampling literature, Eventuality (1) is referred to as *oversampling*—that is, the optimal allocation requires sampling more than 100% in certain strata. Noninteger solutions are rounded off. Eventuality (3) can be easily taken care of by assuming that we sample at least one unit from each stratum, and allocating the rest of $n-L$ units optimally. But noticing that $1/n_i$ is strictly convex in each i, we find the objective function to be a strictly convex function if $a_i > 0$, that is, $S_i^2 > 0$ for all i. Then we are interested in minimizing a strictly convex function over a bounded convex region, created by a linear equality and $2L$ upper and lower bound restrictions. When $L=2$, the feasible region and the objective function appear as in Fig. 5.2.1. In Fig. 5.2.1 both N_1 and N_2 are larger than n. Otherwise, we may have the configuration shown in Fig. 5.2.2.

The similarity between this problem and the knapsack problem introduced in Chapter 4 can be easily seen. Here we have a nonlinear objective function and a linear equality restriction

$$n_1 + \cdots + n_L = n,$$

and upper and lower bound restrictions on n_i,

$$1 \leqslant n_i \leqslant N_i, \qquad i=1,\ldots, L, \text{ and each } n_i \text{ is an integer}.$$

Below we describe a procedure for solving this problem.

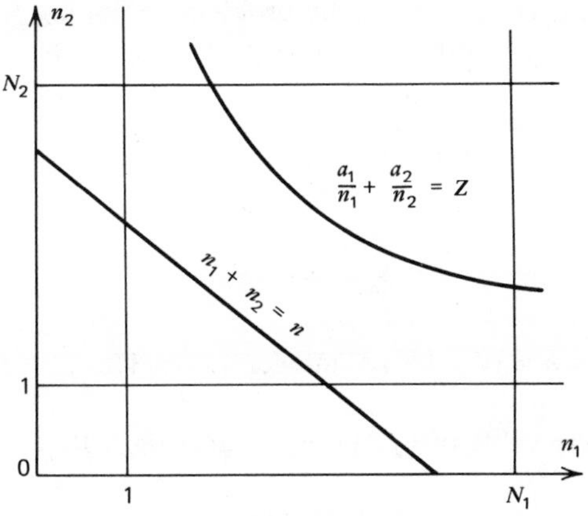

Figure 5.2.1 Feasible region and objective function when N_1 and N_2 are both larger than n.

Let us define $f(k, r)$ to be the minimal value of the objective function, using only the first k strata with total sample size r. That is,

$$\begin{aligned} f(k,r) &= \min \sum_{i=1}^{k} a_i/n_i \\ \text{subject to} \quad & \sum_{i=1}^{k} n_i = r \\ n_i \text{ integer and } & 1 \leqslant n_i \leqslant N_i, \qquad i=1,\ldots,k. \end{aligned} \tag{5.2.6}$$

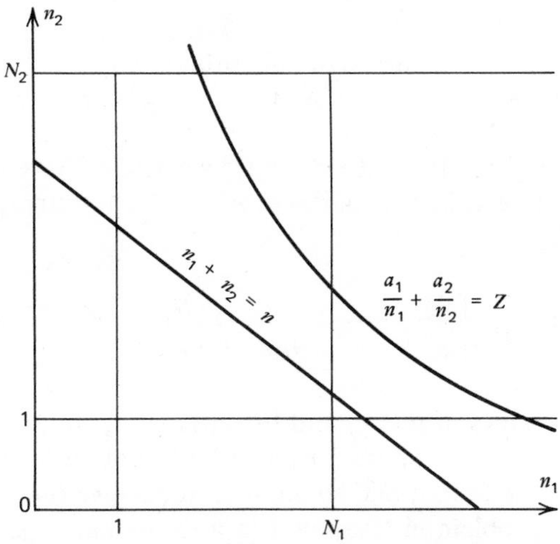

Figure 5.2.2 Feasible region and objective function when only N_2 is larger than n.

Thus Problem A is equivalent to the problem of finding $f(L, n)$. $f(L, n)$ is found recursively, by finding $f(k, r)$ for $k=1,\ldots, L$ and $r=0, 1,\ldots, n$. Now

$$f(k,r)=\min\left[\frac{a_k}{n_k}+\sum_{i=1}^{k-1} a_i/n_i\right]$$

$$\text{subject to}\qquad \sum_{i=1}^{k-1} n_i=r-n_k$$

$$1\leqslant n_i\leqslant N_i,\qquad n_i\ \text{ integer}\qquad i=1,\ldots,k-1.$$

For a fixed integer value of n_k, $1\leqslant n_k\leqslant \min[r, N_k]$, $f(k, r)$ is given by

$$a_k/n_k+\left\{\min\left[\sum_{i=1}^{k-1} a_i/n_i\right]\right.$$

$$\text{subject to}\qquad \sum_{i=1}^{k-1} n_i=r-n_k$$

$$\left.1\leqslant n_i\leqslant N_i,\qquad n_i\ \text{ integer}\qquad i=1,\ldots,k-1\right\}.$$

But by definition, the term in the braces is equal to $f(k-1, r-n_k)$.

Suppose we assume that for a given k, $f(k-1, r)$ is known for all possible $r=0, 1,\ldots, n$. Then

$$f(k,r)=\min_{n_k=1,\ldots,n}\left[a_k/n_k+f(k-1,r-n_k)\right]. \tag{5.2.7}$$

This formula is known as the "dynamic programming recursive formula." Using (5.2.7) for each $k=1,\ldots, L$ and $r=0, 1,\ldots, n$, $f(L, n)$ can be calculated.

Initially we set $f(k, r)=\infty$ if $r<k$ since we wish to have $n_i\geqslant 1$, for each $i=1,\ldots, k$, r must be at least equal to k. Also $f(1, r)=\min[a_1/n_1$ subject to $n_1=r$, $1\leqslant n_1\leqslant N_1]$.

$$f(1,r)=\begin{cases}\infty & \text{for } r>N_1 \quad\text{or}\quad r<1\\ a_1/r & \text{for } 1\leqslant r\leqslant N_1.\end{cases}$$

We tabulate the values of $f(k, r)$ and the optimal n_k, for each k systematically. Then from $f(L, n)$, optimal n_L can be found; from $f(L-1, n-n_L)$ optimal n_{L-1} can be found; and so on, until finally we find optimal n_1. We discuss a similar problem in Chapter 7 in great detail.

Table 5.2.1. Number of inhabitants, in thousands, for the year 1930, in 64 large cities of the U.S. (by stratum)

1			2			3		
90	58	36	31	25	20	14	14	16
82	49	32	27	23	18	17	12	12
78	44	33	28	26	16	15	13	12
81	45	30	25	29	20	14	13	13
67	46	29	27		15	11	10	
124	46	29	21		16	12	11	
57	40	25	26		14	12	11	
63	37	29	21		17	15	11	

EXAMPLE 5.2.1 Table 5.2.1 shows the number of inhabitants, of 64 large cities in the U.S., in thousands, for the year 1930. The cities are grouped into three strata.

There are 16, 20, and 28 cities, respectively, in the first, second, and third stratum. We calculate a_i's as follows: First we compute ΣY, $\overline{Y}$, S_i^2 and W_i for each stratum (see Table 5.2.2). Now a_i is given by $W_i^2 S_i^2$; we find

$$a_1 = 33.7539 \qquad a_2 = 1.4330 \qquad \text{and} \quad a_3 = 1.3885.$$

Suppose we wish to allot optimally a total of 24 samples among the three strata. We obtain the continuous solution given by

$$n_i^* = n\frac{\sqrt{a_i}}{\Sigma\sqrt{a_i}}.$$

We get $n_1^* = 17.0350$, $n_2^* = 3.5100$, and $n_3^* = 3.4549$. The rounded-off integer solutions are 17, 4, and 3. However, Stratum 1 has only 16 cities in all.

Table 5.2.2

Stratum i	N_i	$(\Sigma Y)_i$	$\overline{Y}_i$	S_i^2	W_i
1	16	1007	62.9375	540.0625	0.2500
2	20	552	27.6000	14.6737	0.3125
3	28	394	14.0714	7.2540	0.4375
TOTAL	64	1953			

Therefore this solution is not feasible—i.e., we have the problem of oversampling. So we resort to the dynamic recursive approach.

First we calculate $f(1, r)$. Then we calculate

$$f(2,r) = \min_{[n_2 \text{ feasible}]} \left[a_2/n_2 + f(1, r-n_2)\right]$$

and note down the optimal n_2 for each r. Using $f(2, r)$ we compute $f(3, r)$. Table 5.2.3 gives $f(k, r)$, for $k=1,2$, and $f(3,24)$. Thus we find $n_3=4$, and $f(3,24)=2.8150$. With $r=24-4=20$ and $k=2$, we get $n_2=4$. Finally $r=20-4=16$ and $k=1$, we find $n_1=16$. Therefore 16,4,4 turns out to be optimal.

Table 5.2.3

r	$f(1,4)$	n_1	$f(2,4)$	n_2	$f(3,r)$	n_3
1	33.7539	1	—			
2	16.8769	2	35.1869	1		
3	11.2513	3	18.3099	1		
4	8.4385	4	12.6843	1		
5	6.7508	5	9.8715	1		
6	5.6257	6	8.1838	1		
7	4.8220	7	7.0587	1		
8	4.2192	8	6.2550	1		
9	3.7504	9	5.5385	2		
10	3.3754	10	4.9357	2		
11	3.0685	11	4.4669	2		
12	2.8128	12	4.0919	2		
13	2.5964	13	3.7850	2		
14	2.4110	14	3.5293	2		
15	2.2403	15	3.2905	3		
16	2.1096	16	3.0741	3		
17			2.8887	3		
18			2.7280	3		
19			2.5873	3		
20			2.4679	4		
21			2.3962	5		
22			2.3484	6		
23			2.3143	7		
24			2.2887	8	2.8150	4

The variance of the estimate of the population mean in this allocation is

$$V^{(1)} = V(\bar{y}_{st}) = \sum_{i=1}^{3} W_i^2 S_i^2 \left(\frac{1}{n_i} - \frac{1}{N_i} \right)$$

$$= \sum \frac{a_i}{n_i} - \sum \frac{a_i}{N_i}$$

$$= 2.8150 - 2.2309$$

$$V^{(1)} = 0.5841$$

For comparison the rounded-off integer solution, (17,4,3), has a variance of the estimate equal to

$$V^{(2)} = 2.807 - 2.2309 = 0.5757.$$

(However, this solution is not feasible). For equal allocation—i.e., 8,8,8, the variance of the estimate is

$$V^{(3)} = 4.5719 - 2.2309 = 2.3410.$$

And finally for the proportional allocation (6,8,10) we have the variance of the estimate given by

$$V^{(4)} = 5.9436 - 2.2309 = 3.7127.$$

This approach can be easily extended to the problem of optimally allocating the sample size, subject to budget restriction, instead of the restriction on the total sample size, namely, n.

Suppose the cost per sample differs for the different strata. Let c_i be the cost per sample in the ith stratum. Let C be the total budget available. Then we wish to

$$\text{Minimize} \quad \sum_{i=1}^{L} a_i / n_i \tag{5.2.8}$$

$$\text{subject to} \quad \sum c_i n_i = C$$

$$\text{and } 1 \leqslant n_i \leqslant N_i, \qquad n_i \text{ integer} \qquad \text{for } i = 1, \ldots, L.$$

We have the recursion formula given by

$$f(k,c)=\min_{n_k \text{ feasible}}\left[a_k/n_k+f(k-1,c-c_kn_k)\right]$$

$$\text{where}\quad f(k,c)=\min\sum_{i=1}^{k} a_i/n_i$$

$$\text{subject to}\qquad \sum c_in_i\leqslant c$$

$$1\leqslant n_i\leqslant N_i,\qquad n_i\ \text{integer}\qquad i=1,\ldots,k.$$

for all c feasible—i.e., $\sum_{i=1}^{k} c_i \leqslant c \leqslant C$. The method of finding optimal n_i is exactly as in the earlier example.

Problem B

We can also treat similarly the problem of minimizing the total cost of sampling, subject to certain restrictions on the allowable loss in precision. We have the problem stated as follows:

$$\begin{aligned}&\text{Minimize}\quad \sum_{i=1}^{L} c_in_i\\ &\text{subject to}\quad \sum_{i=1}^{L} a_i/n_i\leqslant v\\ &1\leqslant n_i\leqslant N_i,\qquad n_i\ \text{integer}\qquad \text{for}\quad i=1,\ldots,L.\end{aligned}\tag{5.2.9}$$

So far we have considered only one characteristic for study. But if we have to do a multivariate survey—i.e., we wish to study several characteristics, the problem of optimal allocation does not yield to such a simple approach. In the next section we consider the problem of minimizing the total cost so as to achieve prescribed precision of the estimates of several population characteristics.

5.3 OPTIMAL ALLOCATION OF SAMPLE SIZES IN MULTIVARIATE STRATIFIED RANDOM SAMPLING

We here assume there are p characteristics under study. Let Y_j be the jth characteristic considered. As earlier we have L strata, and N_i units in the

ith stratum,

$$\sum_{i=1}^{L} N_i = N.$$

Assume that the n_i samples are drawn independently from each stratum. Also assume that $\bar{y}_{ij}$ is an unbiased estimate of $\bar{Y}_{ij}$, that is,

$$\bar{y}_{ij} = \frac{1}{n_i} \sum_{h=1}^{n_i} y_{ijh}$$

where y_{ijh} is the value observed for Y_j in the ith stratum for the hth sample unit. An unbiased estimate of the population mean $\bar{Y}_j$ is given by

$$\bar{y}_j(st) = \frac{1}{N} \sum_{i=1}^{L} N_i \bar{y}_{ij} \qquad \text{for} \quad j = 1, \ldots, p. \tag{5.3.1}$$

Now we consider this estimate in the optimal allocation problem. Precision of this estimate is measured by the variance of the estimate of the population characteristic, for each characteristic. As noted in the previous section,

$$V_j = V(\bar{y}_j(st)) = \sum_{i=1}^{L} W_i^2 S_{ij}^2 x_i \tag{5.3.2}$$

where

$$W_i = N_i / N; \; S_{ij}^2 = \frac{1}{N_i - 1} \sum_{h=1}^{N_i} \left(y_{ijh} - \bar{Y}_{ij}\right)^2$$

and

$$x_i = \frac{1}{n_i} - \frac{1}{N_i}.$$

Let $a_{ij} = W_i^2 S_{ij}^2$. Let C_i be the cost of sampling all the p characteristics on a single unit in the ith stratum. The total variable cost of the survey, assuming linearity, is

$$K = \sum_{i=1}^{L} C_i n_i.$$

Assume $a_{ij}, C_i > 0$, for $i = 1, \ldots, L, j = 1, \ldots, p$.

The problem of allocation can now be stated as Problem C.

Problem C

$$\text{Minimize} \quad \sum_{i=1}^{L} C_i n_i \tag{5.3.3}$$

$$\text{subject to} \quad \sum_{i=1}^{L} a_{ij} x_i \leqslant v_j, \qquad j=1,\dots,p. \tag{5.3.4}$$

$$0 \leqslant x_i \leqslant 1 - \frac{1}{N_i}, \qquad i=1,\dots,L \tag{5.3.5}$$

$$x_i = \frac{1}{n_i} - \frac{1}{N_i}, \qquad n_i \text{ integer}, \qquad i=1,\dots,L \tag{5.3.6}$$

where v_j is the allowable error in the estimate of the jth characteristic. Problem C is a integer linear-programming problem but for the Restriction 5.3.6, which is nonlinear. When the new variables $X_i = 1/n_i$, $i=1,\dots,L$ are introduced, Problem C can be equivalently stated as Problem D.

Problem D

$$\text{Minimize} \quad \sum_{i=1}^{L} C_i / X_i \tag{5.3.7}$$

$$\text{subject to} \quad \sum_{i=1}^{L} a_{ij} X_i \leqslant b_j, \qquad j=1,\dots,p \tag{5.3.8}$$

$$\frac{1}{N_i} \leqslant X_i \leqslant 1, \qquad i=1,\dots,L \tag{5.3.9}$$

where $b_j = v_j + \sum_{i=1}^{L} a_{ij}/N_i$, $j=1,\dots,p$.

REMARK 5.3.1 The objective function (5.3.7) in Problem D is a strictly convex function, because C_i/X_i is strictly convex for $C_i > 0$.

REMARK 5.3.2 The restrictions (5.3.8 and 5.3.9) provide a bounded convex feasible region for the problem, formed by linear inequalities. The region is nonempty as

$$\mathbf{X} = \left(\frac{1}{N_1}, \dots, \frac{1}{N_L} \right)$$

is feasible. Thus an optimum $\mathbf{X}=(X_1^*,\ldots, X_L^*)$ exists. Strict convexity also implies uniqueness of the optimal solution.

REMARK 5.3.3 The optimum is attained at a boundary of the convex set.

Problem D is a convex programming problem like the type discussed in Section 5.2. There we developed the necessary and sufficient conditions for an $\mathbf{X}$ to be optimal. There are several methods for solving such problems: the convex-simplex method, feasible direction method, gradient projection method, cutting plane method, linearization method, and so on.

However, all these methods find an $\mathbf{X}$ which may correspond to a noninteger n_i, $i=1,\ldots, p$. Rounding-off yields in those cases a near-optimal solution. But if we wish to find integer optimal solutions to Problem C, we have to resort to some branch-and-bound scheme in which several problems of the type of Problem D may have to be solved, for calculating the bounds. (Branch-and-bound schemes were discussed in Section 2.28.) Similar procedures can be devised in this case.

REMARK 5.3.4 The optimal solution to Problem D provides a lower bound on the value of the optimal solution to Problem C. On the other hand, a rounded-off integer solution that is feasible for Problem D turns out to be an upper bound on the optimal objective function value to Problem C. Thus the deviation from the optimum to Problem C can be measured, before we go to the branch-and-bound procedure. Also, these bounds can help in terminating the branch-and-bound procedure at an intermediate stage, as soon as the upper and lower bounds are sufficiently close, for all practical purposes, as too much computer storage and time are required for problems with a large number of variables.

Geometrical Interpretation of the Problem

We consider the case when L (number of strata) equals 2. The objective function

$$Z=C_1/X_1+C_2/X_2$$

is equivalent to

$$Z=\frac{C_1X_2+C_2X_1}{X_1X_2}.$$

From this,

$$X_1X_2=\frac{C_1X_2}{Z}+\frac{C_2X_1}{Z}$$

or

$$X_1X_2-\frac{C_1X_2}{Z}-\frac{C_2X_1}{Z}=0.$$

This yields the equivalent form for the objective function in terms of X_1, X_2, and Z as

$$(X_1-C_1/Z)(X_2-C_2/Z)=\frac{C_1C_2}{Z^2}, \tag{5.3.10}$$

which is a rectangular hyperbola with center $(C_1/Z, C_2/Z)$. As Z varies, the center $(C_1/Z, C_2/Z)$ lies on the line

$$X_2/X_1=C_2/C_1 \tag{5.3.11}$$

and the vertex of the rectangular hyperbola

$$([C_1+\sqrt{C_1C_2}]/Z,[C_2+\sqrt{C_1C_2}]/Z)$$

lies on the line

$$X_2/X_1=\frac{[C_2+\sqrt{C_1C_2}]}{[C_1+\sqrt{C_1C_2}]}. \tag{5.3.12}$$

Now consider the restrictions of (5.3.8) and (5.3.9). We have the feasible region in the nonnegative orthand, as $a_{ij}X_1+a_{2j}X_2=b_j$ has negative slope and positive X_2-intercepts in the X_1X_2-plane, and the upper and lower bounds on X_1 and X_2 are positive.

To obtain the optimum allocation we have to find the rectangular hyperbola 5.3.10 for some value of Z such that it touches the boundary of the feasible region. See Fig. 5.3.1. In general when we have L-strata we have the following results.

RESULT 5.3.1 The point of contact of the hyperplane

$$\sum a_iX_i=b \qquad (a_i, b>0),$$

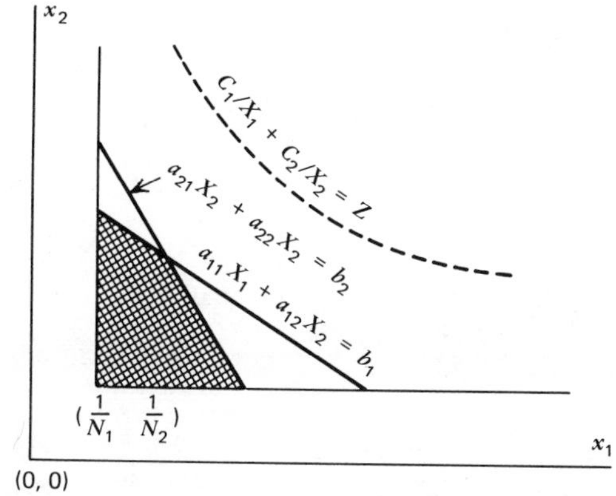

Figure 5.3.1 Objective function and the feasible region.

with the objective function

$$Z = \sum_{i=1}^{L} C_i/X_i$$

is given by $\mathbf{X} = (X_1, \ldots, X_L)$, where

$$X_i = \left[b\sqrt{C_i a_i} \right] \Big/ \left[a_i \sum_{i=1}^{L} \sqrt{C_i a_i} \right], \qquad i = 1, \ldots, L \qquad (5.3.13)$$

Proof The objective function can be written as

$$f(X_1, \ldots, X_L) = \sum_{i=1}^{L} C_i \prod_{\substack{h=1 \\ h \neq i}}^{L} X_h - Z \prod_{h=1}^{L} X_h. \qquad (5.3.14)$$

Let $f_x(\mathbf{X}^{(1)})$ denote $(\partial F/\partial X_1, \ldots, \partial F/\partial X_L)$ evaluated at $\mathbf{X}^{(1)}$. Let $f_x(\mathbf{X}^{(1)})_i$ denote the ith coordinate of $f_x(\mathbf{X}^{(1)})$. Then

$$f_x(\mathbf{X}^{(1)})_i = \sum_{\substack{k=1 \\ i \neq k}}^{L} C_k \prod_{\substack{h=1 \\ k \neq h \neq i}}^{L} X_h^{(1)} - Z \prod_{\substack{h=1 \\ h \neq i}}^{L} X_h^{(1)}.$$

Thus we have the equation for the hyperplane touching the objective function at $\mathbf{X}^{(1)}$, given by

$$\sum_{i=1}^{L}(X_i - X_i^{(1)})f_x(\mathbf{X}^{(1)})_i = 0 \tag{5.3.15}$$

or

$$\sum_{i=1}^{L} X_i\left[\sum_{\substack{k=1\\ i\neq k}}^{L} C_k \prod_{\substack{h=1\\ k\neq h\neq i}}^{L} X_h^{(1)} - Z\prod_{\substack{h=1\\ h\neq i}}^{L} X_h^{(1)}\right]$$

$$+\left\{-\sum_{i=1}^{L} X_i^{(1)}\left[\sum_{\substack{k=1\\ i\neq k}}^{L} C_k \prod_{\substack{h=1\\ k\neq h\neq i}}^{L} X_h^{(1)} + Z\prod_{\substack{h=1\\ h\neq i}}^{L} X_h^{(1)}\right]\right\} = 0.$$

Since the term in the braces is equal to

$$Z\prod_{h=1}^{L} X_h^{(1)} \qquad \text{after simplification.}$$

Thus we have

$$\sum_{i=1}^{L} X_i\left[\sum_{\substack{k=1\\ i\neq k}}^{L} C_k \prod_{\substack{h=1\\ k\neq h\neq i}}^{L} X_h^{(1)} - Z\prod_{\substack{h=1\\ h\neq i}}^{L} X_h^{(1)}\right] + Z\prod_{h=1}^{L} X_h^{(1)} = 0. \tag{5.3.16}$$

This hyperplane will represent the hyperplane $\Sigma_{i=1}^{L} a_i X_i = b$ in case

$$\left[\sum_{\substack{k=1\\ k\neq i}}^{L} C_k \prod_{\substack{h=1\\ k\neq h\neq i}}^{L} X_h^{(1)} - Z\prod_{\substack{h=1\\ h\neq i}}^{L} X_h^{(1)}\right]\Big/ a_i = -Z\prod_{h=1}^{L} X_h^{(1)}/b. \tag{5.3.17}$$

The implication is that

$$\frac{1}{a_i}\sum_{\substack{k=1\\ i\neq k}}^{L} C_k \prod_{\substack{h=1\\ k\neq h\neq i}}^{L} X_h^{(1)} = Z\prod_{\substack{h=1\\ h\neq i}}^{L} X_h^{(1)}\left[\frac{1}{a_i} - \frac{X_i^{(1)}}{b}\right]. \tag{5.3.18}$$

Dividing both sides of 5.3.18 by $\prod_{\substack{h=1\\ h\neq i}}^{L} X_h^{(1)}$, we get

$$\frac{1}{a_i}\sum_{\substack{k=1\\ i\neq k}}^{L} \frac{C_k}{X_k^{(1)}} = Z\frac{\left(b - a_i X_i^{(1)}\right)}{a_i b}. \tag{5.3.19}$$

Cancelling out $1/a_i$ and adding and subtracting $C_i/X_i^{(1)}$ in the left-hand side of (5.3.19) we get

$$\sum_{k=1}^{L} \frac{C_k}{X_k^{(1)}} - C_i/X_i^{(1)} = Z\left[\frac{b - a_i X_i^{(1)}}{b}\right].$$

But

$$\sum_{k=1}^{L} \frac{C_k}{X_k^{(1)}} = Z.$$

Hence, after substitution and simplification we get

$$X_i^{(1)} = \sqrt{\frac{C_i b}{a_i Z}}, \qquad i = 1, \ldots, L. \tag{5.3.20}$$

Now

$$Z = \sum \frac{C_i}{X_i^{(1)}} = \sum \frac{C_i}{\sqrt{\dfrac{C_i b}{a_i Z}}}.$$

The implication is that

$$\sqrt{Z} = \frac{1}{\sqrt{b}} \cdot \sum_{i=1}^{L} \sqrt{C_i a_i}\,. \tag{5.3.21}$$

Eliminating the Z in Expression 5.3.20 we finally obtain

$$X_i^{(1)} = b\sqrt{C_i a_i} \Bigg/ \left[a_i \sum_{i=1}^{L} \sqrt{C_i a_i}\right] \qquad i = 1, \ldots, L$$

as required. Introducing the subscript j for the different characteristics, we have the corresponding result for the jth hyperplane.

We now can describe a procedure which is efficient in case for a certain j; the $X_{ij}^{(1)}$ discussed in Result 5.3.1 for the characteristic j satisfies all the constraints

$$\sum a_{ij} X_{ij}^{(1)} \leqslant b_j, \qquad j = 1, \ldots, p$$

and

$$\frac{1}{N_i} \leqslant X_{ij}^{(1)} \leqslant 1, \qquad i=1,\ldots, L.$$

Step 0 We discard from the set of constraints (5.3.8) those which are not binding—i.e., we find the intercepts $(b_j/a_{1j},\ldots, b_j/a_{Lj})$ for each j and discard those j for which the vector of intercepts strictly dominates the corresponding vector for any other j. Assume that I_1 is the set of binding constraints among the constraints 5.3.8.

Step 1 Compute $\mathbf{X}_j=(X_{1j},\ldots, X_{Lj})$ for each characteristic $j\in I_1$, using Result 5.3.1, that is,

$$X_{ij}=b_j\sqrt{C_i a_{ij}} \Bigg/ \left[a_{ij} \sum_{i=1}^{L} \sqrt{C_i \cdot a_{ij}} \right].$$

Step 2 Find j^* such that $\Sigma_{i=1}^{L} 1/X_{ij^*}$ is maximum for $j\in I_1$. That is for j^* the total sample size is a maximum.

Now if j^* satisfies all the constraints then X_j^* is feasible and the optimal solution is X_j^*. However, if some of the constraints $1/N_i \leqslant X_{ij} \leqslant 1$ are violated, we proceed as follows:

Let $I=\{i \mid \text{either } X_{ij^*}<1/N_i \text{ or } X_{ij^*}>1\}$. Fix $X_{ij}=1/N_i$ or $X_{ij}=1$, as the case may be, for $i\in I$, and eliminate these strata from consideration. For the remaining strata find X_j for all $j\in I_1$ and repeat the process, using Result 5.3.1.

A general procedure along this line is possible, that considers the intersection of some of the hyperplane, finds the point of contact of the objective function with them, and proceeds until all the constraints are satisfied. However, this approach may turn out to be computationally not efficient if several intersections and their contact with the objective function have to be found.

5.4 INTEGRATION OF SURVEYS WITH PRESCRIBED PROBABILITIES OF SELECTION OF SAMPLES

When two or more sample surveys are conducted on the same set of units it may be stipulated that the sample units are to have different probabilities for the different surveys. Such stipulations arise naturally as the characteristics under study in the different surveys may fall into different classes. For example, if crop and demographic surveys are to be conducted on the same set of rural areas, we consider area under cultivation a suitable measure to assign probability of selection of a unit for the crop survey, but probably not for the demographic survey. We might choose number of inhabitants in each unit as an appropriate-size measure for the demographic survey.

Even if one fixes the sample size n to be same for each of the surveys, a sample drawn for one survey may not in general satisfy the restrictions on the probability of selection of units for the other survey. Thus we may have to draw different samples for the different surveys. This in turn may mean increased total cost of the surveys.

Thus we wish to exploit the reduction in cost of surveys, if the same unit appears in the samples drawn for the different surveys, without violating the probability restrictions on selection of samples. Hence the problem of optimizing the total cost of the surveys with restriction of the probabilities of selection of samples for the different surveys.

There are different objectives for consideration. The simplest one is maximization of the expected number of common units in the selected samples for the different surveys.

In the sample-survey literature there are a few selection methods with different objectives. We review some of these methods below and use mathematical-programming models to compare and criticize them, and propose more appropriate models for attacking these problems.

Let $U=\{U_1,\ldots,U_N\}$ be the set of N units, and assume that we are interested in conducting two surveys. A sample of size n is chosen from U for each survey, with prescribed probability of inclusion of a unit in a sample for each survey. Here we assume $n<N$. If $n=N$ the problem is trivial.

Let p_i and p_i' be the probabilities with which the unit U_i has to be included in a sample for the first and the second survey, respectively. We say there is an *overlap* if the same unit U_i is selected in the samples chosen for both the surveys. The number of overlaps can be at most n, and it has to be a nonnegative integer.

A. Keyfitz's Selection Scheme

Keyfitz has suggested the following scheme, which aims at maximizing the expected number of overlaps:

1 Select a sample of n distinct units for Survey I, with the prescribed probabilities, p_i.

2 Suppose unit U_i is in the sample chosen for Survey I. If $p_i' > p_i$, the unit U_i is included in the sample for Survey II.

3 Otherwise—i.e., if $p_i' < p_i$, include unit U_i with probability p_i'/p_i and do not include the unit with probability $(1-p_i'/p_i)$.

4 If $p_i' < p_i$ and the unit U_i is not included in the sample for Survey II, select a unit from those units with $p_j' > p_j$ with probability proportional to $(p_j' - p_j)$ and include it.

This scheme provides a solution to the problem that satisfies the probability restrictions.

RESULT 5.4.1 Unit U_k is selected with the prescribed probability p_k, p_k' for Survey I and Survey II, respectively, in Keyfitz's selection scheme.

Proof In Keyfitz's scheme the sample for the first survey is chosen so that unit U_k is selected in the sample with probability p_k. So the probability restriction for Survey I is satisfied.

For Survey II, we shall show that unit U_k is selected with probability p_k' for each k. Let $P(U_k)$ denote the probability of U_k being included in a sample for Survey II, under this scheme.

CASE 1 $p_k' < p_k$. In this case unit U_k is included in a sample for Survey II with probability p_k'/p_k if U_k is included in the sample for Survey I, that is,

$$P(U_k) = p_k \cdot p_k'/p_k = p_k' \text{ as required.}$$

CASE 2 $p_k' \geqslant p_k$. U_k is included in a sample for Survey II with probability 1 if it is included in the sample for Survey I or if some Unit U_i, $(i \neq k)$, with $p_i' < p_i$ is included in the sample for Survey I and is not included with probability $(1 - p_i'/p_i)$ in the sample for Survey II, and then U_k is included in the sample for Survey II as per step 4 of Keyfitz's scheme, with probability $(p_k' - p_k)/\Sigma_{p_j' > p_j}(p_j' - p_j)$, that is, in this case,

$$P(U_k) = p_k + \left[\sum_{p_i' < p_i} p_i(1 - p_i'/p_i)\right] \frac{p_k' - p_k}{\sum_{p_j' > p_j} (p_j' - p_j)}.$$

But

$$\sum_{p_i' < p_i} (p_i - p_i') = \sum_{p_j' > p_j} (p_j' - p_j).$$

Therefore $P(U_k) = p_k + p_k' - p_k = p_k'$ as desired.

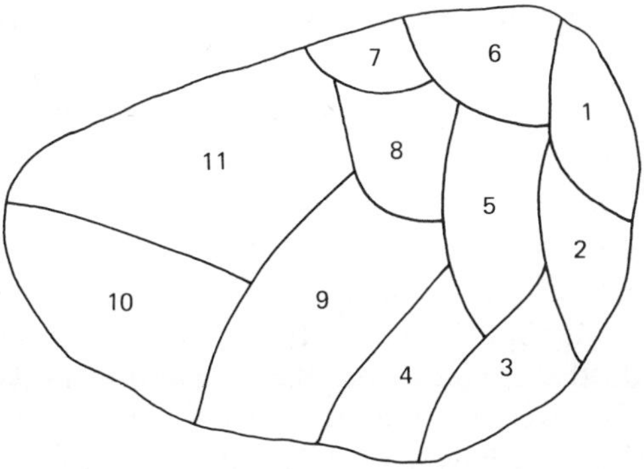

Figure 5.4.1 Lahiri's selection scheme: serpentine ordering of the units.

B. Lahiri's Selection Scheme

Lahiri's scheme of selection needs a serpentine ordering of the units in U as shown in Fig. 5.4.1 so that geographically contiguous units occur next to each other in the sampling frame.

The sample units for the two surveys are selected with the same set of n random numbers chosen from 0 to 1, but with independent cumulative totals of the probabilities p_i and p'_i. For instance, if r is chosen randomly between 0 and 1, then unit U_i is chosen for Survey I if

$$\sum_{j=1}^{i-1} p_j < r \leqslant \sum_{j=1}^{i} p_j$$

and U_k is chosen for Survey II if

$$\sum_{j=1}^{k-1} p'_j < r \leqslant \sum_{j=1}^{k} p'_j.$$

It is expected that $|i-k|$ will be as small as possible, in a large number of cases, because of the serpentine arrangement of the frame. Notice that the probability restrictions are satisfied for both surveys.

C. Roy Choudhury's Selection Scheme

Roy Choudhury offers a third scheme of selection.

1 Select one unit for Survey I, say U_i, with the probability p_i.

2 Select one unit for Survey II other than U_i, say, U_j, with the probability $p'_j / \Sigma_{l \neq i} p'_l$.

3 Let $U_0 = (U_i, U_j)$. $U - \{U_i, U_j\}$ has $N-2$ elements; with U_0 we have $N-1$ elements in all.

4 Draw a sample of size $n-1$ from these $N-1$ units by simple random sampling without replacement.

5 If U_0 is selected in this sample of size $n-1$, we select the unit U_j for Survey I and the unit U_i for Survey II; that is, both U_i and U_j are in the samples for both the surveys and the remaining $n-2$ units are also common. Otherwise, we have the set S of $n-1$ distinct elements from U other than U_i and U_j. We have $\{U_i\} \cup S$ for the first survey and $\{U_j\} \cup S$ for the second survey.

D. Raj's Mathematical Formulations

Let P_{ij} be the probability with which unit U_i and unit U_j are selected for Surveys I and II, respectively. Therefore, the restriction that the unit U_i is included in a sample for Survey I with probability p_i can be written as

$$\sum_{j=1}^{N} P_{ij} = p_i \tag{5.4.1}$$

This equation has to be satisfied for all $i = 1, \ldots, N$.

Similarly, the unit U_j is included in a sample for Survey II with probability p'_j written as

$$\sum_{i=1}^{N} P_{ij} = p'_j. \tag{5.4.2}$$

This equation has to be satisfied for all $j = 1, \ldots, N$. We also require that

$$P_{ij} \geqslant 0 \qquad \text{for all} \quad i, j = 1, \ldots, N \tag{5.4.3}$$

and

$$P_{ij} \leqslant 1 \qquad \text{for all} \quad i, j = 1, \ldots, N. \tag{5.4.4}$$

However, any set of nonnegative P_{ij} satisfying (5.4.1) and (5.4.2) for all i and j automatically satisfies the restrictions $P_{ij} \leqslant 1$, as p_i and p'_j are probabilities. So we drop Restrictions 5.4.4 from further consideration.

We shall show that Keyfitz's scheme provides an optimal solution to Problem 5.4.1, given below:

Problem 5.4.1

$$\text{Maximize} \quad \sum_{i=1}^{N} P_{ii}$$

$$\text{subject to} \quad \sum_{j=1}^{N} P_{ij}=p_i, \qquad i=1,\ldots,N$$

$$\sum_{i=1}^{N} P_{ij}=p'_j, \qquad j=1,\ldots,N$$

$$P_{ij}\geqslant 0, \qquad i,j=1,\ldots,N.$$

Problem 5.4.1 is a linear-programming problem. However, the problem has a special constraint matrix and such matrices are known as *transportation matrices*. Before we discuss the transportation matrices, in Section 5.5, a few observations on Problem 5.4.1 are in order,

RESULTS 5.4.2 In any optimal solutions $P^*=((P^*_{ij}))$ to the Problem 5.4.1,

$$P^*_{ii}=\min(p_i, p'_i) \qquad \text{for all} \quad i=1,\ldots,N.$$

Proof In any feasible solution to Problem 5.4.1, no P_{ii} can be greater than $\min(p_i, p'_i)$, as that would violate the constraints. Therefore in an optimal solution P^*, P^*_{ii} can be less than or equal to $\min(p_i, p'_i)$. Suppose for some k, $P^*_{kk}<\min(p_k, p'_k)$. Then there exist $P^*_{ks}>0$ and $P^*_{rk}>0$ for some s and r, $s\neq k\neq r$.

Let $\theta=\min(P^*_{ks}, P^*_{rk})$. Then the solution P, given by

$$P_{ij}=P^*_{ij}, \qquad r\neq i\neq k$$

$$s\neq j\neq k, \quad i,j=1,\ldots,N$$

$$P_{rk}=P^*_{rk}-\theta$$

$$P_{ks}=P^*_{ks}-\theta$$

$$P_{rs}=P^*_{rs}+\theta$$

and

$$P_{kk} = P^*_{kk} + \theta.$$

Now, P is feasible and the objective function value corresponding to P is at least $\Sigma_{i=1}^{N} P^*_{ii} + \theta$. As $\theta > 0$, this leads to a contradiction as P^* is assumed to be optimal. Hence the result.

This result implies that the optimal objective function value is given by $\Sigma \min(p_i, p'_i)$. So we have

RESULT 5.4.3 Any feasible solution P to Problem 5.4.1 with $P_{ii} = \min(p_i, p'_i)$ for all i, is an optimal solution to the problem. Thus $P_{ii} = \min(p_i, p'_i)$ for all i is both necessary and sufficient for optimality of P as long as P is feasible.

REMARK 5.4.1 In Keyfitz's scheme we select U_i for both the surveys, with the probability $P_{ii} = \min(p_i, p'_i)$. Since in Case 1, $p'_i > p_i$, we include U_i with probability 1 in the sample selected for Survey II. That is, $P_{ii} = p_i \cdot 1 = p_i = \min(p'_i, p_i)$. In Case 2, $p'_i \leqslant p_i$, so we include U_i with probability p'_i/p_i in the sample selected for Survey II. That is, $P_{ii} = p_i \cdot p'_i/p_i = \min(p'_i, p_i)$. Also, for $i \neq j$ in Keyfitz's scheme we have

$$P_{ij} = \begin{cases} (p_i - p'_i)\left[(p'_j - p_j) / \sum_{j \in J} (p'_j - p_j)\right] & \text{for} \quad i \in I, \quad j \in J \\ 0 & \text{otherwise} \end{cases} \tag{5.4.5}$$

where $J = \{j \mid (p'_j - p_j) > 0\}$ and $I = \{i \mid (p_i - p'_i) > 0\}$. Thus Keyfitz's scheme provides an optimal solution to the Problem 5.4.1.

EXAMPLE 5.4.1 Consider four villages with probabilities of being chosen for crop and demographic surveys, as given in Table 5.4.1.

Table 5.4.1

	Village			
Survey	1	2	3	4
Crop p_i	0.5	0.2	0.1	0.2
Demography p'_i	0.3	0.1	0.4	0.2

Table 5.4.2 P_{ij}*'s*

	j				
i	1	2	3	4	p_i
1	0.3		0.2		0.5
2		0.1	0.1		0.2
3			0.1		0.1
4				0.2	0.2
p'_j	0.3	0.1	0.4	0.2	1.0

Keyfitz's scheme provides the following P_{ij}'s for Problem 5.4.1: $P_{11}=\min(0.5,0.3)=0.3$; $P_{22}=\min(0.2,0.1)=0.1$; $P_{33}=\min(0.1,0.4)=0.1$; and $P_{44}=\min(0.2,0.2)=0.2$.

Other P_{ij}'s are worked out according to (5.4.5). For example,

$$J=\{j\mid p'_j-p_j>0\}=\{3\}$$

$$I=\{i\mid p_i-p'_i>0\}=\{1,2\}.$$

Hence $P_{12}=0$; $P_{13}=(0.5-0.3).(0.4-0.1)/(0.4-0.1)=0.2$; $P_{14}=0$; and so on. Nonzero P_{ij}'s are shown in Table 5.4.2. Presenting the P_{ij}'s in this tabular form is a convenient way of exhibiting any feasible solution to the problem. We shall follow this form in the later discussion as well.

Problem 5.4.2

$$\text{Minimize} \quad \sum_{i=1}^{N}\sum_{j=1}^{N} P_{ij}\cdot|j-i|$$

$$\text{subject to} \quad \sum_{j=1}^{N} P_{ij}=p_i, \qquad i=1,\ldots,N$$

$$\sum_{i=1}^{N} P_{ij}=p'_j, \qquad j=1,\ldots,N$$

$$P_{ij}\geqslant 0, \quad i,j=1,\ldots,N.$$

Problem 5.4.2 has the same feasible region as Problem 5.4.1. Only the objective function is different. In general the objective function coefficients are denoted by C_{ij}. Here $C_{ij}=|j-i|$.

We shall show that Lahiri's scheme provides an optimal basic feasible solution to Problem 5.4.2.

As the problem is a linear-programming problem, we can consider basic feasible solutions. For this purpose we discuss the properties of the constraint matrix and bases.

5.5 CONSTRAINT MATRIX, BASES, PROPERTIES, AND CHARACTERIZATION

Let A denote the $2N\times N^2$ coefficient matrix corresponding to the constraints of Problem 5.4.2. The column vector in A corresponding to the variable P_{ij} contains only two nonzero entries, a "1" in Row i and another "1" in Row $N+j$. Such a matrix A for $N=4$ is shown below:

$$A=\begin{bmatrix} 1&1&1&1&0&0&0&0&0&0&0&0&0&0&0&0\\ 0&0&0&0&1&1&1&1&0&0&0&0&0&0&0&0\\ 0&0&0&0&0&0&0&0&1&1&1&1&0&0&0&0\\ 0&0&0&0&0&0&0&0&0&0&0&0&1&1&1&1\\ 1&0&0&0&1&0&0&0&1&0&0&0&1&0&0&0\\ 0&1&0&0&0&1&0&0&0&1&0&0&0&1&0&0\\ 0&0&1&0&0&0&1&0&0&0&1&0&0&0&1&0\\ 0&0&0&1&0&0&0&1&0&0&0&1&0&0&0&1 \end{bmatrix}.$$

Notice that the sum of the first N rows of such a matrix A is equal to the sum of the last N rows of A. Hence the rows of A are linearly dependent. Therefore the rank of A is less than $2N$. So any one of the rows of A can be dropped as redundant. Let $\bar{A}$ denote the matrix obtained from A by dropping the $2N$th row. Now, the rank of $\bar{A}\leqslant 2N-1$. Suppose the rank of $\bar{A}$ is strictly less than $2N-1$. Then there exist $(\alpha_1,\ldots,\alpha_{2N-1})$ not all zero such that

$$\alpha_1\bar{A}^1+\cdots+\alpha_{2N-1}\bar{A}^{2N-1}=0 \tag{5.5.1}$$

where $\bar{A}^i$ denotes the ith row of $\bar{A}$.

Now $\alpha_1,\ldots,\alpha_N$ must all be zero as the equation obtained in (5.5.1) corresponding to P_{iN} has a "1" in the ith row, and the rest of the entries are all zero in these columns as we have dropped the last row. This is so for all $i=1,\ldots,N$. However, the column corresponding to any variable P_{ij}, $j\neq N$, has a single nonzero element among the rows $N+1,\ldots,2N-1$. Thus if $\alpha_1=\ldots=\alpha_N=0$, (5.5.1) also implies that $\alpha_{N+1},\ldots,\alpha_{2N-1}$ are all zeroes. But the fact is that not all α_i's are zeroes. Hence Rows $1,\ldots,2N-1$ are linearly independent or the rank of $\bar{A}$ is $2N-1$. So is the rank of A.

Thus in any basic feasible solution to the problem, we can have at most $2N-1$ nonzero P_{ij}'s. We present these facts as Result 5.5.1.

RESULT 5.5.1 The coefficient matrix A of Problem 5.4.2 has rank $2N-1$. Hence, any basic feasible solution to Problem 5.4.2 can have at most $2N-1$ nonzero P_{ij}'s.

Consider any $(2N-1)\times(2N-1)$ submatrix B of $\bar{A}$, with the rank of B equal to $2N-1$. Observe that B cannot have $2(2N-1)$ or more nonzero entries. Suppose this were not true, then every column of B would have 2 nonzero elements. Summing the first N rows and the last $N-1$ rows of B will yield the same vector, contradicting the fact that B is nonsingular. Since the rank of B is $2N-1$, there must exist a row of B with a single nonzero entry. The submatrix of B obtained by dropping this row and the corresponding column where this single "1" appears is also nonsingular. So it contains a single "1" in some row. This way we proceed until at last we are left with a single "1". This property of B is known as *triangularity*. Or we say that B is *triangular*. Therefore to find the basic solution corresponding to B, we need not invert B. We can first find the row i_1 in which B has a single "1" in the j_1 column and solve for $P_{i_1 j_1}$. We eliminate $P_{i_1 j_1}$ from further consideration by substituting its value in the other constraints. Similarly $P_{i_2 j_2}$ is found, and so on. This process is known as *back substitution*.

This observation greatly reduces the computation as we do not need to calculate and store the usual simplex table entries. As mentioned earlier it is convenient to give the solution P_{ij} in tabular form. So it is advantageous to identify basic feasible solutions without having to consider the columns corresponding to the positive P_{ij}'s and proving them to be linearly independent. With this end in view, we proceed to discuss certain graph theoretical concepts. Notice that each P_{ij} has a cell in the table of P_{ij}'s (see Example 5.4.4). Let a subset of these N^2 cells be denoted by $\bar{B}$.

DEFINITION 5.5.1 A *graph* $\mathcal{G}$ is a pair of sets $(\mathfrak{N}, \mathcal{E})$, where $\mathfrak{N}$ is a finite set of *points* and $\mathcal{E}$, called the set of *edges*, is a subset of the set of unordered pairs of distinct points in $\mathfrak{N}$. An element of $\mathcal{E}$ is denoted by $(i; j)$ where $i\neq j$, $i, j\in\mathfrak{N}$. The elements of $\mathfrak{N}$ are also called vertices or nodes. If $(i; j)\in\mathcal{E}$, $(j; i)$ also refers to the same edge (i, j), Edge (i, j) is said to be incident at i and j.

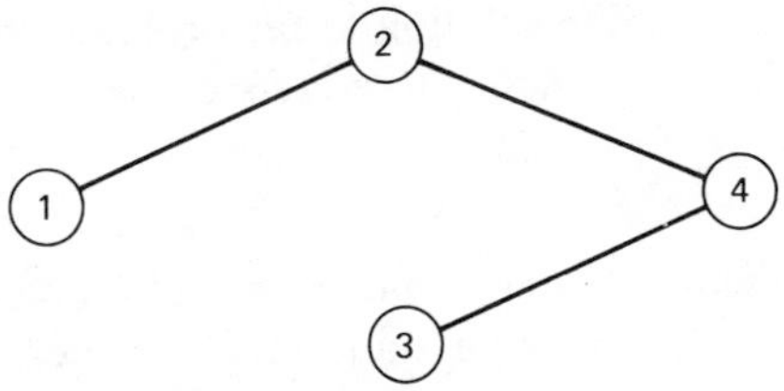

Figure 5.5.1 A connected graph.

DEFINITION 5.5.2 A *path* from i_0 to $i_k, i_0, i_k \in \mathfrak{N}$ in graph $\mathcal{G}$ is an alternating sequence of nodes and edges of the form

$$i_0, (i_0; i_1), i_1, (i_1; i_2), i_2, \ldots, i_{k-1}, (i_{k-1}; i_k), i_k$$

where no edge appears more than once.

EXAMPLE 5.5.1 Consider $\mathcal{G} = (\mathfrak{N}, \mathcal{E})$ where $\mathfrak{N} = \{1,2,3,4\}$ and $\mathcal{E} = \{(1;2),(3;4),(2;4)\}$. Figure 5.5.1 shows a graphical model of $\mathcal{G}$.

DEFINITION 5.5.3 A graph is *connected* if there exists a path from any node of the graph to any other node.

In Fig. 5.5.1 we have a connected graph.

DEFINITION 5.5.4 A *closed path* in a graph is a path from some node i back to itself.

In Fig. 5.5.1 there is no closed path. However, if $\mathcal{E}$ includes the edge $(2;3)$ then $2,(2;4),4,(4;3),3,(3;2),2$ is a closed path.

DEFINITION 5.5.5 A connected graph without any closed path is called a *tree*. In Example 5.5.1 we have a tree. Any node of this graph is designated as a "root" of the tree. Node 2 can be taken as a root in Example 5.5.1.

Now consider the P_{ij} table. The rows of this table are associated with the numbers from 1 to N. The columns of this table are associated with the numbers $N+1$ to $2N$. Let $\mathfrak{N} = \{1, \ldots, N, N+1, \ldots, 2N\}$. Consider a subset $\overline{B}$ with cells corresponding to some P_{ij}'s. Each cell can be associated with two points from $\mathfrak{N}$. Thus, each cell can be represented as an edge. Thus, given a subset $\overline{B}$ of cells, we have an associated graph $\mathcal{G} = (\mathfrak{N}, \mathcal{E})$, where $(i; N+j) \in \mathcal{E}$ if and only if the cell $(i, j) \in \overline{B}$.

Table 5.5.1

	j			
i	1	2	3	4
1		$\surd$		
2		$\surd$	$\surd$	
3			$\surd$	
4				$\surd$

EXAMPLE 5.5.2 Consider the P_{ij} table given in Table 5.5.1, where $i, j=1,\ldots,4$.

Let $\bar{B}=\{(1,2),(2,2),(2,3),(3,3),(4,4)\}$. These cells are marked with a $\surd$ in Table 5.5.1. The corresponding graph $\mathcal{G}$ is given in Fig. 5.5.2. Notice that in such a graph an edge $(i;\,j)$ is such that either $i \leqslant N$ and $j \geqslant N+1$ or $i \geqslant N+1$ and $j \leqslant N$.

RESULT 5.5.2 A subset $\bar{B}$ of columns of $\bar{A}$ is linearly dependent if and only if the associated graph contains a closed path.

Proof Suppose the associated graph contains a closed path, say, $i_1,(i_1;\,i_2),\ldots,(i_{k-1};\,i_1),\,i_1$. Consider the columns corresponding to the edges in this path by alternately assigning coefficients "+1" and "−1" for these columns, and notice that these columns have a "+1" in the i_lth and i_{l+1}th place and that the rest are all zeroes. The implication is that this linear combination yields **0**, and thus that the columns are linearly dependent. On the other hand, suppose the columns in the subset are linearly dependent. Then there exists a nonempty subset $\hat{\bar{B}}$ of $\bar{B}$ such that all the coefficients used while expressing **0** as a linear combination of the columns of $\hat{\bar{B}}$ are nonzero. Hence, each row of $\hat{\bar{B}}$ has at least two "1"s and there are

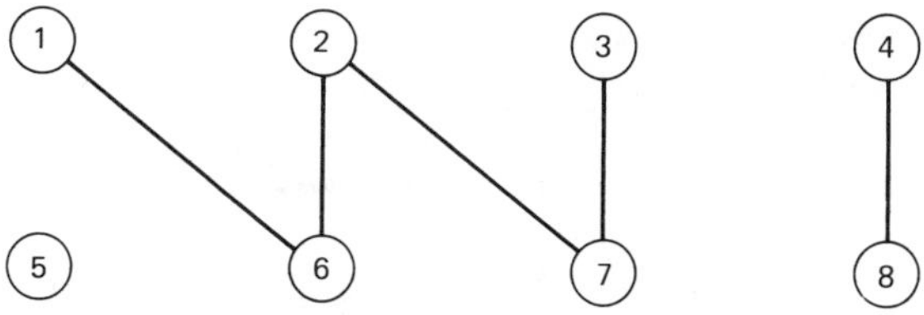

Figure 5.5.2 Graph $\mathcal{G}$ corresponding to Table 5.5.1.

$2N-1$ such rows in $\hat{\bar{B}}$. Let $\hat{B}$ be obtained from $\hat{\bar{B}}$ by taking the corresponding columns of matrix A, that is, including the last row of A. Now consider the path $i_1,(i_1;i_2),i_2,(i_2;i_3),\ldots,(i_{k-1};i_k),i_k$ obtained by starting with Row i_1 Set $l=1$, continuing as follows:

Find a column of $\hat{B}$ which has a "1" in the i_lth row. This column corresponds to the edge $(i_l;i_{l+1})$. Delete this column from further consideration. Set $l=l+1$ and repeat this process, until $i_k=i_1$ for some k.

The path so obtained is a closed path, as there are only finitely many columns in $\bar{B}$, we must repeat row i_1 at some stage. Hence the result.

EXAMPLE 5.5.3 Consider the subset of columns of A corresponding to the subset of cells $\{(1,1),(1,3),(2,2),(2,4),(3,3),(3,4),(5,1),(5,2)\}$ in a 5×5 transportation table. In Table 5.5.2 these columns are marked with $\surd$. These columns of A form the following submatrix.

Node Number

$$\begin{array}{r}1\\2\\3\\4\\5\\6\\7\\8\\9\\10\end{array}\begin{bmatrix}1&1&0&0&0&0&0&0\\0&0&1&1&0&0&0&0\\0&0&0&0&1&1&0&0\\0&0&0&0&0&0&0&0\\0&0&0&0&0&0&1&1\\1&0&0&0&0&0&1&0\\0&0&1&0&0&0&0&1\\0&1&0&0&1&0&0&0\\0&0&0&1&0&1&0&0\\0&0&0&0&0&0&0&0\end{bmatrix}$$

Table 5.5.2

	j				
i	1	2	3	4	5
1	√		√		
2		√		√	
3			√	√	
4					
5	√	√			

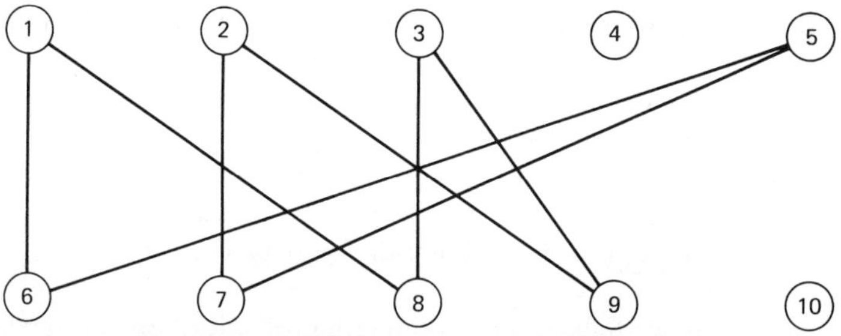

Figure 5.5.3 Graph $\mathcal{G}$ corresponding to Table 5.5.2.

Start with row $i_1=1$. The column corresponding to Cell (1,1) has a "1" in Row 1. This column corresponds to the edge (1;6). We delete this column, and consider Row $i_2=5+1=6$, where there is a "1" in the column corresponding to Cell (1,1). In this row there is a "1" in the column corresponding to Cell (5,1). We delete this column and consider Row $i_3=5$. In Row 5 there is a "1" in the column corresponding to Cell (5,2). We delete this colunn, and so on. The closed path we obtain is given by

$$1,(1;6),6,(6;5),5,(5;7),7,(7;2),2,(2;9),9,(9;3),3,(3;8),8,(8;1),1.$$

This path is shown in the table by dashed lines connecting these cells. Also the graph is shown in Fig. 5.5.3.

Thus we have characterized the linear dependence of columns of $\bar{A}$, by the presence of a closed path in the associated graph. So if we have a subset B of $2N-1$ columns of $\bar{A}$, such that the associated graph has no closed path in it, then these columns form a basis for $\bar{A}$.

RESULT 5.5.3 A graph $\mathcal{G}$ with n edges and at most $n+1$ nodes, having no closed path, is connected and hence is a tree.

Proof (By induction on n) The graph $\mathcal{G}$ consisting of a single edge is connected, so the result is true for $n=1$. For $n>1$, since there is no closed path in $\mathcal{G}$, there exists a node which is incident with only one edge, the one connecting this node with the rest of the nodes of $\mathcal{G}$. Deleting this node and this edge we have a graph $\mathcal{G}'$ with $n-1$ edges and at most n nodes, and has no closed path. The induction hypothesis is that such a graph is connected; hence $\mathcal{G}$ is also connected. By definition a connected graph without any closed path is a tree. Hence the result.

RESULT 5.5.4 The graph $\mathcal{G}$ associated with a subset $\bar{B}$ of $2N-1$ linearly independent columns of $\bar{A}$, is a tree.

Proof $\mathcal{G}$ contains no closed path. $\mathcal{G}$ has $2N-1$ edges and $2N$ nodes. From Result 5.5.3 it is a tree. Hence the result.

Obtaining a Basic Feasible Solution

We describe below a general procedure for obtaining a basic feasible solution to a problem with a constraint matrix that is a transportation matrix.

At each stage, we have a set of remaining cells. Initially, all the cells are remaining. One of these cells is selected according to a specified rule. For different selection rules we obtain different basic feasible solutions. However, the steps of the procedure at each stage are the same otherwise. Set $k=1$ (stage index).

Step 1 According to the selection rule under consideration, choose a cell (r, s) among the remaining cells; include it among the set of basic cells. Set $P_{rs}=\text{minimum}\ (p_r, p'_s)$.

Step 2 If $p_r<p'_s$, delete the cells of the rth row from the set of remaining cells. Set $p'_s=p'_s-p_r$. Go to Step 3. If $p_r>p'_s$, delete the cells of the sth column from the set of remaining cells. Set $p_r=p_r-p'_s$. Go to Step 3. If $p_r=p'_s$ and the remaining cells lie in two or more rows, delete the cells in the rth row from the set of remaining cells. Set $p'_s=0$. Go to Step 3. Otherwise, that is, if $p_r=p'_s$ and the remaining cells lie in a single row, delete the cells of the sth column from the set of remaining cells. Set $p_r=0$. Go to Step 3.

Step 3 If $k=2N-1$, Stop. Otherwise, set $k=k+1$. Go to Step 1.

In this procedure, we choose $2N-1$ basic cells. The corresponding columns of the $\bar{A}$ matrix form a basis, as the associated graph $\mathcal{G}$ has no closed path, as we never return to a row or column of the P_{ij} table, once we have selected a cell from there. Notice that the value of P_{ij}'s corresponding to the selected cells is always greater than or equal to zero. Hence this procedure obtains a basic feasible solution.

Selection rules We describe two selection rules here.

1. **Northwest corner rule (NWCR)** Select the cell (r, s) such that $(r+s)$ is a minimum among the remaining cells.

2. **Matrix Minima rule (MMR)** Select the cell (r, s) such that the coefficient of P_{ij} in the objective function of the problem is at a minimum among the remaining cells.

Table 5.5.3

	j				
i	1	2	3	4	p_i
1	0	1	2	3	0.5
2	1	0	1	2	0.2
3	2	1	0	1	0.1
4	3	2	1	0	0.2
p'_j	0.3	0.1	0.4	0.2	1.0

EXAMPLE 5.5.4 Consider the data given in Example 5.4.1. Suppose we wish to find a basic feasible solution to Problem 5.4.2. We have Table 5.5.3 giving the objective function coefficients $C_{ij} = |j-i|$ and the p_i, p'_j for $i, j = 1,\ldots,4$.

We find it convenient to give the values of the P_{ij}'s selected as basic variables in Table 5.5.3 within a box. Let us consider NWCR as the selection rule. In Stage $k=1$, we select the cell (1, 1), as $r+s=2$ and is the

Table 5.5.4

	j				
i	1	2	3	4	p_i
1	0 [0.3]	1 [0.1]	2 [0.1]	3	0.5
2	1	0	1 [0.2]	2	0.2
3	2	1	0 [0.1]	1	0.1
4	3	2	1	0 [0.2]	0.2
p'_j	0.3	0.1	0.4	0.2	1.0

Table 5.5.5

i	j = 1	2	3	4	p_i
1	0 [0.3]	1	2 [0.2]	3	0.5
2	1	0 [0.1]	1 [0.1]	2	0.2
3	2	1	0 [0.1]	1	0.1
4	3	2	1	0 [0.2]	0.2
p'_j	0.3	0.1	0.4	0.2	1.0

minimum among the remaining cells. Now as per Step 1, P_{11} = minimum (0.5, 0.3) = 0.3. We delete the first-column cells from the set of remaining cells, according to Step 2, and set $P_1 = 0.5 - 0.3 = 0.2$. In Stage $k = 2$ we select the cell (1,2), as $r+s=3$ and is the minimum among that of the remaining cells. Now as per Step 1, P_{12} = minimum (0.2, 0.1) = 0.1. We delete Column 2 from the set of remaining cells, and so on. The basic feasible solution obtained is shown in Table 5.5.4.

Suppose we use the matrix minima rule for selection of the basic cells. In stage 1, we can select any one of the cells (1,1), (2,2), (3,3), or (4,4) as all of them have $C_{ij} = 0$ and zero is the minimum of the C_{ij}'s for all the cells. Suppose we select (1,1). Then P_{11} = minimum(0.5, 0.3) = 0.3. Delete the cells in Column 1. Set $p_1 = 0.2$. In Stage 2, we can select cell (2,2) as it has the minimum C_{ij} value among the remaining cells. Now, P_{22} = min (0.1, 0.2) = 0.1 and so on. The solution obtained is shown in Table 5.5.5.

REMARK 5.5.1 The discussions and results can be easily seen to hold for the general transportation problem, with obvious modifications. The general transportation problem can be stated as follows:

$$\text{Minimize} \quad \sum_{j=1}^{n} \sum_{i=1}^{m} C_{ij} P_{ij}$$

$$\text{subject to} \quad \sum_{j=1}^{n} P_{ij} = p_i, \qquad i=1,\ldots,m$$

$$\sum_{i=1}^{m} P_{ij} = p'_j, \qquad j=1,\ldots,n \tag{5.5.2}$$

$$P_{ij} \geqslant 0.$$

The number of rows in the constraint matrix A here is $m+n$ instead of $2N$, as in Problem 5.4.1 or 5.4.2. This brings in the obvious changes in the rank of the matrix and the number of columns in a basis for the problem. $2N-1$ must be read as $(m+n-1)$ in these results.

In the next section, we prove the optimality of NWCR for Problem 5.4.2.

5.6 NORTHWEST CORNER RULE (NWCR), LAHIRI'S SELECTION SCHEME, AND OPTIMAL BASIC SOLUTIONS

Consider Problem 5.4.2. Let $P=((P_{ij}))$. Let $C(P)$ denote the objective function value corresponding to the solution P. Using NWCR we select one basic cell in each stage, until we have $2N-1$ basic cells in all. Let the cell selected at the lth stage be designated as the lth basic cell. Let the adjusted values of p_i, p'_j at the lth stage be denoted by $p_i(l)$ and $p'_j(l)$, respectively.

RESULT 5.6.1 Let the indices s, l, r, and k be such that $s<l$ and $r<k$, with $P_{ks}>0$ and $P_{rl}>0$ for a feasible solution P to Problem 5.4.2. Then there exists $P'=((P'_{ij}))$ such that $C(P')\leqslant C(P)$ and $P'_{rs}>P_{rs}$.

Proof: Consider $P'=((P'_{ij}))$ given by

$$P'_{ij}=P_{ij}, \qquad i\neq r,k, \qquad j\neq s,l$$

$$P'_{rs}=P_{rs}+\varepsilon, \qquad P'_{kl}=P_{kl}+\varepsilon, \qquad P'_{rl}=P_{rl}-\varepsilon, \qquad \text{and} \quad P'_{ks}=P_{ks}-\varepsilon.$$

where $\varepsilon=\min(P_{ks}, P_{rl})>0$. We wish to show that $C(P')=\sum\sum P'_{ij}|j-i|\leqslant$

$C(P)=\Sigma\Sigma P_{ij}|j-i|$ or $\Sigma\Sigma(P'_{ij}-P_{ij})|j-i|\leqslant 0$. By definition of P'_{ij}, we have to show that

$$(P'_{rs}-P_{rs})|s-r|+(P'_{kl}-P_{kl})|l-k|$$
$$+(P'_{rl}-P_{rl})|l-r|+(P'_{ks}-P_{ks})|s-k|\leqslant 0,$$

that is,

$$\varepsilon[|s-r|+|l-k|-|l-r|-|s-k|]\leqslant 0.$$

Let Δ denote the term inside the square bracket in the expression above. Thus we are required to show that $\Delta\leqslant 0$ as $\varepsilon>0$.

CASE 1 $s\leqslant r$, $l\leqslant k$, and $r\leqslant l$.

$$\Delta=(r-s)+(k-l)-(k-s)-(l-r)=2r-2l\leqslant 0.$$

CASE 2 $s\leqslant r$, $l\leqslant k$, and $l<r$.

$$\Delta=(r-s)+(k-l)-(k-s)-(r-l)=0.$$

CASE 3 $s\leqslant r$, $l>k$, and $r\leqslant l$.

$$\Delta=(r-s)+(l-k)-(k-s)-(l-r)=2(r-k)<0.$$

CASE 4 $s>r$, $l\leqslant k$, and $r\leqslant l$.

$$\Delta=(s-r)+(k-l)-(k-s)-(l-r)=2(s-l)<0.$$

CASE 5 $s>r$, $l>k$, and $r\leqslant l$.

$$\Delta=(s-r)+(l-k)-|k-s|-(l-r)$$
$$=(s-k)-|k-s|\leqslant 0.$$

This completes the proof, as either $s\leqslant r$ or $s>r$ and $r\leqslant l\leqslant k$, $l<r$ or $l>k$, covers all the cases and $s>r$ and $l<r$ is not possible as $s<l$.

RESULT 5.6.2 A basic feasible solution obtained by NWCR is optimal for Problem 5.4.2.

Proof Let P^* denote the matrix corresponding to the basic feasible solution obtained by NWCR.

Suppose P^* is not optimal. Let there exist a $P \neq P^*$ such that $C(P) < C(P^*)$. Now let l be the smallest positive integer such that the first $l-1$ Northwest corner cells of P^* agree with P. Let the next Northwest corner cell determined by the NWCR be the cell (r, s), as the cell (r, s) is such that $P^*_{rs} = \min\{p_r(l), p'_s(l)\}$. If $P_{rs} \neq P^*_{rs}$ it must be $P_{rs} < P^*_{rs}$. Therefore, there exist k, l such that $r < k$ and $s < l$ such that $P_{ks} > 0$ and $P_{rl} > 0$. Now applying Result 5.6.1 we can show that there exists a P' as good as P and $P'_{rs} > P_{rs}$. If $P'_{rs} = P^*_{rs}$, we have shown, if there exists a P agreeing with P^* up to the first $l-1$ NWC cells then there exists a P' agreeing with P^* up to the first l Northwest corner cells. If $P'_{rs} < P^*_{rs}$ we can repeat with $P = P'$ till P^*_{rs} equals the new P_{rs} obtained. Hence the result.

REMARK 5.6.1 Comparison of NWCR and Lahiri's scheme makes it clear that they are one and the same. Thus we have shown the optimality of Lahiri's scheme for Problem 5.4.2.

5.7 SOME COMMENTS ON THE SELECTION SCHEMES CONSIDERED FOR INTEGRATING SURVEYS

Consider Lahiri's method of numbering the units in a serpentine manner (see Fig. 5.4.1). We find that two units geographically contiguous to a third unit may not be so considered, as they receive different numbers. In fact in the mathematical model considered by Raj the costs are proportional to $|i-j|$. For instance, in Fig. 5.4.1, units U_2 and U_6 are adjacent to unit U_1 but we give a cost 5 for unit U_1 to U_6 and only 1 for unit U_1 to U_2.

Instead we may use any C_{ij} that reflects the actual cost of traveling from U_i to U_j. In that case we have a general transportation problem.

Let us now consider Problem 5.4.1. The objective function in Problem 5.4.1 is

$$\text{Maximize} \quad \sum_{i=1}^{N} P_{ii}.$$

The purpose is maximization of the expected number of overlaps. Overlaps between the samples chosen for the two surveys can happen, even when we choose i for the first survey and j for the second survey, with probability P_{ij}; and similarly j is chosen for the first survey and i is chosen for the second survey, with probability P_{ji}.

First we note that in an optimal solution given to Problem 5.4.1 only overlaps of the kind $(i_1, i_1), \ldots, (i_n, i_n)$ are possible.

RESULT 5.7.1 For Problem 5.4.1, in an optimal solution given by $P_{ii}=\min(p_i, p'_i)$, $i=1,2,\ldots,N$ and P_{ij}'s are chosen so as to satisfy the constraints of the problem; only overlaps of the kind $(i_1,i_1),\ldots,(i_n,i_n)$ are possible.

Proof Consider $(i_1,i_2,\ldots,i_n)$ indices of a sample chosen for Survey I. To have an overlap other than the one given by Result 5.7.1, that is, $(i_1,j_1),\ldots,(i_n,j_n)$ without all $i_k=j_k$ and $(j_1,\ldots,j_n)$ a permutation of $(i_1,i_2,\ldots,i_n)$, we require $P_{i_kj_k}>0$ for all $k=1,\ldots,n$. Now, for any i_k,

$$P_{i_ki_k}=\min\{p_{i_k}, p'_{i_k}\}.$$

If $p'_{i_k}>p_{i_k}$, then $P_{i_kr}=0$ for all $r\neq i_k$. Hence j_k has to be i_k. If $p'_{i_k}\leqslant P_{i_k}$, the $P_{ri_k}=0$, for all $r\neq i_k$. We can also come to the same conclusion in this case. Hence, a correct objective for maximizing the expected number of overlaps is not

$$\text{Maximize}\quad \sum_{i=1}^{N} P_{ii}.$$

However if the sample size $n=1$, then this is appropriate for a problem of maximizing the expected overlap. In Section 5.8 a transportation model is considered that actually solves this problem of maximizing expected overlap in general.

With Roy Choudhury's scheme, it can be shown that a particular sample s is selected for Survey I with probability proportional to the total size of the sample, that is,

$$p(s)=Q\sum_{i\in s} p_i$$

where

$$Q=\left[{}^{N-1}c_{n-1}\right]^{-1}.$$

Such is the case because for any sample $s=\{i_1,\ldots,i_n\}$, if i_j is chosen in the first selection for Survey I with probability p_{i_j}, whatever appears in the second draw, in the subsequent sample of size $n-1$ we can select the remaining $(n-1)$ elements of s with probability Q. Therefore,

$$p(s)=\sum_{j=1}^{n} p_{i_j}Q.$$

So $p(s)$ is proportional to the total size of the sample s. However, for the second survey the sample is not necessarily selected with probability proportional to the total size, the size measure being according to Survey II.

Consider the case in which $n=2$, $n<N$. Let the sample be $\{i, j\}$, $i \neq j$. Then we choose this sample for Survey II, when

(1) $i(j)$ is chosen for the first survey, $j(i)$ is chosen for the second survey, and the unit $\{i, j\}$ appears in the sample subsequently chosen from the $N-1$ units. (Recall Roy Choudhury's scheme, given in Section 5.4.1.) Or,

(2) k, different from i and j is chosen for Survey I, $j(i)$ is chosen for the second survey, and $i(j)$ is chosen in the subsequent sampling.

The probability p of $\{i, j\}$ being as selected for Survey II is given by

$$p=(p_i p_j'/1-p_i')Q+(p_j p_i'/1-p_j')Q$$

$$+\sum_{i\neq k\neq j}(p_k p_i'/1-p_k')Q+\sum_{i\neq k\neq j}(p_k p_j'/1-p_k')Q.$$

Now, combining the first two terms and the last two terms, we have

$$p=(p_i p_j'/1-p_i'+p_j p_i'/1-p_j')Q$$

$$+\left[\sum_{i\neq k\neq j}(p_k p_i'/1-p_k')+\sum_{i\neq k\neq j}(p_k p_j'/1-p_k')\right]Q.$$

Adding, subtracting

$$(p_i p_i'/1-p_i'+p_j p_j'/1-p_j')Q$$

from the second term in the expression above, and simplifying, we get

$$p=Q\sum_r(p_r/1-p_r')(p_i'+p_j')-Q(p_i p_i'/1-p_i'+p_j p_j'/1-p_j').$$

Thus p is not proportional to $(p_i'+p_j')$, as claimed.

However, in this scheme at most one unit is different in the two samples selected for the two surveys. In Section 5.10 we give another transportation model for which an optimal solution chooses two samples with at most one unit different. The probabilities of selecting these samples are proportional to the total size of each of the samples.

5.8 INTEGRATION OF SURVEYS MINIMIZING THE EXPECTED COST WHILE UNITS ARE INCLUDED IN THE SAMPLE WITH PROBABILITY PROPORTIONAL TO SIZE

Let the sample size be n. Let $S=\{s_1,\ldots,s_{N_{c_n}}\}$ be the set of all possible samples from U. Let $p(s_i)$ denote the probability of selecting sample s_i for Survey I, while the units are included in the sample with probability p_i proportional to size. Similarly let $p'(s_i)$ denote that of Survey II.

Let C_{ij} be the cost of selecting sample s_i for Survey I and sample s_j for Survey II. Let P_{ij} be the probability with which we select samples s_i and s_j for two surveys, respectively. Let $N_{c_n}=t$.

Problem 5.8.1

$$\text{Minimize} \quad \sum_i \sum_j C_{ij} P_{ij}$$

$$\text{subject to} \quad \sum_j P_{ij}=p(s_i), \qquad i=1,\ldots,t$$

$$\sum_i P_{ij}=p'(s_j), \qquad j=1,\ldots,t$$

$$P_{ij}\geqslant 0, \qquad i,j=1,\ldots,t.$$

The problem of maximizing the expected number of overlaps is a special case of Problem 5.8.1, as is evident when the objective function coefficients are specified as follows:

Let $C_{ij}=n-\#(s_i\cap s_j)$, where $\#(s_i\cap s_j)$ denotes the cardinality of $s_i\cap s_j$.

The problem of maximizing the probability of complete overlap can also be seen as Problem 5.8.1 with $C_{ii}=-1$ and $C_{ij}=0$, $i\neq j$. A simple-solution procedure along the same line as that of Problem 5.4.1 is possible with this special objective function.

If C_{ij} reflects the cost of conducting the two surveys with samples s_i and s_j for Surveys I and II, respectively, depending on the units selected in the samples, we have a more general objective than Lahiri's.

Problem 5.8.1 is also a transportation problem, and we describe a method to find an optimal solution to the problem. We have already described a procedure that finds a basic feasible solution to a transportation problems. As in the linear-programming simplex procedure, we derive the criteria for exit of a basic column from the basis and entry of a nonbasic column into the basis, and also provide an optimality criterion for termination of the procedure. However, we make use of the observations on the special properties of the bases, and the duality results of linear programming, to modify the simplex iteration so as to achieve considerable reduction in computational burden.

5.9 DUAL OF THE PROBLEM, OPTIMALITY CRITERION, AND IMPROVEMENT PROCEDURES

Let u_i, v_j denote the dual variables associated with the first and second set of constraints in Problem 5.8.1. The dual of this problem can now be written as

$$\begin{aligned} &\text{Maximize} \quad \sum_{i=1}^{t} p_i u_i + \sum_{j=1}^{t} p'_j v_j \\ &\text{subject to} \quad u_i + v_j \leqslant C_{ij}, \quad i, j = 1,\ldots,t. \end{aligned} \tag{5.9.1}$$

u_i, v_j unrestricted in sign for $i, j = 1,\ldots,t$.

Notice that the rank of the constraint matrix A of Problem 5.8.1 is $2t-1$, and we can drop the last constraint of the primal problem and obtain matrix $\bar{A}$, which has $2t-1$ rows and t^2 columns. In the dual problem it corresponds to setting $v_t = 0$. Thus, we eliminate this dual variable from further consideration. There are, then, $2t-1$ variables and t^2 constraints in the dual problem.

Let $C_{\bar{B}}$ denote the objective function coefficients of Problem 5.8.1, corresponding to a feasible basis $\bar{B}$ of $\bar{A}$. We know from the duality theorem of linear programming that $C_{\bar{B}}\bar{B}^{-1}$ provides the corresponding dual solution. Let this dual solution be denoted as the vector $\lambda' = (u_1,\ldots,u_t, v_1,\ldots,v_{t-1})$. Then

$$C_{\bar{B}}\bar{B}^{-1} = \lambda' \tag{5.9.2}$$

or equivalently

$$C_{\bar{B}} = \lambda'\bar{B}.$$

Finding λ' does not require inverting $\bar{B}$, for the structure of $\bar{B}$ is such that we can obtain λ' by back-substitution. The proof is along the same lines as that of the proof given for obtaining a basic feasible solution to Problem 5.4.2 in Section 5.4, when the basis is given. Corresponding to a basic cell

(r, s) the column in the matrix $\bar{A}$ has a "1" in the rth and $t+s$th coordinates for $s \neq t$. If $s=t$, then as we have dropped the $2t$th row from matrix A, we have in a column of $\bar{A}$ corresponding to the cell (r, s), a single "1" in the rth coordinate. Thus the equation $C_B = \lambda'\bar{B}$ yields for a basic cell (r, s)

$$C_{rs} = \begin{cases} u_r + v_s & \text{for} \quad s \neq t, \\ u_r & \text{for} \quad s = t. \end{cases} \tag{5.9.3}$$

u_i, v_j's can be calculated using (5.9.3) and back-substitution in a unique manner.

We have brought in the discussion of the dual solution to the problem, as the solution can be used in determining the vector to enter the basis. In the simplex procedure, we calculate $C_j - Z_j$'s for all nonbasic columns and choose the vector accordingly. As the variables in Problem 5.8.1 are double-subscripted, we have "Z_{ij}" in place of "Z_j." By definition

$$Z_{ij} = C_{\bar{B}}\bar{B}^{-1}\bar{\mathbf{a}}_{ij},$$

where $\bar{\mathbf{a}}_{ij}$ is the column corresponding to P_{ij} in $\bar{A}$. However,

$$C_{\bar{B}}\bar{B}^{-1} = \lambda'.$$

Therefore,

$$Z_{ij} = \lambda'\bar{\mathbf{a}}_{ij} \qquad i = 1, \ldots, t$$

$$= u_i + v_j \qquad j = 1, \ldots, t-1 \tag{5.9.4}$$

from the nature of $\bar{\mathbf{a}}_{ij}$. Hence as $v_t = 0$, we can write

$$C_{ij} - Z_{ij} = C_{ij} - (u_i + v_j) \qquad \text{for} \quad i, j = 1, \ldots, t. \tag{5.9.5}$$

Notice that if $C_{ij} - Z_{ij} \geqslant 0$ for i, j then λ' is feasible for the dual problem with $v_t = 0$. Then, from the duality theorem of linear programming, $\bar{B}$ is an optimal basis for Problem 5.8.1. Thus we have the optimality criterion given by

$$\Delta_{ij} = C_{ij} - Z_{ij} \geqslant 0 \qquad \text{for all} \quad i, j = 1, \ldots, t \tag{5.9.6}$$

If (5.9.6) is not satisfied, corresponding to a basis $\bar{B}$, then we find (r, s)

such that

$$C_{rs} - Z_{rs} = \min\{C_{ij} - Z_{ij} | C_{ij} - Z_{ij} < 0\} \tag{5.9.7}$$

and introduce (r, s) as a basic cell. The cell to be removed from the set of basic cells corresponding to $\bar{B}$ is determined by the following rule:

Find a closed path in the graph associated with the basic cells corresponding to $\bar{B}$ and the cell (r, s). Suppose

$$r, (r; s), s, (s; i_2), \ldots, (i_{k-1}; r), r$$

is a closed path in the graph. Partition the edges in the closed path into two sets E^+ and E^- as follows:

> Include $(r; s)$ in E^+. Then alternatingly include the edges along the path in E^- and E^+. Finally, $(i_{k-1}; r)$ will be included in E^-.

Now calculate $\varepsilon = \min\{P_{ij}$ corresponding to the edges in $E^-\}$. If the minimum is attained for $(l; k) \in E^-$, then the corresponding cell is made nonbasic and (r, s) is made basic. The new basic solution is given by:

$$\hat{P}_{ij} = \begin{cases} P_{ij} + \varepsilon & \text{if} \quad (i, j) \text{ corresponds to an edge } \in E^+ \\ P_{ij} - \varepsilon & \text{if} \quad (i, j) \text{ corresponds to an edge } \in E^- \\ P_{ij} & \text{otherwise.} \end{cases} \tag{5.9.8}$$

Justification follows from the criterion for exit in the simplex procedure, and the property of the basis, $\bar{B}$.

EXAMPLE 5.9.1 Consider the four village, two-survey problem discussed in Example 5.4.1. We have the data as given in Table 5.9.1.

Table 5.9.1

Survey		Village			
		1	2	3	4
I	(p_i)	0.5	0.2	0.1	0.2
II	(p_i')	0.3	0.1	0.4	0.2

Let the sample size n be equal to 2. Suppose we wish to solve the corresponding problem, Problem 5.8.1. We have to obtain $p(s_i)$ and $p'(s_j)$ for all samples s_i, s_j of size 2. For instance, for $s_i=(1,2)$, we have $p(s_1)=0.325$. As we obtain them we notice that we are sampling without replacement; the units are included with probability proportional to size and the units may be included in the sample in any order.

$$p(s_1)=p_1\cdot\frac{p_2}{1-p_1}+p_2\cdot\frac{p_1}{1-p_2}$$

$$=0.5\frac{0.2}{0.5}+0.2\frac{0.5}{0.8}=0.2+0.125=0.325.$$

Similarly the $p(s_i)$ and $p'(s_j)$ are worked out as shown in Table 5.9.2. In this example, C_{ij}'s are so chosen that $C_{ii}=0$ and the other C_{ij}'s are arbitrary. The C_{ij} are shown at the upper right corner of the cells in Table 5.9.2. Also we have a initial basic solution given, which is shown in Table 5.9.2 within the basic cells. We wish to see whether this solution is optimal. Otherwise we apply the improvement procedure described earlier. Here $t=4c_2=6$. We first set $v_6=0$. Then using (5.9.3), u_3, u_5, and u_6 are calculated first, as (3,6), (5,6), and (6,6) are the basic cells in Column 6. Now $u_3=C_{36}=1$; $u_5=C_{56}=3$, and $u_6=C_{66}=0$. Since $u_3=1$, $v_2+u_3=C_{32}$ as (3; 2) is a basic cell. So $v_2=1-1=0$. Similarly $v_3=0-1=-1$. Now $v_2=0$ yields $u_1=1$, $u_2=0$. As $u_1=1$, v_1 turns out to be -1 to satisfy $v_1+u_1=C_{11}=0$. Similarly $v_4=0$ is obtained, yielding $u_4=0$. We also have $u_5=3$, yielding $v_5=-3$. Thus we have obtained all the u_i's and v_j's; they are shown in Table 5.9.2.

We proceed to calculate $\Delta_{ij}=C_{ij}-(u_i+v_j)$ for all nonbasic (i,j)'s. For instance, for cell (1,3), we have $\Delta_{13}=1-(1-1)=1$; for cell (5,3), we have $\Delta_{53}=1-(3-1)=-1$. We show these Δ_{ij}'s at the lower left corner of the cells. As only negative Δ_{ij}'s are needed for identifying the cell to enter the basis, only these Δ_{ij}'s are shown in the table.

Cells (5,1), (5,3), and (5,4) have negative Δ_{ij}. As the minimum of these Δ_{ij}'s is -2, attained for cell (5,4), we choose (5,4) for entry.

The closed path in the corresponding graph is shown in Fig. 5.9.1. Recall that the nodes of the graph are numbered from 1 to $2t$ where $t=6$. So we have the edge (5; 10) corresponding to the cell (5,4), and so on.

Now

$$E^+=\{(5;10),(1;8),(3;12)\}$$

$$E^-=\{(10;1),(8;3),(12;5)\}.$$

Table 5.9.2 C_{ij}/P_{ij}

i	j → / s_i ↓ s_j →	1 (1,2)	2 (1,3)	3 (1,4)	4 (2,3)	5 (2,4)	6 (3,4)	$p(s_i)$	u_i
1	1,2	0 $\boxed{0.076}$	1 $\boxed{0.184}$ +	1	1 $\boxed{0.065}$ −	1	2	0.325	1
2	1,3	1	0 $\boxed{0.156}$	1	1	3	1	0.156	0
3	1,4	1	1 $\boxed{0.031}$ −	0 $\boxed{0.161}$	2	1	1 $\boxed{0.133}$ +	0.325	1
4	2,3	1	1	2	0 $\boxed{0.047}$	1	1	0.047	0
5	2,4	1	3	1	1	0 $\boxed{0.047}$	3 $\boxed{0.053}$ −	0.100	3
6	3,4	−1 2	1	−1 1	−2 + 1	3	0 $\boxed{0.047}$	0.047	0
$p'(s_j)$		0.076	0.371	0.161	0.112	0.047	0.233	1.000	—
v_j		−1	0	−1	0	−3	0	—	—

Corresponding cells in the table are marked + and −, respectively. Therefore,

$$\varepsilon = \min(P_{14}, P_{32}, P_{56})$$

$$= \min(0.065, 0.031, 0.053)$$

$$= 0.031.$$

So cell (3,2) leaves the basis. The new solution is obtained according to

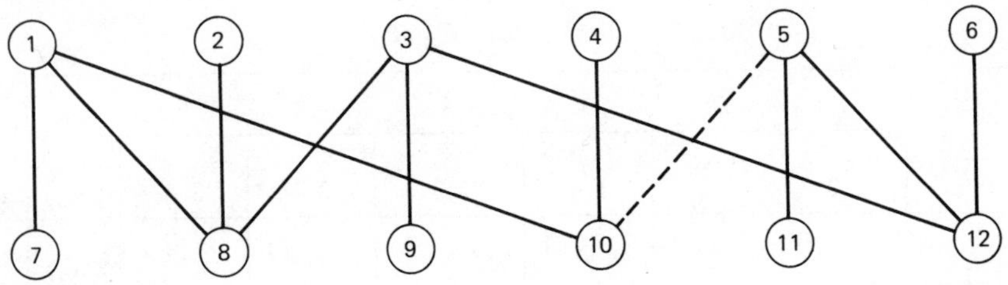

Figure 5.9.1 Graph $\mathcal{G}$ corresponding to Table 5.9.2. The closed path is 5,(5;10), 10,(10;1), 1,(1;8), 8,(8;3), 3,(3;12), 12,(12;5), 5.

Table 5.9.3 C_{ij}/P_{ij}

i	s_i \ j, s_j	1 1,2	2 1,3	3 1,4	4 2,3	5 2,4	6 3,4	$p(s_i)$	u_i
1	1,2	0 [0.076]	1 [0.215]	1 −1	1 [0.034]	1	2 −1	0.325	3
2	1,3	1	0 [0.156]	1	1	3	1 −1	0.156	2
3	1,4	1	1	0 [0.161] −	2	1	1 [0.164] +	0.325	1
4	2,3	1	1	2	0 [0.047]	1	1 −1	0.047	2
5	2,4	1	3	1 −1 +	1 [0.031]	0 [0.047]	3 [0.022] −	0.100	3
6	3,4	2	1	1	1	3	0 [0.047]	0.047	0
$p'(s_j)$		0.076	0.371	0.161	0.112	0.047	0.233	1.000	—
v_j		−3	−2	−1	−2	−3	0	—	—

(5.9.8). Table 5.9.3 gives this solution. We have

$$\hat{P}_{54}=0.031;\ \hat{P}_{14}=0.065-0.031=0.034$$

$$\hat{P}_{12}=0.184+0.031=0.215, \text{ and so on.}$$

With the new basis we again calculate u_i's and v_j's, starting with $v_6=0$ (Table 5.9.3). We calculate Δ_{ij}'s. Cells with negative Δ_{ij}'s have the value of Δ_{ij} at the lower left-hand corner. Minimum of these Δ_{ij} is -1. We choose cell (5,3) for entry. There are other choices, though. The closed path in the associated graph is shown in Fig. 5.9.2. We have

$$E^{+}=\{(5;9),(3;12)\}$$

$$E^{-}=\{(9;3),(12;5)\}$$

Then $\varepsilon=$minimum $(0.161, 0.022)=0.022$, and cell (5,6) leaves the basis. The new solution is shown in Table 5.9.4. We calculate the u_i's and v_j's and find that all the Δ_{ij}'s are nonnegative. We stop. The solution shown in Table 5.9.4 is optimal.

REMARK 5.9.1 For maximizing the probability of complete overlap, we use the objective function $C_{ii}=-1$ and $C_{ij}=0$ for $i\neq j$. And for such a problem we have to allot as much as possible for all the P_{ii}'s. The procedure is to have $P_{ii}=\min(p(s_i), p'(s_i))$ for $i=1,\ldots,t$ and the rest of the P_{ij}'s should be such that P is feasible for the problem. In this sense, for the objective of maximizing the probability of complete overlap we have an optimal solution given in Table 5.9.2: $\Sigma P_{ii}=0.534$. However, from the solution obtained in Problem 5.4.1 this probability of complete overlap works out to be only 0.430.

Thus solution to Problem 5.4.1 does not in general optimize the probability of complete overlap.

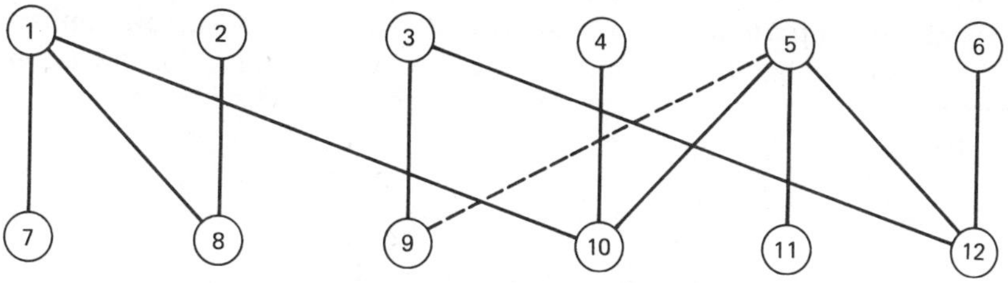

Figure 5.9.2 Graph $\mathcal{G}$ corresponding to Table 5.9.3. The closed path is 5,(5;9), 9,(9;3), 3,(3;12), 12,(12;5), 5.

Table 5.9.4 C_{ij}/P_{ij}

i	s_i \ j, s_j	1 1,2	2 1,3	3 1,4	4 2,3	5 2,4	6 3,4	$p(s_i)$	u_i
1	1,2	0 / 0.076	1 / 0.215	1	1 / 0.034	1	2	0.325	2
2	1,3	1	0 / 0.156	1	1	3	1	0.156	1
3	1,4	1	1	0 / 0.139	2	1	1 / 0.186	0.325	1
4	2,3	1	1	2	0 / 0.047	1	1	0.047	1
5	2,4	1	3	1 / 0.022	1 / 0.031	0 / 0.047	3	0.100	2
6	3,4	2	1	1	1	3	0 / 0.047	0.047	0
$p'(s_j)$		0.076	0.371	0.161	0.112	0.047	0.233	1.000	—
v_j		−2	−1	−1	−1	−2	0	—	—

5.10 INTEGRATION OF SURVEY WITH PROBABILITY PROPORTIONAL TO TOTAL SIZE OF THE SAMPLES

Consider the set of N_{c_n} samples $S=\{s_1, s_2, \ldots, s_{N_{c_n}}\}$. Let $p(s_i)$ be the probability proportional to the total size of s_i for the first survey. Similarly $p'(s_i)$ is the probability proportional to the total size of s_i for the second survey.

Let

$$C_{ij}=\begin{cases}0 & \text{if } \; s_i=s_j \\ 1 & \text{if } \; \#(s_i \cap s_j)=n-1 \\ M & \text{otherwise, } M \text{ is a large positive.}\end{cases} \tag{5.10.1}$$

With these changes we can show that Problem 5.8.1 solves the problem of integrating surveys with probability proportional to the total size, such that the samples chosen differ by at most one unit. That is, Choudhury's objective of having at least $n-1$ units common to the two samples is achieved, without violating the restriction on the probability of selection of sample for Survey II.

Let $F=\{(i, j) | C_{ij} \leqslant 1\}$.

RESULT 5.10.1 For Problem 5.8.1 with C_{ij} as given by (5.10.1) there exists a feasible solution with the property $P_{ij}=0$ for $(i, j) \notin F$ if and only if for no i,

$$p(s_i) > \sum_{\substack{j \ni \\ (i,j) \in F}} p'(s_j)$$

and for no j,

$$p'(s_j) > \sum_{\substack{i \ni \\ (i,j) \in F}} p(s_i).$$

Proof

"If" part Existence of a feasible solution follows by using the procedure available for finding a basic feasible solution to the problem, with the set of remaining cells in the beginning being F.

"Only if" part Suppose there exists a feasible solution for the problem with $P_{ij}=0$, $(i, j) \notin F$. Then

$$\sum_{\substack{i \ni \\ (i,j) \in F}} P_{ij} = p'(s_j) \qquad \text{for all } j$$

$$\sum_{\substack{j \ni \\ (i,j) \in F}} P_{ij} = p(s_i) \qquad \text{for all } i$$

$$\sum p(s_i) = \sum p'(s_j) = 1 = \sum_{\substack{i \\ (i,j) \in F}} \sum_{j} P_{ij}$$ implying that there is no i and j such that

$$p(s_i) > \sum_{\substack{j \ni \\ (i,j) \in F}} p(s_j) \quad \text{and} \quad p'(s_j) > \sum_{\substack{i \ni \\ (i,j) \in F}} p(s_i).$$

Hence the result.

RESULT 5.10.2 For the problem (5.8.1) with $p(s_i)$, $p'(s_j)$ proportional to total size of s_i, s_j, respectively, and C_{ij} as given by (5.8.1), we have

$$p(s_i) \leqslant \sum_{\substack{j \ni \\ (i,j) \in F}} p'(s_j), \qquad \text{for all} \quad i=1,\ldots,t$$

and

$$p'(s_j) \leqslant \sum_{\substack{i \ni \\ (i,j) \in F}} p(s_i), \qquad \text{for all} \quad j=1,\ldots,t$$

when $F=\{(i,j)|C_{ij} \leqslant 1\}$.

Proof Let $s_i=(U_{i_1},\ldots,U_{i_n})$, for some i. For j such that $(i,j)\in F$ for this given i, either

$$s_i = s_j$$

or

$$\#(s_i \cap s_j) = n-1, \qquad \text{that is,}$$

$$s_j = \{U_{i_1},\ldots,U_{i_{k-1}},U_{i_{k+1}},\ldots,U_{i_n}\} \cup \{U_r\}$$

for some $U_r \in N - s_i$. Therefore

$$\begin{aligned}
\sum_{\substack{j \ni \\ (i,j) \in F}} p'(s_j) &= p(s_i) + Q \sum_{k=1}^{n} \sum_{U_r \notin s_i} \left[p'(s_i) - p'_{i_k} + p'_r \right] \\
&= Q\left\{ \sum_{k=1}^{n} p'_{i_k} + n(N-n) \sum_{k=1}^{n} p'_{i_k} - N - \sum_{k=1}^{n} p'_{i_k} \right. \\
&\qquad \left. + n\left(1 - \sum_{k=1}^{n} p'_{i_k}\right)\right\} \\
&= Q\left\{ [1 + n(N-n) - N] \sum_{k=1}^{r} p'_{i_k} + n \right\} \\
&= Q\left\{ (n-1)(N-n-1) \sum_{k=1}^{n} p'_{i_k} + n \right\} \geqslant nQ.
\end{aligned}$$

Because

$$n \geqslant 1,\ N > n \qquad \text{and} \qquad \sum_{k=1}^{n} p'_{i_k} \geqslant 0.$$

Thus we have,

$$\sum_{\substack{j \ni \\ (i,j) \in F}} p'(s_j) \geqslant nQ \geqslant Q \sum_{k=1}^{n} p_{i_k} = p(s_i).$$

This proves the result for all the rows. Similarly, we can prove the result for columns. Hence the result.

From these results it follows that we have a feasible solution to this problem with $P_{ij}=0$ for $(i, j) \notin F$. Hence we can find an optimal solution to the problem with the samples chosen for the two surveys having at least $(n-1)$ units in common.

EXAMPLE 5.10.1 If we apply the transportation method to find an optimal solution to Example 5.4.1 with $p(s_i)$, $p'(s_i)$ as defined in this section, an optimal solution is obtained in Table 5.10.1.

Here the probability of overlap is as high as 0.8. If we apply Roy Choudhury's scheme we get the probability of overlap at only 0.333. Further, the column totals are not proportional to total size of the sample, as seen in Table 5.10.2.

Table 5.10.1

	Survey II						
Survey I	1,2	1,3	1,4	2,3	2,4	3,4	$p(s_i)$
1,2	0.133	0.032		0.068		—	0.233
1,3		0.200			—		0.200
1,4			0.168	—		0.066	0.234
2,3			—	0.099		0.001	0.100
2,4		—			0.100	0.033	0.133
3,4	—					0.100	0.100
$p'(s_i)$	0.133	0.232	0.168	0.167	0.100	0.200	1.000

Table 5.10.2

	Survey II						
Survey I	1,2	1,3	1,4	2,3	2,4	3,4	$p(s_i)$
1,2	0.045	0.030	0.015	0.095	0.048	—	0.233
1,3	0.006	0.111	0.011	0.024	—	0.048	0.200
1,4	0.008	0.033	0.074	—	0.024	0.095	0.234
2,3	0.017	0.022	—	0.035	0.011	0.015	0.100
2,4	0.025	—	0.022	0.033	0.023	0.030	0.133
3,4	—	0.025	0.017	0.008	0.006	0.044	0.100
$\Sigma_i P_{ij}$	0.101	0.221	0.139	0.195	0.112	0.232	1.000
$p'(s_i)$	0.133	0.232	0.168	0.167	0.100	0.200	1.000

REMARK 5.10.1 The $p(s_i)$ and $p'(s_j)$ can be any two probability measures over the set of all possible samples of size n from the population consisting of N units. Still we can apply the transportation method. However, we may not be able to ensure that the samples have the property of the Roy Choudhury's scheme, namely, the samples chosen for the two surveys to differ by at most one unit.

Integration of multiple surveys can also be approximated in a stage by stage manner using the transportation method.

5.11 ESTIMATING POPULATION PROPORTIONS WITH RESTRICTION ON MARGINALS

The Gallup polls in the United States, and other survey organizations have beyond doubt shown that when scientific methods, as opposed to subjective procedures, are used, overall public opinion can be gauged very accurately. Success of such approaches depends heavily on a well-defined sampling frame, accuracy of the collected information, and periodic updating of the information. In many underdeveloped countries we have yet to have such systematic information collected periodically, for planning and framing laws that reflect the people's will.

Considering the known factors that are likely to influence the opinion of the people, in order to predict an accurate picture of the situation under study, samples have to be such that there is proportional representation from all possible different factor combination groups (FCG's). For example, we can list a few of the factors like age, sex, income, educational background, occupation, the density of the population in the region, and even the geographical location of residence.

The exact proportion of people in each FCG is easily found or a good estimate is easily made if well-organized data are readily available. In such cases the sampling scheme is quite simple. Here we mainly consider the situation in which we have information about the factors, but we do not have a complete picture of the population—that is, the proportions of the different FCG's are not known.

We consider the problem of estimating population proportions when marginals are known about the factors. We give a linear-programming formulation of the problem of minimizing the observed χ^2 value, to estimate the parameters of interest. An alternate formulation of the problem is also discussed. We approach the problem from the point of view of maximum likelihood estimation as well, and bring out the connection between these approaches.

Let $U=\{U_1,\ldots,U_N\}$ be the total population. Let $F_1, F_2,\ldots, F_r$ be the r factors under consideration. Each factor F_j is grouped under k_j, mutually exclusive and collectively exhaustive classes, say, $1,2,\ldots, k_j$. Let p_{jl} be the proportion of units in the lth class in the factor F_j, such that

$$\sum_{l=1}^{k_j} p_{jl}=1.$$

When there are only two factors we use $p_{i.}$ and $p_{.j}$ for the proportion of the units in the ith class in Factor 1 and jth class in Factor 2, respectively.

Let $(l_1, l_2,\ldots, l_r)$ denote the FCG in which class l_j is considered for Factor F_j. There are $K=\prod_{j=1}^{r}k_j$ such FCG's in all. Let these FCG's be numbered, $1,2,\ldots, K$ in a fixed order. Let $p(s)$ be the proportion of units in the population in the FCG's given by $s=(l_1,\ldots, l_r)$.

Let C be the cost of the sampling scheme. Let C_0 be the overhead cost, which is a constant for certain broad ranges of the total sample size n. Let C_s be the average cost of surveying one unit in the sth FCG, which may depend on number of units in the FCG.

Then,

$$C=C_0+\sum_{s=1}^{K} n_s C_s. \tag{5.11.1}$$

When no information other than the $p(s)$ is available a given sample size n may be allocated proportional to $p(s)$, assuming that the sampling variance in the smaller factor combinations is less than in the larger one. Proportional allocation may be desirable in many situations, as discussed earlier.

Let n_s denote the number of units to be sampled from the sth FCG's for $s=1,\ldots, K$. For proportional allocation

$$n_s = np(s). \tag{5.11.2}$$

For this allocation

$$C = C_0 + \sum_{s=1}^{K} np(s)C_s$$

or for a given cost $C=C^*$, we have

$$n = (C^* - C_0) \Big/ \sum_{s=1}^{k} p(s)C_s. \tag{5.11.3}$$

If a reasonable estimate of $p(s)$ is available we use it in (5.11.2) and (5.11.3) to get n_s and n. We next consider the problem of estimation of $p(s)$.

Estimating $p(s)$ When Only Marginals Are Known

For simplicity, first we consider the case when $r=2$. (Later we shall deal with the case when $r>2$.) We assume $p_{i.}$ and $p_{.j}$ are known.

Suppose now we have reasons to believe that $p((i, j))$ is of the form $p_{i.}p_{.j}$. We can test for this independence of the factors.

We have

$$H_0: p((i, j)) = p_{i.}p_{.j} \quad i=1,2,\ldots, k_1 \qquad j=1,2,\ldots, k_2 \tag{5.11.4}$$

$$H_1: \text{Not all the equations given under } H_0 \text{ are satisfied}$$

where

$$\sum_j p((i, j)) \quad \text{adds up to } p_{i.} \text{ and}$$

$$\sum_i p((i, j)) \quad \text{adds up to } p_{.j}.$$

We can use the approximate χ^2 procedures to test the hypothesis. The procedure rejects h_0 if

$$\sum_{i=1}^{k_1}\sum_{j=1}^{k_2}\frac{(x'_{ij}-t_{i.}\cdot t_{.j}/n)^2}{t_{i.}\cdot t_{.j}/n}>\chi^2_{(k_1-1)(k_2-1);\,(1-\alpha)} \tag{5.11.5}$$

where α is the significance level, x'_{ij} is the number of units falling in the FCG, (i, j) out of the n units sampled for testing the hypothesis, and

$$t_{i.}=\sum_j x'_{ij};$$

similarly,

$$t_{.j}=\sum_i x'_{ij}.$$

Thus, if the H_0 is not rejected we have the estimates

$$\hat{p}((i, j))=p_{i.}\cdot p_{.j}$$

where $p_{i.}$ and $p_{.j}$ are known.

What shall we do if the hypothesis is rejected? We approach the problem in the next section.

5.12 ESTIMATING $p(s)$, WHEN INDEPENDENCE HYPOTHESIS IS REJECTED

When Hypothesis 5.11.4 is rejected, we let

$$\hat{p}((i, j))=p_{i.}\cdot p_{.j}+d_{ij} \tag{5.12.1}$$

where

d_{ij} is unrestricted in sign but is subject to certain restriction.
$\hat{p}((i, j))$ is taken as an estimate of $p((i, j))$.

We require

$$\sum_j \hat{p}((i, j))=p_{i.},\qquad i=1,\ldots,k_1 \tag{5.12.2}$$

and

$$\sum_i \hat{p}((i, j))=p_{.j},\qquad j=1,\ldots,k_2 \tag{5.12.3}$$

RESULT 5.12.1 If (5.12.1) holds,

$$(5.12.2) \quad \Leftrightarrow \sum_j d_{ij}=0, \qquad i=1,\ldots,k_1$$

$$(5.12.3) \quad \Leftrightarrow \sum_i d_{ij}=0, \qquad j=1,\ldots,k_2.$$

Proof

$$(5.12.2) \quad \Leftrightarrow \sum_j p_{i.}\cdot p_{.j}+\sum_j d_{ij}=p_{i.}, \quad \text{since}$$

$$\sum_j p_{i.}\cdot p_{.j}=p_{i.}$$

$$\Leftrightarrow \sum_j d_{ij}=0.$$

Similarly

$$(5.12.3) \quad \Leftrightarrow \sum_i d_{ij}=0.$$

We also require that $\hat{p}((i,j))$ be probabilities, so we have

$$\hat{p}((i,j)) \geqslant 0 \tag{5.12.4}$$

$$\hat{p}((i,j)) \leqslant 1. \tag{5.12.5}$$

But (5.12.5) is redundant, when (5.12.4) holds with (5.12.2) and (5.12.3).
Notice that (5.12.4) implies that $d_{ij} \geqslant -p_{i.}\cdot p_{.j}$; also

$$\hat{p}((i,j)) \leqslant \min(p_{i.}\cdot p_{.j})$$

or

$$d_{ij} \leqslant \min(p_{i.}-p_{i.}\cdot p_{.j}, p_{.j}-p_{i.}\cdot p_{.j}),$$

which implies

$$d_{ij} \leqslant \min\big(p_{i.}(1-p_{.j}), p_{.j}(1-p_{i.})\big).$$

Let

$$U_{ij}=\min\big(p_{i.}(1-p_{.j}), p_{.j}(1-p_{i.})\big)$$

and

$$L_{ij} = -p_{i.} \cdot p_{.j}.$$

Let

$$t_{ij} = t_{i.} \cdot t_{.j}/n \quad \text{and} \quad x_{ij} = x'_{ij} - t_{ij}.$$

We wish to increase the chances of the new hypothesis being accepted. We resort to minimizing the observed χ^2 value.

Problem 5.12.1

$$\text{Minimize} \quad \sum_{i=1}^{k_1} \sum_{j=1}^{k_2} \frac{1}{t_{ij}} (x_{ij} - nd_{ij})^2$$

$$\left.\begin{aligned} \text{subject to} \quad & \sum_j d_{ij} = 0, \qquad i = 1, \ldots, k_1 \\ & \sum_i d_{ij} = 0, \qquad j = 1, \ldots, k_2 \\ & U_{ij} \geqslant d_{ij} \geqslant L_{ij} \end{aligned}\right\} \qquad (5.12.6)$$

all d_{ij} unrestricted in sign.

Now let $y_{ij} = (x_{ij} - nd_{ij})$. Then Problem 5.12.1 is equivalent to,

$$\text{Minimize} \quad \sum_{i=1}^{k_1} \sum_{j=1}^{k_2} \frac{y_{ij}^2}{t_{ij}}$$

$$\text{subject to} \quad (5.12.6), \text{ and} \qquad (5.12.7)$$

$$y_{ij} = x_{ij} - nd_{ij}$$

$$y_{ij} \quad \text{unrestricted in sign.}$$

Observe that we have a quadratic-programming problem. The objective function is separable and all constraints are linear. Also the objective function is convex in each y_{ij}. An algorithm for solving such problems was discussed in Chapter 2. However, the separability and convexity of the objective function can be exploited to develop an approximate linear-programming-equivalent problem. Firstly, we approximate the nonlinear separable objective function by a piecewise linear separable function.

Consider the variable y_{ij}. As

$$y_{ij}=x_{ij}-nd_{ij} \quad \text{and} \quad L_{ij}\leqslant d_{ij}\leqslant U_{ij}$$

and $L_{ij}\leqslant 0$ and $U_{ij}\geqslant 0$, we have

$$x_{ij}-nU_{ij}\leqslant y_{ij}\leqslant x_{ij}+n. \tag{5.12.8}$$

That is, each y_{ij} is bounded. Let

$$y_{1ij}=x_{ij}-nU_{ij} \quad \text{and} \quad y_{tij}=x_{ij}+n, \tag{5.12.9}$$

where t is a positive integer chosen so as to divide the range of y_{ij} into $t-1$ intervals. Choose $y_{2ij},\ldots, y_{t-1,ij}$ such that the piecewise linear approximation of y_{ij}^2 is sufficiently good (Fig. 5.12.1).

Let w_{kij} be such that,

$$y_{ij}=\sum_{k=1}^{t} y_{kij}w_{kij} \tag{5.12.10}$$

with

$$\sum_{k=1}^{t} w_{kij}=1 \quad \text{and} \quad w_{kij}\geqslant 0, \quad \text{for } k=1,\ldots,t. \tag{5.12.11}$$

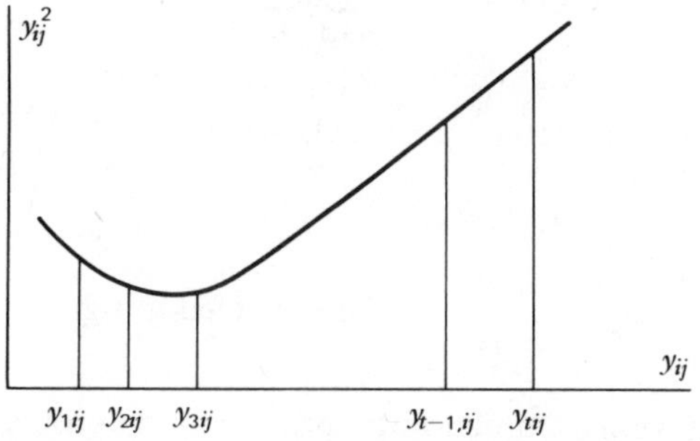

Figure 5.12.1 A piecewise linear approximation of y_{ij}^2.

Now y_{ij}^2, when $y_{r-1,ij} \leqslant y_{ij} \leqslant y_{rij}$, can be approximated as

$$y_{ij}^2 = y_{r-1,ij}^2 w_{r-1ij} + y_{rij}^2 w_{rij}$$

with

$$w_{r-1ij} + w_{rij} = 1 \qquad w_{kij} \geqslant 0.$$

The implication is that $w_{kij} = 0$, $k \neq r$, $r-1$. Therefore, we put the additional restrictions on w_{kij}'s that

(1) At most two of the w_{kij} are allowed to be positive

and (5.12.12)

(2) the positive w_{kij}'s must be such that they correspond to adjacent grid points $y_{r-1,ij}$ and y_{rij} for some r.

Now we can write $y_{ij}^2 = \Sigma_k y_{kij}^2 w_{kij}$, with w_{kij} satisfying (5.12.11) and (5.12.12).

This procedure can be done for each variable y_{ij}, and y_{ij} can be replaced by $\Sigma y_{kij} w_{kij}$ in all the constraints. The approximate problem now turns out to be Problem 5.12.2.

Problem 5.12.2

$$\text{Minimize} \quad \sum_{i,j,k} (y_{kij}^2 / t_{ij}) w_{kij}$$

$$\text{subject to} \quad \sum_j d_{ij} = 0, \qquad i = 1, \ldots, k_i, \qquad (5.12.13)$$

$$\sum_i d_{ij} = 0, \qquad j = 1, \ldots, k_2 \qquad (5.12.14)$$

$$L_{ij} \leqslant d_{ij} \leqslant U_{ij}$$

$$d_{ij} \qquad \text{unrestricted in sign,}$$

$$\sum_k y_{kij} w_{kij} + n d_{ij} = x_{ij}$$

$$\sum_k w_{kij}=1, \quad i=1,\ldots,k_1;$$

$$j=1,\ldots,k_2.$$

$$w_{kij} \geqslant 0, \qquad \text{for all} \quad k, i, \text{and } j$$

and (5.12.12) is satisfied by the w_{kij}'s for all i, j.

This problem is a linear-programming problem but for the restrictions of (5.12.12). However, because of the convexity of the objective function in each variable and the fact that we are seeking a minimum of the objective function in Problem 5.12.1, we can show that it is sufficient to solve Problem 5.12.2 as a linear-programming problem, dropping the restrictions of (5.12.12).

Thus we in fact have an approximate equivalent problem that is a linear-programming problem. But the size of the problem largely depends on the number of grid points, t, chosen for each variable. After the problem is solved, optimal d_{ij}'s are used to find an estimate of $p((i, j))$, as follows

$$\hat{p}((i, j)) = p_{i.} \cdot p_{.j} + d_{ij}.$$

When $r>2$, we can go step by step, taking first, Factors F_1 and F_2, and then after obtaining estimates of $p((i, j))$'s we go after the next factor, combining the first two factors and treating them as a single factor.

For example, FCG's $((1,1),\ldots,(k_1,k_2))$ will be renumbered from 1 to, say, some k_1'; the third factor, F_3, is now taken in place of the second factor, and we go through the routine that obtains estimates of $p((i, j))$'s.

As more and more factors are considered the problem size explodes. So it is better to judiciously restrict the factors and classes in each factor to a small number. And subdivisions of a class or factor can be taken up at a later stage.

EXAMPLE 5.12.1 Consider the following data on marital status of a large sample from the population (10 years of age and over), by sex and age for the city of Somnan, Iran, given in Table 5.12.1.

Suppose we want to draw a sample of size 200 from this population such that the different FCG's are proportionately represented in the sample.

Here we have the population proportions of the different FCG's. For instance, among men ages 30–49, there are 4408 married. This proportion works out to 0.1569. Therefore in the sample of size 200, we should include 31 from the FCG of married men ages 30–49.

Table 5.12.1 Marital Status of the Population, 10 Years of Age and Over, by Sex and Age: Somnan, Iran

Age (years)	Married	Widowed	Divorced	Never Married	Total
		MEN			
10–14	28	2	1	2736	2767
15–19	202	2	3	2590	2797
20–24	829	2	4	1510	2345
25–29	1237	2	12	413	1664
30–34	1340	4	5	170	1519
35–39	1213	8	8	62	1291
40–44	1088	6	3	33	1130
45–49	767	9	3	18	797
50–54	650	20	3	24	697
55–59	397	15	1	11	424
60–64	351	29	3	7	390
Over 65	529	107	3	21	660
TOTAL	8631	206	49	7595	16481
		WOMEN			
10–14	75	0	2	2030	2107
15–19	854	4	7	1055	1920
20–24	1360	3	3	360	1726
25–29	1112	10	7	68	1197
30–34	921	18	14	21	974
35–39	746	29	7	9	791
40–44	611	59	11	14	695
45–49	418	86	10	7	521
50–54	358	153	10	6	527
55–59	144	90	6	12	252
60–64	127	209	7	10	353
Over 65	121	424	4	6	555
TOTAL	6847	1085	88	3598	11618

Next, to illustrate the approach discussed so far, let us assume we know only the proportion of men and women in the population and the proportions of married, widowed, divorced, and never married (Tables 5.12.2 and 5.12.3). We need to obtain the proportions for men who are married, widowed, divorced, or never married, and similar proportions for women.

First we test whether the factors are independent. Suppose a random sample of size 200 drawn from the population gives rise to the x'_{ij} as given in Table 5.12.4.

Table 5.12.2

Sex	Proportion
Men $p_{1.}$	0.5865
Women $p_{2.}$	0.4135

Table 5.12.3

Married $p_{.1}$	Widowed $p_{.2}$	Divorced $p_{.3}$	Never Married $p_{.4}$
0.5508	0.0459	0.0049	0.3984

Table 5.12.4

	Marital Status (x'_{ij})				
Sex	Married	Widowed	Divorced	Never Married	Total
Men	57	3	0	51	111
Women	49	11	1	28	89
TOTAL	106	14	1	79	200

As we reject the hypothesis on the independence of the factors after performing the χ^2 test we go on to solve the linear-programming problem described earlier as Problem 5.12.2.

We obtain the optimal d_{ij}'s as given in Table 5.12.5. The $p_{i.} \cdot p_{.j}$'s are as shown in Table 5.12.6. The alternative estimates obtained using d_{ij}'s are:

$$\hat{p}((i, j)) = p_{i.} p_{.j} + d_{ij}$$

as given in Table 5.12.7.

Table 5.12.5

d_{ij}			
-0.00915	-0.02385	-0.00278	0.03578
0.00915	0.02385	0.00278	-0.03578

Table 5.12.6

	$p_{i.} \cdot p_{.j}$		
0.32304	0.02692	0.00287	0.23366
0.22775	0.01898	0.00203	0.16473

Now a test of the kind

$$H_0: p((i, j)) = \hat{p}((i, j)) \text{ for all } i, j$$

$$H_2: \text{not all equations in } H_0 \text{ are holding}$$

accepts the hypothesis. Thus an acceptable alternative is found by the procedure.

So, using the estimates of $p((i, j))$'s we decide to include members from different FCG's in a sample of size 200, as indicated in Table 5.12.8.

EXTREME CASES The estimates obtained by the procedure are dependent on the x'_{ij} of the sample chosen. Hence it would be interesting to see what happens if extremely impossible samples were chosen in the first place.

Table 5.12.7

0.31389	0.00307	0.00009	0.26944
0.23690	0.04283	0.00481	0.12895

Table 5.12.8

	Marital Status				
Sex	Married	Widowed	Divorced	Never Married	Total
Men	63	1	0	53	117
Women	47	9	1	26	83
TOTAL	110	10	1	79	200

Suppose we have obtained the following samples (Table 5.12.9):

Table 5.12.9

	Marital Status (x_{ij})				
Sex	Married	Widowed	Divorced	Never Married	Total
Men	39	2	1	85	127
Women	65	2	1	5	73
Total	104	4	2	90	200

If we test for independence of the factors, we find the hypothesis is required. We go for the linear-programming solution and find the d_{ij}'s as given in Table 5.12.10.

Therefore the estimates of $p((i, j))$ work out, as shown in Table 5.12.11.

And so our decision would be to include $200 \times \hat{p}((i, j))$ from each FCG in a sample of size 200.

It is easy to see that these proportions are far from the actual proportions in the population. But this should not alarm us, as the probability of getting such an odd sample from the given population is, approximately, as small as 0.76464×10^{-19}. Thus we can conclude that unless we start with a reasonably good sample from the population, we are bound to estimate the population FCG proportions in a faulty manner.

Table 5.12.10

d_{ij}			
−0.13520	−0.00270	−0.00135	0.13925
0.13520	0.00270	0.00135	−0.13925

Table 5.12.11

$\hat{p}(i, j)$			
0.18780	0.02422	0.00152	0.37291
0.36295	0.02168	0.00338	0.02548

Table 5.12.12

	Marital Status				
Sex	Married	Widowed	Divorced	Never Married	Total
Men	0.31	0.01	0.00	0.27	0.59
Women	0.24	0.04	0.00	0.13	0.41
Total	0.55	0.05	0.00	0.40	1.00

On the other hand, suppose we have obtained a sample with proportions exactly as in the population. We would like to know whether the procedure obtains the actual proportions. But for the round-off errors, it does. The actual proportions up to two decimal places are given in Table 5.12.12 for the same population discussed earlier.

If the sample drawn is exact to these proportions we find in a sample of size 200, x'_{ij} as given in Table 5.12.13.

The d_{ij}'s are found to be as given in Table 5.12.14, and the estimates turn out to be the same as the actuals after rounding off to two decimal places, indicating the ability of the method to estimate correct proportions.

EXAMPLE 5.12.2 In Example 5.12.1 we have considered two factors, sex and marital status. Suppose we consider age also as an important factor.

Table 5.12.13

	Marital Status (x_{ij})			
Sex	Married	Widowed	Divorced	Never Married
Men	62	2	0	54
Women	48	8	0	26

Table 5.12.14

d_{ij}'s			
−0.014	−0.020	0.000	0.034
0.014	0.020	0.000	−0.034

We introduce that factor after estimating the population proportions for the first two factor combinations. Age is grouped into three groups, 10–29, 30–49, and 50 and above. We have 8 classes corresponding to sex and marital status; Male Married, Male Widowed, Male Divorced, Male Never Married, Female Married, Female Widowed, Female Divorced, and Female Never Married. In a sample of size 200 as drawn we have the x'_{ij}, $i=1,2,\ldots,8$, $j=1,2,3$, as given in Table 5.12.15. As the χ^2 rejects the independence hypothesis we go to the linear-programming solution.

The d_{ij}'s obtained are as in Table 5.12.16. The estimates thus obtained are given in Table 5.12.17.

This example illustrates how to consider three factors. Similarly we can bring in as many factors as we desire; the problem is the same.

Table 5.12.15

Class j / i	Age group 1	2	3	Total
1	14	30	13	57
2	0	1	2	3
3	0	0	1	1
4	48	3	0	51
5	27	17	5	49
6	0	3	8	11
7	0	0	1	1
8	27	0	0	27
TOTAL	116	54	30	200

Table 5.12.16

i \ j	d_{ij}: 1	2	3
1	−0.10701	0.08180	0.02521
2	−0.00182	−0.00085	0.00267
3	−0.00006	−0.00003	0.00009
4	0.09084	−0.05385	−0.03699
5	−0.00710	0.01885	−0.01175
6	−0.02517	−0.00917	0.03434
7	−0.00282	−0.00132	0.00414
8	0.05314	−0.03543	−0.01771

Table 5.12.17

	$\hat{p}((i, j))$		
i \ j	1	2	3
1	0.07757	0.16802	0.06831
2	0.00000	0.00000	0.00310
3	0.00000	0.00000	0.00010
4	0.24925	0.02015	0.00000
5	0.13338	0.08447	0.02105
6	0.00000	0.00259	0.04022
7	0.00000	0.00000	0.00480
8	0.12900	0.00000	0.00000

5.13 AN ALTERNATE FORMULATION OF THE PROBLEM

We can reformulate the problem discussed in Section 5.12 in the following way.

Let p_{ij} be the estimate of $p((i, j))$ which we want to determine. Then we require

$$\sum_j p_{ij} = p_{i.}, \qquad i = 1, \ldots, k_1 \tag{5.13.1}$$

$$\sum_i p_{ij} = p_{.j}, \qquad j = 1, \ldots, k_2 \tag{5.13.2}$$

$$p_{ij} \geqslant \varepsilon \qquad \text{for all} \quad i, j \tag{5.13.3}$$

where ε is a very small given positive number. The reason for Restriction 5.13.3 will be made clear soon.

Given the x'_{ij}'s from a sample of size n drawn from the given population, let us consider the expression for the observed χ^2, while the hypothesis is

$$H_0: p((i, j)) = p_{ij} \qquad \text{for all} \quad i, j.$$

It is given by

$$\sum_i \sum_j \frac{(x'_{ij} - np_{ij})^2}{np_{ij}}$$

We can therefore try to minimize this quantity subject to Restrictions 5.13.1–3.

Now

$$\sum_i \sum_j \frac{(x'_{ij}-np_{ij})^2}{np_{ij}} = \sum_i \sum_j \frac{x'^2_{ij}}{np_{ij}} + \text{a constant}.$$

Let $C_{ij} = x'^2_{ij}/n$.
Thus we have the problem,

$$\begin{aligned} &\text{Minimize} \quad \sum_i \sum_j C_{ij}/p_{ij}. \\ &\text{subject to} \quad \sum_j p_{ij} = p_{i.}, \qquad j = 1,2,\dots,k_1 \\ &\qquad\qquad\quad \sum_i p_{ij} = p_{.j}, \qquad i = 1,2,\dots,k_2 \\ &\qquad\qquad\quad p_{ij} \geqslant \varepsilon. \end{aligned} \tag{5.13.4}$$

The constraints of the problem are that of a transportation problem, with the difference that the variables are restricted to be greater than or equal to ε, so as to have a meaningful objective function as $1/p_{ij}$ appears in there.

The objective function of the problem is a convex function; in fact, it is convex in each variable, as $C_{ij} > 0$.

$$\text{Now let} \quad p'_{ij} = p_{ij} - \varepsilon\ . \tag{5.13.5}$$

$$\text{Let} \quad p'_{i.} = p_{i.} - \varepsilon k_2; \qquad p'_{.j} = p_{.j} - \varepsilon k_1.$$

Then we have the equivalent problem.

$$\begin{aligned} &\text{Minimize} \quad \sum_i \sum_j C_{ij}/(p'_{ij}+\varepsilon) \\ &\text{subject to} \quad \sum_j p'_{ij} = p'_{i.} \\ &\qquad\qquad\quad \sum_i p'_{ij} = p'_{.j} \\ &\qquad\qquad\quad p'_{ij} \geqslant 0 \qquad \text{for all} \quad i, j. \end{aligned} \tag{5.13.6}$$

Now we have a convex transportation problem to be solved, as $C_{ij}/(p'_{ij}+\varepsilon)$ is a convex function for $C_{ij}>0$. There are special procedures to solve such transportation problems with convex objective functions.

This formulation is an exact formulation of the problem under consideration.

5.14 MAXIMUM LIKELIHOOD ESTIMATION

The problem considered in the foregoing sections can be approached from a different point of view, that of maximization of the likelihood function. The population with the different proportions corresponding to the different FCG's constitutes a multinomial distribution. If a sample of size n drawn from this population yields x'_{ij} from the FCG, (i, j), then the probability of obtaining x'_{ij}, given $p((i, j))$ is

$$L(p)=\frac{n}{\prod_{i,j} x'_{ij}} \prod_{i,j} [p((i, j))]^{x'_{ij}}$$

The true $p((i, j))$ are unknown but we know they must satisfy (5.13.1), (5.13.2), and $p((i, j))>\varepsilon$, where $\varepsilon>0$ and small, to avoid taking the logarithm of zero. The object is to obtain an estimate of $p((i, j))$ on the basis of the sample of size n drawn from the population.

Observing that maximizing $L(p)$ is the same as maximizing the logarithm of $L(p)$, and the portion corresponding to $(n/\Pi_{i,j}x'_{ij})$ is a constant for a given sample of size n, we can formulate the problem as follows:

$$\text{Maximize} \quad \sum_{i,j} x'_{ij} \log p_{ij}$$

$$\text{subject to} \quad (5.13.1), (5.13.2), \text{and } (5.13.3)$$

where p_{ij} is the estimate of $p((i, j))$, which we want to determine.

Here $\varepsilon<p_{ij}<\min(p_{i.}, p_{.j})$. Therefore, $\log p_{ij}$ is also bounded and is concave in each p_{ij}. Thus we have a separable objective function that is concave in each variable. Along the same lines as in Section 5.13, we can reformulate the problem as a concave maximization problem with a transportation matrix as the constraint matrix.

It can be noticed that the objective function in the minimum χ^2 problem and maximum-likelihood problem have the following properties:

1 The function is a separable function with convexity in each variable in the minimum χ^2 problem, whereas the corresponding one is a separable function with concavity in each variable.

2 In each variable, the functions C_{ij}/p_{ij} and $x'_{ij}\log p_{ij}$ attain their minimum and maximum, respectively, at the same point.

The distributional properties of the mathematical-programming solution for the problem need to be explored for studying the properties of the estimates obtained.

BIBLIOGRAPHICAL NOTES

5.1 The problem of choosing optimal sample size can be found in Cochran (1977), Kish (1967), Murthy (1967), and Raj (1969). When the cost function is strictly convex, an integer solution can be found by using a method described in Washburn (1974).

5.2 Optimum allocation in stratified sampling has been considered by Dalenius (1957), Folks and Antle (1965), Stock and Frankel (1939), Ghosh (1965), and Kokan and Khan (1967), among others. Example 5.2.1 is taken from Cochran (1977). Allocation of total sample size among different strata, when the sample means are required to have the sampling variance as much as possible in a given ratio, so as to reflect different degrees of importance in the various data, is considered by Chaddha et al. (1971). They formulate the problem as a minimax problem and suggest integer, nonlinear, and dynamic programming methods to solve it. The type of nonlinear knapsack problem encountered in this section has a strictly convex separable objective function. Michaeli and Pollastschek (1977) consider a similar problem and derive a necessary and sufficient condition for an integer solution to be optimal for the problem. Also they show that an integer optimal solution $\mathbf{x}^0$ and the continuous solution $\mathbf{x}^*$ are such that either $x_i^0 \geq [x_i^*]$ for each i or $x_i^0 \leq [x_i^*]+1$ for each i or both. Thus the problem is reduced to a 0–1 integer programming problem.

5.3 The discussion on optimal allocation in multivariate stratified sampling closely follows that of Kohan and Khan (1967), Bruvold and Murphy (1978), and Pfanzagl (1966), who consider similar formulations of certain related problems. Algorithms for solving convex programming can be found in Zangwill (1969), Fiacco and McCormick (1968), Collatz and Wetterling (1975), and Hadley (1964), Bazaraa and Shetty (1979).

5.4 Integration of surveys is discussed in Murthy (1967) and Des Raj (1969). Selection schemes considered are from Lahiri (1954), Keyfitz (1951), and Raj (1956).

5.5 Discussion on the transportation problem appears in almost every book on linear programming. The tree structure of the basic cells is used for efficient storage and computation when electronic computers are used. Discussion on this point appears in Murty (1976).

5.8 The transportation problem appears in Arthanari and Dodge (1977).

5.11 Arthanari and Dodge (1977) consider this estimation problem. A closely related problem in the context of contingency tables is considered by Melnick and Yechiali (1976).

5.12 Approximating nonlinear separable objective functions by piecewise linear functions and formulating the problem as a restricted linear-programming problem are discussed under

"Separable Programming" in Hadley (1964). Also a proof of the sufficiency of solving the problem as a linear-programming problem when the objective function is convex appears in Hadley (1964).

5.13 The convex transportation problem was first considered by Beale (1959), who describes an algorithm for solving such problems. See Meyer (1979) for two segmented programming approach.

5.14 Alldredge and Armstrong (1974) apply geometric programming for solving a maximum-likelihood-estimation problem arising in estimation of overlap sizes created by interlocking sampling frames.

REFERENCES

Alldredge, J. R., and Armstrong, D. W. (1974). "Maximum Likelihood Estimation for the Multinomial Distribution Using Geometric Programming." *Technometrics* **16**, 585.

Arthanari, T. S., and Dodge, Y. (1977). *Integration of Surveys with Prescribed Probabilities of Selection*. Technical Report **1**, School of Planning and Computer Applications, Tehran.

Arthanari, T. S., and Dodge, Y. (1977). *Estimating Population Proportions with Restriction on Marginals*. Technical Report **2**, School of Planning and Computer Applications, Tehran.

Bard, Y. (1974). *Nonlinear Parameter Estimation*. Academic Press, New York.

Bazaraa, M. S. and Shatty, C. M. (1979). *Nonlinear Programming and Algorithms*. Wiley, New York.

Beale, E. M. L. (1959). "On Quadratic Programming." *Naval Res. Log. Quart.* **6**, 227.

Beale, E. M. L. (1959). "An Algorithm for Solving the Transportation Problem When the Shipping Cost over each Route is Convex." *Naval Res. Log. Quart.* **6**, 43.

Bruvold, N. T., and Murphy, R. A. (1978). "Sample Sizes for Comparison of Proportions." *Technometrics* **20**, 437.

Cassel, C. M., Sarndel, C. E., and Wretman, J. H. (1977). *Foundations of Inference in Survey Sampling*. Wiley-Interscience, New York.

Chaddha, R. L., et al. (1971). "Allocation of Total Sample Size When Only the Stratum Means Are of Interest." *Technometrics* **13**, 817.

Cochran, W. G. (1977). *Sampling Techniques* (ed. 3). Wiley, New York.

Collatz, L., and Wetterling, W. (1975). *Optimization Problems*. Springer-Verlag, New York.

Dalenius, T. (1957). "Sampling in Sweden-Contributions to the Methods and Theories of Sample Survey Practice." Almquist and Wissell, Stockholm.

Fiacco, A. V., and McCormick, G. P. (1968). *Non Linear Programming: Sequential Unconstrained Minimization Techniques*. Wiley, New York.

Folks, J. L., and Antle, C. E. (1965). "Optimum Allocation of Sampling Units to Strata When There Are R Responses of Interest." *J. Am. Stat. Assoc.* **60**, 225.

Ford, L. R., and Fulkerson, D. R. (1956). "Solving Transportation Problem." *Manag. Sci.* **3**, 24.

Ghosh, S. P. (1958). "A Note on Stratified Random Sampling with Multiple Characters." *Bull. Calcutta Stat. Assoc.* **8**, 81.

Ghosh, S. P. (1965). "Optimum Allocation in Stratified Sampling with Replacement." *Metrika* **9**, 212.

Hadley, G. (1964). *Nonlinear and Dynamic Programming*. Addison-Wesley, Reading, Mass.

Hartly, H. O., and Rao, J. N. K. (1968). "A New Estimation Theory for Sample Surveys." *Biometrica* **55**, 547.

Ireland, C. T., and Kullback, S. (1968). "Contingency Tables with Given Marginals." *Biometrika* **55**, 179.

Keyfitz, N. (1951). "Sampling with Probability Proportional to Size Adjustment for Changes in Size." *J. Am. Stat. Assoc.* **46**, 105.

Kish, L. (1967). *Survey Sampling*. 2nd edition. Wiley, New York.

Kokan, A. R. (1963). "Optimum Allocation in Multivariate Surveys." *J. R. Stat. Soc.* **A126**, 557.

Kokan, A. R., and Khan, S. (1967). "Optimum Allocation in Multivariate Surveys: An Analytical Solution." *J. R. Stat. Soc.* **B29**, 115.

Kullback, S. (1974). "Loglinear Models in Contingency Table Analysis." *Am. Stat.* **28**, 115.

Lahiri, D. H. (1954). "Technical Paper on Some Aspects of the Development of the Sample Design." Indian National Sample Survey, Report 5, repr. *Sankhya*, **14**.

Lee, T. C., Judge, G. G., and Zellner, A. (1968). "Maximum Likelihood and Bayes Estimation of Transition Probabilities." *J. Am. Stat. Assoc.* **63**, 1162.

Melnick, E. L., and Yechiali, U. (1976). "A Mathematical Programming Formulation of Estimation Problems Related to Contingency Tables." *Manage. Sci.* **22**, 701.

Meyer, R. R. (1979). "Two-Segmented Separable Programming." *Manag. Sci.*, Appl. and Theory, **5**, 385.

Michaeli, I., and Pollastschek, M. A. (1977). "On Some Nonlinear Knapsack Problems." In *Studies in Integer Programming*. P. L. Hammer, E. L. Johnson, B. H. Korte, and G. L. Nemhauser, Eds. *Ann. Discrete Mathematics* **1**, North-Holland Publishing, Amsterdam, pp. 403–14.

Murthy, K. G. (1976). *Linear and Combinatorial Programming*. Wiley, New York.

Murthy, M. N. (1967). *Sampling Theory and Method*. Statistical Publishing Society, Calcutta.

Pfanzagl, J. (1966). *Allemeine Methodenlehre der Statistik* (Bd. II, 2). Auflage, Berlin.

Pilot Census Annexes. (1977). Statistical Center of Iran, Tehran.

Raj, D. (1956). "On the Method of Overlapping Maps in Sample Surveys." *Sankhya* **17**, 89.

Raj, D. (1969). *Sampling Theory*. McGraw-Hill, New York.

Roy Choudhury, D. K. (1956). "Integration of Several PPS Surveys." *Sci. Cult.* **22**, 119.

Stock, J. S., and Frankel, L. R. (1939). "The Allocations of Sampling Among Several Strata." *Ann. Math. Stat.* **10**, 288.

Washburn, A. (1975). "A Note on Integer Maximization of Unimodel Functions." *Oper. Res.* **23**, 358.

Wolfe, P. (1959). "The Simplex Method for Quadratic Programming." *Econometrica*.

Zangwill, W. I. (1969). *Nonlinear Programming: A Unified Approach*. Prentice-Hall, Englewood Cliffs, N.J.

CHAPTER 6

Design and Analysis of Experiment

6.1 INTRODUCTION

Scientists and engineers need to assess the effects of certain controllable conditions on the experiments they conduct. Unless the experiments are planned in a scientific manner, the data collected may not yield any statistically valid information. So the experiment under consideration must be designed appropriately.

A set of possible conditions is available to the experimenter, under which to conduct an experiment. This set is called the *factor space* or domain of interest. For each possible condition in the factor space, specification of the number of observations to be made is called a *design*.

After choosing the design we need to make assumptions on the relationship between the controllable conditions and the outcome. A random variable of interest is the *response* an observable variable. The response variable is related to controllable variables by some functional relationship, which is assumed to hold. When the functional structure between the response variable and the controllable variables, and the distribution of errors for distinct runs of the experiment, are specified we have a *model*.

The choice of both the design and the model influence the conclusions drawn from the experiment. Thus problems of optimally choosing the design and the model are introduced. Also methods of analyzing the data are required for estimating the unknown parameters in the model. There may be technological, budget, or other restrictions on the number of observations that can be made.

We can consider application of mathematical programming related to optimization problems in the design and analysis of experiments. There are four major areas of interest: (1) analysis of the outcome of the experiment, (2) choosing the model, (3) constructing designs with desired properties, and (4) optimally choosing the design among the available designs, with respect to a given criterion. In this chapter we illustrate a few applications of mathematical programming in some of these areas.

6.2 SOME PRELIMINARIES FROM DESIGN AND ANALYSIS OF EXPERIMENTS

A particular set of experimental conditions is called a *treatment*. A unit to which a single treatment is applied is called an *experimental unit*. When a treatment is repeated with more than one experimental unit, we have a replication. A *block* is a set of t experimental units in which t treatments are applied, one in each of the t experimental units. When identical treatments are applied to identical experimental units we like to observe identical responses; however, such is not the case, because of variables that are not under consideration. Thus the difference between responses in such a case is called the *experimental error*.

If we wish to investigate simultaneously the effect of several types of conditions on a given process, each treatment consists of all combinations that can be formed from different factors. The possible conditions of a factor are called the *levels* for that factor. Hence a treatment is a combination of the factors at different levels. Such an experiment is called a *factorial experiment*.

If the treatments are assigned to the units completely at random, then the design is called a *completely randomized design*. There is no restriction on the allocation of treatments to experimental units. In such a setting homogeneous units should be available. Any number of replicates may be used.

The response obtained from an experimental unit with treatment k differs from the response with treatment l, by a constant, $\tau_k - \tau_l$. The objective of the experiment is to estimate such differences.

Associated with the design above we assume a linear model of the form

$$Y_{ij} = \mu + \tau_i + \varepsilon_{ij}, \quad i = 1, \ldots, t$$
$$j = 1, \ldots, n_i \tag{6.2.1}$$

where μ is the true mean effect, τ_i is the true effect of the ith treatment, and ε_{ij} is the experimental error arising when the treatment i is applied to experimental unit j. Notice that we are also assuming that the response obtained on one unit is unaffected by the treatment applied to another unit. In Model 6.2.1 in addition we assume that

$$\sum_{i=1}^{t} \tau_i = 0.$$

We then have what is called a fixed one-way classification model (Model

I). In the event that we assume τ_i in the experiment are randomly selected from a population of treatments, then we have a random effect model (Model II). In such a case a usual assumption is that τ_i follows normal distribution with mean zero and variance σ_τ^2.

In the matrix notation, (6.2.1) can be written as

$$\mathbf{Y}=\mathbf{X}\boldsymbol{\theta}+\boldsymbol{\varepsilon} \tag{6.2.2}$$

where **Y** is the vector of responses, and **X** is the matrix of zeros and ones, called the *design matrix*,

$$\boldsymbol{\theta}=(\mu,\tau_1,\ldots,\tau_t)', \text{ and } \boldsymbol{\varepsilon}=(\varepsilon_{11},\ldots,\varepsilon_{tn_t})'.$$

If the homogeneous units are divided into blocks so that the number of units in each block is same as the number of treatments, and treatments are alloted with equal probability to each unit within each block, we have a *randomized complete block design*.

The model associated with the randomized block design can be written as

$$Y_{ijk}=\mu+\tau_i+\beta_j+\varepsilon_{ijk} \qquad i=1,\ldots,a$$
$$j=1,\ldots,b$$
$$k=1,\ldots,n_{ij} \tag{6.2.3}$$

where Y_{ijk}, μ, τ_i, and ε_{ijk} are same as in (6.2.1) and β_j is the true effect of the jth block. In addition

$$\sum_{j=1}^{b}\beta_j=0.$$

As in (6.2.1) we may assume Model I or II with respect to τ_i's. Model (6.2.3) is also called a *two-way classification* model.

The linear model for (6.2.3) can be written in the form

$$\mathbf{Y}=\mathbf{X}\boldsymbol{\theta}+\boldsymbol{\varepsilon}$$

where $\boldsymbol{\theta}=(\mu,\tau_1,\ldots,\tau_a,\beta_1,\ldots,\beta_b)'$. The least-squares estimate $\hat{\boldsymbol{\theta}}$ for $\boldsymbol{\theta}$ is obtained as

$$\hat{\boldsymbol{\theta}}=(\mathbf{X}'\mathbf{X})^{-}\mathbf{X}'\mathbf{Y}$$

where $(\mathbf{X}'\mathbf{X})^{-}$ is the generalized inverse of $\mathbf{X}'\mathbf{X}$. $\mathbf{X}'\mathbf{X}$ is called the *information matrix*.

In the following section we consider a problem in analyzing the outcome of the experiment.

We have already mentioned in Chapter 2 the problems arising with the least-squares estimation method when the normality assumptions are not valid. Also we have noted the MINMAD estimation as a possible alternative to the least-squares estimation.

Now we go on to consider using the MINMAD method for estimating the parameters in the two-way classification model.

6.3 MINMAD ESTIMATES FOR A TWO-WAY CLASSIFICATION MODEL

Consider the two-way classification model (Model 6.2.3). Suppose we employ MINMAD regression to estimate the parameters. We have the following problem:

Problem 6.3.1

$$\underset{\mu,\,\tau_i,\,\beta_j}{\text{Minimize}} \sum_{i,\,j,\,k} |Y_{ijk} - (\mu + \tau_i + \beta_j)|.$$

Let $\alpha_i = \mu + \tau_i$, $i = 1, \ldots, a$. The equivalent linear-programming problem in matrix notation can be given as follows:

Problem 6.3.2

$$\begin{array}{ll} \text{Minimize} & \mathbf{e}'\mathbf{d}_1 + \mathbf{e}'\mathbf{d}_2 \\ \text{subject to} & \mathbf{X}\boldsymbol{\lambda} + \mathbf{I}\mathbf{d}_1 - \mathbf{I}\mathbf{d}_2 = \mathbf{Y} \\ \boldsymbol{\lambda} \quad \text{unrestricted in sign,} & \mathbf{d}_1 \geqslant \mathbf{0},\ \mathbf{d}_2 \geqslant \mathbf{0}, \end{array}$$

where $\mathbf{X}$ is an $n \times m$ matrix, $n = \sum_{i=1}^{a} \sum_{j=1}^{b} n_{ij}$ and $m = a + b$; $\boldsymbol{\lambda} = (\alpha_1, \ldots, \alpha_a, \beta_1, \ldots, \beta_b)$; $\mathbf{Y}$ is the column vector of Y_{ijk}'s with n components; $\mathbf{e}'$ is an n-component row vector with components "1"s, $\mathbf{d}_1 = (d_{1111}, \ldots, d_{111n_{11}}, \ldots, d_{1ab1}, \ldots, d_{1abn_{ab}})'$ and $\mathbf{d}_2 = (d_{2111}, \ldots, d_{211n_{11}}, \ldots, d_{2ab1}, \ldots, d_{2abn_{ab}})'$.

We have considered a general problem of this kind in the MINMAD regression, in Section 2.7. Here we study the special structure of the matrix $\mathbf{X}$, introducing to the MINMAD problem additional simplifications of the Barrodale and Roberts modified simplex method.

Each row of $\mathbf{X}$ has only two nonzero elements. There is a "1" in the ith component and $(a+j)$th component of any row of $\mathbf{X}$. Thus $\mathbf{X}$ is the transpose of a transportation matrix, if $n_{ij}=1$ for all i, j. Assuming $n_{ij} \geqslant 1$ for all i, j, then the rank of $\mathbf{X}$ is $m-1$.

Consider a basis $\mathbf{B}$ of order n for Problem 6.3.2. At most $m-1$ of the λ_i's can be basic.

Given a basis, we can partition the set of n indices of the rows into three sets as follows:

$$\begin{aligned} J_0 &= \{(i,j,k) \mid d_{1ijk}, d_{2ijk} \text{ are both nonbasic}\} \\ J_1 &= \{(i,j,k) \mid d_{1ijk} \text{ is basic}\} \\ J_2 &= \{(i,j,k) \mid d_{2ijk} \text{ is basic}\} \end{aligned} \tag{6.3.2}$$

where (ijk) is the index of the row $\alpha_i+\beta_j+d_{1ijk}-d_{2ijk}=Y_{ijk}$. Therefore, after the rows are rearranged so that the rows corresponding to J_0 appear at the top and columns corresponding to basic λ_i's appear before the other columns in the basis, we obtain

$$\mathbf{B}=\begin{pmatrix} \mathbf{X}_{(1)} & \mathbf{0} \\ \mathbf{X}_{(2)} & \mathbf{D} \end{pmatrix} \tag{6.3.3}$$

where $\mathbf{X}_{(1)}$ is the submatrix of order $m-1$ and is nonsingular, $\mathbf{0}$ is a matrix with $(m-1)$ rows and $n-m+1$ columns, and $\mathbf{D}$ is a diagonal matrix of order $n-m+1$, with the diagonal elements $+1$ or -1. Notice that the submatrix $\mathbf{X}_{(1)}$ is triangular.

Then $\mathbf{B}^{-1}$ can be written as

$$\mathbf{B}^{-1}=\begin{bmatrix} \mathbf{X}_{(1)}^{-1} & \mathbf{0} \\ -\mathbf{D}\mathbf{X}_{(2)}\mathbf{X}_{(1)}^{-1} & \mathbf{D} \end{bmatrix}. \tag{6.3.4}$$

The condition for optimality of $\mathbf{B}$ can be written as

$$-\mathbf{e}'_{m-1} \leqslant \mathbf{e}'\mathbf{D}\mathbf{X}_{(2)}\mathbf{X}_{(1)}^{-1} \leqslant \mathbf{e}'_{m-1}. \tag{6.3.5}$$

The feasibility of $\mathbf{B}$ is easily shown, as follows: Once $\mathbf{X}_{(1)}$ is known, as $\mathbf{X}_{(1)}$ is triangular, the values of λ_i's in the basis can be determined by back-substitution. With these λ_i's, d_{1ijk} and d_{2ijk} can be determined, such that $d_{1ijk}\cdot d_{2ijk}=0$, and $d_{1ijk}, d_{2ijk} \geqslant 0$, such that we obtain a feasible solution corresponding to $\mathbf{B}$.

Also given a set J_0 of $m-1$ linearly independent rows from $\mathbf{X}$, we can build a corresponding basis $\mathbf{B}$. The submatrix of $\mathbf{X}$ corresponding to J_0 has

m columns and, any one of these can be shown to be linearly dependent on the others. So we drop say the last column from this submatrix. Let the resultant matrix be $\mathbf{X}_{(1)}$. The corresponding λ_i's will be basic. For the rows in J_0 we set $d_{1ijk} = d_{2ijk} = 0$.

By back substitution these λ_i can be worked out. With these λ_i's, we can determine the other basic variables from among the d_{1ijk}'s and d_{2ijk}'s for $(i, j, k) \notin J_0$, depending on the residuals for these rows. In this way we obtain a basis $\mathbf{B}$. We explain these with an example.

EXAMPLE 6.3.1 In an agricultural experimental station in Ahwaz, Iran, the experimenter studied the yield of three varieties of wheat when two different types of fertilizers were used. The responses are given in Table 6.3.1.

Table 6.3.1

	Variety		
Fertilizer	Karoon	Shoeleh	Mexican
A	21	7	46
	19	11	
B	16	8	38
	15		40
			42

Entries are kilograms per 100 sq. m.

Suppose we consider a two-way classification model and use MINMAD regression to estimate the parameters of the model. Here we have $a=2$, $b=3$, and $n_{11}=n_{12}=n_{21}=2$, $n_{13}=n_{22}=1$, and $n_{23}=3$. Therefore $n=\Sigma_{i=1}^{2}\Sigma_{j=1}^{3}n_{ij}=11$. The matrix $\mathbf{X}$ and $\mathbf{Y}$ corresponding to Problem 6.3.2 are as follows:

$$\mathbf{X}=\begin{bmatrix} 1&0&1&0&0\\ 1&0&1&0&0\\ 1&0&0&1&0\\ 1&0&0&1&0\\ 1&0&0&0&1\\ 0&1&1&0&0\\ 0&1&1&0&0\\ 0&1&0&1&0\\ 0&1&0&0&1\\ 0&1&0&0&1\\ 0&1&0&0&1 \end{bmatrix} \quad \text{and} \quad \mathbf{Y}=\begin{bmatrix} 21\\ 19\\ 7\\ 11\\ 46\\ 16\\ 15\\ 8\\ 38\\ 40\\ 42 \end{bmatrix}.$$

The rank of $\mathbf{X}$ is 4.

Suppose we consider the rows corresponding to the indices $\{111,122,221,233\}=J_0$. We have a 4×5 submatrix as given below:

$$\begin{bmatrix} 1 & 0 & 1 & 0 & 0 \\ 1 & 0 & 0 & 1 & 0 \\ 0 & 1 & 0 & 1 & 0 \\ 0 & 1 & 0 & 0 & 1 \end{bmatrix}.$$

Notice that the last column is linearly dependent on the other columns. Let us choose λ_1, λ_2, λ_3, and λ_4 to be in the basis. Then

$$\mathbf{X}_{(1)} = \begin{bmatrix} 1 & 0 & 1 & 0 \\ 1 & 0 & 0 & 1 \\ 0 & 1 & 0 & 1 \\ 0 & 1 & 0 & 0 \end{bmatrix} \text{ with rank equal to 4.}$$

For the rows corresponding to J_0, we set $d_{1ijk}=d_{2ijk}=0$, yielding the equations

$$\begin{aligned} \lambda_1 \quad +\lambda_3 \qquad &=21 \\ \lambda_1 \qquad\qquad +\lambda_4 &=11 \\ \lambda_2 \qquad +\lambda_4 &=8 \\ \lambda_2 \qquad\qquad &=42. \end{aligned}$$

These equations readily yield $\lambda_2=42$. Substitution in the third equation above gives $\lambda_4=-34$. Similarly we find $\lambda_1=45$ and $\lambda_3=-24$. λ_i can be calculated in the response table itself, as with the transportation problem, by setting $\lambda_5=0$. By using these λ_i's with $\lambda_5=0$, which is nonbasic, we can work out the residuals, and the d_{1ijk}'s, d_{2ijk}'s can be obtained from the residuals depending on the sign of the residuals (Table 6.3.2). For instance, residual $r(112)=8$; therefore $d_{1,111}=8$; and residual $r(121)=-4$, so $d_{2,121}=4$. Correspondingly we obtain the following basis $\mathbf{B}$ for Problem 6.3.2.

$$\mathbf{B}=\left[\begin{array}{cccc|ccccccc} 1 & 0 & 1 & 0 & & & & & & & \\ 1 & 0 & 0 & 1 & & & & \mathbf{0} & & & \\ 0 & 1 & 0 & 1 & & & & & & & \\ 0 & 1 & 0 & 0 & & & & & & & \\ \hline 1 & 0 & 1 & 0 & 1 & 0 & 0 & 0 & 0 & 0 & 0 \\ 1 & 0 & 0 & 1 & 0 & -1 & 0 & 0 & 0 & 0 & 0 \\ 1 & 0 & 0 & 0 & 0 & 0 & 1 & 0 & 0 & 0 & 0 \\ 0 & 1 & 1 & 0 & 0 & 0 & 0 & -1 & 0 & 0 & 0 \\ 0 & 1 & 1 & 0 & 0 & 0 & 0 & 0 & -1 & 0 & 0 \\ 0 & 1 & 0 & 0 & 0 & 0 & 0 & 0 & 0 & -1 & 0 \\ 0 & 1 & 0 & 0 & 0 & 0 & 0 & 0 & 0 & 0 & -1 \end{array}\right].$$

Table 6.3.2

	Variety						
	Karoon		Shoeleh		Mexican		
Fertilizer	Y_{ijk}	residual	Y_{ijk}	residual	Y_{ijk}	residual	λ
	21	⓪	7	−4	46	1	
A							45
	29	8	11	⓪			
	16	−2	8	⓪	38	−4	
B					40	−2	42
	15	−3			42	⓪	
λ	−24		−34		0		

We can calculate $\mathbf{e}'\mathbf{D}\mathbf{X}_{(2)}\mathbf{X}_{(1)}^{-1}$ to find whether this $\mathbf{B}$ is optimal or not. However, as in the transportation problem we do this determination via the dual variables. The circled cells correspond to J_0; the objective function value $=24$.

Let us now look at the dual of Problem 6.3.2.

Problem 6.3.3

$$\text{Maximize} \quad \mathbf{Y}'\boldsymbol{\pi}$$

$$\text{subject to} \quad \mathbf{X}'\boldsymbol{\pi}=\mathbf{0} \tag{6.3.6}$$

$$-\mathbf{e}' \leqslant \boldsymbol{\pi}' \leqslant \mathbf{e}' \tag{6.3.7}$$

where $\boldsymbol{\pi}'=(\pi_{111},\dots,\pi_{11n_{11}},\dots,\pi_{ab1},\dots,\pi_{abn_{ab}})$. The constraints (6.3.6) are those of a transportation problem. Thus we have an $m\times n$ transportation problem, with the additional restrictions that the π_{ijk}'s should be bounded from below by -1 and above by $+1$. Such problems are called *capacited* transportation problems.

From the duality theory we know that $\mathbf{C_B}\mathbf{B}^{-1}$ gives the dual variables, corresponding to a basis to Problem 6.3.2. $\mathbf{C_B}\mathbf{B}^{-1}$ satisfies the dual constraints of (6.3.6); however it may not satisfy in general the restrictions of (6.3.7), unless $\mathbf{B}$ is optimal. This precisely is the optimality condition of (6.3.5). So when the optimality condition of (6.3.5) is not satisfied, we have only an infeasible solution to the dual, corresponding to a primal basis.

The optimality condition implies that the $C_{1ijk}-Z_{1ijk}, C_{2ijk}-Z_{2ijk}$ corresponding to the $(i,j,k)\in J_0$ given as follows are nonnegative.

$$C_{1ijk}-Z_{1ijk}=1+\mathbf{e}'\mathbf{D}\mathbf{X}_{(2)}\mathbf{X}_{(1)}^{-1}\mathbf{e}_p \qquad \text{for } d_{1ijk},(i,j,k)\in J_0$$

$$C_{2ijk}-Z_{2ijk}=1-\mathbf{e}'\mathbf{D}\mathbf{X}_{(2)}\mathbf{X}_{(1)}^{-1}\mathbf{e}_p \qquad \text{for } d_{2ijk},(i,j,k)\in J_0 \quad (6.3.8)$$

where $\mathbf{e}_p$ is the pth column of $\mathbf{I}$ of order $m-1$ and p is such that the pth row of $\mathbf{X}_{(1)}$ corresponds to index (i,j,k). From duality theory we know that $\mathbf{e}'\mathbf{D}\mathbf{X}_{(2)}\mathbf{X}_1^{-1}$ is actually the row vector of dual variables, π_{J_0}, corresponding to the row indices in J_0. As

$$\mathbf{e}'\mathbf{D}\mathbf{X}_{(2)}=\mathbf{e}'\mathbf{D}\mathbf{X}_{(2)}\mathbf{X}_{(1)}^{-1}\mathbf{X}_{(1)}=\pi_{J_0}\mathbf{X}_{(1)}. \qquad (6.3.9)$$

As $\mathbf{X}_{(1)}$ is triangular, π_{J_0} can be calculated by back-substitution if we know $\mathbf{e}'\mathbf{D}\mathbf{X}_{(2)}$. However $\mathbf{D}$ is a diagonal matrix with ± 1 along the diagonal, depending on the sign of the residuals. Hence $\mathbf{e}'\mathbf{D}$ gives the vector of ± 1, depending on the sign of the residuals. And any columns of $\mathbf{X}_{(2)}$ corresponding to an α_l in the basis have a "1" in the rows such that (l,j,k), for all j and k such that $(l,j,k)\notin J_0$. Similarly if a column of $\mathbf{X}_{(2)}$ corresponds to a β_l in the basis, there is a "1" in the rows such that (i,l,k), for all i and k such that $(i,l,k)\notin J_0$. Thus computing $\mathbf{e}'\mathbf{D}\mathbf{X}_{(2)}$ requires only additions and subtractions. The process of computing $\mathbf{e}'\mathbf{D}\mathbf{X}_{(2)}$ continues with Example 6.3.1.

EXAMPLE 6.3.1 (CONTINUED) From Table 6.3.2, we can note the sign of the residuals. Let $\mathbf{e}'\mathbf{D}\mathbf{X}_{(2)}$ be denoted by $\mathbf{T}=(T_1,T_2,\ldots,T_{m-1})$. T_1 is obtained by considering the cells corresponding to α_1, in a systematic way. Corresponding to α_1 we have the cells after deleting the circled cells $\{(1,1,2), (1,2,1), \text{and } (1,3,1)\}$. The signs of the residuals corresponding to these cells are $(1,-1,1)$. Therefore $T_1=1-1+1=1$. Similarly T_2 is worked out as $T_2=-1-1-1-1=-4$. For T_3, the corresponding cells are $\{(1,1,2), (2,1,1), \text{and } (2,1,2)\}$. Hence $T_3=1-1-1=-1$, and so on. Now we have, from (6.3.9),

$$\mathbf{T}=(1,-4,-1,-1)=(\pi_{111},\pi_{122},\pi_{221},\pi_{233})\cdot\begin{bmatrix}1&0&1&0\\1&0&0&1\\0&1&0&1\\0&1&0&0\end{bmatrix}$$

corresponding to the equations,

$$\begin{aligned}
\pi_{111}+\pi_{122} \qquad\qquad &= 1\\
\pi_{221}+\pi_{233} &= -4\\
\pi_{111} \qquad\qquad &= -1\\
\pi_{122}+\pi_{221} \qquad &= -1.
\end{aligned}$$

Readily we have, $\pi_{111}=-1$, $\pi_{122}=2$, $\pi_{221}=-3$, and $\pi_{233}=-1$.

Notice that π_{ijk}'s for $(i, j, k)\in J_0$ can be calculated using the circled cells and T_i as in the transportation problem. We add an additional column and row corresponding to **T** for this purpose. (See Table 6.3.3.) Now these π_{ijk} can be used to calculate the reduced costs $C_{1ijk}-Z_{1ijk}$ and $C_{2ijk}-Z_{2ijk}$, for $(i, j, k)\in J_0$. For instance,

$$C_{1,111}-Z_{1,111}=C_{1,111}+\pi_{111}=1+(-1)=0$$

$$C_{2,111}-Z_{2,111}=C_{2,111}-\pi_{111}=1-(-1)=2.$$

Similarly other reduced costs are worked out and are shown in Table 6.3.4. Notice that it is enough to calculate either $C_{1ijk}-Z_{1ijk}$ or $C_{2ijk}-Z_{2ijk}$, as the sum of them is always equal to 2, as observed in Section 2.8. We choose $d_{1,221}$ to enter the basis.

Table 6.3.3

	Variety							
	Karoon		Shoeleh		Mexican			
Fertilizer	Y_{ijk}	residual	Y_{ijk}	residual	Y_{ijk}	residual	λ	**T**
	21	⓪	7	−4	46	1		
A							45	1
	29	8	11	⓪				
	16	−2	8	⓪	38	−4		
B					40	−2	42	−4
	15	−3			42	⓪		
λ	−24		−34		0			
T	−1		−1		−1			

Table 6.3.4

$(i,j,k)\in J_0$	(1,1,1)	(1,2,2)	(2,2,1)	(2,3,3)
$C_{1ijk}-Z_{1ijk}$	0	3	-2	0
$C_{2ijk}-Z_{2ijk}$	2	-1	$+4$	2

To summarize the discussions so far, we have had:

1 The MINMAD regression problem corresponding to a two-way classification model has a special structure. That is, the dual problem is a capacited transportation problem.

2 Any $(m-1)$th order nonsingular submatrix $\mathbf{X}_{(1)}$ of the design matrix is triangular, and so the values of the corresponding basic variables α_i or β_j can be computed easily by back-substitution. Once they are known, the residuals can be worked out. Depending on the sign of the residuals, d_{1ijk} and d_{2ijk} can be determined.

3 The reduced cost coefficients need only be calculated for (i, j, k) such that both d_{1ijk} and d_{2ijk} are nonbasic. Also these computations can be done with use of the corresponding dual variables.

4 Dual variables π_{ijk} are again computed, with use of triangularity property of $\mathbf{X}_{(1)}$.

5 The vector to enter the basis is chosen according to the simplex criterion: the one with the most negative reduced cost is selected for entry.

Vector to Leave the Basis

At this stage we want to bring in the advantages derived from the structure of the general MINMAD regression problem. We noticed in Section 2.8 that the usual simplex exchange steps can be bypassed because of the structure of the matrix corresponding to the MINMAD regression problem: we choose the vector to leave the basis according to the minimum-ratio criterion. We increase the value of the entering variable beyond the point where it would cause some of the basic variables restricted in sign to become negative. However, if $d_{1ijk}(d_{2ijk})$ in the basis becomes negative in such a context, we introduce $d_{2ijk}(d_{1ijk})$ into the basis. New reduced cost is worked out for the entering variable. If the reduced cost is still negative, we continue such bypassing if possible. When the reduced cost becomes positive, we cannot bypass.

We introduce the variable chosen for entry and remove the appropriate variable from the basis, according to the simplex exit criterion.

The modifications of Barrodale and Roberts to the simplex method, discussed in Section 2.8, enter here as well.

The minimum-ratio criterion chooses the minimum of

$$\frac{|r(s)|}{\pm d_{ss}\mathbf{X}_{(2)(s)}\mathbf{X}_{(1)(p)}^{-1}}, \quad \text{over } s \text{ such that the denominator is positive} \tag{6.3.10}$$

where $r(s)$ is the residual corresponding to the sth row of $\mathbf{X}_{(2)}$, d_{ss} is the sth diagonal element of $\mathbf{D}$, $\mathbf{X}_{(2)(s)}$ is the sth row of $\mathbf{X}_{(2)}$, and $\mathbf{X}_{(1)(p)}^{-1}$ is the pth column of $\mathbf{X}_{(1)}^{-1}$ where p corresponds to the entering variable. If it is d_{1ijk} we take the "−" in the denominator, and if it is d_{2ijk} we take the "+" in the denominator.

If the minimum of the ratios corresponds to $d_{1tuv}(d_{2tuv})$ we bring into the basis $d_{2tuv}(d_{1tuv})$ and revise the reduced cost corresponding to the entering variable, as follows:

$$\text{new reduced cost} = (\text{old reduced cost}) + 2|\mathbf{X}_{(2)(s)}\mathbf{X}_{(1)(p)}^{-1}|. \tag{6.3.11}$$

If this result is negative we consider the next minimum of the ratios and proceed until (6.3.11) is positive, when we make a usual simplex exchange.

EXAMPLE 6.3.1 (CONTINUED) To determine the vector to leave the basis we require $\mathbf{X}_{(2)(s)}\mathbf{X}_{(1)(p)}^{-1}$. Let

$$\mathbf{X}_{(2)(s)}\mathbf{X}_{(1)}^{-1} = \mathbf{L}.$$

Then,

$$\mathbf{X}_{(2)(s)} = \mathbf{L}\mathbf{X}_{(1)}.$$

By back-substitution $\mathbf{L}$ can be obtained. The pth element of $\mathbf{L}$ gives

$$\mathbf{X}_{(2)(s)}\mathbf{X}_{(1)(p)}^{-1}.$$

Consider $s=1$, corresponding to $(1,1,2)\notin J_0$.

$$\mathbf{X}_{(2)(s)} = (1,0,1,0), \quad r(s) = 8.$$

The variable determined to enter the basis is d_{1221}. The index $(2,2,1)\in J_0$ corresponds to the third row of $\mathbf{X}_{(1)}$; hence $p=3$, $d_{11}=1$ in $\mathbf{D}$. Now

$$(L_1,\ldots,L_4)\cdot\begin{bmatrix}1&0&1&0\\1&0&0&1\\0&1&0&1\\0&1&0&0\end{bmatrix} = (1,0,1,0)$$

yielding $\mathbf{L}=(1,0,0,0)$. So $\mathbf{X}_{(2)(s)}\mathbf{X}_{(1)(p)}^{-1}=0$. $(1,1,2)$ cannot be considered for removal.

Consider $s=2$, corresponding to $(1,2,1)$.

$$\mathbf{X}_{(2)(s)}=(1,0,0,1),\ |r(s)|=|-4|=4 \text{ and } d_{22}=-1.$$

Now the corresponding $\mathbf{L}=(0,1,0,0)$, so $L_3=0$. Hence $(1,2,1)$ also cannot be removed from the basis.

Consider $s=3$, corresponding to $(1,3,1)$. $\mathbf{X}_{(2)(s)}=(1,0,0,0)$, $r(s)=1$, and $d_{33}=1$. Now the corresponding $\mathbf{L}=(0,1,-1,1)$, so $L_3=-1$. Therefore, $-d_{33}\cdot\mathbf{X}_{(2)(s)}\cdot\mathbf{X}_{(1)(p)}^{-1}=1>0$. $\theta(3)=\frac{1}{1}=1$.

Similarly for $s=4$. $\mathbf{X}_{(2)(s)}=(0,1,1,0)$, $|r(s)|=|-2|=2$, and $d_{44}=-1$. $\mathbf{L}=(1,-1,1,0)$, so $L_3=1$. Therefore $-d_{44}\mathbf{X}_{(2)(s)}\mathbf{X}_{(1)(p)}^{-1}=1>0$. $\theta(4)=\frac{2}{1}=2$. For $s=5$ we have $\theta(5)=\frac{3}{1}=3$. We find for $s=6,7$, $L_3=0$. Thus $\theta=\min_s\theta(s)=1$, for $s=3$. $\mathbf{L}$ can also be calculated as in the transportation problem, by using the circled cells and $\mathbf{X}_{(2)(s)}$.

In fact, for any row s with the corresponding $(i,j,k)\notin J_0$ is such that $(i,j,r)\in J_0$ for some $r\neq p$ we will have $\mathbf{X}_{(2)(s)}\mathbf{X}_{(1)(p)}^{-1}=0$ and if $r=p$ then $\mathbf{X}_{(2)(s)}\mathbf{X}_{(1)(p)}^{-1}=1$. Hence we need not consider such s for computing $\mathbf{L}$.

Also, any coordinate of $\mathbf{X}_{(2)(s)}\mathbf{X}_{(1)}^{-1}=\mathbf{L}$ is either ±1 or 0. We find a loop connecting the cell (i,j,k) corresponding to s and the circled cells which are in J_0. We start with a "+1" in cell (i,j,k) and alternatingly give "-1" and "+1" to the cells in the loop. If the cell corresponding to the variable to enter receives ±1 we have $\mathbf{X}_{(2)(s)}\mathbf{X}_{(1)(p)}^{-1}=\mp1$, respectively. Hence there is no need of solving $\mathbf{L}$ explicitly by back substitution.

Thus, the minimum-ratio criterion indicates in the example, the exit of $d_{1,131}$ from the basis $\mathbf{B}$. As yet we do not decide to exchange this with $d_{1,221}$. We revise the reduced cost for $d_{1,221}$ as given by (6.3.11).

$$\begin{aligned}\text{New reduced cost} &= (-2)+2|\mathbf{X}_{(2)(3)}\mathbf{X}_{(1)(3)}^{-1}|\\ &= -2+2=0.\end{aligned}$$

As the new reduced cost is no more negative, we now do the usual exchange of $d_{1,131}$ and $d_{1,221}$, obtaining a new basis, as shown in Table 6.3.5. λ_i's are again found as usual using the circled cells corresponding to rows in $\mathbf{X}_{(1)}$. The objective function value is 21.

We proceed with the new $\mathbf{X}_{(1)}$ and calculate the vector $\mathbf{T}$. Now, $T_1=1-1=0$, $T_2=-1-1+1-1-1=-3$, $T_3=1-1-1=-1$, and $T_4=-1+1=0$. From the circled cells in J_0 and the values of $\mathbf{T}$, we find

$$\pi_{111}=-1,\qquad \pi_{122}=0,\qquad \text{so}\quad \pi_{131}=1\qquad \text{and}\quad \pi_{233}=-3.$$

Next we find the variable to enter the basis as $d_{1,233}$.

Table 6.3.5

	Variety							
	Karoon		Shoeleh		Mexican			
Fertilizer	Y_{ijk}	residual	Y_{ijk}	residual	Y_{ijk}	residual	λ	T
	21	⓪	7	−4	46	⓪	46	
A								0
	29	8	11	⓪				
	16	−1	8	1	38	−4		
B					40	−2		−3
	15	−2			42	⓪	42	
λ	−25		−35		0			
T	−1		0		−2			

We do one more iteration and obtain a basis for which optimality conditions are satisfied.

The variable to leave the basis is determined as $d_{2,211}$. The optimal solution obtained is shown in Table 6.3.6. We find $\pi_{111}=1$, $\pi_{122}=0$, $\pi_{131}=-1$, and $\pi_{211}=-1$; as all four are between $+1$ and -1, we have optimality. The objective function values is 20. The estimates of μ, τ_1, τ_2, β_1, β_2, and β_3 are obtained as follows.

Table 6.3.6

	Variety							
	Karoon		Shoeleh		Mexican			
Fertilizer	Y_{ijk}	residual	Y_{ijk}	residual	Y_{ijk}	residual	λ	T
	21	⓪	7	−4	46	⓪		
A							46	0
	29	8	11	⓪				
	16	⓪	8	2	38	−3		
B					40	−1	41	−1
	15	−1			42	1		
λ	−25		−35		0			
T	0		0		−1			

Let the optimal solution obtained for the problem be denoted by α_1^*, α_3^*, and β_1^*, β_2^*, and β_3^* where $\boldsymbol{\lambda}^*=(\alpha_1^*, \alpha_2^*; \beta_1^*, \beta_2^*, \beta_3^*)$.

Let $\bar{\alpha}^*=\Sigma_{i=1}^{a}\alpha_i^*/a$, $\bar{\beta}^*=\Sigma_{j=1}^{b}\beta_j^*/b$. Then,

$$\hat{\tau}_i=\alpha_i^*-\bar{\alpha}^*, \qquad i=1,\ldots,a$$

$$\hat{\beta}_j=\beta_j^*-\bar{\beta}^*, \qquad j=1,\ldots,b$$

and

$$\hat{\mu}=\bar{\alpha}^*+\bar{\beta}^*.$$

This insures that $\Sigma\tau_i=0$, $\Sigma\hat{\beta}_j=0$. Thus we have $\hat{\tau}_1=2.5$; $\hat{\tau}_2=-2.5$; $\hat{\beta}_1=-5$; $\hat{\beta}_2=-15$, and $\hat{\beta}_3=20$. Also $\hat{\mu}=23.5$.

It is helpful to present the tree corresponding to the rows of **B** for which both d_{1ijk} and d_{2ijk} are nonbasic.

For instance, in Example 6.3.1, the tree corresponding to the initial J_0 is given below in Figure 6.3.1, with the root of the tree as Node 5 corresponding to $\lambda_5=\beta_3$. If $(i,j,k)\in J_0$ then we have an edge between nodes i and $a+j$ and the number along the edge corresponds to the k (as there is more than one arc between the nodes when $n_{ij}>1$). For instance $(1,1,1)\in J_0$ we have an edge between node 1 and 3 and we indicate 1 along ①—③.

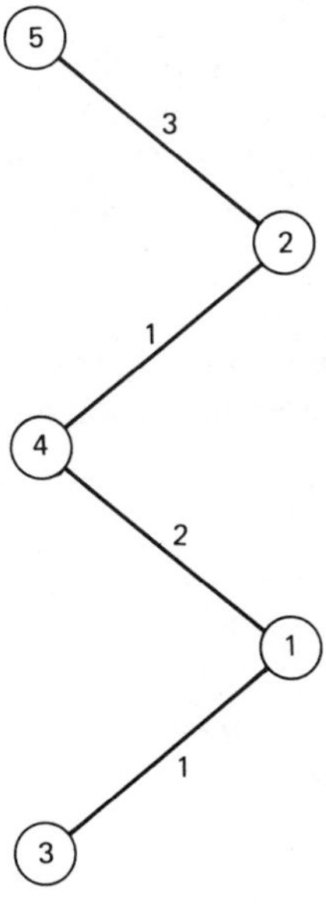

Figure 6.3.1 Tree corresponding to the first J_0 in Example 6.3.1.

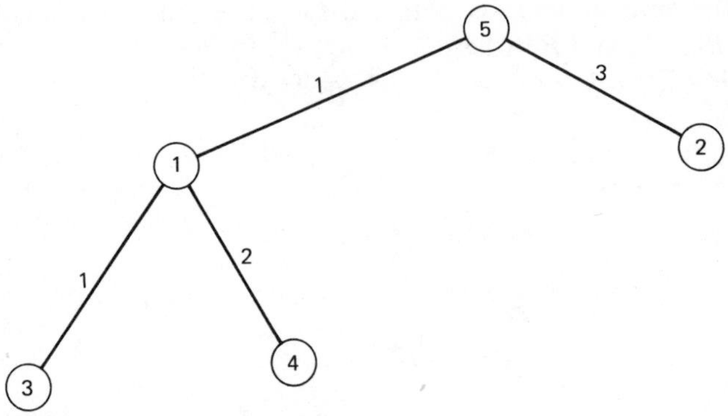

Figure 6.3.2 Tree corresponding to $J_0 = \{(1,1,1), (1,2,2), (1,3,1), (2,3,3)\}$.

According to the simplex criterion the vector to enter the basis was determined as $d_{1,221}$. In the subsequent tree we remove the edge $(2,4)$ with $k=1$.

Since $(2,2,1)$ no longer belongs to J_0, when the corresponding edge is removed the tree is no longer connected and we have two subtrees. If we wish to have a tree corresponding to the next basis, the new edge added to the tree, corresponding to the d_{1ijk} or d_{2ijk} leaving the basis, should be such that the two subtrees form a tree. We need to calculate only the minimum ratio corresponding to such edges. For instance, it is enough to consider edges $(5;1)$, $k=1$ and $(2;3)$, $k=1,2$. Corresponding to $J_0 = \{(1,1,1),(1,2,2),(1,3,1),(2,3,3)\}$ we have the tree given in Fig. 6.3.2.

But for the notational complexity introduced by the four subscripts in the variables, the ideas are exactly same as those encountered in Section 2.8, and in the discussions of transportation problem in Chapter 5.

However, the algorithm implemented as a computer program is available written in FORTRAN, and such a program can be used to obtain estimates in these problems. Also the additional computational advantages derived by the tree representation of the basic sets are incorporated in the program. And computationally, this primal approach is superior to solving the dual-capacited transportation problem.

6.4 CONSTRUCTION OF BIB DESIGNS

Complete block designs require that each treatment be assigned to each block, where a block is a collection of experimental units—plots or rats or runs. Suppose there are v treatments. In each block we require at least v experimental units. If the design has b blocks we require in all bv experimental units. But in many practical situations, for various reasons the experimentor may not be able to choose bv experimental units for his experiment. So naturally there is a need for minimizing the number of experimental units required for the experiment, subject to restrictions on the precision of the estimates obtained from the analysis of the experimental data. Balanced incomplete block designs were introduced with this problem in mind.

Let $V=\{1,2,\ldots,v\}$ be the set of treatments and let $v\Sigma k$ be the set of all distinct subsets of size k based on V. Let vCk denote the cardinality of $v\Sigma k$. Let b denote the total number of blocks in the design.

A *balanced incomplete block design*, *d*, with parameters v, b, r, k, and λ denoted by BIB (v, b, r, k, λ) is a collection of b elements of $v\Sigma k$ (not necessarily distinct), called blocks, with these properties: (1) each element of V occurs in exactly r blocks, and (2) each pair of distinct elements of V appears in exactly λ blocks.

Thus a BIB design, BIB (v, b, r, k, λ) is a combinatorial arrangement of v treatments in b blocks, containing k experimental units in each, and these v treatments occur in such a way that each treatment does not occur more than once in any block, each treatment occurs on r experimental units, and each pair of treatments occurs λ times.

It is necessary for v, b, k, r, λ to satisfy the following relations.

1 $bk=vr$. The reason is that each treatment appears r times in the design and k treatments appear in each block. Thus both of them give the total number of experimental units.

2 $\lambda(v-1)=r(k-1)$. Since we have $bk(k-1)=\lambda v(v-1)$ and from (1), $bk=vr$,

$$vr(k-1)=\lambda v(v-1), \text{ or}$$

$$\lambda(v-1)=r(k-1).$$

3 From (1) and (2), we have $\lambda(v-1)\equiv 0 \bmod (k-1)$ and $\lambda v(v-1)\equiv 0 \bmod (k(k-1))$.

4 $b\geqslant v$. This inequality, due to Fisher, can be shown by noticing that the rank of the treatment-block-incidence matrix N associated with an incomplete block design where N has v rows and b columns, and $N=((n_{ij}))$ is such that $n_{ij}=1$ if the ith treatment appears in the jth block; $n_{ij}=0$ otherwise. Here N has rank v.

Now $v\leqslant \min(v, b)$. Hence $b\geqslant v$.

These conditions can be shown to be not sufficient for the existence of a BIB(v, b, r, k, λ). Therefore we have to find ways to construct such designs. Classically such designs are obtained by using results from orthogonal Latin squares, finite geometrics, and difference sets, among other possibilities. Here we do not go into the discussion of these.

Notice that we have not explicitly restricted the b blocks, should all be distinct. So some of the blocks corresponding to certain elements of $v\Sigma k$, may occur more than once. BIB designs in which some blocks are repeated are called BIB designs with *repeated blocks*.

Such designs are of practical significance as they allow us to restrict certain treatment combinations being excluded from the experiment, for various considerations. We restrict our attention to study the problem of constructing BIB designs with repeated blocks.

A BIB design, d, is said to be a *uniform BIB design* if the distinct elements of $v\Sigma k$ appearing in the blocks b are such that they are repeated the same number of times in b. A BIB design with $b = vCk$ is denoted by $d(v, k)$ called the *trivial BIB design*. A BIB design with $b < vCk$ is said to be a *reduced BIB design*.

DEFINITION 6.4.1 A *support* of a BIB design, d, is a collection of distinct blocks in d, denoted by d^*. The support size is the cardinality of d^* and is denoted by b^*.

Before we consider the BIB designs and the problem of construction of such designs, we consider some related problems in integer programming.

6.5 INTEGER PROGRAMMING PROBLEMS WITH 0-1 MATRICES AND CONSTANT RIGHT-HAND-SIDE VECTOR

Let $\mathbf{A}$ be an $m \times n$ matrix. Let $\mathbf{b}$ be an $m \times 1$ vector. Let $\mathbf{C}$ be a $1 \times n$ vector. Then the problem of finding $\mathbf{x}$, an $n \times 1$ integer vector, satisfying the following conditions is known as an integer-programming problem.

$$\begin{aligned} \text{Minimize} \quad & \mathbf{Cx} \\ \text{subject to} \quad & \mathbf{Ax} \geqslant \mathbf{b} \\ & \mathbf{x} \geqslant \mathbf{0} \\ & \mathbf{x} \quad \text{integer.} \end{aligned}$$

When we have the additional information that the elements of $\mathbf{A}$ are either zeroes or ones, we can solve the IP problem above using special methods developed for this purpose. Such problems have received the special attention of the mathematical programmers as these problems arise in many practical applications.

We come to the problem known as the set representation problem (SRP). Let $\Gamma=\{1,\ldots,n\}$ and $F=\{F_1,\ldots,F_m\}$, a family of m nonempty subsets of Γ. The SRP is to select a subset S such that

$$S\subseteq\Gamma$$
$$S\cap F_i\neq\varnothing,\qquad i=1,\ldots,m$$

and $\Sigma_{j\in S}C_j$ is minimized, where C_j is the cost of including element j in the set S, for $j=1,\ldots,n$. This problem can be formulated as an integer-programming problem with a 0-1 matrix. Let $\mathbf{A}=((a_{ij}))$ with a_{ij} equal to 1 if $j\in F_i$ and 0 otherwise. Let x_j be a zero-one variable (binary), such that x_j is 1 if $j\in S$ and 0 otherwise. Let $\mathbf{e}=(1,\ldots,1)'$.

Now we have an SRP equivalent to Problem 6.5.1.

Problem 6.5.1

$$\begin{aligned}&\text{Minimize} && \mathbf{Cx}\\ &\text{subject to} && \mathbf{Ax}\geqslant\mathbf{e}\\ &&& x_i,\ 0,\text{ or }1.\end{aligned}\tag{6.5.1}$$

Another problem related to SRP is the one in which the inequality is replaced by an equality in (6.5.1). This problem is known as the set partitioning problem (SP) and is given by Problem 6.5.2.

Problem 6.5.2

$$\begin{aligned}&\text{Minimize} && \mathbf{Cx}\\ &\text{subject to} && \mathbf{Ax}=\mathbf{e}\\ &&& x_i,\ 0,\text{ or }1.\end{aligned}\tag{6.5.2}$$

With a view to showing the connection between the solutions to the two problems we in fact prove certain results in a more general setup, with the right-hand sides of (6.5.1) and (6.5.2) replaced by a constant vector $\lambda\mathbf{e}$, where λ is a positive integer.

Let $\mathcal{Q}$ be the set of all $m\times n$, matrices with elements 0 or 1.

In this discussion whenever $\mathbf{A}$ is used it is understood that $\mathbf{A}\in\mathcal{Q}$.

DEFINITION 6.5.1 Any subset M of the index set of columns from matrix $\mathbf{A} \in \mathcal{Q}$ is called a *represent* of $\mathbf{A}$ if and only if

$$\mathbf{A}_M \mathbf{e} \geqslant \mathbf{e} \tag{6.5.3}$$

where $\mathbf{A}_M$ is the submatrix obtained from $\mathbf{A}$ corresponding to the elements in M.

In the language of linear programming M is a feasible solution to Problem 6.5.1, since setting x_j

$$x_j = \begin{cases} 1 & \text{for } j \in M \\ 0 & \text{otherwise} \end{cases}$$

satisfies all the constraints of Problem 6.5.1. Represents are also called *coverings*.

DEFINITION 6.5.2 A represent M of $\mathbf{A}$ is *minimal* if no M', $M' = M - \{j\}$, for any $j \in M$, is a represent of $\mathbf{A}$.

A minimal represent can also be defined equivalently as follows:

For any row r, let $\alpha(M, r)$ denote the number of columns with $a_{rj} = 1$ among $j \in M$.

DEFINITION 6.5.3 A represent M of $\mathbf{A}$ is minimal if for every $j \in M$, there exists an $r(j)$ such that $\alpha(M, r(j)) = 1$ and $a_{r(j)j} = 1$.

The equivalence of these definitions follows from the fact that if for some $j \in M$, for all r such that $a_{rj} = 1$, $\alpha(M, r) > 1$, then that j can be deleted from M and the remaining elements of M will still be a represent.

Let the cardinality of M be denoted by $n(M)$. Let $R(j)$ denote the set of indices of the rows for which $a_{rj} = 1$. Let $(\mathbf{Ax})_r$ denote the rth row of $\mathbf{Ax}$. For any M, let $\overline{M}$ denote the complement of M.

EXAMPLE 6.5.1 Consider the matrix $\mathbf{A}$ given below

$$\mathbf{A} = \begin{bmatrix} 1 & 0 & 1 & 0 & 1 \\ 0 & 0 & 0 & 1 & 1 \\ 1 & 1 & 0 & 0 & 1 \end{bmatrix}, \qquad M = \{1,2,3,4,5\}$$

as a represent of $\mathbf{A}$. However M is not a minimal represent, as $M' = M - \{5\}$ is also a represent of $\mathbf{A}$. But $M'' = \{2,3,4\}$ is a minimal represent of $\mathbf{A}$, since $\alpha(M'', r) = 1$, for $r = 1,2,3$. Here $r(2) = 3$, $r(3) = 1$, and $r(4) = 2$.

For $M' = \{1,2,3,4\}$, $R(1) = \{1,3\}$, $R(2) = \{3\}$, $R(3) = \{1\}$, and $R(4) = \{2\}$. Notice that M' is not a minimal represent of $\mathbf{A}$.

RESULT 6.5.1 Any represent M of $\mathbf{A}$ is a feasible solution to Problem 6.5.2 if and only if $\alpha(M,r)=1$ for all r.

Proof If we have a feasible solution to Problem 6.5.2, that satisfies all the constraints, it is automatically feasible for Problem 6.5.1, and any solution to Problem 6.5.2 has to have exactly one 1 in each row of the submatrix corresponding to the solution. On the other hand, suppose $\alpha(M,r)=1$ for all r. Then M is feasible for Problem 6.5.2. Thus the result is proven.

Stating that M is feasible for Problem 6.5.2 is the same thing as saying $\mathbf{A}_M\mathbf{e}=\mathbf{e}$. Here the $\mathbf{e}$ multiplying $\mathbf{A}_M$ has $n(M)$ coordinates and the $\mathbf{e}$ on the right hand side has m coordinates.

Now consider $\mathbf{Ax}=\lambda\mathbf{e}$, with $\mathbf{x}$ a nonnegative integer vector, and λ a positive integer. Let for any $M, \mathbf{x}_M$ denote the vector of the corresponding x_j's.

RESULT 6.5.2 Let M be a minimal represent of $\mathbf{A}$. There exist $\mathbf{x}_M$ such that $\mathbf{A}_M\mathbf{x}_M=\lambda\mathbf{e}$ for some λ positive integer if and only if $\mathbf{A}_M\mathbf{e}=\mathbf{e}$.

Proof If $\mathbf{A}_M\mathbf{e}=\mathbf{e}$, we can take $x_j=\lambda$ for $j\in M$ and as a solution to $\mathbf{A}_M\mathbf{x}_M=\lambda\mathbf{e}$. Suppose $\mathbf{A}_M\mathbf{e}>\mathbf{e}$ with $(\mathbf{A}_M\mathbf{e})_r>1$ for some r. By definition M has $\alpha(M,r(j))=1$ for all $j\in M$, for some $r(j)$. The implication is that x_j has to be equal to λ for all $j\in M$ in any solution $\mathbf{x}_M$ to $\mathbf{A}_M\mathbf{x}_M=\lambda\mathbf{e}$; so $\mathbf{x}_M=\lambda\mathbf{e}$. But for row r we have $(\mathbf{A}_M\mathbf{e})_r>1$ or $(\mathbf{A}_M\mathbf{x}_M)_r>\lambda$. Hence there does not exist any solution to $\mathbf{A}_M\mathbf{x}_M=\lambda\mathbf{e}$. This completes the proof.

Notice that Result 6.5.2 says that if M is a minimal represent of $\mathbf{A}$, then either we have an integer solution to $\mathbf{A}_M\mathbf{x}_M=\lambda\mathbf{e}$ for every positive integer λ or for none.

RESULT 6.5.3 Consider any M, a represent of $\mathbf{A}$. Suppose $\alpha(M,r)=1$ for some $r\in R(j)$, for some $j\in M$. If there exists an $\mathbf{x}_M$ such that $\mathbf{x}_M>\mathbf{e}$ and $\mathbf{A}_M\mathbf{x}_M=\lambda\mathbf{e}$, then $\alpha(M,s)=1$ for all $s\in R(j)$.

Proof As $\lambda(M,r)=1$ and $r\in R(j)$, x_j must be equal to λ. Suppose for some $s\in R(j)$, $\alpha(M,s)>1$. Then there exists an l such that $a_{sl}=1$, $l\neq j$. Now as $\mathbf{x}_M>\mathbf{e}$, $x_l>1$. Hence $(\mathbf{A}_M\mathbf{x}_M)_s>\lambda$, leading to a contradiction. Hence the result.

RESULT 6.5.4 Suppose M is a represent of $\mathbf{A}$. Then there exists an $\mathbf{x}_M>\mathbf{e}$, $\mathbf{A}_M\mathbf{x}_M=\lambda\mathbf{e}$ with $x_j<\lambda$ for some $j\in M$, only if for each $r\in R(j)$, $\alpha(M,r)>1$.

Proof The proof follows the same lines as the proof of Result 6.5.3.

Let g_j be the cardinality of $R(j)$ for any column j of **A**.

RESULT 6.5.5 Consider **A** such that for any represent M of **A**, $\Sigma_{j\in M} g_j > m$, the number of rows in **A**. Then no minimal represent of **A** can be feasible for Problem 6.5.2.

Proof Let M be a minimal represent of **A**. As $\Sigma_{j\in M} g_j > m$, we have at least one row r, with $\alpha(M,r) > 1$. From Result 6.5.1, M is not feasible for Problem 6.5.2 as $\mathbf{A}_M\mathbf{e}=\mathbf{e}$ is not satisfied. Hence the result.

RESULT 6.5.6 Consider **A** such that $g_j = g$ for all j, columns of **A**. Let $s = m/g$. If s is not an integer then no minimal represent of **A** can be feasible for Problem 6.5.2.

This result follows from Result 6.5.5, and the observation that if s is not an integer we have $\Sigma_{j\in M} g_j > m$.

DEFINITION 6.5.4 Given a represent M of **A**, we say that $\mathbf{A}_M$ is row-balanced if for every $j \in M$, $\alpha(M,r)$ is same for all $r \in R(j)$.

EXAMPLE 6.5.2 Consider the matrix **A** given below:

$$\mathbf{A} = \begin{bmatrix} 1 & 0 & 1 & 0 & 1 \\ 0 & 0 & 0 & 1 & 1 \\ 1 & 1 & 0 & 0 & 1 \end{bmatrix}$$

Consider $M = \{1,2,3,4\}$.

We find that for $j=1$, $R(j) = \{1,3\}$, and both these rows have two 1's in $\mathbf{A}_M$. For $j=2$, 3, and 4, as $R(j)$ is a singleton set, the condition is satisfied automatically. Thus $\mathbf{A}_M$ is row-balanced.

But if we consider $M = \{2,5\}$ we find the condition not satisfied for $j=1$, as row 1 has a single 1 in $\mathbf{A}_M$ but row 3 has two 1's in $\mathbf{A}_M$.

RESULT 6.5.7 If $\mathbf{A}_M$ is row-balanced then there exists $\mathbf{x}_M > \mathbf{e}$ such that $\mathbf{A}_M\mathbf{x}_M = \lambda\mathbf{e}$ for some positive λ.

Proof We can put the elements of M into equivalence classes according to $\alpha(M,r)$, that is, for any j in the qth class, for all $r \in R(j)$ we have $\alpha(M,r) = n_q$. Let l be the least common multiple of n_q's. Consider x_j defined as follows from $j \in M$.

$x_j = l/n_q$ if j is in the qth class. Now it can be verified that this **x** satisfies $\mathbf{A}_M\mathbf{x}_M = l\mathbf{e}$ and $\mathbf{x}_M > \mathbf{e}$. Hence the result.

Notice that we have no restriction on the matrix **A**, excepting $\mathbf{A}\in\mathcal{Q}$. For instance, in Example 6.5.2 for $M=\{1,2,3,4\}$ we have two classes $\{1,2,3\}$ and $\{4\}$; $n_1=2$ and $n_2=1$. So we have $l=2$ and $x_1=x_2=x_3=1$ and $x_4=2$.

RESULT 6.5.8 Consider **A** with $g_j=g$ for all j column of **A**. If there exists a represent for **A**, then the cardinality of M, $n(M)$ for any minimal represent M of **A** is such that

$$s=m/g\leqslant n(M)\leqslant \min(m,n).$$

Proof As each column of **A** can represent at most g rows, the minimum number of columns required to represent **A** is therefore equal to $s=m/g$. And further, $n(M)$ can be equal to s only when s is an integer, as seen from Result 6.5.6, establishing the lower bound for $n(M)$. On the other hand, as **A** has a represent, every row of **A** has at least one "1". Therefore, to represent all the rows of **A**, at most we require m columns of **A**. So no minimal represent can have cardinality more than m. Since **A** has a represent, if $n<m$, no minimal represent has to have cardinality more than n. Hence the result.

RESULT 6.5.9 If M is such that $\mathbf{A}_M\mathbf{x}_M=\lambda\mathbf{e}$ for some λ and $\mathbf{x}_M>\mathbf{e}$ then there exists an M' such that $\mathbf{A}_{M'}\mathbf{X}_M{}'=\lambda'\mathbf{e}$ for some λ' and $\mathbf{x}_{M'}>\mathbf{e}$ and $n(\mathbf{A}_{M'})\leqslant \text{rank }(\mathbf{A})\leqslant \min(m,n)$.

Proof Suppose $n(\mathbf{A}_M)\leqslant \text{rank }(\mathbf{A})$ then $M'=M$ does this. On the other hand suppose $n(\mathbf{A}_M)>\text{rank }(\mathbf{A})$. Then the columns of $\mathbf{A}_M$ are linearly dependent. So we can find $\mathbf{h}\neq\mathbf{0}\ni\mathbf{h}'\mathbf{A}_M=\mathbf{0}$. Let

$$\frac{(\mathbf{x}_M)_r}{h_r}=\min\left\{\frac{(\mathbf{x}_M)_j}{h_j},\qquad h_j>0\right\}.$$

Then consider

$$M'=M-\left\{j\,\middle|\,\frac{(\mathbf{x}_M)_j}{h_j}=\frac{(\mathbf{x}_M)_r}{h_r}\right\}$$

with

$$(\mathbf{x}_{M'})_j=\begin{cases}(\mathbf{x}_M)_j-(\mathbf{x}_M)_r\cdot\dfrac{(\mathbf{x}_M)_j}{h_r}, & j\neq r\\ 0 & j=r\end{cases}.$$

Now $(\mathbf{x}_{M'})_j$ may not be all integers. Let t be the smallest integer such that $t(\mathbf{x}_{M'})_j$ are all integers. Then $\mathbf{A}_{M'}t\mathbf{x}_{M'} = t\lambda\mathbf{e}$. Thus $n(M') < n(M)$ and we have $\lambda' = t\lambda$. We can repeat this process until $n(M') \leqslant \text{rank}\ (\mathbf{A})$. Hence the result.

We shall now look back at the main problems, namely, those related to the construction of BIB designs with repeated blocks.

6.6 CONSTRUCTION OF BIB DESIGNS AND RELATED INTEGER PROGRAMMING PROBLEMS

We give here an interesting and useful formulation of the problem of construction of BIB designs. Then we prove certain results relating to the supports of BIB designs and represents defined in Section 6.5.

Let $v\Sigma k$ be the set of all possible blocks, the distinct subsets of size k from V.

DEFINITION 6.6.1 F is called a frequency vector corresponding to a BIB design if $\mathbf{F} = (f_1, \ldots, f_{vCk})$ is such that f_i is the frequency of the ith element of $v\Sigma k$.

Now let $\Sigma_{i=1}^{vCk} f_i = b$ and b^* is the number of nonzero entries in $\mathbf{F}$. Let the elements of $v\Sigma 2$ be numbered from 1 to $vC2$, and similarly let the elements of $v\Sigma k$ be numbered from 1 to vCk.

DEFINITION 6.6.2 Matrix $\mathbf{P}$ is called the pair inclusion matrix for a given v and k, in case $P_{ij} = 1$ if the ith element of $v\Sigma 2$ is contained in the jth element of $v\Sigma k$ and $P_{ij} = 0$ otherwise. Similarly we call the vector $\mathbf{P}_j = (P_{1j}, \ldots, P_{vC2,j})'$ the pair inclusion vector associated with $j \in v\Sigma k$.

Thus $\mathbf{P} = [\mathbf{P}_1, \ldots, \mathbf{P}_{vCk}]$. Let $\mathbf{e}$ denote the vector $(1, \ldots, 1)'$ with $vC2$ elements. We have the following result.

RESULT 6.6.1 Any frequency vector **F** corresponding to a BIB design of $\text{BIB}(v, \ldots, k, \lambda)$ satisfies

$$\mathbf{PF} = \lambda\mathbf{e} \tag{6.6.1}$$

$$\mathbf{F} \geqslant \mathbf{0}, \text{ integer vector} \tag{6.6.2}$$

and any integer **F** satisfying (6.6.1) and (6.6.2) is a frequency vector corresponding to a BIB $(v, \ldots, k, \lambda)$.

Proof The first part is obvious. The second part follows from the fact that $\Sigma_j P_{ij} f_j$ is the number of times the ith pair appears in the design.

Thus the problem of constructing BIB designs based on v, k, and λ is formulated as the problem of finding feasible solutions to an integer programming problem. In fact, if we consider $\mathbf{PF}-\lambda\mathbf{e}=\mathbf{0}$ and find integer solution to this homogeneous system with $(\mathbf{F},\lambda)$ integers we have a BIB design. Observe that whenever we have a rational solution to the system we have an integer solution as well. We find such a solution by taking the least common multiple (*lcm*) of the denominators of the f_j's and multiplying f_j's by the *lcm*; for this modified $\mathbf{F}$, λ will be an integer.

Next we turn our attention to the characterization of the supports of BIB designs.

We observe that the integer program obtained in Result 6.6.1 is a special case of the problems discussed in Section 6.5. Here we have the following correspondence: $\mathbf{A}=\mathbf{P}$, $m=v(v-1)/2$, $n=vCk$, $g_j=k(k-1)/2$ for all j, columns of $\mathbf{A}$, $\mathbf{x}$ is denoted by $\mathbf{F}$. For convenience, we have already denoted the right-hand-side elements by $\lambda\mathbf{e}$, a constant vector, instead of the usual linear-programming notation "**b**." With this correspondence, we can observe that (1) the existence of $\mathbf{x}_M > \mathbf{e}$ such that $\mathbf{A}_M\mathbf{x}_M=\lambda\mathbf{e}$, is equivalent to saying that M is a support of a BIB design; (2) $\mathbf{A}_M\mathbf{e}=\mathbf{e}$ is equivalent to saying that M is itself a BIB design with $\lambda=1$.

Results 6.6.2 and 6.6.3 connect minimal represents and support of BIB designs.

EXAMPLE 6.6.1 Consider $v=5$, $k=3$. The $\mathbf{P}$ matrix corresponding to these parameters is given in Table 6.6.1.

Table 6.6.1

	Block									
Pair	1,2,3	1,2,4	1,2,5	1,3,4	1,3,5	1,4,5	2,3,4	2,3,5	2,4,5	3,4,5
1,2	1	1	1							
1,3	1			1	1					
1,4		1		1		1				
1,5			1		1	1				
2,3	1						1	1		
2,4		1					1		1	
2,5			1					1	1	
3,4				1			1			1
3,5					1			1		1
4,5						1			1	1

RESULT 6.6.2 If M is a minimal represent of $\mathbf{P}$, then M is the support of a BIB design if and only if M is itself a BIB design for all λ, positive integers.

Proof If M is a BIB design then it is a support of the same design. On the other hand, if M is a support of a BIB design, from Result 6.5.3 M is a BIB design with $\lambda=1$. This proves the result.

RESULT 6.6.3 BIB designs with repeated blocks for all λ exist, if and only if there exists a minimal represent of $\mathbf{P}$ which is the support of a BIB design.

Proof This result follows from Result 6.6.2 and the fact that existence of a BIB design for $\lambda=1$ is equivalent to $\mathbf{P}_M\mathbf{e}=\mathbf{e}$ for some M; this M has to be a minimal represent. Hence the result.

We also obtain bounds on the minimal support size $b^*_{\min}$ using the preceding discussion and Result 6.5.8, and Result 6.5.9. We state this as Result 6.6.4.

RESULT 6.6.4 The bounds on the minimal support size denoted by $b^*_{\min}$, for a BIB design, BIB $(v,.,.,k,.)$, is given by

$$\frac{v(v-1)}{k(k-1)} \leqslant b^*_{\min} \leqslant vC2.$$

Also from Result 6.5.6, we see that the lower bound will be attained only possibly when $s=\{v(v-1)\}/\{k(k-1)\}$ is an integer. Further, from the necessary condition for existence of BIB designs we have $\lambda(v-1)\equiv 0 \bmod(k-1)$. If a minimal represent is the support of a BIB design, then for $\lambda=1$ we have a BIB design. Therefore $q=(v-1)/(k-1)$ also has to be an integer. Thus when s and q are integers we have hope to get a BIB design with minimal support size equal to s.

EXAMPLE 6.6.2 Consider $v=36$ and $k=6$. We have

$$s=\frac{36\times 35}{6\times 5}=42$$

and

$$q=\tfrac{35}{5}=7.$$

Both are integers. However, there does not exist a BIB design with $b=s=42$, $\lambda=1$. This example establishes the fact that s, q integers are not sufficient for the existence of a BIB design with minimal support size equal to s.

6.7 SOME MINIMAL SUPPORTS FOR BIB DESIGNS WITH $k=3$

In this section we give a procedure to find the minimal support in case s and q are integers. In general, we pose the problem as that of finding a Latin square with certain additional restrictions. A Latin square is an arrangement of v elements, each repeated v times, in a square matrix of order v, in such a manner that each element appears exactly once in each row and in each column.

Let L be a Latin square. Let L_{ij} represent the ijth element of L, $i, j=1,\dots,v$. Assume that $L_{ii}=i$ for all $i=1,2,\dots,v$. Call such a Latin square a Latin square with natural diagonal.

EXAMPLE 6.7.1 Consider $v=4$.

$$\mathbf{L}=\begin{pmatrix} 1 & 2 & 3 & 4 \\ 4 & 1 & 2 & 3 \\ 3 & 4 & 1 & 2 \\ 2 & 3 & 4 & 1 \end{pmatrix}$$

L as given is a Latin Square. L as given below has the natural diagonal,

$$\mathbf{L}=\begin{pmatrix} 1 & 4 & 2 & 3 \\ 3 & 2 & 4 & 1 \\ 4 & 1 & 3 & 2 \\ 2 & 3 & 1 & 4 \end{pmatrix}$$

RESULT 6.7.1 A BIB $(v, k=3, \lambda=1)$ with support size s exists if there is a Latin square with a natural diagonal satisfying the condition $L_{ji}=L_{ij}=r$. $L_{ir}=L_{ri}=j$ and $L_{rj}=L_{jr}=i$, for all i, j, and $r=1,\dots,v$.

Proof Suppose there is a BIB design with support size s and $\lambda=1$. Notice that $s=v(v-1)/6$. As $\lambda=1$ if (i, j, r) is a block in the support, then the blocks (i, r, j), (r, i, j), (r, j, i), (j, r, i), and (j, i, r) are not in the support. Thus for each block in the support we can determine six distinct elements of the square matrix satisfying the conditions required. In all $v(v-1)$ elements will be obtained from the s blocks in the design, excluding the diagonal elements which can be fixed as $L_{ii}=i$, $i=1,\dots,v$. Now we have to show that we have obtained a Latin square, which also follows from the fact that $\lambda=1$.

Thus we have a natural Latin square with the conditions specified.

On the other hand, first we discard the diagonal, then we proceed as follows: Select an element $L_{i_1 j_1} = r_1$ and delete all five of the other related elements. We have the block (i_1, j_1, r_1). Next we select another element $L_{i_2 j_2} = r_2$ and repeat the process. This way we will get s blocks as required. And we can see that each pair appears exactly once and each block covers three distinct pairs. Hence the result.

Observe that, because of symmetry, it is sufficient to construct the upper triangle of the natural Latin square. Also observe that this problem can be posed as finding feasible solutions to an integer programming problem.

EXAMPLE 6.7.2 Consider $v=7$, $k=3$. We have $s=7$, $\lambda=1$, $b=7$. We have a Latin square with a natural diagonal satisfying the restrictions.

	1	2	3	4	5	6	7
1	1	3	2	5	4	7	6
2		2	1	6	7	4	5
3			3	7	6	5	4
4				4	1	2	3
5					5	3	2
6						6	1
7							7

From this square we obtain the design:

123	246	347
145	257	356
167		

EXAMPLE 6.7.3 $v=9$, $k=3$. We have $s=12$, $\lambda=1$, and $b=12$. The corresponding Latin square is:

	1	2	3	4	5	6	7	8	9
1	1	6	5	7	3	2	4	9	8
2		2	4	3	8	1	9	5	7
3			3	2	1	9	8	7	6
4				4	9	8	1	6	5
5					5	7	6	2	4
6						6	5	4	3
7							7	3	2
8								8	1
9									9

We have the design:

126	234	369	459	567
135	258	378	468	
147	279			
189				

6.8 GENERAL APPROACH FOR FINDING BIB DESIGNS WITH MINIMAL SUPPORT SIZE WHEN s AND q ARE INTEGERS

We have observed earlier that when s and q are integers we can possibly have a minimal represent which is also the minimal support of a BIB design. This problem can be formulated as follows:

Problem 6.8.1

$$\begin{aligned} &\text{Minimize} \quad \mathbf{e}'\mathbf{F} \\ &\text{subject to} \quad \mathbf{PF}=\mathbf{e} \\ &\qquad\qquad f_j=0 \text{ or } 1. \end{aligned}$$

This problem is a set partitioning problem formulated as Problem 6.5.2 earlier, with $\mathbf{C}=\mathbf{e}$. Most available methods for solving this problems are search procedures which implicitly search for certain solutions, using systematic branching of the set of all possible solutions. Basically, the efficiency of these methods depends very much on the sharpness of the bounds and the storage and related computational requirements. In this special form the problem is known as a *minimum cardinality set partitioning problem*. There are special algorithms developed for solving 0-1 integer programming problems of this kind. We discuss one such method here for solving this problem.

Let the linear-programming problem corresponding to Problem 6.8.1 be as given by Problem 6.8.2 when the 0-1 integer restriction on f_j is removed.

Problem 6.8.2

$$\begin{aligned} &\text{Minimize} \quad \mathbf{e}'\mathbf{F} \\ &\text{subject to} \quad \mathbf{PF}=\mathbf{e} \\ &\qquad\qquad \mathbf{F}\geqslant\mathbf{0} \end{aligned}$$

REMARK 6.8.1 In no feasible solution to Problem 6.8.2 can f_j be greater than 1, since that would violate some constraint of the problem as P_{ij}'s are 0 or 1 and $f_j \geqslant 0$. Also notice that if there is a feasible solution to Problem 6.8.1, it must correspond to an extreme point of the feasible region of Problem 6.8.2. Implicitly, each $f_j \leqslant 1$. Therefore, corresponding to an optimal basic feasible solution of the problem if the variables take values 0 or 1 we have in fact a solution to Problem 6.8.1. Thus if **P** is of moderate size, solving Problem 6.8.2 can possibly yield a solution to the problem of finding a minimal support to the corresponding BIB design. However, the number of rows and columns of the problem rapidly increases as v and k increase. But the way the columns and rows of **P** are numbered, it is sufficient to keep a formula that works out the blocks when the index of the column is given. There is no need to keep the columns of **P** in the storage of the computer while solving the problem with a computer. The columns in the basis can also be kept track of by storing the indices of the corresponding columns. Further, if the special structure of the columns—namely, that each column has same number of 1's and that the rest of the elements are zeroes—can be exploited to find the inverse of the basis, computation can be further reduced.

We indicate another approach, where we avoid solving the linear-programming problem, Problem 6.8.2. The following results are true for any A, 0-1 matrix; recall that A is an $m\times n$ matrix.

A graph $\mathcal{G}=(\mathcal{N},\mathcal{Q})$ is called a directed graph if the edges are ordered pairs of elements from $\mathcal{N}$; we call such edges *arcs*.

Let

$$\mathcal{N}_A=\{1,\ldots,n\}$$

and

$$\mathcal{Q}=\{(i,k)|\text{There exists a } j \text{ such that}$$

$$a_{ij}=1,\ a_{kj}=1,\ a_{qj}=0,\quad i<q<k\}.$$

Consider $\mathcal{G}_A=(\mathcal{N}_A,\mathcal{Q})$. Now, $\mathcal{G}_A$ is a directed graph. Also $\mathcal{G}_A$ does not contain a closed directed path, one in which the edges are directed in the same direction. While considering directed paths it is sufficient to give the nodes in the order in which they appear in the directed path, $p=(i_1,\ldots,i_k)$.

EXAMPLE 6.8.1 Consider

$$A=\begin{bmatrix}1&0&1&0\\1&1&0&0\\0&1&0&1\\0&0&0&1\end{bmatrix}$$

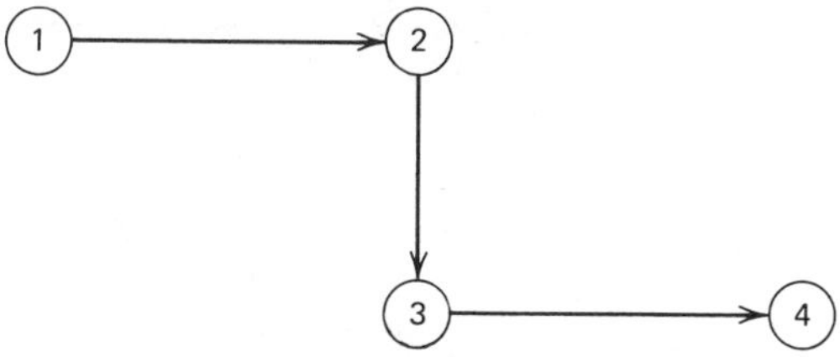

Figure 6.8.1 Graph corresponding to A in Example 6.8.1. The path is directed from 1 to 4.

The graph corresponding to A is given in Fig. 6.8.1, where $p=(1,2,3,4)$ is a directed path from 1 to 4.

Now, a *chain* in graph $\mathcal{G}_A$ is defined as a directed path or an isolated node.

We have the following result, Result 6.8.1.

RESULT 6.8.1 If $R(j)=\{i|a_{ij}=1\}=\{i_1,\ldots,i_{g_j}\}$ where g_j is the cardinality of $R(j)$. Then $C=(i_1,\ldots,i_{g_j})$ is a chain in $\mathcal{G}_A$, where $i_1<i_2<,\ldots,<i_{g_j}$.

Proof By construction of the graph $(i_r,i_{r+1})\in\mathcal{Q}$. $r=1,\ldots,g_{j-1}$. Hence the result.

We say $\mathcal{G}_A$ is *decomposed* into chains if the nodes are partitioned into certain number of chains. Naturally, the question arises of finding a partition with a minimum number of chains. This problem can be formulated as a linear-programming problem.

RESULT 6.8.2 Let $\mathbf{F}^*$ be a feasible solution to Problem 6.8.1, and $M=\{j|f_j^*=1\}$. Then $\{C(j)\}, j\in M$, is a chain decomposition of $\mathcal{G}_{\mathbf{P}}$, where $C(j)$ corresponds to the chain obtained from Column j.

Proof Notice that for any column the corresponding $C(j)$ is a chain. Since $\mathbf{F}^*$ is feasible to Problem 6.8.1, $\cup_{j\in M}R(j)=\{1,\ldots,vC2\}=$ the indices of the set of rows of $\mathbf{P}$, and if j and $k\in M, j\neq k$, then $C(j)$ and $C(k)$ cannot have 1's in the same row; therefore we get a set of mutually disjointed chains. So M decomposes $\mathcal{G}_{\mathbf{P}}$. Hence the result.

Thus the minimum number of chains in a decomposition by $\mathcal{G}_{\mathbf{P}}$ is a lower bound for the optimal objective function value of Problem 6.8.1. Also notice that the objective function value of Problem 6.8.2 is a lower bound to the minimal support size, as we have relaxed the integer restriction on $\mathbf{F}$. However, it can be shown that the optimal objective function value of Problem 6.8.2 is greater than or equal to that of the chain decomposition problem, with $\mathcal{G}_{\mathbf{P}}$. Yet we can use this weaker lower bound, as it is computationally superior.

Now consider $\mathcal{G}_{\mathbf{P}} = (\mathcal{N}_{\mathbf{P}}, \mathcal{Q})$. Consider the graph obtained from $\mathcal{G}_{\mathbf{P}}$ as follows:

$$\mathcal{G} = (V, \mathcal{Q}')$$

where

$$V = \mathcal{N}_{\mathbf{P}}^1 \cup \mathcal{N}_{\mathbf{P}}^2,$$

$$\mathcal{Q}' = \{(i, j) | i \in \mathcal{N}_{\mathbf{P}}^1, j \in \mathcal{N}_{\mathbf{P}}^2, \text{ and } (i, j) \in \mathcal{Q}\}$$

where $\mathcal{N}_{\mathbf{P}}^1$ and $\mathcal{N}_{\mathbf{P}}^2$ are copies of $\mathcal{N}_{\mathbf{P}}$. Such a graph is called a bipartite graph. By using duality theory we can show that the minimum number of chains in a decomposition of $\mathcal{G}_{\mathbf{P}}$ is equal to the maximum number of independent sets of edges in the graph $\mathcal{G}$, where a set of edges is said to be independent if no two of them are incident with a common node.

Finding the maximal independent set of edges in a bipartite graph can be solved efficiently by a method known as *labeling procedure*. A simplified version of this procedure appears in Chapter 7.

Thus a lower bound found by solving the minimum chain decomposition of $\mathcal{G}_{\mathbf{P}}$ can be used in a branch-and-bound scheme, to solve Problem 6.8.1. As noted in Chapter 2, branch-and-bound schemes depend very much on the bounds, and the branching strategy used, for their efficiency.

Several such schemes are available in the literature and their computational performances have been empirically compared.

Another way of finding lower bounds is through relaxing the constraints by introducing Lagrangian multipliers and solving the relaxed problem by subgradient optimization procedures. Such an approach is explained in Chapter 7.

6.9 OBTAINING DESIGNS WITH REDUCED SUPPORT SIZE FROM OTHER DESIGNS

Given a BIB design, $d = \text{BIB}(v, b, r, k, \lambda)$ with the number of distinct blocks equal to b^*, we may be interested in finding designs with reduced support size with a subset of blocks, d.

From the theory of linear programming we know that any feasible solution to the linear-programming problem can be reduced to a basic feasible solution.

So if the set of columns corresponding to the distinct blocks in d is linearly dependent, we can systematically reduce the number of distinct blocks in d. However, if these columns are linearly independent, no other design with reduced support size is possible from d.

Let d^* denote the set of distinct blocks in d. Let $\mathbf{P}_{d^*}$ refer to the submatrix corresponding to d^*.

If the columns in $\mathbf{P}_{d^*}$ are not linearly independent, then $\mathbf{F}_{d^*}$ is a feasible solution to $\mathbf{P}_{d^*}\mathbf{F}_{d^*}=\lambda\mathbf{e}$ and is not a basic feasible solution in the language of linear programming. Now let $\mathbf{h}$ be a nonzero rational vector such that $\mathbf{P}_{d^*}\mathbf{h}=\mathbf{0}$. From the theory of linear programming we know that we can find a column r among the columns of $\mathbf{P}_{d^*}$ and make the corresponding $(\mathbf{F}_{d^*})_r=0$. We can choose either

$$\frac{f_r}{h_r}=\min\left\{\frac{f_j}{h_j},\ h_j>0\right\} \tag{6.9.1}$$

or

$$\frac{f_r}{h_r}=\max\left\{\frac{f_j}{h_j},h_j<0\right\}. \tag{6.9.2}$$

The new f_j's are given by

$$\hat{f}_j=\begin{cases} f_j-f_r\dfrac{f_j}{h_r} & j\neq r\\ 0 & j=r\end{cases}$$

and they are nonnegative.

Now if we require integer solutions we can find the smallest t such that $t\hat{f}_j$ is an integer for all columns in d^*. As $\mathbf{P}_{d^*}\hat{\mathbf{F}}_{d^*}=\lambda\mathbf{e}$, $t\hat{f}_j$ will satisfy $\mathbf{P}_{d^*}\hat{\mathbf{F}}_{d^*}=t\lambda\mathbf{e}$. Thus $\hat{f}_j=t\hat{f}_j$ corresponding to the columns of d^* excepting perhaps those columns which have $f_r/h_r=f_j/h_j$. The implication is that we have a support for another design d_l. d_l^* is a proper subset of d^*, as at least one of the $\hat{f}_j=0$. A table of designs for $v=8$, $k=3$ with support sizes 22 to 55 is given at the end of the chapter.

6.10 AN APPROACH TO CERTAIN PROBLEMS RELATED TO CONSTRUCTION OF BIB DESIGNS

The problem of determining whether there is any design with a support size less than or equal to a given positive integer is first identified as an LP problem with restriction on positive cardinality. This approach can also be used to determine the minimal support size. Moreover if we wish to know whether there exists a design of a given support size, at most two problems must be solved.

We have already seen that a BIB design exists (with repeated blocks or otherwise) if we can find an integer solution to

$$\mathbf{PF} = \lambda \mathbf{e}$$
$$\mathbf{F} \geqslant \mathbf{0}$$

for some positive integer λ, given v and k.

The first problem posed, namely, does there exist a BIB design with support size $\leqslant l$, can be formulated as

Problem 6.10.1

$$\mathbf{PF} = \lambda \mathbf{e}$$
$$\mathbf{F} \geqslant \mathbf{0}$$
$$|\mathbf{F}|^{+} \leqslant l$$
$$\mathbf{F} \quad \text{integer},$$

where $|\mathbf{F}|^{+}$ is the cardinality of the set of positive f_j's.

This is a cardinality-constrained linear-programming (CCLP) problem without the objective function, but we do have the additional restriction that **F** has to be an integer.

Now we shall show that the algorithm mentioned in Section 2.27 for the cardinality-constrained linear-programming problem can be used in this case as well, except that we will not be introducing the cuts generated by the objective function.

RESULT 6.10.1 The algorithm described for the cardinality-constrained linear-programming problem, when applied to Problem 6.10.1, produces only integer solutions to the problem.

Proof: We have

$$\mathbf{PF} = \lambda \mathbf{e}$$
$$\mathbf{F} \geqslant \mathbf{0}$$
$$\mathbf{F} \quad \text{integer}.$$

If we denote by $\mathbf{D}=[\mathbf{P}, -\mathbf{e}]$, we have $\mathbf{w}=\begin{pmatrix}\mathbf{F}\\ \lambda\end{pmatrix}$ and

$$\mathbf{Dw}\geqslant\mathbf{0}, \quad \mathbf{w}\geqslant\mathbf{0}$$

and also

$$\mathbf{w} \quad \text{integer}.$$

Recall Algorithm 2.27.1. Now consider applying the algorithm to the **D** above. We see initially that all the elements of **D** are either 0, +1, or −1. And new columns generated are obtained in general by adding two columns of **Q** with integer coefficients. Therefore the resulting columns are also integer vectors, and we always have integer solutions to the problem if any solutions exist.

Next we consider the problem of finding BIB designs with minimal support size, and formulate it as Problem 6.10.2.

Problem 6.10.2

$$\begin{aligned} \text{Minimize} \quad & |\mathbf{F}|^{+} \\ \text{subject to} \quad & \mathbf{PF}=\lambda\mathbf{e} \\ & \mathbf{F}\geqslant\mathbf{0}. \end{aligned}$$

The problem can also be solved using the algorithm with a slight modification.

Any time an extreme point of $\mathbf{PF}=\lambda\mathbf{e}, \mathbf{F}\geqslant\mathbf{0}$ is generated we find the number of positive elements in **F**, say r, and then in the process only generated extreme points with positive components less than or equal to r. If none exists, the incumbent solution is the optimal one.

The third problem we consider is that of finding a BIB design with a specified support size.

This problem is more difficult, as we cannot solve it only by exploring the set of extreme points. So we may have to solve a second-stage problem obtained from the extreme-point solutions to the first-stage problem.

Problem 6.10.3

$$\mathbf{PF}=\lambda\mathbf{e}$$

$$\mathbf{F}\geqslant\mathbf{0}, \qquad \mathbf{F} \quad \text{integer}$$

and

$$|\mathbf{F}|^+ = l.$$

We first consider $\mathbf{Dw} \geqslant \mathbf{0}$, $\mathbf{w} \geqslant \mathbf{0}$, and $|\mathbf{F}|^+ \leqslant l$, and find all the extreme points of this convex set, using the algorithm as if we had Problem 6.10.1. Then if there is an extreme point of $\mathbf{P}$ with $|\mathbf{F}|^+ = l$ we stop. Otherwise we consider the set V_l of extreme points of $\mathbf{PF} = \lambda\mathbf{e}$ with $|\mathbf{F}|^+ \leqslant l$, $\mathbf{F} \geqslant \mathbf{0}$. Consider the matrix $\mathbf{A}$ of zeroes and ones obtained by replacing each positive coordinate by 1 in any element of V_l.

Let $\mathbf{x}$ be 0 or 1 vector.

Now if there is a solution to Problem 6.10.3 then $\mathbf{Ax} = \mathbf{b}$. $\mathbf{0} \leqslant \mathbf{x} \leqslant \mathbf{e}$, $|\mathbf{b}|^+ = l$ has a solution for some $\mathbf{b}$. Also, if there is a solution to the latter system, there is a solution $\mathbf{F}$ with $|\mathbf{F}|^+$ equal to l for Problem 6.10.3.

Also note that if there is a solution to

$$\mathbf{Ax} = \mathbf{b}, \qquad \mathbf{0} \leqslant \mathbf{x} \leqslant \mathbf{e} \tag{6.10.1}$$

$$|\mathbf{b}|^+ = l$$

then there is an extreme point of $\{\mathbf{x} | \mathbf{0} \leqslant \mathbf{x} \leqslant \mathbf{e}\}$ which is also a solution to the problem. The reason is that when there is a solution $\mathbf{x}$ to the system above and it is not an extreme point of $\mathbf{0} \leqslant \mathbf{x} \leqslant \mathbf{e}$, then by replacing $\mathbf{x}$ by $\bar{\mathbf{x}}$, given by

$$\bar{x}_i = \begin{cases} 1 & \text{if } x_i > 0 \\ 0 & \text{otherwise} \end{cases}$$

we have an extreme point of $\mathbf{0} \leqslant \mathbf{x} \leqslant \mathbf{e}$ and for some $\mathbf{b} = \bar{\mathbf{b}}$ that satisfies $\mathbf{Ax} = \bar{\mathbf{b}}$ with $|\bar{\mathbf{b}}|^+ = l$.

Thus in the second stage we consider

$$[\mathbf{A} - \mathbf{I}]\begin{pmatrix} \mathbf{x} \\ \mathbf{b} \end{pmatrix} = \mathbf{0}$$

$$\mathbf{0} \leqslant \mathbf{x} \leqslant \mathbf{e}$$

$$|\mathbf{b}|^+ \leqslant l.$$

If there is no extreme point of this problem with $|\mathbf{b}|^+$ equal to l then there is no solution to Problem 6.10.3.

We have given these formulations only as an alternative way of looking at these problems. Computational experience with cardinality-constrained linear-programming algorithms is not encouraging at present. The problems related to construction of designs are very huge, which further worsens the situation. For instance, if we wish to settle the existence or nonexistence of a BIB design with $v=46$, $b=69$, $r=1$, $k=6$, and $\lambda=1$, we must consider a **P** matrix of 1035 rows and 9366819 columns.

So these formulations should not be taken too seriously and one should not assume the questions raised in the beginning are settled. In fact these formulations have to be studied with the properties of the **P** matrix to devise special algorithms best suited for this situation.

6.11 CONSTRUCTING PAIRS OF ORTHOGONAL LATIN SQUARES

In this section we consider an integer-programming formulation of the problem of constructing Latin squares that are orthogonal. Two Latin squares L_1, L_2 of the same order p are said to be *orthogonal* if the ordered pairs (l_{ij}^1, l_{ij}^2) are all distinct for all i, j.

EXAMPLE 6.11.1 Notice that L_1, L_2 given below are orthogonal.

$$L_1 = \begin{bmatrix} 2 & 3 & 5 & 1 & 4 \\ 3 & 1 & 4 & 5 & 2 \\ 4 & 2 & 1 & 3 & 5 \\ 1 & 5 & 2 & 4 & 3 \\ 5 & 4 & 3 & 2 & 1 \end{bmatrix} \qquad L_2 = \begin{bmatrix} 4 & 1 & 5 & 3 & 2 \\ 5 & 2 & 1 & 4 & 3 \\ 3 & 5 & 4 & 2 & 1 \\ 1 & 3 & 2 & 5 & 4 \\ 2 & 4 & 3 & 1 & 5 \end{bmatrix}.$$

Interest in constructing pairs of orthogonal Latin squares arose because of a conjecture of Euler's, which says that there is no pair of orthogonal Latin squares of order p, $p\equiv 2\,\mathrm{mod}(4)$. It was verified that this conjecture is true for $p=6$. However, later it was shown that such pairs exist for all $p>6$, $p\equiv 2\,\mathrm{mod}(4)$.

Here we are formulating this problem as an integer-programming problem with 0-1 variables.

Let

$$x_{ijkl} = \begin{cases} 1 & \text{if the ordered pair } (i, j) \text{ appears in the location } k, l \\ 0 & \text{otherwise} \end{cases}$$

Now consider the problem:
Find x_{ijkl} 0 or 1 such that

$$\sum_k \sum_l x_{ijkl} = 1, \qquad i, j = 1, \ldots, p \tag{6.11.1}$$

$$\sum_i \sum_j x_{ijkl} = 1, \qquad k, l = 1, \ldots, p \tag{6.11.2}$$

$$\sum_j \sum_k x_{ijkl} = 1, \qquad i, l = 1, \ldots, p \tag{6.11.3}$$

$$\sum_i \sum_k x_{ijkl} = 1, \qquad j, l = 1, \ldots, p \tag{6.11.4}$$

$$\sum_j \sum_l x_{ijkl} = 1, \qquad i, k = 1, \ldots, p \tag{6.11.5}$$

$$\sum_i \sum_l x_{ijkl} = 1, \qquad j, k = 1, \ldots, p \tag{6.11.6}$$

Here, (6.11.1) implies that each order pair (i, j) is assigned to only one location; (6.11.2) insures that exactly one pair (i, j) is assigned to each location; (6.11.3) and (6.11.4) represent the conditions that element i appear only once in the first Latin square in column l and that element j appear only once in the second Latin square in row k; similarly (6.11.5) and (6.11.6) insures that i appear only once in the first Latin square in row k and element j appear only once in the second Latin square in column l.

Thus we have a matrix **A** corresponding to this problem with elements zeroes and ones, and the right-hand sides are all 1's. We have a set partitioning problem for which we have to find a feasible solution.

The approach was discussed in Section 6.8.

6.12 CONNECTED DESIGNS AND TRANSPORTATION MATRICES

Consider a two-way classification model. Let Y_{ijk} be a collection of independent random variables with a common unknown variance, σ^2, and each having expectation of the form

$$E(Y_{ijk}) = \mu + \alpha_i + \beta_j \qquad i = 1, \ldots, a, j = 1, \ldots, b$$

$$k = 1, \ldots, n_{ij}$$

$$n_{ijk} \geqslant 0 \qquad \text{for all} \quad i, j, k.$$

If $n_{ijk}=0$ no random variable with subscript ijk occurs in the collection. Thus, we are working with a two-way classification model with arbitrary pattern.

Let N be the incidence matrix associated with the model above—that is, $N=((n_{ij}))$. We assume that

$$n_{i.}=\sum_j n_{ij}\neq 0, \qquad i=1,\dots,a$$

$$n_{.j}=\sum_i n_{ij}\neq 0, \qquad j=1,\dots,b.$$

That is, there is at least one observation in each row and columns.

Here a natural question arises: Is the design matrix $\mathbf{X}$ of full rank? Equivalently, are all elementary contrasts estimable?

In this regard we discuss connected designs.

DEFINITION 6.12.1 A linear combination of the parameters which occur in the expectation of the Y_{ijk}'s is called a linear parametric function. Such a function is said to be *estimable* if it can be written as a linear combination of the cell expectations $\mu+\alpha_i+\beta_j$ for which $n_{ij}\geqslant 1$.

DEFINITION 6.12.2 A linear parametric function is said to be a *contrast* if the sum of the coefficients of the parameters is zero.

An *α-contrast* is a contrast involving only α_i and α_j, $i\neq j$. Similarly, a *β-contrast* is a contrast involving only β_i and β_j, $i\neq j$.

DEFINITION 6.12.3 A design is said to be *connected* if and only if every contrast is estimable.

EXAMPLE 6.12.1 Consider the model mentioned earlier. Let A and B be at two levels. Suppose we have $n_{11}, n_{12}, n_{21}>0$. We have

	B	
A	0	1
0	x	x
1	x	

The cells marked x have $n_{ij}>1$.

Now

$$\alpha_0-\alpha_1=\mu+\alpha_0+\beta_0-(\mu+\alpha_1+\beta_0)$$
$$=\alpha_0-\alpha_1$$

and

$$\beta_0-\beta_1=\mu+\alpha_0+\beta_0-(\mu+\alpha_0+\beta_1)$$
$$=\beta_0-\beta_1.$$

Hence both α-contrast and β-contrast are estimable.

Notice that if we have less than three $n_{ij}\geqslant 1$, all contrasts will not be estimable in this case. So we require a minimum of three cells with $n_{ij}\geqslant 1$. Such a design is called a *minimal connected design*.

In this context we have a combinatorial equivalent definition for connected designs.

DEFINITION 6.12.4 A treatment and block are said to be *associated* if the treatment is contained in the block. Two treatments, two blocks, or a treatment and a block are said to be connected if it is possible to pass from one to the other by means of a chain consisting alternatingly of blocks and treatments such that any two members of a chain are associated. A design (or a portion of the design) is said to be a connected design (or a connected portion of a design) if every block or treatment of the design (or a portion of the design) is connected to every other.

In regard to the transportation problem discussed in Section 5.5, we have seen that given a subset of cells, $\overline{B}$, from $a\times b$, the transportation table, we can draw an associated graph $\mathcal{G}=(\mathfrak{N},\mathcal{Q})$, where $\mathfrak{N}=\{1,\ldots,a+b\}$ and $(i;a+j)\in\mathcal{Q}$ if and only if the cell $(i,j)\in\overline{B}$ and the rows are associated with $1,\ldots,a$ and the columns with $a+1,\ldots,a+b$. If the associated graph is a connected graph we have a path from any node to any other node.

Also we noted that the associated graph can be written, looking at the cells in $\overline{B}$, by the following process:

Start at Node $(a+b)$ corresponding to Column b. In Column b consider the cells in $\overline{B}$. Consider the corresponding rows and connect the corresponding nodes to Node $(a+b)$ by edges. Then in each of those rows we search for cells in $\overline{B}$. If Column j has a cell in Row r, we connect Nodes r and $(a+j)$ by an edge. And this process is continued until no more connections can be incorporated among the nodes. We illustrate this process by an example.

EXAMPLE 6.12.2 Consider an 8×8 design with 48 missing cells, having the following pattern:

	β_1	β_2	β_3	β_4	β_5	β_6	β_7	β_8
α_1	1							
α_2		1	1					
α_3			1	1			1	
α_4				1	1			
α_5					1	1		
α_6						1	1	
α_7							1	1
α_8	1							1

Let us for convenience name the nodes $\alpha_1, \ldots, \alpha_8$ and $\beta_1, \ldots, \beta_8$, corresponding to rows and columns, so that we need not give numbers from 1 to 16. We start with Column 8, corresponding to β_8. Now Column 8 has cells in Rows 7 and 8. So we connect Node α_7 and α_8 to β_8. For Row 8 we find there is a cell in Column 1, so we connect β_1 to α_8. In Column 1 there is a cell in Row 1, so we connect α_1 to β_1. But there is no other cell in Row 1, so using this row we can not make any more connections. So we go back to Row 7. Row 7 has a cell in Column 7, so β_7 is connected to α_7. And we proceed further, obtaining the connected graph shown in Fig. 6.12.1.

We also noted if the connected graph obtained has no closed path among the nodes then the graph is a tree, corresponding to a linearly independent set of columns from the design matrix, or the transportation matrix. Then the corresponding design is minimally connected. So proceeding as above, we find that if the connected graph is a tree, it becomes apparent that the root of the tree is at Node $(a+b)$ or β_b. Thus whether the design is connected can be easily decided by obtaining the associated graph.

There are other ways of checking whether a design is connected. One such is described below:

Let N be the incidence matrix of a two-way classification model, with arbitrary pattern. Then a process is applied to N to obtain the final matrix M. If M has all entries of 1, the design is connected; otherwise it is not. The process is as follows:

1. Set

$$m_{ij} = \begin{cases} 1 & \text{if } n_{ij} \geqslant 1 \\ 0 & \text{otherwise} \end{cases}$$

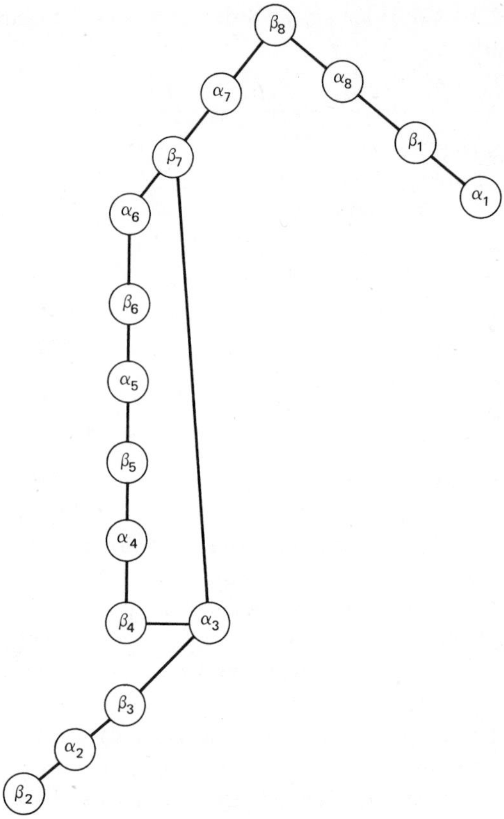

Figure 6.12.1 Connected graph corresponding to Example 6.12.2.

2. Change any $m_{ij}=0$, to 1 if there exist k and l such that $m_{il}=m_{kl}=m_{kj}=1$.

3. Continue Step 2, with the transformed M, until no more m_{ij}'s can be changed.

In the transportation terminology this process proceeds by forming elementary loops, and the cells connected thus are included in the subset of original cells. The process is repeated until no more cells can be included.

Drawing the associated graph directly by joining Nodes i and $a+j$ if $(i, j)\in\overline{B}$ yields the result much more quickly. However, the process above makes it apparent whether the connected graph is a tree or not.

We now consider the problem of finding minimal connected designs for 2^3 factorial layout. There are three factors, at two levels each. Here we are addressing ourselves to the problem of finding all 2^3 minimally connected designs.

Consider the matrix obtained by transposing the design matrix of a 2^3 factorial layout. We have

$$\mathbf{A}_3 = \left[\begin{array}{cccc|cccc} 1 & 1 & 1 & 1 & 0 & 0 & 0 & 0 \\ 0 & 0 & 0 & 0 & 1 & 1 & 1 & 1 \\ 1 & 1 & 0 & 0 & 1 & 1 & 0 & 0 \\ 0 & 0 & 1 & 1 & 0 & 0 & 1 & 1 \\ 1 & 0 & 1 & 0 & 1 & 0 & 1 & 0 \\ 0 & 1 & 0 & 1 & 0 & 1 & 0 & 1 \end{array}\right].$$

Note that

$$\mathbf{A}_3 = \left[\begin{array}{c|c} \mathbf{e} & \mathbf{0} \\ \mathbf{0} & \mathbf{e} \\ \mathbf{A}_2 & \mathbf{A}_2 \end{array}\right], \qquad \triangleq [\mathbf{A}_{31}|\mathbf{A}_{32}]$$

where $\mathbf{A}_2$ denotes the transportation matrix for a 2×2 transportation problem or, equivalently the transpose of the design matrix corresponding to a 2^2 factorial layout,

$$\mathbf{e}=(1,1,1,1) \quad \text{and} \quad \mathbf{0}=(0,0,0,0).$$

We know from the discussion on the transportation problem that the rank of an $m\times n$ transportation matrix is $m+n-1$. So $\mathbf{A}_2$ has the rank 3. Also we find that any column of $\mathbf{A}_{31}$ is linearly independent of columns in $\mathbf{A}_{32}$. Therefore $\mathbf{A}_3$ has a rank greater than or equal to 4. However it can be easily shown that the rank of $\mathbf{A}_3=4$. Therefore a basis for $\mathbf{A}_3$ has to have four linearly independent columns from $\mathbf{A}_3$. And a basis gives a minimally connected design. Note that any three columns of $\mathbf{A}_{31}$ are linearly independent. Such is the case with $\mathbf{A}_{32}$.

Therefore a basis for $\mathbf{A}_3$ can be any of these three forms: (1) 3 columns from $\mathbf{A}_{31}$ and 1 column from $\mathbf{A}_{32}$; (2) 2 columns from $\mathbf{A}_{21}$ and 2 columns from $\mathbf{A}_{32}$; or (3) 1 column from $\mathbf{A}_{31}$ and 3 columns from A_{32}.

In Choices (1) and (3), it is easy to verify that the columns are linearly independent. However in Choice (2) it is possible that two columns of $\mathbf{A}_{31}$ and two columns from $\mathbf{A}_{32}$ form a linearly dependent set.

Table A6.1 BIB Designs with $v=8$, $k=3$; Support Sizes 22–55

b^*	22	23	24	25	26	27	28	29	30	31	32	33	34	35	36	37	38	39	40	41	42	43	44	45	46	47	48	49	50	51	52	53	54	55
b	56	56	56	56	56	56	56	56	56	56	56	56	56	56	56	56	56	56	56	56	56	56	56	56	56	56	56	56	56	112	56	112	112	112
123	—	—	—	3	1	—	—	2	—	—	3	—	—	—	2	—	3	1	3	—	1	1	2	1	1	1	1	1	1	2	1	2	2	2
124	—	—	—	—	—	2	2	—	—	3	—	—	—	—	—	—	—	1	3	1	1	1	—	1	1	1	1	1	1	2	1	2	2	1
125	1	1	1	1	—	—	—	—	4	—	1	3	—	1	—	1	1	2	—	3	2	1	1	—	2	—	1	1	1	2	1	—	1	2
126	—	—	—	—	4	2	2	3	—	2	—	2	—	—	2	—	—	—	—	1	—	1	1	3	—	2	1	1	1	2	1	3	3	2
127	2	2	2	—	1	2	2	1	—	1	—	—	3	1	1	1	1	1	—	—	1	1	1	1	1	2	2	1	1	2	1	3	2	2
128	3	3	3	2	—	—	—	—	2	—	2	1	3	4	1	4	1	1	—	1	1	1	1	—	1	—	—	1	1	2	1	2	2	3
134	—	—	—	2	—	—	—	1	1	—	2	1	1	1	1	1	1	1	—	1	1	1	1	1	1	1	1	1	1	2	1	1	1	3
135	—	—	—	1	3	4	4	2	—	3	1	—	1	1	2	1	1	1	2	1	1	1	1	1	1	1	2	1	1	3	1	3	3	3
136	5	3	5	—	2	1	1	1	2	1	—	2	1	3	1	2	1	1	1	1	1	—	1	1	1	1	—	1	1	1	1	2	2	2
137	1	3	1	—	—	1	1	—	2	2	—	2	1	1	—	1	—	1	—	2	1	2	1	1	1	1	1	1	1	2	1	2	2	2
138	—	—	—	—	—	—	—	—	1	—	—	1	2	—	—	1	—	1	—	1	1	1	—	1	1	1	1	1	1	2	1	2	2	—
145	5	5	5	—	3	—	—	2	2	—	—	3	3	3	2	3	—	2	—	1	2	1	1	1	1	1	1	1	1	4	2	2	2	1
146	1	1	1	—	—	3	3	—	2	2	—	1	2	2	—	2	—	—	2	1	—	2	—	1	1	1	2	1	1	3	1	2	2	2
147	—	—	—	3	1	—	—	1	—	—	3	—	—	—	2	—	2	1	1	1	1	—	1	1	1	1	—	1	—	—	—	2	2	2
148	—	—	—	1	2	1	1	2	1	1	1	1	—	—	1	—	3	1	—	1	1	1	3	1	1	1	1	1	2	1	1	3	3	3
156	—	—	—	4	—	—	—	2	—	1	4	—	1	—	2	—	3	1	1	—	1	—	2	1	1	1	1	1	1	2	1	3	2	2
157	—	—	—	—	—	—	—	—	—	—	—	—	—	1	—	1	1	—	2	—	—	1	1	1	1	1	—	1	1	1	1	2	2	2
158	—	—	—	—	—	2	2	—	—	2	—	—	1	—	—	—	—	—	1	1	—	2	—	2	—	2	1	1	1	—	—	2	2	2
167	—	—	—	1	—	—	—	—	2	—	1	1	2	1	—	2	1	2	—	2	2	2	1	—	1	—	1	1	2	2	1	1	2	2
168	—	2	—	1	—	—	—	—	—	—	1	—	—	—	1	—	1	2	2	1	2	1	1	—	2	1	1	1	—	2	1	1	1	2
178	3	1	3	2	4	3	3	4	2	3	2	3	—	2	3	1	1	1	3	1	1	—	1	2	1	1	2	1	1	5	2	2	2	2
234	4	4	4	3	2	2	2	2	3	1	2	3	1	3	1	3	2	—	—	2	—	2	2	1	—	1	1	2	1	2	1	3	3	2
235	—	—	—	—	—	2	1	—	—	2	—	—	3	—	—	—	—	2	—	—	1	1	—	2	1	2	1	—	1	1	1	1	1	2
236	—	—	—	—	—	—	—	—	3	—	1	2	1	1	—	1	1	—	1	3	1	1	1	—	1	—	1	2	1	2	1	3	3	2
237	2	2	2	—	—	—	—	—	—	—	—	1	1	2	1	2	—	1	1	1	1	—	—	—	1	—	1	1	—	3	1	2	2	2
238	—	—	—	—	3	2	3	2	—	3	—	—	—	—	2	—	—	2	1	—	2	1	1	2	2	2	1	—	2	2	1	1	1	2
245	1	1	1	1	2	—	—	2	—	—	1	—	2	1	2	1	1	—	2	—	1	—	1	1	1	1	1	1	1	2	1	2	1	3

246	1	1	1	2	1	—	—	1	—	—	1	—	1	1	1	1	1	4	—	—	3	1	1	—	3	1	1	—	1	2	1	2	2	2
247	—	—	—	—	—	2	2	—	2	2	1	1	—	1	1	1	1	1	—	1	1	1	1	2	1	1	1	1	1	2	1	1	2	3
248	—	—	—	—	1	—	—	1	1	—	1	2	2	—	1	—	1	—	1	2	—	1	1	1	—	1	1	1	1	2	1	2	2	1
256	2	2	2	2	1	2	3	1	—	2	1	—	1	3	2	3	1	1	1	—	1	1	1	1	1	1	1	1	1	3	1	2	2	2
257	2	2	2	1	3	1	1	3	1	1	2	2	—	1	2	1	1	—	1	2	—	1	2	1	—	1	1	1	2	2	1	4	4	1
258	—	—	—	1	—	1	1	—	1	1	1	1	—	—	—	—	2	1	2	1	1	2	1	1	1	1	1	2	—	2	1	3	3	2
267	—	—	—	2	—	—	—	—	2	1	2	1	2	—	—	—	2	1	3	1	1	2	1	1	1	1	—	1	1	1	1	—	—	2
268	3	3	3	—	—	2	1	1	1	1	1	1	1	1	1	1	1	—	1	1	—	—	1	1	—	1	2	1	1	2	1	2	2	2
278	—	—	—	3	2	1	1	2	1	1	1	1	—	1	1	1	1	2	1	1	2	1	1	1	2	1	1	1	1	2	1	2	2	2
345	—	—	—	—	—	—	1	—	—	1	—	—	—	—	1	—	1	—	1	1	—	1	1	1	1	1	—	1	1	—	—	2	2	1
346	—	2	—	1	1	—	—	1	—	1	2	—	—	—	1	1	1	2	1	—	2	1	1	1	1	1	1	—	1	3	1	2	2	2
347	2	—	—	—	3	—	—	2	—	—	—	—	3	—	1	—	1	2	2	—	1	1	1	1	1	1	1	—	2	3	2	2	2	1
348	—	—	2	—	—	4	3	—	2	3	—	2	1	2	1	1	—	1	2	2	2	—	—	1	2	1	2	2	—	2	1	2	2	3
356	—	—	—	—	—	—	—	—	—	—	—	1	1	—	—	—	—	1	1	1	1	1	1	—	1	—	1	1	1	2	1	1	1	2
357	—	—	2	5	—	—	—	—	4	—	4	3	1	2	—	2	3	2	1	2	3	2	2	1	2	1	1	2	1	2	1	2	2	2
358	6	6	4	—	3	—	—	4	2	—	1	2	—	3	3	3	1	—	1	1	—	—	1	1	—	1	1	1	1	4	2	3	3	2
367	1	1	1	—	3	5	5	4	—	4	—	—	—	1	4	1	—	—	1	—	—	—	—	3	1	3	2	1	1	2	1	2	2	2
368	—	—	—	5	—	—	—	—	1	—	3	1	3	1	—	1	3	2	1	1	1	3	2	1	1	1	1	1	1	2	1	2	2	2
378	—	—	—	1	—	—	—	—	—	—	2	—	—	—	—	—	2	—	1	1	—	1	2	—	—	—	—	1	1	—	—	2	2	3
456	—	—	—	—	—	2	1	—	3	1	1	2	1	—	—	—	2	—	1	3	—	2	1	2	—	2	1	2	1	—	1	2	3	2
457	—	—	—	—	1	4	4	2	1	4	—	1	—	1	1	1	—	1	1	1	1	1	—	1	1	1	2	1	1	3	1	2	2	3
458	—	—	—	5	—	—	—	—	—	—	4	—	—	1	—	1	2	3	1	—	2	1	2	—	2	—	1	—	1	3	1	2	2	2
467	1	1	3	3	1	—	—	1	1	—	2	3	1	2	1	1	2	—	1	2	1	—	3	—	1	—	1	2	1	2	1	3	2	2
468	3	1	1	—	3	1	2	3	—	2	—	—	1	1	3	1	—	—	1	—	—	—	—	2	—	1	—	1	1	2	1	1	1	2
478	3	5	3	—	—	—	—	—	2	—	—	1	2	2	—	3	—	1	1	1	1	3	—	1	1	2	1	1	1	2	1	2	2	1
567	4	4	2	—	2	—	—	1	—	—	—	—	1	1	1	1	—	2	1	—	1	1	—	1	1	1	1	—	—	2	1	2	2	2
568	—	—	2	—	3	2	2	2	3	2	—	3	1	2	1	2	—	1	1	2	2	1	1	1	2	1	1	1	2	2	1	2	2	2
578	—	—	—	—	—	1	1	—	—	1	—	—	4	—	2	—	1	1	—	1	1	—	1	1	1	1	1	1	1	1	1	—	—	2
678	—	—	—	—	—	1	1	—	1	1	1	1	—	1	—	1	1	1	—	1	1	1	1	1	1	1	1	1	1	2	1	4	4	2

Note: For each support size the number of blocks b, is a minimum.

Table A6.2 All Possible 2^3 Minimally Connected Designs

1	1	1	1	1	1	1	1	1	1	1	1	1	1	1	1	1	1
1	2	1	1	2	1	1	2	1	1	2	1	1	2	1	1	2	1
2	1	1	2	1	1	2	1	1	2	1	1	2	1	2	2	1	2
2	2	2	1	2	2	1	1	2	2	1	2	1	1	2	1	2	2
1	1	1	1	1	1	1	1	1	1	1	1	1	1	1	1	1	1
1	2	1	1	2	1	1	2	1	1	2	1	1	2	1	1	2	1
2	1	2	2	2	2	2	2	2	2	2	2	2	2	1	2	2	1
2	2	1	1	2	2	1	1	2	2	2	1	1	1	2	1	2	2
1	1	1	1	1	1	1	1	1	1	1	1	1	1	1	1	1	1
1	2	2	1	2	2	1	2	2	1	2	2	1	2	2	1	2	2
1	1	2	1	1	2	1	1	2	1	1	2	2	2	1	2	2	1
2	2	2	2	1	2	2	1	1	2	2	1	2	1	1	2	1	2
1	1	1	1	1	1	1	1	1	1	1	1	1	1	1	1	1	1
1	2	2	1	1	2	1	2	2	1	2	2	1	1	2	1	1	2
2	2	1	2	1	1	2	1	2	2	2	2	2	1	2	2	1	1
2	2	2	2	2	1	2	1	1	2	1	2	2	2	2	2	2	2
1	1	1	1	1	1	1	1	1	1	1	1	1	1	2	1	1	1
1	1	2	2	1	1	2	1	1	2	1	1	2	2	1	2	1	2
2	1	2	2	2	2	2	2	1	2	2	1	1	2	2	2	2	2
2	2	1	2	1	2	2	2	2	2	1	2	2	2	2	2	2	1
1	1	2	1	1	2	1	1	2	1	1	2	1	1	2	1	1	2
2	2	1	2	2	1	2	2	1	2	2	1	2	2	1	2	2	1
1	2	2	1	2	1	1	2	1	1	2	2	2	2	2	2	2	2
2	1	2	2	2	2	2	1	1	1	2	1	2	1	2	2	1	1
1	1	2	1	1	2	1	1	2	1	1	2	1	1	2	1	1	2
2	2	1	2	2	2	2	2	2	2	2	2	2	2	2	2	2	2
2	1	2	2	1	1	2	1	1	2	1	1	1	2	1	2	1	2
2	1	1	2	1	2	1	2	1	1	2	2	1	2	2	1	2	1
1	1	2	1	1	2	1	1	2	1	1	2	1	2	2	1	2	2
2	1	2	2	1	1	2	1	2	2	1	2	2	2	2	2	2	2
1	2	2	1	2	1	1	2	1	1	2	2	2	1	1	2	1	1
1	2	1	1	2	2	2	1	1	2	1	1	2	1	2	1	2	1
1	2	2	1	2	2	1	2	2	1	2	2	1	2	2	1	2	2
2	2	2	2	2	1	2	2	1	2	2	1	2	2	1	2	2	1
2	1	2	2	2	2	1	2	1	1	2	1	2	1	2	2	2	2
1	2	1	2	1	2	2	1	2	2	1	1	2	1	1	2	1	1
1	2	1	1	2	1	1	2	1	1	2	1						
2	2	1	2	2	2	2	2	1	2	2	1						
2	1	1	2	1	2	2	1	2	2	1	1						
2	2	2	2	1	1	2	2	2	2	1	2						

For instance, Columns 1 and 2 from $\mathbf{A}_{31}$ and $\mathbf{A}_{32}$ form a dependent set, which can be shown in the form of incidence matrices:

		γ			
		1		2	
		β		β	
		1	2	1	2
α	1	1	1	1	1
	2				

In all there are 10 such possibilities in which we get dependent sets of columns, two taken from $\mathbf{A}_{31}$ and two from $\mathbf{A}_{32}$.

In Choice (1) we have $4\times4=16$ minimally connected designs; such is also the case with Choice (3). In Choice (2) if we eliminate the 10 dependent sets we have $36-10=26$ minimally connected designs.

Thus for a 2^3 factorial layout, there are 58 minimally connected designs. In fact we can build the minimally connected designs for a 2^4 factorial layout by choosing any of these 58 minimal connected designs with δ at Level 1 and adding one of the 8 cells with δ at Level 2, or one of the 8 cells with δ at Level 1 and any of the 58 minimally connected designs with $\delta=2$. Thus easily we can have $2\times8\times58=928$ minimally connected designs for a 2^4 factorial layout, and this building process can be continued similarly for higher-order factorial layouts.

Table A6.2 is a table of minimally connected designs for a 2^3 factorial layout.

BIBLIOGRAPHICAL NOTES

6.1 Cox (1958) and John (1971) provide sufficient background in design and analysis of experiments. Also see Cochran and Cox (1977), Searle (1971), Fisher (1974), Anderson and McLean (1974), and Gill (1978).

6.3 This material in this section is due to Armstrong, Elam and Hultz (1977); however, the example does not appear in their paper.

6.4 BIB designs with repeated blocks have received the attention of various researchers. See Parker (1963), Seiden (1963), Mann (1969), Van Lint and Ryser (1972), Van Lint (1973, 1974), and Wynn (1975). For a brief survey up to 1976 of constructing designs with specified number of distinct blocks see Foody and Hedayat (1977).

6.5 See Arthanari and Dodge (1978).

6.6 Foody and Hedayat (1977) appear to be the pioneers in formulating the problem of construction of BIB designs as an integer programming problem. They also give a proof of rank of P is $vC2$. So the upper bound in Result 6.6.4 can not be improved.

6.7 See Arthanari and Dodge (1978). For detailed applications of Latin squares see Denes and Keedwell (1974). A FORTRAN program for the method described appears in Emmami (1979).

6.8 See Nemhauser, Trotter, and Nauss (1974). There are many different approaches to solving Problem 6.8.1 and related problems such as set covering, set representation, and so on. Garfinkel and Nemhauser (1969) give a survey of these methods. Salkin and Konkal (1970) discuss a cutting-plane approach for these types of problems. Lemke et al. (1971) give an implicit enumeration approach using linear programming. Balas and Padberg (1972a, b), Etcheberry (1974), Roth (1969), among others, have considered such problems and their algorithms.

6.9 See Foody and Hedayat (1977).

6.10 CCLP is explained in Section 2.27. See Chernikova (1964), (1965) and Rubin (1975).

6.11 Gale has shown that construction of orthogonal Latin squares can be formulated as an integer programming problem (see Dantzig (1963)).

6.12 The concept of connectedness was first introduced by Bose (1949). See Weeks and Williams (1964), Srivastava and Anderson (1970), Eccleston and Hedayat (1974), Birkes, Dodge, and Seely (1976), Shah and Khatri (1973), Shah and Dodge (1977), and Raghavarao and Federer (1975). Example 6.12.1 is from Dodge and Majumdar (1979). A graphic presentation of three-way connectedness is given by Wynn (1977); Hamada and Tamari (1978) approach the construction of fractional factorials using linear programming.

Table A6.1 is from Foody and Hedayat (1977). Table A6.2 is from Dodge (1976), generated in 1979.

REFERENCES

Anderson, V. L., and McLean, R. A. (1974). *Design of Experiments*. Marcel Dekker, New York.

Armstrong, R. D., Elam, J. J., and Hultz, J. W. (1977). "Obtaining Least Absolute Value Estimates for a Two-Way Classification Model." *Commun. Stat.-Simul. Comput.* **B6**(4), 365.

Arthanari, T. S., and Dodge, Y. (1978). *Mathematical Programming and Construction of BIB Designs*. Technical Report 4, School of Planning and Computer Applications, Tehran.

Balas, E., and Padberg, M. (1972a). "On the Set Covering Problem." *Oper. Res.* **20**, 1152.

Balas, E., and Padberg, M. (1972b). *On the Set Covering Problem. II: An Algorithm*. Management Science Research Report 295, Graduate School of Industrial Administration, Carnegie-Mellon Univ., Pittsburgh.

Berge, C. (1973). *Graphs and Hypergraphs*. North-Holland, Amsterdam.

Birkes, D., Dodge, Y., and Seely, J. (1976). "Spanning Sets for Estimable Contrasts in Classification Models." *Ann. Stat.* **4**, 86.

Bose, R. C. (1949). "*Least Square Aspects of Analysis of Variance*." Institute of Statistics, Memo Ser. 9, University of North Carolina, Chapen Hill, N.C.

Chakrabarti, M. C. (1963). "On the C-Matrix in Design of Experiments." *J. Indian Stat. Assoc.* **1**, 8.

Chernikova, N. V. (1964). "Algorithm for Finding a General Formula for the Non-Negative Solutions of a System of Linear Equations." *USSR Comput. Math. Math. Phys.* **4**, 151.

Chernikova, N. V. (1965). "Algorithm for Finding a General Formula for the Non-Negative Solutions of a System of Linear Inequalities." *USSR Comput. Math. Math. Phys.* **5**, 228.

Cochran, W. G., and Cox, G. M. (1977). Experimental Designs. Second edition. John Wiley, New York.

Cox, D. R. (1958). *Planning of Experiments.* Wiley, New York.

Dantzig, G. B. (1963). *Linear Programming and Extensions.* Princeton Univ. Press, Princeton, N.J.

Denes, J., and Keedwell, A. D. (1974). *Latin Squares and Their Applications.* Academic Press, New York.

Dodge, Y. (1976). *Estimability Considerations for 2^n factorial Experiments with Missing Observations.* Technical Report 7606, Indian Statistical Institute, New Delhi.

Dodge, Y., and Majumdar, D. (1979). "An Algorithm for Finding Least Square Generalized Inverses for Classification Models with Arbitrary Patterns." *J. Stat. Comput. Simul.* **9**, 1.

Eccleston, J. A., and Hedayat, A. (1974). "On the Theory of Connected Designs: Characterization and Optimality." *Ann. Math. Stat.* **2**, 1238.

Emmami, S. (1979). *A FORTRAN Program for Construction of BIB Designs Using Natural Diagonal Latin Squares.* B.S. Report, School of Planning and Computer Applications, Tehran.

Etcheberry, J. (1974). "The Set Representation Problem." Unpublished doctoral dissertation, University of Michigan, Ann Arbor, Mich.

Fisher, R. A. (1974). *The Design of Experiments.* Macmillan, New York.

Foody, W., and Hedayat, A. (1977). "On Theory and Application of BIB Designs with Repeated Blocks." *Ann. Math. Stat.* **5**, 932.

Garfinkel, R., and Nemhauser, G. L. (1969). "The Set-Partitioning Problem: Set Covering with Equality Constraints." *Oper. Res.* **17**, 848.

Garfinkel, R., and Nemhauser, G. L. (1972). *Integer Programming.* Wiley, New York.

Gill, J. L. (1978). *Design and Analysis of Experiments in the Animal and Medical Sciences* (Vol. 1–3). Iowa State Univ. Press, Ames, Iowa.

Hamada, N., and Tamari, F. (1978). "*Construction of Optimal Linear Codes and Optimal Fractional Factorial Designs Using Linear Programming.*" *Technical Rep. Hiroshima University, Japan.*

Hedayat, A. (1979). "Sampling Designs with Reduced Support Sizes." In *Optimizing Methods in Statistics.* J. S. Rustagi, ed., Academic Press, New York, pp. 273–88.

Held, M., Wolfe, P., and Crowder, H. (1974). "Validation of Subgradient Optimization." *Math. Program.* **6**, 68.

John, P. W. M. (1971). *Statistical Design and Analysis of Experiments.* Macmillan, New York.

Lemke, C., Salkin, H., and Spielberg, K. (1971). "Set Covering by Single Branch Enumeration with Linear Programming Subproblems." *Oper. Res.* **19**, 978.

Mann, H. B. (1969). "A Note on Balanced Incomplete Block Designs." *Ann. Math. Stat.* **40**, 679.

Murty, K. (1972). "A Fundamental Problem in Linear Inequalities with Applications to the Traveling Salesman Problem." *Math. Program.* **2**.

Nemhauser, G. L., Trotter, L. E. Jr., and Nauss, R. M. (1974). "Set Partitioning and Chain Decomposition." *Manage. Sci.* **20**, 1413.

Neuhardt, J. B., Bradley, H. E., and Henning, R. W. (1973). "Computational Results in Selecting Multifactor Experimental arrangements." *J. Am. Stat. Assoc.* **68**, 608.

Parker, E. T. (1963). "Remarks on Balanced Incomplete Block Designs." *Proc. Am. Math. Soc.* **14**, 731.

Raghavarao, D. (1971). *Construction and Combinatorial Problems in Design of Experiments*. Wiley, New York.

Raghovarao, D., and Federer, W. T. (1975). "On Connectedness in Two-Way Elimination of Heterogeneity." *Ann. Stat.* **3**, 730.

Roth, R. (1969). "Computer Solutions to Minimum Cover Problems." *Oper. Res.* **17**, 455.

Rubin, D. S. (1975). "Vertex Generation and Cardinality Constrained Linear Programs." *Oper. Res.* **23**, 555.

Salkin, H., and Konkal, R. (1970). *A Pseudo Dual All-Integer Algorithm for the Set Covering Problem*. Technical Memo 204, Operations Research Dept., Case Western Univ.

Searle, S. R. (1971). *Linear Models*. Wiley, New York.

Sedransk, J. (1967). "Designing Some Multi-Factor Analytical Studies." *J. Am. Stat. Assoc.* **62**, 1121.

Seiden, E. (1963). "A Supplement to Parker's Remarks on Balanced Incomplete Block Designs." *Proc. Am. Math. Soc.* **14**, 731.

Shah, K. R., and Khatri, C. G. (1973). "Connectedness in Row-Column Designs." *Commun. Stat.* **2**, 571.

Shah, K. R., and Dodge, Y. (1977). "On Connectedness of Designs." *Sankhya* **39**, 284.

Srivastava, J. N., and Anderson, D. A. (1970). "Some Basic Properties of Multidimensional Partially Balanced Designs." *Ann. Math. Stat.* **41**, 1438.

Neuhardt, J. B., and Bradley, H. E. (1971). "On the Selection of Multi-Factor Experimental Arrangements with Resource Constraints." *J. Am. Stat. Assoc.* **66**, 618.

Van Lint, J. H. (1973). "Block Design with Repeated Blocks and $(b, r, \lambda) = 1$, *J. Combin. Theory* **A 15**, 288.

Van Lint, J. H., (1974). *Combinatorial Theory Seminar*. Eindhoven University of Technology Lecture Notes in Mathematics 382. Springer-Verlag, New York.

Van Lint, J. H., and Ryser, H. J. (1972). "Block Designs with Repeated Blocks." *Disc. Math.* **3**, 381.

Weeks, D. L., and Williams, D. R. (1964). "A Note on the Determination of Connectedness in an n-Way Cross Classification." *Technometrics* **6**, 319; Errata, *Technometrics* **7**, 281.

Wynn, H. P. (1977). "The Combinatorial Characterization of Certain Connected $2 \times J \times K$ Three-Way Layouts." *Commun. Stat. Theor.* **A(6)10**, 945.

CHAPTER 7

Cluster Analysis

7.1 INTRODUCTION

One of the problems in multivariate data analysis is the problem of cluster analysis. Problems which involve grouping a certain number of entities into a certain number of groups, in some sense optimally, arise in various fields of scientific inquiry. Such problems are encountered in the sciences of zoology, biology, botany, sociology, psychology, medicine, astronomy, artificial intelligence, information retrieval, and criminology, in business, industry, and marketing, and in education. Because it is so all-embracingly present in so many totally different contexts, the problem has been christened with several different names: cluster analysis, *Q*-analysis, grouping, clumping, classification, typology, and numerical taxonomy, among others. The vastness of the areas in which such problems require solutions can be compared only to the vastness of the various approaches available for solving them.

Unfortunately, stating the problem in its generality is itself a problem, as clusters mean different things to different people. However, with an intuitive understanding of the problem it can be stated as follows:

Given a set N, of n entities, we wish to partition this set into a number of subsets such that it is optimal with respect to a certain chosen criterion function, defined on the set of all partitions.

In general, if we have to talk about a clustering problem we must suppose that for each element belonging to the set N there is a vector giving the measurements on the characteristics or numerical codes for attributes that we wish to use as the input information for grouping. Thus we can assume that $\mathbf{X}_i=(X_{1i},\ldots,X_{zi})\in R^z$ is available for all $i\in N$. Next, we also require that a *distance* be defined between any two elements of N.

DEFINITION 7.1.1 A real valued function d_{ij} of i and $j \in N$ is said to be a *metric* or *distance* defined in N if it satisfies the following conditions:

$$\begin{aligned} &(1) \quad d_{ij} \geqslant 0,\ d_{ij}=0 \text{ if } i=j \\ &(2) \quad d_{ij}=d_{ji} \\ &(3) \quad d_{ik}+d_{kj} \geqslant d_{ij}, \qquad i, j, k \in N. \end{aligned} \tag{7.1.1}$$

We consider a few distances that are commonly used.

1 $d_{ij}=[\sum_{k=1}^{z}|X_{ik}-X_{jk}|^{r}]^{1/r}$, with r a positive integer, is a function of i and j which is called a *Minkowski metric*. If we insert $r=1$ we get what is known as an *absolute* or *city block metric* or *L_1-norm*. If $r=2$ we have what is known as a *Euclidean metric* or *L_2-norm*.

2 $d_{ij}=[\sum_{k=1}^{z}W_k|X_{ik}-X_{jk}|^{r}]^{1/r}$, r a positive integer, is called *weighted Minkowski distance*. When $r=1,2$ we have a corresponding weighted absolute and Euclidean metric, respectively.

3 $d_{ij}=(\mathbf{X}_i-\mathbf{X}_j)'\mathbf{P}(\mathbf{X}_i-\mathbf{X}_j)$ where $\mathbf{P}$ is a $z \times z$ symmetric positive semidefinite matrix. This metric is called a *general Euclidean metric*, and gives weights to pairs of characteristics under study. A special case is $d_{ij}=(\mathbf{X}_i-\mathbf{X}_j)'\Sigma_z^{-1}(\mathbf{X}_i-\mathbf{X}_j)$ where $\mathbf{\Sigma}_z$ is the variance-covariance matrix corresponding to the z characteristics under study. This metric is more general than the weighted Euclidean metric, where we have $\mathbf{\Sigma}_z=\mathbf{I}_z$. This distance is known as Mahalanobis D^2.

Let d_{ij} denote the *distance* between i and j for $i, j \in N$. The matrix of d_{ij} is also called the *dissimilarity matrix*.

Given a subset A of N, we define a real valued function of the elements belonging to A denoted by $\tau(A)$. In many cases, the function $\tau(A)$ is a function of the d_{ij}'s. One such criterion is given below:

$$\tau(A)=\begin{cases} \dfrac{1}{\lambda} \displaystyle\sum_{\substack{i<j \\ i,j \in A}} d_{ij}^2 & \text{for } \lambda \geqslant 2 \\ 0 & \text{otherwise} \end{cases} \tag{7.1.2}$$

where λ is the number of elements in A.

Given $\tau(.)$ and a real number h called "threshold level," a cluster $A \subset N$ can be defined as $\tau(A) \leqslant h$. A is maximal in the sense that $\tau(A \cup \{i\}) > h$ for any $i \notin A$, $i \in N$. Then we may be interested in finding all clusters, for a given $\tau(.)$ and h. Usually, we wish to find disjointed clusters. In some cases this disjointness restriction on clusters is relaxed. Then we are said to be seeking overlapping clusters. In these problems the number of clusters is not prespecified.

However, if we specify the number of clusters m for set N, we have the problem of choosing the best partition of N into m clusters, according to certain objective criteria. Such problems have the general form shown in Problem 7.1.1.

Let C be a real-valued function defined over R^m. Let $\tau_J = \{\tau(J_1),\ldots,\tau(J_m)\}$, where $J_1,\ldots, J_m$ are the m clusters such that $N=\cup_{i=1}^m J_i$ and $J=\{J_1,\ldots, J_m\}$. Let $\mathcal{J}=\{J \mid J \text{ is a partition of } N\}$.

Problem 7.1.1

$$\underset{J\in\mathcal{J}}{\text{Optimize}} \qquad C(\tau_J)$$

For instance, when $C(\tau_J)$ has the form $\Sigma_{l=1}^m \tau(J_l)$ for any partition J of N, with $\tau(.)$ as given in (7.1.2), we have the problem of *minimizing the total of the average within-group sum of squares* of the distances. Suppose $C(\tau_J)= \Sigma_{l=1}^m \tau(J_l)$ for any partition J of N, with $\tau(A)=\Sigma_{\substack{i,j\in A\\ i<j}} d_{ij}^2$. We have the problem of *minimizing the total within-group sum of squares* of the distances. If $C(\tau_J)=\max_{1<l<m}\tau(J_l)$, for any partition J of N, with $\tau(\cdot)$ given by $\tau(A)=\max_{\substack{i<j\\ i,j\in A}} d_{ij}$, then we have the problem of *minimizing the maximum of the within-group* distances, where $\tau(A)$ is called the within-group distance for A.

Suppose $C(\tau_J)=\Sigma_{l=1}^m \tau(J_l)$, for any partition J of N with $\tau(\cdot)$ given by $\tau(A)=\min_{j\in A}\Sigma_{i\in A} d_{ij}$. Then we have the problem of *minimizing the total sum* of the distances from all the elements in a cluster to its median, where r is called the median of A in case $\Sigma_{i\in A} d_{ir}=\min_{j\in A}\Sigma_{i\in A} d_{ij}$. In this case the problem is known as the *cluster-median* problem.

7.2 CLUSTERING METHODS

Methods available for clustering can be divided broadly into two categories, those that use probability distribution assumptions on the characterization vectors and those that do not. Methods like discriminant analysis and factor analysis come under the former category and we do not go into the discussion of those methods here. We consider a few methods in the second category that are simple but may not produce an optimal solution. Discussion of these methods is useful (as we shall see later, that the optimal clustering problems are computationally very complex), although they trade off optimality for computational ease.

Hierarchical Clustering Methods

The methods to be considered here have a common structure. Initially we consider that there are n clusters corresponding to the n elements in N. We define the distance $d_{J_iJ_j}$ between any two clusters J_i and J_j in a certain fashion and use this distance to clump two clusters into a single cluster. We clump J_i and J_j at any stage, if

$$d^2_{J_iJ_j} = \min_{r,s} d^2_{J_rJ_s}$$

We continue this process until there is a single cluster; that is N.

Depending on the way $d_{J_iJ_j}$ is defined, we obtain different clustering procedures. When

$$d^2_{J_iJ_j} = \min_{\substack{p \in J_i \\ q \in J_j}} d^2_{pq}$$

we have the *nearest neighbor method*. When

$$d^2_{J_iJ_j} = \max_{\substack{p \in J_i \\ q \in J_j}} d^2_{pq}$$

we have the *farthest neighbor method*. Let

$$\overline{\mathbf{X}}_{J_i} = \frac{1}{n_i} \sum_{p \in J_i} \mathbf{X}_p \quad \text{and} \quad \overline{\mathbf{X}}_{J_j} = \frac{1}{n_i} \sum_{q \in J_j} \mathbf{X}_q.$$

They are the vectors of the average of the characteristic measurements for J_i and J_j. Then

$$d^2_{J_iJ_j} = \text{the square of the Euclidean distance between } \overline{\mathbf{X}}_{J_i} \text{ and } \overline{\mathbf{X}}_{J_j}$$

and is called the centroid distance between J_i and J_j: That is

$$d^2_{J_iJ_j} = \left(\overline{\mathbf{X}}_{J_i} - \overline{\mathbf{X}}_{J_j}\right)'\left(\overline{\mathbf{X}}_{J_i} - \overline{\mathbf{X}}_{J_j}\right).$$

In this case we have the *centroid method*. When

$$d^2_{J_iJ_j} = \frac{n_1 n_2}{n_1 + n_2}\left(\overline{\mathbf{X}}_{J_i} - \overline{\mathbf{X}}_{J_j}\right)'\left(\overline{\mathbf{X}}_{J_i} - \overline{\mathbf{X}}_{J_j}\right),$$

we have Ward's distance, corresponding to the increase in the within-group sum of squares when J_i and J_j are clumped. Using this definition of $d^2_{J_iJ_j}$ we have the corresponding *Ward's method*. The clumping of the clusters at the various stages, $l=1,\ldots,n-1$, can be represented graphically. For instance, suppose we have five elements in N and in the first stage we clump Elements 1 and 2 together, as shown in Fig. 7.2.1. In the next stage we can clump 4 and 5 together, as shown in Fig. 7.2.2. In the third stage clusters 1, 2, and 3 are clumped, as shown in Fig. 7.2.3. In the fourth stage, we clump 1, 2, and 3 with 4 and 5, which yields N. This way we obtain a graphical representation of the formation of clusters, known as a *dendogram* (Fig. 7.2.4).

We are led to study these methods from a graph theory point of view. It is possible to define graphs corresponding to the different stages with the hierarchical clustering methods. In this graph N is the set of nodes and when i, j belong to the same cluster we introduce an edge between i and j.

Figure 7.2.1 First-stage clumping of elements.

Figure 7.2.2 Second-stage clumping of elements.

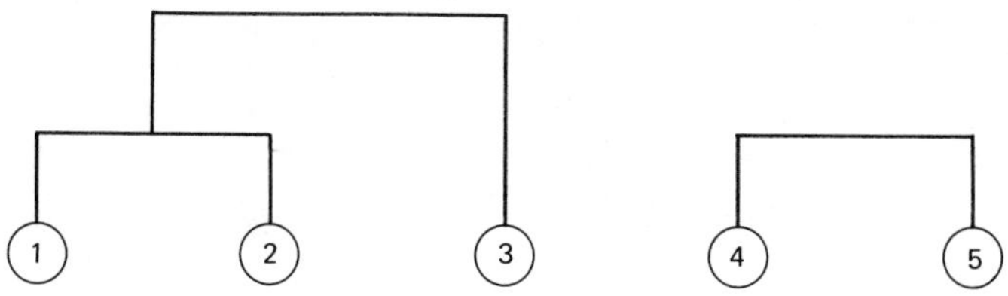

Figure 7.2.3 Third-stage clumping of elements.

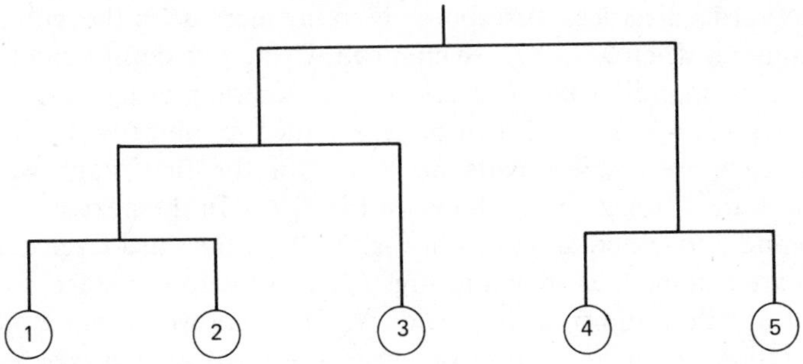

Figure 7.2.4 Dendogram showing final clumping of elements in Fig. 7.2.1–7.2.3.

Graph theoretical structure related to these methods is also studied with a view to obtain procedures with desirable properties.

EXAMPLE 7.2.1 Consider the clustering problem, with $N=\{1,2,\ldots,8\}$ and the matrix of d_{ij}^2 as given below (only the upper triangular portion is shown).

$$((d_{ij}^2))=\begin{bmatrix} 0 & 1 & 4 & 9 & 100 & 9 & 9 & 36 \\ & 0 & 1 & 4 & 121 & 4 & 16 & 49 \\ & & 0 & 9 & 144 & 1 & 25 & 64 \\ & & & 0 & 81 & 16 & 4 & 25 \\ & & & & 0 & 169 & 49 & 16 \\ & & & & & 0 & 36 & 81 \\ & & & & & & 0 & 9 \\ & & & & & & & 0 \end{bmatrix}$$

Suppose we employ the nearest-neighbor method to cluster the elements of N. Initially we have 8 clusters, $\{1\},\ldots,\{8\}$.

$l=1$: Elements 1 and 2 or 2 and 3 or 3 and 6 can be clumped to form a cluster, as $\min d_{ij}^2=1=d_{12}^2=d_{23}^2=d_{36}^2$. Suppose we clump 1 and 2 and form the cluster $\{1,2\}$. At this stage we have the clusters $J_1=\{1,2\}$, $J_2=\{3\},\ldots,J_7=\{8\}$.

Now,

$$d_{J_1J_i}^2=\min_{\substack{p\in J_1\\ q\in J_i}}\{d_{pq}^2\}, \qquad i=2,3,\ldots,7$$

are to be calculated. Other $d_{J_iJ_j}^2$, for $i>1$, are the same as before, since

J_i, $i>1$ has only one element. For instance

$$d^2_{J_1J_2}=\min\{d^2_{13}, d^2_{23}\}$$
$$=\min\{4,1\}=1.$$

Similarly other $d^2_{J_1J_i}$ can be found. The new squared distances are denoted by $d^2_{ij(l)}$, $l=1$; i and j refer the ith and jth clusters. They are given below:

$$\begin{array}{cccccccc} & \{1,2\} & \{3\} & \{4\} & \{5\} & \{6\} & \{7\} & \{8\} \\ ((d^2_{ij(1)}))= & \left[\begin{array}{c} 0 \\ \\ \\ \\ \\ \\ \\ \end{array}\right. & \begin{array}{c} 1 \\ 0 \\ \\ \\ \\ \\ \\ \end{array} & \begin{array}{c} 4 \\ 9 \\ 0 \\ \\ \\ \\ \\ \end{array} & \begin{array}{c} 100 \\ 144 \\ 81 \\ 0 \\ \\ \\ \\ \end{array} & \begin{array}{c} 4 \\ 1 \\ 16 \\ 169 \\ 0 \\ \\ \\ \end{array} & \begin{array}{c} 9 \\ 25 \\ 4 \\ 49 \\ 36 \\ 0 \\ \\ \end{array} & \left.\begin{array}{c} 36 \\ 64 \\ 25 \\ 16 \\ 81 \\ 9 \\ 0 \end{array}\right] \end{array}$$

$l=2$: Now we find, $\min d^2_{ij(1)}=d^2_{12(1)}=d^2_{25(1)}=1$. That is, we can clump $\{3\}$ with $\{1,2\}$ or $\{3\}$ with $\{6\}$. Suppose we clump $\{3\}$ with $\{6\}$. Then we have the following clusters: $J_1=\{1,2\}$, $J_2=\{3,6\}$, $J_3=\{4\}$, $J_4=\{5\}$, $J_5=\{7\}$, and $J_6=\{8\}$. We have to compute $d^2_{ij(2)}$. As mentioned earlier, it is sufficient to find $d^2_{ij(2)}$ for $i=2$, the newly formed cluster, as $d^2_{ij(2)}=d^2_{ij(1)}$, $i\neq 2$. We have

$$\begin{array}{ccccccc} & \{1,2\} & \{3,6\} & \{4\} & \{5\} & \{7\} & \{8\} \\ ((d^2_{ij(2)}))= & \left[\begin{array}{c} 0 \\ \\ \\ \\ \\ \\ \end{array}\right. & \begin{array}{c} 1 \\ 0 \\ \\ \\ \\ \\ \end{array} & \begin{array}{c} 4 \\ 9 \\ 0 \\ \\ \\ \\ \end{array} & \begin{array}{c} 100 \\ 144 \\ 81 \\ 0 \\ \\ \\ \end{array} & \begin{array}{c} 9 \\ 25 \\ 4 \\ 49 \\ 0 \\ \\ \end{array} & \left.\begin{array}{c} 36 \\ 64 \\ 25 \\ 16 \\ 9 \\ 0 \end{array}\right] \end{array}$$

$l=3$: We find $\min d^2_{ij(2)}=d^2_{12(2)}=1$. We clump $\{1,2\}$ and $\{3,6\}$ and form the cluster 1,2,3,6. Now $J_1=\{1,2,3,6\}$, $J_2=\{4\}$, $J_3=\{5\}$, $J_4=\{7\}$, and $J_5=\{8\}$. $d^2_{ij(3)}$ are as follows:

$$\begin{array}{cccccc} & \{1,2,3,6\} & \{4\} & \{5\} & \{7\} & \{8\} \\ ((d^2_{ij(3)}))= & \left[\begin{array}{c} 0 \\ \\ \\ \\ \\ \end{array}\right. & \begin{array}{c} 4 \\ 0 \\ \\ \\ \\ \end{array} & \begin{array}{c} 100 \\ 81 \\ 0 \\ \\ \\ \end{array} & \begin{array}{c} 9 \\ 4 \\ 49 \\ 0 \\ \\ \end{array} & \left.\begin{array}{c} 36 \\ 25 \\ 16 \\ 9 \\ 0 \end{array}\right] \end{array}$$

$l=4$: Observe that {4} and {7} or {1,2,3,6} and {4}, can be clumped. Suppose we clump {4} and {7}. Now $J_1=\{1,2,3,6\}$, $J_2=\{4,7\}$, $J_3=\{5\}$, and $J_4=\{8\}$. $d^2_{ij(4)}$ are as given below:

$$\begin{array}{cccc} \{1,2,3,6\} & \{4,7\} & \{5\} & \{8\} \end{array}$$

$$((d^2_{ij(4)}))=\begin{bmatrix} 0 & 4 & 100 & 36 \\ & 0 & 49 & 9 \\ & & 0 & 16 \\ & & & 0 \end{bmatrix}$$

Proceeding this way, we find the dendogram corresponding to the clusters formed as given in Fig. 7.2.5.

The primary charm of these methods is their computational ease. However, these methods suffer from a basic deficiency: they generally search for locally optimal clustering; once a cluster is formed we do not remove any element or elements to see whether that can produce different clusters at different threshold levels.

For instance, when Ward's definition of distance between any two clusters is used, and in Stage $n-2$ we obtain 2 clusters, there is no guarantee that these two clusters are optimal for the corresponding optimization problem with the objective of minimizing the total within-group sum of squares, with $m=2$.

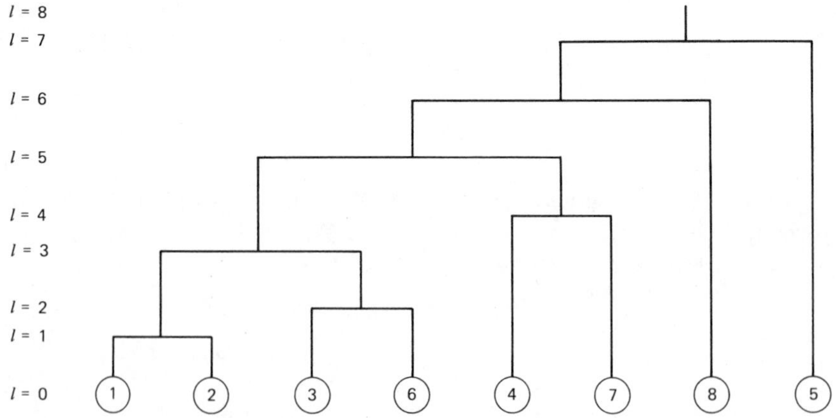

Figure 7.2.5 Dendogram showing the different stages of cluster formation; the elements are rearranged for neat presentation.

These advantages and disadvantages are shared by other similar procedures available for solving clustering problems. Thus we should consider methods that can produce optimal solutions to clustering problems.

Clustering techniques which are used to form clusters by optimizing a clustering criterion have to face the *curse* of mathematical programming—the *local* optimum may not be the *global* optimum. Obviously all possible partitions of the original set of objects or elements must be considered before we can conclude that the local optimum at hand is really the overall best. However, the enormously large number of partitions that are possible prohibits the use of complete enumeration for this purpose. The need for algorithms that can find the global optimal cluster has been filled to some extent by the applications of integer and dynamic programming methods and partial-enumeration techniques, known as branch-and-bound methods.

We shall first consider the integer programming approach for the problems of clustering.

7.3 INTEGER PROGRAMMING AND CLUSTER-MEDIAN PROBLEM

Let $N=\{1,2,\ldots,n\}$ be the set of objects or elements that are to be clustered into m clusters. We are given for each $i\in N$, a vector $\mathbf{X}_i=(X_{1i},\ldots,X_{zi})\in R^z$.

Let n_k denote the number of elements in the kth cluster, $k=1,2,\ldots,m$. So

$$\sum_{k=1}^{m} n_k=n.$$

Let m_0 denote the given upper bound on n_k. If there is no such restriction, $m_0=n$. Even though we wish to have only m clusters we define n clusters fictitiously, $n-m$ of which will have no elements at all. Further, for each nonempty cluster we have a median. Thus there should be m medians in all. We call the cluster for which element j is the median, j-cluster. Let

$$x_{ij}=\begin{cases}1 & \text{if the } i\text{th element belongs to the } j\text{-cluster}\\ 0 & \text{otherwise,}\end{cases}$$

for $i, j=1,\ldots,n$.

Let $((d_{ij}))$ be any distance matrix. Consider Problem 7.3.1 given below.

Problem 7.3.1

$$\text{Minimize} \qquad \sum_{i=1}^{n}\sum_{j=1}^{n} x_{ij}d_{ij} \tag{7.3.1}$$

$$\text{subject to} \qquad \sum_{j=1}^{n} x_{ij}=1, \qquad i=1,\dots,n \tag{7.3.2}$$

$$\sum_{j=1}^{n} x_{jj}=m \tag{7.3.3}$$

$$x_{jj} \geqslant x_{ij}, \qquad \begin{matrix} i=1,\dots,n \\ j=1,\dots,n \end{matrix} \tag{7.3.4}$$

each x_{ij} is either 0 or 1.

REMARK 7.3.1 The objective function (7.3.1) minimizes the total sum of distances in a cluster to the cluster median. (7.3.2) means that an element cannot belong to more than one cluster. (7.3.3) implies the restriction that we require exactly m nonempty clusters. (7.3.4) ensures that j-cluster is formed only when the corresponding element is a median; if $x_{jj}=1$, $n_j \neq 0$. For example, if $x_{99}=1$ and we have the cluster $\{2,9,25\}$, then $x_{2,9}=x_{9,9}=x_{25,9}=1$ and $x_{9,2}=x_{9,25}=x_{25,2}=x_{2,25}=0$ and also $x_{2,2}=x_{25,25}=0$.

Now observe that (7.3.4) can be reformulated as

$$nx_{jj} \geqslant \sum_{i=1}^{n} x_{ij}, \qquad j=1,\dots,n. \tag{7.3.5}$$

Thus we can reduce the number of constraints in the problem by using (7.3.5) in place of (7.3.4), which has n^2 constraints.

The cluster-median problem can thus be solved as a 0-1 integer programming problem. However, the complexity of the problem is such that as the size of the problem increases, the computational time and storage required for the available optimum-producing algorithms, grow exponentially. And also, as the studies done on the complexity of such combinatorial problems indicate, the possibility is dim of obtaining algorithms which are "good" in the sense that the computational time required is a polynomial of the size parameters of the problem.

So a reasonable approach is to find algorithms that can quickly produce solutions very close to the optimal solution, and also can determine how close the present solution is to the optimal. The partial search procedures, such as the branch-and-bound algorithms described in Chapter 2, and algorithms based on Lagrangian relaxation (mentioned in Chapter 6) have these properties.

We describe one such algorithm for solving the cluster-median problem. Consider constraints of (7.3.2) in Problem 7.3.1. With use of the Lagrangian multiplier, these constraints can be absorbed into the objective function, yielding Problem 7.3.2, for a *given* set of Lagrangian multipliers $\{u_1,\ldots,u_n\}$. Let $\mathbf{u}=(u_1,\ldots,u_n)$.

Problem 7.3.2

$$\text{Minimize} \quad \sum\sum d_{ij}x_{ij}+\sum u_i\left(1-\sum x_{ij}\right) \qquad (7.3.6)$$

$$\text{subject to} \quad \sum x_{jj}=m$$

$$x_{jj}\geqslant x_{ij},\ i,\ j=1,\ldots,n$$

$$x_{ij}=0 \text{ or } 1,\ i,\ j=1,\ldots,n.$$

Let $L(\mathbf{u})$ = optimal objective function value of Problem 7.3.2 for a given $\mathbf{u}$. Let F denote the feasible region for Problem 7.3.2.

RESULT 7.3.1 $L(\mathbf{u})\leqslant z^*$, where z^* is the optimal objective function value for Problem 7.3.1, for all $\mathbf{u}\in R^n$.

Proof Let $\mathbf{x}^*=((x^*_{ij}))$ be optimal for Problem 7.3.1. Then

$$z^*=\sum\sum d_{ij}x^*_{ij}.$$

Now corresponding to $\mathbf{x}^*$ the objective function (7.3.6) is

$$z^*+\sum u_i\left(1-\sum_j x^*_{ij}\right)\geqslant L(\mathbf{u})$$

by definition of $L(\mathbf{u})$.

Since $\mathbf{x}^*$ is optimal for Problem 7.3.1, $1-\Sigma_j x^*_{ij}=0$ as $\mathbf{x}^*$ is feasible. This implies $z^*\geqslant L(\mathbf{u})$ as required. For this reason Problem 7.3.2 is called a bounding problem. And the constraints are said to be relaxed. Problem 7.3.2 can equivalently be stated as

$$L(\mathbf{u})=\underset{\mathbf{x}\in F}{\text{minimum}}\left\{\sum u_i+\sum\sum(d_{ij}-u_i)x_{ij}\right\}. \qquad (7.3.7)$$

REMARK 7.3.2 Given a vector $\bar{\mathbf{u}} \in R^m$, $L(\bar{\mathbf{u}})$ can be easily worked as

$$L(\bar{\mathbf{u}}) = \min_{\mathbf{x} \in F} \left\{ \sum \bar{u}_i + \sum \sum (d_{ij} - \bar{u}_i) x_{ij} \right\}$$

$$= \sum \bar{u}_i + \min_{\mathbf{x} \in F} \left\{ \sum \sum (d_{ij} - \bar{u}_i) x_{ij} \right\}$$

Let

$$\bar{C}_{ij} = \begin{cases} d_{ij} - \bar{u}_i, & \text{if } d_{ij} - \bar{u}_i < 0 \\ 0 & \text{otherwise.} \end{cases}$$

Let $\bar{C}_j = \Sigma_i \bar{C}_{ij}$. $\bar{C}_j$ guides in assigning optimally values to x_{ij}, so as to minimize (7.3.7). Let $\bar{C}_{j_1} \leqslant \bar{C}_{j_2} \leqslant \cdots \leqslant \bar{C}_{j_m} \cdots \leqslant \bar{C}_{j_n}$. Then we set $x_{j_1 j_1} = \cdots = x_{j_m j_m} = 1$ and other x_{jj} are set to 0. Let $\{j_1, \ldots, j_m\}$ be called M, the set of medians.

The remaining x_{ij} are assigned values optimally as follows:

$$x_{ij} = \begin{cases} 1 & \text{if } j \in M \text{ and } \bar{C}_{ij} < 0 \\ 0 & \text{otherwise.} \end{cases}$$

This choice of x_{ij} can easily be seen to minimize $L(\bar{\mathbf{u}})$. However, as $\Sigma x_{ij} = 1$ is relaxed, the corresponding x_{ij}'s as defined above may not satisfy this constraint, and so may not be feasible for Problem 7.3.1.

REMARK 7.3.3 A feasible solution to Problem 7.3.1 can be found corresponding to the set of medians M selected, as in Remark 7.3.2, for a given $\bar{\mathbf{u}}$, as follows:

Set

$$x_{jj} = \begin{cases} 1 & \text{if } j \in M, \\ 0 & \text{otherwise} \end{cases}$$

and for other x_{ij}, $i \neq j$ we get

$$x_{ij} = \begin{cases} 1 & \text{if } d_{ij} = \min_{r \in M} d_{ir} \\ 0 & \text{otherwise.} \end{cases}$$

This solution insures that $\Sigma_j x_{ij}=1$ for all i, if we assign $x_{ij}=1$ for only one of the j such that $d_{ij}=\min_{r\in M} d_{ir}$ if there are ties. Other constraints are automatically satisfied.

Such a solution provides an upper bound for the optimal objective function value of Problem 7.3.1.

As we noticed in result 7.3.1, $L(\bar{\mathbf{u}})$ is a lower bound for z^* for all $\bar{\mathbf{u}}\in R^m$. Therefore the best lower bound for z^* is obtained by maximizing $L(\mathbf{u})$ over all $\mathbf{u}\in R^m$. We have Problem 7.3.3.

Problem 7.3.3

$$\underset{\mathbf{u}\in R^m}{\text{Maximize}}\, L(\mathbf{u})$$

With a view of obtaining the best lower bound for z^*, we turn our attention to solving Problem 7.3.3. Notice that $L(\mathbf{u})$ is a piecewise linear concave function in $\mathbf{u}$, as discussed earlier in a similar context. To solve Problem 7.3.3, we will be using a subgradient optimization algorithm which produces a sequence of $\mathbf{u}_1,\mathbf{u}_2,\ldots$ such that $L(\mathbf{u}_i)$ goes to $L(\mathbf{u}^*)$, where $\mathbf{u}^*$ is optimal for the problem. We require some preliminaries.

DEFINITION 7.3.1 Given $\bar{\mathbf{u}}\in R^m$, a vector $\mathbf{w}\in R^n$ is called a *subgradient* of the concave function $L(\mathbf{u})$ at $\bar{\mathbf{u}}$ if $L(\mathbf{u})\leqslant L(\bar{\mathbf{u}})+\mathbf{w}(\mathbf{u}-\bar{\mathbf{u}})$ for all $\mathbf{u}\in R^m$.

REMARK 7.3.4 The set of feasible solutions F, without the integer restriction on x_{ij}'s, is a linear-programming feasible region. Let us denote this region by F_c, that is,

$$F_c=\left\{\mathbf{x}=((x_{ij}))\,|\sum_j x_{jj}=m;\ x_{jj}\geqslant x_{ij}, 0\leqslant x_{ij}\leqslant 1, i, j=1,\ldots,n\right\}$$

F_c is a closed, bounded, convex region. Let the extreme points of this region be denoted by $\mathbf{x}^1,\mathbf{x}^2,\ldots,\mathbf{x}^r$. Let $F_c^*=\{\mathbf{x}^1,\mathbf{x}^2,\ldots,\mathbf{x}^r\}$.

Now Problem 7.3.3, with F replaced by F_c in the expression for $L(\mathbf{u})$, can be stated as follows:

Problem 7.3.4

$$\underset{\mathbf{u}\in R^m}{\text{Maximize}}\, L(\mathbf{u})$$

$$\text{subject to } L(\mathbf{u})=\underset{1\leqslant k\leqslant r}{\text{minimum}}\left\{y^k+\mathbf{u}\mathbf{v}^k\right\}$$

where

$$y^k = \sum_i \sum_j d_{ij} x_{ij}^k \text{ is a real number}$$

$$\mathbf{v}^k = \left(1 - \sum_j x_{ij}^k\right) \text{ is a row vector where the } i\text{th element is}$$

$1 - \sum_j x_{ij}^k$, and $\mathbf{x}^k = ((x_{ij}^k))$ is the kth extreme point in F_c^*.

REMARK 7.3.5 Problem 7.3.4 has the same structure as that of the general subgradient optimization problem. We state an algorithm for solving the general subgradient optimization problem, specialized to solve Problem 7.3.4.

Finding Initial Solution to Problem 7.3.1

Let $\bar{L}$ be equal to the objective function value corresponding to a feasible solution to Problem 7.3.1, obtainable by any of the available near-optimal methods, which are computationally simple.

Suppose we have m clusters thus obtained. We can find the median for each of these clusters and then we can allocate the other elements to the cluster whose median is closest to the element. Then the new cluster can be taken as the starting cluster and the median for each can be found again. Reallocate the other elements to the cluster with the nearest median. And this process can be repeated until no more reallocation results.

EXAMPLE 7.3.2 Consider the distance matrix given in Example 7.2.1. Suppose $m=2$, we have obtained the clusters $J_1 = \{1,2,3,6\}$, $J_2 = \{4,5,7,8\}$. Here we have given the d_{ij}'s instead of the d_{ij}^2's given in Example 7.2.1.

$$
\begin{array}{cc}
 & \begin{array}{cccccccc} 1 & 2 & 3 & 4 & 5 & 6 & 7 & 8 \end{array} \\
((d_{ij})) = & \left[\begin{array}{cccccccc}
0 & 1 & 2 & 3 & 10 & 3 & 3 & 6 \\
 & 0 & 1 & 2 & 11 & 2 & 4 & 7 \\
 & & 0 & 3 & 12 & 1 & 5 & 8 \\
 & & & 0 & 9 & 4 & 2 & 5 \\
 & & & & 0 & 13 & 7 & 4 \\
 & & & & & 0 & 6 & 9 \\
 & & & & & & 0 & 3 \\
 & & & & & & & 0
\end{array}\right]
\end{array}
$$

Now the median for J_1 can be chosen as 2 or 3, as both of them have $\min_j \Sigma_{i\in J_1} d_{ij}=4$. Let us choose 2 as the median for J_1. Similarly, the median for J_2 is 7. Now, 2 and 7 are the medians selected at this stage. As Element 1 is closer to 2 than to 7, 1 belongs to the cluster for which 2 is median; similarly, 3 and 6 are in the right cluster. Also we find that 5, 4, and 8 belong to the right cluster. So no reallotment was required in this example. Thus we have $J_1=\{1,2,3,6\}$ and $J_2=\{4,5,7,8\}$, with 2 and 7 as the respective medians.

Corresponding to this partition the objective function value of Problem 7.3.1 is $4+12=16$. That is, $z=\Sigma\Sigma d_{ij}x_{ij}=16$, when $x_{i2}=1$, $i\in J_1$ and $x_{i7}=1$, $i\in J_2$, and other x_{ij}'s are zero. So we take $\bar{L}=16$, corresponding to this solution.

The subgradient method proceeds by finding $\mathbf{u}^0,\mathbf{u}^1,\ldots,\mathbf{u}^p,\ldots$ a sequence of vector $\mathbf{u}$ such that $\mathbf{u}^p\to\mathbf{u}^*$, the optimal solution to Problem 7.3.4. However, there is no guarantee that $L(\mathbf{u}^p)$ monotonically increases to $L(\mathbf{u}^*)$.

Let $2\geqslant\rho_0\geqslant\rho_1\geqslant\rho_2\geqslant\cdots$ be a sequence of positive numbers, such that $\rho_p\to 0$ as $p\to\infty$. Let $k(p)$ be the index of the extreme point corresponding to the optimal solution $\mathbf{x}$ for the Problem 7.3.2, with $\mathbf{u}=\mathbf{u}^p$. This solution is obtained as given in Remark 7.3.2. Let

$$t^p=\rho_p\left(\left[\bar{L}-L(\mathbf{u}^p)\right]/\|\mathbf{v}^{k(p)}\|^2\right) \tag{7.3.8}$$

where

$$\mathbf{v}^{k(p)}=\left(1-\sum_j x_{ij}^{k(p)}\right).$$

Suppose that, starting with some $\mathbf{u}^0$, we have found $\mathbf{u}^1,\mathbf{u}^2,\ldots,\mathbf{u}^p$ so far. Then $\mathbf{u}^{p+1}$ is found as follows:

$$\mathbf{u}^{p+1}=\mathbf{u}^p+t^p\cdot\mathbf{v}^{k(p)}. \tag{7.3.9}$$

The subgradient algorithm guarantees, $\mathbf{u}^p\to\mathbf{u}^*$. We can stop after some p steps if the difference between $\bar{L}$ and $L(\mathbf{u}^p)$ is close enough for all practical purposes, by specifying an allowable error ε, that is the ratio $(\bar{L}-L(\mathbf{u}^p))/L(\mathbf{u}^p)\leqslant\varepsilon$.

REMARK 7.3.6 A good choice for $\rho_p, p=1,2,3,\ldots$ is $\rho_0=2$ and $\rho_{p+1}=\rho_p/2$ if there is not considerable improvement between $L(\mathbf{u}^{p+1})$ and $L(\mathbf{u}^p)$ in q steps, where q is an integer less than or equal to 10. Empirical studies indicate that $q=5$ is a good choice. Notice that $\mathbf{v}^{k(p)}$ is a subgradient for $L(\mathbf{u})$ at $\mathbf{u}=\mathbf{u}^p$.

We illustrate the steps of the method with an example.

EXAMPLE 7.3.3 Consider the data and the clusters given in Example 7.3.2. We have $\bar{L}=16$ with 2 and 7 as medians, $J_1=\{1,2,3,6\}$ and $J_2=\{4,5,7,8\}$.

$p=0$: suppose $\mathbf{u}^0=\mathbf{0}$.

Then as per Remark 7.3.2 the corresponding optimal $\mathbf{x}^{k(0)}$ can be found as follows:

Recall that,

$$\bar{C}_{ij}=\begin{cases} d_{ij}-u_i,\ d_{ij}-u_i<0 \\ 0 \qquad \text{otherwise for } i,\ j=1,\ldots,n \end{cases}$$

As $\mathbf{u}^0=\mathbf{0}$, $\bar{C}_{ij}=0$ for all i, j at this stage, or $\bar{C}_j=0$ for all j. Any j_1, j_2 can form the set M as $m=2$. Therefore, suppose we choose $M=\{2,7\}$, $x_{11}^{k(0)}=1$, $x_{22}^{k(0)}=1$ other $x_{jj}^{k(0)}=0, j=3,\ldots,8$ and $x_{ij}^{k(0)}=0$ as no $\bar{C}_{ij}<0$. This yields

$$\mathbf{v}^{k(0)}=(1,0,1,1,1,1,0,1).$$

Now,

$$\|\mathbf{v}^{k(0)}\|^2=36.$$

$L(\mathbf{u}^0)=\mathbf{0}$. Taking $\rho_0=2$ we get $t^0=2([16-0]/36)=0.889$. This means, we choose

$$\mathbf{u}^1=\mathbf{u}^0+0.889\ \mathbf{v}^{k(0)}$$

$$\mathbf{u}^1=(0.889,0,0.889,\ldots,0,0.889)$$

$p=1$:

We go on to compute $\mathbf{v}^{k(1)}$.

Now $((d_{ij}-u_i^1))$ is such $d_{ii}-u_i^1=-0.889$ for $i=1,3,4,5,6$, and 8. The rest of the $d_{ij}-u_i^1$ are greater than or equal to zero.

Hence $\bar{C}_{ij}<0$ for C_{11}, C_{33}, C_{44}, C_{55}, C_{66} and C_{88} and the rest are zeros. Also $\bar{C}_2=\bar{C}_7=0$ and the other $\bar{C}_j=-0.889$. We can choose $x_{33}=1$, $x_{55}=1$ and $x_{jj}=0$ for all other j's and the x_{ij} are zero, $i\neq j$. Now

$$\mathbf{v}^{k(1)}=(1,1,0,1,0,1,1,1)$$

and

$$\|\mathbf{v}^{k(1)}\|^2=36. \quad L(\mathbf{u}^1)=3.556.$$

Taking $\rho^1=2$ we get

$$t^1=2([16-3.556]/36)$$
$$=0.691$$

and so

$$\mathbf{u}^2=\mathbf{u}^1+0.691\mathbf{v}^{k(1)}$$
$$=(1.58, 0.691, 0.889, 1.58, 0.889, 1.58, 0.691, 1.58)$$

$p=2$:

Now $((d_{ij}-u_i^2))$ is such $d_{11}-u_1^2=-1.58$ and $d_{12}-u_1^2=-0.58$; $d_{22}-u_2^2=-0.691$; $d_{33}-u_3^2=-0.889$; $d_{44}-u_4^2=-1.58$; $d_{55}-u_5^2=-0.889$; $d_{66}-u_6^2=-1.58$ and $d_{63}-u_6^2=-0.58$; $d_{77}-u_7^2=-0.691$ and $d_{88}-u_8^2=-1.58$. The rest are positive. Therefore $\overline{C}_{jj}<0$ for all j. $\overline{C}_{63}<0$, $\overline{C}_{12}<0$ and all zeros. Also $\overline{C}_1=-1.58$, $\overline{C}_2=-1.271$, $\overline{C}_3=-1.469$, $\overline{C}_4=-1.58$, $\overline{C}_5=-0.899$, $\overline{C}_6=-1.58$, $\overline{C}_7=-0.691$, and $\overline{C}_8=-1.58$. Here

$$\overline{C}_1=\overline{C}_4=\overline{C}_6=\overline{C}_8<\overline{C}_3<\overline{C}_2<\overline{C}_5<\overline{C}_7.$$

So we choose $\{1,6\}$ as M or $x_{11}=1$ and $x_{66}=1$ and $x_{jj}=0$ otherwise, and

$$x_{ij}=0 \qquad \text{for } i\neq j \text{ as no } \overline{C}_{ij} \text{ for } j\in M \text{ is negative.}$$

Now

$$\mathbf{v}^{k(2)}=(0,1,1,1,1,0,1,1)$$

and

$$\|\mathbf{v}^{k(2)}\|^2=36. \quad L(\mathbf{u}^2)=6.32$$

We proceed with the computation until convergence to the optimal solution is indicated.

REMARK 7.3.7 In place of $\overline{L}$ we can use any feasible solution obtained according to Remark 7.3.3 if that is a better solution to Problem 7.3.1. And while doing this we may choose the medians judiciously for a better set of medians than that offered by Remark 7.3.2. Such may be necessary to bring both the upper bound $\bar{L}$ and the lower bound $L(\mathbf{u}^p)$ closer.

The notations and the procedure are clarified with the Example 7.3.2. However, a computer implementation of the algorithm is required to solve large problems. Encouraging experience is reported on the applicability of this method for practical problems with moderate size, obtaining solutions within 1% of the optimum.

7.4 MINIMIZATION OF TOTAL WITHIN-CLUSTER SUM OF SQUARES: ONE-DIMENSIONAL CASE

Here we consider the clustering problem with the criterion of minimizing the total within-cluster sum of squares when there is only one characteristic under study.

Assume without loss of generality that the elements are numbered such that

$$X_1 \leqslant X_2 \leqslant \cdots \leqslant X_n. \tag{7.4.1}$$

Let G_j denote the cluster in which j is such that X_j is the smallest element.

Let

$$x_{ij} = \begin{cases} 1 & \text{if } i \text{ belongs to } G_j \\ 0 & \text{otherwise.} \end{cases}$$

Notice that for $i<j$, x_{ij} cannot be equal to 1, as that would mean j cannot be the index of the smallest element in G_j. Hence, X_i cannot belong to $G_{i+1},\ldots,G_m$. It can also be expressed in terms of the constraint

$$\sum_{i=1}^{n} \sum_{j=i+1}^{n} x_{ij} = 0. \tag{7.4.2}$$

Now the problem can be stated as Problem 7.4.1.

Problem 7.4.1

Minimize $$\sum_{i=1}^{n}\sum_{j=1}^{n} x_{ij}(X_i-\bar{X}_j)^2$$

subject to $$\bar{X}_j=\sum_{i=1}^{n} x_{ij}X_i/n_j \qquad j=1,\ldots,n$$

$$\sum_{i=1}^{n} x_{ij}=n_j \qquad j=1,\ldots,n \tag{7.4.3}$$

$$\sum_{j=1}^{n} x_{jj}=m \tag{7.4.4}$$

$$\sum_{i=1}^{n}\sum_{j=i+1}^{n} x_{ij}=0, \tag{7.4.5}$$

each x_{ij} 0 or 1.

REMARK 7.4.1 The objective function minimizes the total within-cluster sum of squares. The set of constraints of (7.4.3) implies that there are n_j elements in G_j. The constraints of (7.4.4) and (7.4.5) insure that there are m clusters in all, and that X_j is the smallest element in G_j, respectively. Note that we have a nonlinear integer-programming problem.

Before solving this problem as such we prove a necessary condition for the optimal clustering which transfers this nonlinear integer problem into a linear integer-programming problem.

DEFINITION 7.4.1 A partition of the elements of N, such that each subset in the partition satisfies the condition: if $i<j<k$ and elements i and k are in the same subset, then j must also belong to that subset, is called a *contiguous clustering*.

For example, consider $N=\{1,2,\ldots,8\}$, $m=3$ with $X_1 \leqslant X_2 \leqslant \cdots \leqslant X_8$: $\{1,2,3\}$, $\{4,5,6\}$, $\{7,8\}$ is a contiguous clustering. But $\{1,3\}$, $\{2,4,5\}$ $\{5,7,8\}$ is not a contiguous clustering.

RESULT 7.4.1 While minimizing the total within-cluster sum of squares, it is sufficient to consider contiguous clusterings.

Proof We prove this result by considering $G_j=\{X_j, X_{j+1},\ldots, X_{j+s-1}\}$ and showing that it causes less damage to include the element X_{j+s} in this set than including any other X_r for $r>j+s$, in this set G_j.

It can be noticed that the increase in the within-cluster sum of squares when X_{j+s} is included in G_j is

$$\frac{s}{s+1}\left(X_{j+s}-\frac{1}{s}\sum_{i=0}^{s-1}X_{j+i}\right)^2.$$

Similarly, if X_r for $r>j+s$ is included in G_j we have the increment given by

$$\frac{s}{s+1}\left(X_r-\frac{1}{s}\sum_{i=0}^{s-1}X_{j+i}\right)^2.$$

But $X_r \geqslant X_{j+s}$, according to the numbering. Hence the result.

This result leads us to impose a restriction on the x_{ij} so that the partition obtained is contiguous. We require

$$x_{jj} \geqslant x_{j+1j} \geqslant \cdots \geqslant x_{nj}, \qquad j=1,\ldots,n \tag{7.4.6}$$

to insure contiguous partitioning, as when $x_{jj}=1$ and $x_{j+1j}=0$, x_{ij} cannot be 1 for $i>j+1$. In fact, if we look at the solution matrix $((x_{ij}))$ we have an uninterrupted string of 1's below the diagonal on its columns if this constraint is satisfied. This constraint is therefore called the *string property constraint*.

To summarize, the x_{ij}'s have to satisfy the constraints above in addition to the other constraints of the problem.

We now try to express the nonlinear objective function total within-cluster sum of squares by a linear one, by defining d_{ij}'s corresponding to x_{ij}'s as follows:

$$d_{ij}=\begin{cases} X_i^2-(X_j+\cdots+X_i)^2/(i-j+1) & \\ \quad +(X_j+\cdots+X_{i-1})^2/(i-j) & i>j \\ 0 & \text{otherwise.} \end{cases} \tag{7.4.7}$$

We claim that d_{ij} for $i>j$ given by (7.4.7) is in fact the net increment in the within-cluster sum of squares achieved by including Element i in the cluster G_j. As we are required to consider contiguous partitions only, i cannot belong to Cluster C_j unless all the elements between j and i are in cluster G_j, that is, unless $j, j-1,\ldots,i-1$ already belong to cluster G_j. From this fact the claim is proved.

RESULT 7.4.2 If $((x_{ij}))$ is a feasible solution to the problem satisfying (7.4.6), and $((d_{ij}))$ is defined as in (7.4.7), then

$$\sum_{i=1}^{n}\sum_{j=1}^{n} d_{ij}x_{ij} = \sum_{i=1}^{n}\sum_{j=1}^{n} x_{ij}\left(X_i - \bar{X}_j\right)^2. \tag{7.4.8}$$

Proof Consider G_l for some l, with $n_l \neq 0$. Now $G_l = \{X_l, \ldots, X_{l+n_l-1}\}$ because of (7.4.6).

The right-hand side of (7.4.8) corresponding to G_l is

$$\sum_{i=l}^{l+n_l-1} X_i^2 - \left(\sum_{i=l}^{l+n_l-1} X_i\right)^2 \Big/ n_l.$$

However, on the left-hand side, corresponding to G_l, we have

$$d_{ll} + \cdots + d_{l+n_l-1,l}.$$

Substituting for these d_{ij}'s from (7.4.7) and simplifying, we have the desired result.

Thus, we have formulated the one-dimensional problem of forming m clusters when the criterion is to minimize the within-cluster sum of squares, as an integer-programming problem. We have:

$$\begin{aligned}
&\text{Minimize} && \sum_{i=1}^{n}\sum_{j=1}^{n} d_{ij}x_{ij} \\
&\text{subject to} && \sum_{i=1}^{n}\sum_{j=i+1}^{n} x_{ij}=0 \\
& && x_{jj} \geqslant x_{j+1,j} \geqslant \cdots \geqslant x_{nj}, \quad j=1,\ldots,n \\
& && \sum_{j=1}^{n} x_{ij}=1 \qquad i=1,\ldots,n \qquad (7.4.9) \\
& && \sum_{j=1}^{n} x_{jj}=m \\
& && \text{each } x_{ij} \text{ 0 or 1 variable}
\end{aligned}$$

where d_{ij}'s are as given by (7.4.7).

Notice that we can solve this problem by the method described in Section 7.3, by relaxing the constraints of (7.4.9). String property can be easily generalized for higher dimensions. However it is not necessary for optimality.

7.5 DYNAMIC-PROGRAMMING APPROACH TO CLUSTER ANALYSIS

In this section we show that clustering problems can be viewed as dynamic-programming problems.

A general dynamic-programming problem can be explained by assuming that we are observing a *system* and that the system moves from one *state* to another depending on the *action* taken at the beginning of each stage of decision-making. On account of the decision made we have a *reward*, which may be profit made, losses incurred, cost of the action, and so forth. We are interested in finding the optimal action to be taken when the system is in a particular state at a particular stage of decision-making.

Any specification of action depending on the state and the stage is called a *policy* and if the policy optimizes the total reward, it is called an *optimal policy*.

The approach for such problems, due to Bellman, is to obtain a functional equation which assumes that the following principle (*principle of optimality*) holds good:

An optimal policy is such that, whatever the initial states and initial actions are, the remaining actions must constitute an optimal policy with respect to the resultant state from the initial decisions.

The functional equation obtained for a particular problem depends very much on the ingenuity of the researcher. In the literature, we have several examples to show that the stages or states are identified differently; the same problem can be solved with different computational burdens.

Certain terms and notations are relevant for solving the clustering problems through dynamic programming.

DEFINITION 7.5.1 A *network* $G=(V, A)$ is a pair of sets in which V is a set of elements, called nodes, including 0 and *, and A is a subset of the ordered set $V\times V$. Further, $0\in V$ is called a *source*, and $(k,0)\notin A$ for any $k\in V$ and $*\in V$ is called a *sink*; $(*, k)\notin A$ for any $k\in V$. A is called the set of *directed arcs*.

DEFINITION 7.5.2 Given a network $G=(V, A)$ we say $p=(i_1, i_2,\dots, i_k)$, yields a *path* from i_1 to i_k if $(i_l, i_{l+1})\in A$ for all $1\leqslant l\leqslant k-1$.

DEFINITION 7.5.3 Given a network $G=(V, A)$, G is called a k *stage network* if every path from 0 to * has exactly k arcs and $(0, i_1, i_2, \ldots, i_{k-1}, *)$ is such that $i_l \in V_l \subset V$, $1 \leqslant l \leqslant k-1$, and $(V_l, 1 \leqslant l \leqslant k-1)$ is a partition of the set $V-\{0,*\}$.

EXAMPLE 7.5.1 G as given below is a network

$$V=\{0,1,2,3,4,*\}, A=\{(0,1), (0,2), (1,3), (2,4), (2,*), (3,*), (4,*)\}.$$

We can also represent any network in terms of a diagram, as shown in Fig. 7.5.1. Here we have an arrow for each element of A connecting the two ordered nodes, which are the elements of V.

The network in Fig. 7.5.1 has the following paths from 0 to *: $(0,1,3,*)$, $(0,2,4,*)$, and $(0,2,*)$.

Notice that this network is not a k-stage network, as the path $0,2,*$ has two arcs and the other paths have three arcs. The network shown in Fig. 7.5.2 is a three-stage network. Here $V_1=\{1,2,3\}$, $V_2=\{4,5\}$.

Given G and $C_{ij} \geqslant 0$ for $(i, j) \in A$, called the distance from node i to j, the length of a path, $p=(i_1, i_2, \ldots, i_k)$ defined as $C(p)=\sum_{l=1}^{k-1} C_{i_l i_{l+1}}$. The problem in which we wish to find the shortest path from 0 to * is called the *shortest-route* problem.

When the network is a k-stage network, the problem is known as a *stagecoach* problem. Dynamic programming has been successfully applied to such problems.

Consider a k-stage network with C_{ij}'s given. We call the elements of V the states. Let s denote any state. Let n denote the number of stages remaining to reach * from s. Let $f_n(s)$ be the shortest length from s to * of a path having n arcs from s to *. Let the corresponding optimal path from

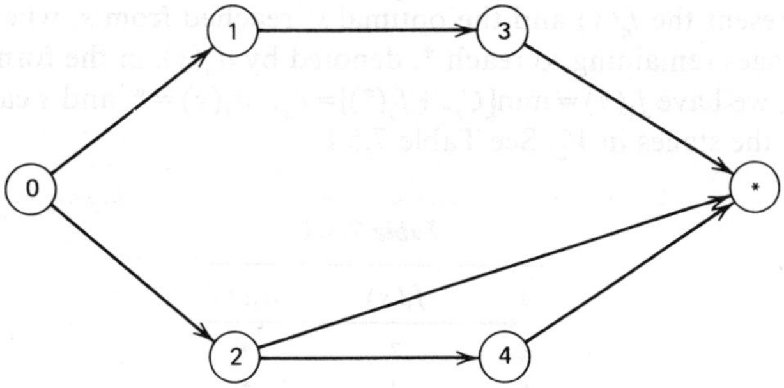

Figure 7.5.1 Diagram of a network that is not a k-stage network.

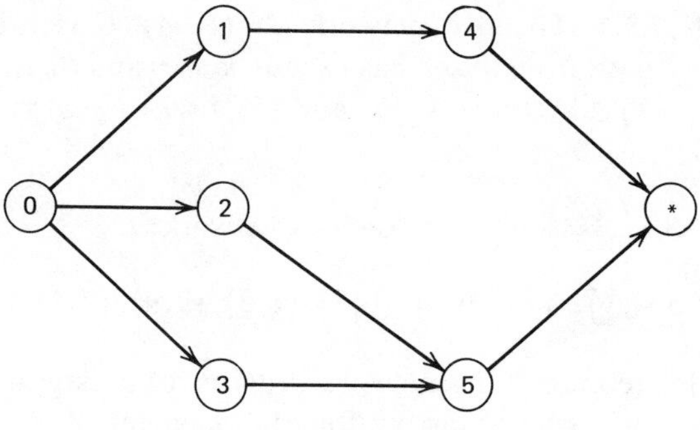

Figure 7.5.2 Diagram of a three-stage network.

s to * be $p_n(s)$. Now, if from s we go to s' such that $(s, s') \in A$, then we have $(n-1)$ stages remaining. Therefore, once we take the action to go to s' from s, assuming we use an optimal path from s' to *, the length of the path $(s, p_{n-1}(s'))$ will be $C_{ss'} + f_{n-1}(s')$. Thus, the minimum length for a path from s to * will be

$$f_n(s) = \min_{(s, s') \in A} \{C_{ss'} + f_{n-1}(s')\}. \tag{7.5.1}$$

Once we are in * there are no more stages remaining, and so $f_0(*) = 0$. Using this value, we can recursively calculate

$$f_i(s), \qquad i = 1, \ldots, k.$$

Equation (7.5.1) is known as the recursive equation.

EXAMPLE 7.5.2 Consider the three-stage network given below. Its $C_{ss'}$ is shown in Fig. 7.5.3; $C_{ss'}$ lies along the arcs.

We present the $f_n(s)$ and the optimal s' reached from s, when there are n more stages remaining to reach *, denoted by $a_n(s)$, in the form of tables. For $n = 1$, we have $f_1(s) = \min[C_{s*} + f_0(*)] = c_{s*}$, $a_1(s) = *$, and s can be either 4, 5, or 6, the stages in V_2. See Table 7.5.1.

Table 7.5.1

s	$f_1(s)$	$a_1(s)$
4	7	*
5	3	*
6	5	*

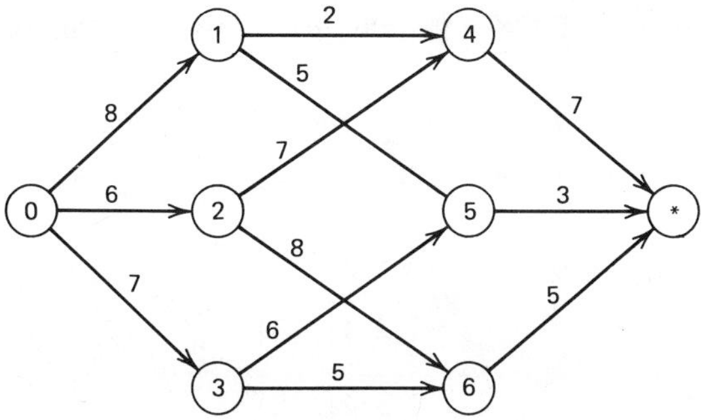

Figure 7.5.3 A three-stage network with $C_{ss'}$ along its arcs.

For $n=2$, we have $f_2(s)=\min_{(s,s')\in A}[C_{ss'}+f_1(s')]$. We have $f_1(s')$ from Table 7.5.1. For $s=1$, we have $f_2(1)=\min[C_{14}+f_1(4), C_{15}+f_1(5)]=\min(2+7,5+3)=8$. In this manner we obtain Table 7.5.2. For $n=3$, $f_3(0)=\min_{(0,s')\in A}[C_{0s'}+f_2(s')]$. We have $f_3(0)=\min[8+8,6+13,7+9]=16$. Here $a_3(0)$ can be either 1 or 3, as both give the minimum.

To obtain an optimal path $p_3(0)$, we proceed from 0 to 1 as $a_3(0)=1$ or 3. Then from 1 we find $a_2(1)=5$; we proceed from 1 to 5. From 5, $a_1(5)=*$. Thus we track a path given by $(0,1,5,*)$. $(0,3,5,*)$ is also an optimal path.

So far we have considered the shortest-route problem in a k-stage network, and given a dynamic-programming formulation of the problem. We shall use these ideas to solve clustering problems.

Let us return to the clustering problem in which n elements ought to be clustered optimally into m nonempty clusters according to some criterion.

DEFINITION 7.5.4 We shall call a vector $\mathbf{P}=(n_1, n_2, \ldots, n_m)$ a *distribution form* if $\sum_{j=1}^{m} n_j=n$ and n_j is a positive integer for all j, with $n_1 \geqslant n_2 \geqslant \cdots \geqslant n_m$. We use $\mathbf{P}$ to denote a distribution form.

Table 7.5.2

s	$f_2(s)$	$a_2(s)$
1	8	5
2	13	6
3	9	5

EXAMPLE 7.5.3 Suppose $n=4$ and $m=2$. We have the following distribution forms:

$$P_1=(3,1) \quad \text{and} \quad P_2=(2,2).$$

The possible clusters having distribution form P_1 are:

$$\{1,2,3\},\ \{4\}$$

$$\{1,2,4\},\ \{3\}$$

$$\{1,3,4\},\ \{2\}$$

$$\{2,3,4\},\ \{1\}.$$

Corresponding to P_2, we have:

$$\{1,2\},\ \{3,4\}$$

$$\{1,3\},\ \{2,4\}$$

$$\{1,4\},\ \{2,3\}.$$

The definition of "distribution form" introduces an order among the clusters formed, to wit, the first cluster has n_1 elements, the second has n_2 elements, and so on, with $n_1 \geqslant n_2 \geqslant \cdots \geqslant n_m$.

Now, the *stages* for the dynamic programming formulation correspond to the formation of Clusters 1, 2, and so on, until it becomes obvious, regarding the clustering of the remaining elements, that we require m nonempty clusters. For instance, when we have clustered $\{1,2\}$ in the first stage, it is obvious that $\{3,4\}$ is the second cluster when we are clustering four elements into two groups.

The number of stages in the problem m_0 is m if $n \geqslant 2m$, and $n-m+1$ if $n<2m$.

The set of *states* in each stage is the different subsets of elements that have been clustered so far, corresponding to the different distribution forms.

EXAMPLE 7.5.4 Consider $n=4$, $m=2$. We have the following states corresponding to Stage 1:

$$\{1,2,3\},\ \{1,2,4\},\ \{1,3,4\},\ \{2,3,4\},$$

$$\{1,2\},\ \{1,3\},\ \{1,4\}.$$

In the second stage, we have only one state, $\{1,2,3,4\}$, as we require only two clusters.

Between successive stages k and $k+1$, we connect the states by an arc if the state s in stage k is a subset of the state s' in stage $k+1$. For example, consider the following states in Stage 1 and Stage 2.

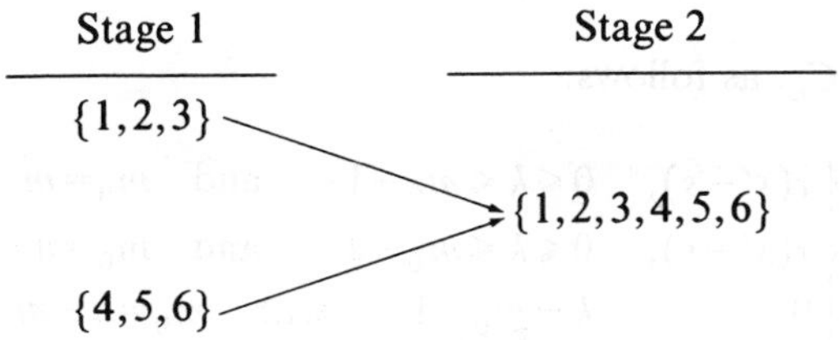

As both the states in Stage 1 are subsets of the state in Stage 2, we have two arcs connecting those states to the state in Stage 2. But we do not obtain different clusters as a result. Both these arcs represent the clusters $\{1,2,3\}$, $\{4,5,6\}$, so we can omit one of them. Such an arc is called a ***redundant*** arc.

Another form of redundancy may also arise. For example, consider $n=7$ and $m=3$.

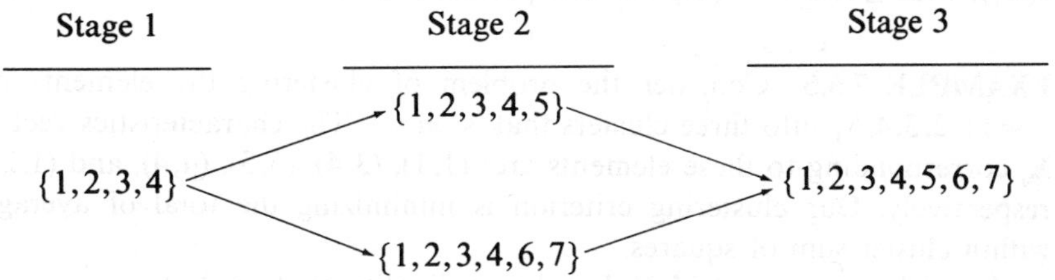

If we consider $\{1,2,3,4\}$ connected to $\{1,2,3,4,5\}$ it means we are forming the two clusters of $\{1,2,3,4\}$ and $\{5\}$. Now connecting $\{1,2,3,4,5\}$ to $\{1,2,3,4,5,6,7\}$ means that the third cluster is $\{6,7\}$. The same clustering is also obtained by $\{1,2,3,4\}\rightarrow\{1,2,3,4,6,7\}\rightarrow\{1,2,3,4,5,6,7\}$. But the second one also satisfies the order restriction on the number of elements in the clusters, namely, $n_1=4$, $n_2=2$, and $n_3=1$, $n_1 \geqslant n_2 \geqslant n_3$. So arc $\{1,2,3,4\}$ to $\{1,2,3,4,6,7\}$ is sufficient.

We add a Stage 0, and the empty cluster as the only state in that stage. This stage is connected to all states in Stage 1. In this fashion we get a network. The final state (State *) is $\{1,2,\ldots,n\}$. Such a network has the property that any path from the initial state 0 to the final state * has the same number of arcs, m_0. That is, we have an m_0-stage network.

Next we consider the reward $C_{ss'}$ arising out of an action taken to go from State s in Stage k to State s' in Stage $k+1$ such that (s, s') is an arc in the network.

Let us assume the clustering objective criterion is the minimization of $\Sigma_{j=1}^{m}\tau(G_j)$ where $\tau(G_j)$ is a function of the elements in the set G_j and $G_j \cap G_i = \emptyset$, $i \neq j$ and $\cup_{j=1}^{m} G_j = \{1,2,\ldots,n\}$, and $\tau(G_j)=0$ if G_j is a singleton set.

We define $C_{ss'}$ as follows:

$$C_{ss'} = \begin{cases} \tau(s'-s), & 0 \leqslant k \leqslant m_0 - 1 \quad \text{and} \quad m_0 = m \\ \tau(s'-s), & 0 \leqslant k \leqslant m_0 - 2 \quad \text{and} \quad m_0 = n - m + 1 \\ 0 & k = m_0 - 1 \quad \text{with} \quad m_0 = n - m + 1 \end{cases} \tag{7.5.2}$$

where s', s, the states, represent the corresponding subset of elements also. For $m_0 < m$, we define as above because once we have formed $m_0 - 1$ clusters, the remaining elements must be singleton clusters, as we require m nonempty clusters.

Thus we have a problem exactly like the *stagecoach* problem. We can make use of the dynamic-programming approach to solve the clustering problem with any criterion to be minimized satisfying the conditions on $\tau(G_j)$, with $\Sigma\,\tau(G_j) = C(G)$, for any partition G of N.

EXAMPLE 7.5.5 Consider the problem of clustering the elements of $N = \{1,2,3,4,5\}$ into three clusters that is, $m=3$. The characteristics vector $\mathbf{X}_i$ corresponding to these elements are: (1, 1), (3, 4), (5, 5), (4, 4), and (1, 2), respectively. Our clustering criterion is minimizing the total of average within-cluster sum of squares.

Let A be any subset of N. Let the cardinality of $A = l$. Let

$$\tau(A) = \begin{cases} \dfrac{1}{l} \displaystyle\sum_{\substack{i<j \\ i,j \in A}} d_{ij}^2 & l > 2 \\ 0 & \text{otherwise} \end{cases}$$

where d_{ij}^2 is the square of the Euclidean distance between the vectors corresponding to i and j. Table 7.5.3 gives the d_{ij}^2. Notice that $\tau(A)$ gives the within-cluster sum of squares.

In the first stage we have 20 states in all, as the distribution forms possible are (3, 1, 1) and (2, 2, 1). We take all possible C_3^5 subsets of N and

Table 7.5.3

i	j = 1	2	3	4	5
1		13	32	18	1
2			5	1	8
3				2	25
4					13

all possible C_2^5 subsets of N; in all there are 20 such subsets. In Stage 2, we can only cluster up to four elements, as in both the distribution forms we get only four elements clustered up to Stage 2 (that is, $3+1, 2+2$). So we take all the C_4^5 subsets from N in Stage 2 as states. Finally, in the last stage we have only one state, $\{1,2,3,4,5\}$.

Notice that:

1 From Stage 2 to Stage 3 we have only a singleton set formed, so the cost, as discussed earlier, will be zero for all arcs connecting states in Stage 2 to $\{1,2,3,4,5\}$.

2 From Stage 1 to Stage 2 for all states with three elements in Stage 1 to any state in Stage 2 that is permissible, the cost is zero, as only a singleton cluster is formed.

3 Only for the two-element states in Stage 1 to any permissible state in Stage 2 do we have to compute $C_{ss'}$. For the cost corresponding to the arcs from 0 to s, states in Stage 1. See Table 7.5.4.

Table 7.5.4

A	$\tau(A)$	A	$\tau(A)$
1,2	6.50	1,2,3	16.67
1,3	16.00	1,2,4	10.67
1,4	9.00	1,2,5	7.33
1,5	0.50	1,3,4	17.33
2,3	2.50	1,3,5	19.33
2,4	0.50	1,4,5	10.67
2,5	4.00	2,3,4	2.67
3,4	1.00	2,3,5	12.67
3,5	12.50	2,4,5	7.33
4,5	6.50	3,4,5	13.33

Table 7.5.5

s	$f_2(s)$	$a_2(s)$
1,2	1.00	(1,2,3,4)
1,3	0.50	(1,2,3,4)
1,4	2.50	(1,2,3,4)
1,5	0.50	(1,2,4,5)
2,3	6.50	(2,3,4,5)
2,4	12.50	(2,3,4,5)
2,5	1.00	(2,3,4,5)

Using the recursive formula of (7.5.1), $f_0(*)=0$ and $f_1(s)=0$, as observed earlier, and $a_1(s)=*$. We need only calculate $f_2(s)$ for s belonging to the set of states in Stage 1 (Table 7.5.5).

For instance, for State $s=(1,2)$ the permissible arcs are between (1,2) and (1,2,3,4), (1,2,3,5) and (1,2,4,5). The corresponding $C_{ss'}$ for $s'=$ (1,2,3,4), (1,2,3,5), and (1,2,4,5) are $\tau(3,4)$, $\tau(3,5)$, and $\tau(4,5)$, respectively. Therefore

$$\begin{aligned} f_2(s) &= \min[\tau(3,4)+0, \tau(3,5)+0, \tau(4,5)+0], \text{ as } f_1(s')=0 \quad \text{for all } s' \\ &= \min\{1.00, 12.50, 6.5\} \\ &= 1.00. \end{aligned}$$

Now for $f_3(0)$ we need to consider all states in Stage 2 as s'. We have

$$f_3(0) = \min_{s'} [C_{0s'} + f_2(s')].$$

Notice that $f_2(s')=0$ for s', a three-element subset. We find $\min[C_{0s'}]$ over 3-element subsets to be 2.67.

$$\begin{aligned} f_3(0) = \min[&2.67;\ 6.50+1.00,\ 16.00+0.50,\ 9.00+2.50, \\ &0.50+0.50,\ 2.50+6.50,\ 0.50+12.50,\ 4.00+1.00]. \end{aligned}$$

So we have

$$f_3(0) = 1.00 \quad \text{and} \quad a_3(0) = (1,5).$$

Thus we have the clustering (1,5), (2,4), and (3). The example illustrates the dynamic-programming approach to the problem. A general formula giving the number of states in each stage, and the number of nonredundant arcs between Stages k and $k+1$, can be worked out with combinatorial arguments.

7.6 LINEAR-PROGRAMMING APPROACH

The clustering problem can be viewed as a linear-programming problem as well as a dynamic-programming problem. The problem is finding the shortest path from 0 to * in a m_0-stage network. Such problems immediately yield to the application of linear programming. To be more precise, they can be viewed as minimal-cost-flow problems. Once we have identified the problem as a shortest-route problem, we can in fact use any of the efficient shortest-route algorithms that are available. Unfortunately, the network we obtain grows very rapidly as the number of partitions increases, and that in turn increases with n and m. However, we discuss this approach as an alternative way of looking at the clustering problem.

The theory of *flows in networks* was developed by Ford and Fulkerson. They consider a problem called minimal cost flow problem, which can be described as follows:

Consider a network $G=(V, A)$. We are given $d(i, j)\geqslant 0$ (called the capacity of the arc (i, j), for all $(i, j)\in A$. Also for each arc we are given C_{ij}, the cost per unit of flow of certain material from i to j, along $(i, j)\in A$.

Let $f_{ij}\geqslant 0$ be the amount of material that flows from i to j along $(i, j)\in A$. The capacity of an arc restricts the amount of flow along the arc. So $d(i, j)\geqslant f_{ij}\geqslant 0$. Suppose we have v units available at the source, node 0. We wish to find the flows along the arcs in the network such that these v units reach the sink, node *, and the cost of doing this is minimized.

If $v>v^*$, the maximal flow that is possible from 0 to *, we have no solution to the problem. This problem can be formulated as a linear-programming problem.

However, in this section we consider a special case of this problem and give a simple procedure for finding an optimal solution using the complementary slackness property of linear programming. This special case is related to the problem of finding a shortest route in an uncapacitated k-stage network.

Consider the m_0-stage network corresponding to a clustering problem. Let $G(V, A)$ be the corresponding network. Let $f_{ij}\geqslant 0$ be the *flow* along the arc $(i, j)\in A$. Let C_{ij} be as defined by 7.5.2. We assume that we have 1 unit available at Node 0, the *source*. Similarly, we require 1 unit at node *, the *sink*. We express these two constraints as follows:

$$\sum_{(0,y)\in A} f_{0y} = 1$$

and

$$\sum_{(y,*)\in A} f_{y*} = 1.$$

Also we do not want any of the material starting from 0 to be left out at any of the other nodes except Node *. This requirement is expressed as

$$\sum_{(x,y)\in A} f_{xy} - \sum_{(y,x)\in A} f_{yx} = 0 \qquad \text{for } 0 \neq x \neq *,\ x \in V.$$

Next we state the cost of the flow of 1 unit from 0 to * as the sum of the products of C_{xy} and f_{xy} for $(x, y) \in A$:

$$\sum_{(x,y)\in A} C_{xy} f_{xy}.$$

We wish to minimize this total cost in a minimal-cost-flow problem. Thus we have the linear-programming problem, Problem 7.6.1.

Problem 7.6.1

$$\begin{aligned}
\text{Minimize} \quad & \sum_{(x,y)\in A} C_{xy} f_{xy} \\
\text{subject to} \quad & \sum_{(0,y)\in A} f_{0y} = 1 \\
& -\sum_{(y,*)\in A} f_{y*} = -1 \\
& \sum_{(x,y)\in A} f_{xy} - \sum_{(y,x)\in A} f_{yx} = 0 \qquad \text{for} \quad 0 \neq x \neq * \\
& f_{xy} \geqslant 0, \quad (x, y) \in A.
\end{aligned}$$

This problem resembles the transportation problem discussed earlier. Notice that we do not have the capacity restrictions on f_{xy}. For each variable f_{xy} we have exactly two nonzero elements in the corresponding column, and they are ± 1. This special structure enables us to devise a method easier than the simplex method. In essence we use the complementary-slackness property of the optimal solutions to primal and dual problems. So next we consider the dual of Problem 7.6.1. For better understanding of the problem we consider a simple four-stage network (not necessarily corresponding to a clustering problem). We give the cost C_{xy} along the arc (x, y) in Fig. 7.6.1. The linear-programming matrix **A** corresponding to this network is given in Table 7.6.1.

Here $\mathbf{f}=(f_{01},\ldots,f_{6*})'$ is the vector corresponding to the flow variables, $\mathbf{b}'=(1,0,0,\ldots,-1)'$, and $\mathbf{C}=(5,4,3,4,\ldots,4)$ corresponding to $\mathbf{f}$. Thus we wish to

$$\begin{aligned}&\text{Minimize} && \mathbf{Cf}\\ &\text{subject to} && \mathbf{Af}=\mathbf{b}\\ &&& \mathbf{f}\geqslant \mathbf{0}.\end{aligned}$$

The dual of this problem is given by

$$\begin{aligned}&\text{Maximize} && \mathbf{b}'\mathbf{w}\\ &\text{subject to} && \mathbf{A}'\mathbf{w}\leqslant \mathbf{C}'\\ &&& \mathbf{w}\text{ unrestricted in sign,}\end{aligned}$$

where $\mathbf{w}=(w_0, w_1,\ldots,w_6, w_*)'$.

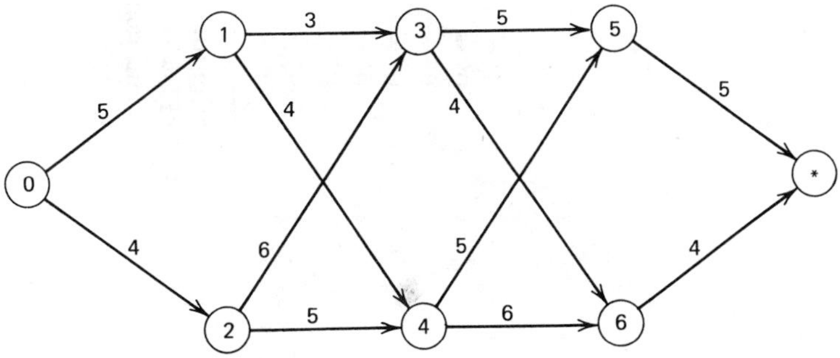

Figure 7.6.1 A four-stage network.

Table 7.6.1

	Cost											
	5	4	3	4	6	5	5	4	5	6	5	4
b	(f_{01})	(f_{02})	(f_{13})	(f_{14})	(f_{23})	(f_{24})	(f_{35})	(f_{36})	(f_{45})	(f_{46})	(f_{5*})	(f_{6*})
$1^a =$	1	1										
$0^b =$	−1		1	1								
$0^b =$		−1			1	1						
$0^b =$			−1		−1		1	1				
$0^b =$				−1		−1			1	1		
$0^b =$							−1		−1		1	
$0^b =$								−1		−1		1
$-1^c =$											−1	−1

[a]Constraint 1.
[b]Flow conservation constraints corresponding to Nodes 1–6.
[c]Constraint 2.

The dual problem corresponding to the example is shown below in expanded form. We have

$$\begin{aligned}
\text{Maximize}\quad & w_0 - w_* \\
\text{subject to}\quad & w_0 - w_1 \leqslant 5 \\
& w_0 - w_2 \leqslant 4 \\
& w_1 - w_3 \leqslant 3 \\
& w_1 - w_4 \leqslant 4 \\
& w_2 - w_3 \leqslant 6 \\
& w_2 - w_4 \leqslant 5 \\
& w_3 - w_5 \leqslant 5 \\
& w_3 - w_6 \leqslant 4 \\
& w_4 - w_5 \leqslant 5 \\
& w_4 - w_6 \leqslant 6 \\
& w_5 - w_* \leqslant 5 \\
& w_6 - w_* \leqslant 4 \\
& \mathbf{w} \quad \text{unrestricted in sign.}
\end{aligned}$$

As we did for the transportation problem, we can find a feasible solution to the dual problem above, as follows:

$$w_0 = \min_{(0,k)\in A} C_{0k} = \min(5,4) = 4.$$

We now take $w_1 = w_0 - 5 = -1$, $w_2 = w_0 - 4 = 0$. Then for $w_3 \geqslant \max(w_1 - 3, w_2 - 6)$, so $w_3 = \max(-4, -6) = -4$, and $w_4 = \max(w_1 - 4, w_2 - 5) = -5$. Now for $w_5 = \max(w_3 - 5, w_4 - 5) = \max(-9, -10) = -9$, and $w_6 = \max(w_3 - 4, w_4 - 6) = -8$. Finally we get $w_* = \max(w_5 - 5, w_6 - 4) = \max(-14, -12) = -12$. Therefore, $w_0 - w_* = 4 - (-12) = 16$.

Now for this set of feasible w_i's, we find that some of the constraints hold as equalities and some are strict inequalities: the slack variables corresponding to some inequalities are zero and they are positive in the other inequalities. Let Δ_{ij} denote the slack variable corresponding to the constraint involving w_i and w_j, $(i, j) \in A$. That is $\Delta_{ij} = C_{ij} - (w_i - w_j) \geqslant 0$.

From the complementary-slackness property, we find that whenever $\Delta_{ij} > 0$, the corresponding f_{ij} must be zero in an optimal solution to the primal problem. The method known as the *out-of-kilter* method makes use of this property of optimal solutions to primal and dual problems to find a solution to the minimal-cost-flow problem.

As f_{ij} must be zero for $(i, j) \in A$, such that $\Delta_{ij} > 0$, we need only allow flow through those arcs in the network for which $\Delta_{ij} = 0$; consider the network in which we deleted the arcs for which $\Delta_{ij} > 0$. The resultant network is called *restricted network*. For instance, with $\mathbf{w} = (4, -1, 0, -4, -5, -9, -8, -12)'$, we have the restricted network given in Fig. 7.6.2 with w_i written above Node i. Now we can find a flow of one unit from 0 to * along the path $(0, 1, 3, 6, *)$. Thus the optimal cost is $5 + 3 + 4 + 4 = 16$, which is the same as the objective function value of the dual problem $w_0 - w_* = 4 - (-12) = 16$.

But in general we might not get a feasible solution to the restricted problem in one step. In such cases we choose another feasible solution to the dual problem, consider the corresponding restricted primal, and find a path from 0 to *. If we succeed, we stop. Otherwise, we repeat this process. When we choose the next feasible solution to the dual problem, we improve the objective function of the dual. This approach is also known as the *primal-dual* method. If the restricted problem has a path from 0 to *, we have a flow of one unit from 0 to *, and we have a feasible solution to the primal problem.

A path from 0 to * in the restricted network can be found via a labeling procedure.

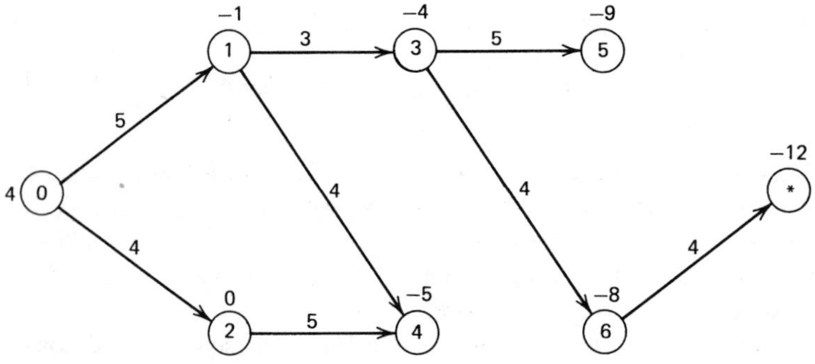

Figure 7.6.2 A restricted network.

Let us denote by $A_{\mathbf{w}}$ the set of those restricted arcs in the restricted network corresponding to **w**, a feasible solution to the dual.

Step 1 0 is always considered to be labeled.

Step 2 Then, if i is labeled, we label j if $(i, j) \in A_w$ is an arc in the restricted network.

This procedure is called "forward labeling." We label j with $(i+)$.

We repeat Step 2 to label as many nodes as possible. If * is labeled, we stop. We have a path from 0 to * that can be traced from the labels. If * is not labeled we proceed as follows:

Let us denote the set of labeled nodes at each stage by L_1, L_2, L_3, and so on. Let

$$\varepsilon = \min_{1 \leqslant k \leqslant m_0 - 1} \left[\underset{\substack{i \in L_k \\ j \in V_{k+1} - L_{k+1}}}{\text{minimum}} (\Delta_{ij}) \right]$$

V_k is the set of nodes in Stage k. Notice that ε is always greater than zero. Suppose $\varepsilon = 0$. This means there is a Node i in some stage k and a Node j in stage $k+1$ such that $\Delta_{ij} = 0$, in which case we will have an arc from i to j in the restricted network. As $i \in L_k$, we will end up labeling j. But $j \in V_{k+1} - L_{k+1}$, that is, j is not labeled, so the supposition is wrong.

Now change the w_i's as follows:

$$\hat{w}_i = \begin{cases} w_i + \varepsilon \text{ if } i \text{ is labeled.} \\ w_i \quad \text{otherwise} \end{cases}$$

We can note that this solution is feasible for the dual problem. We consider four cases.

CASE 1

$$i \in L_k, j \in L_{k+1} \quad \text{for some} \quad k \quad \text{and} \quad (i, j) \in A$$

$$\hat{w}_i - \hat{w}_j = w_i + \varepsilon - (w_j + \varepsilon) = w_i - w_j \leqslant C_{ij}.$$

The corresponding dual constraints are satisfied.

CASE 2

$$i \in L_k, j \in V_{k+1} - L_{k+1} \quad \text{and} \quad (i, j) \in A.$$

$$\hat{w}_i - \hat{w}_j = w_i + \varepsilon - w_j; \quad \text{it must be} \leqslant C_{ij}.$$

But

$$\varepsilon \leqslant \underset{\substack{i \in L_k \\ j \in V_{k+1} - L_{k+1}}}{\text{minimum}} (C_{ij} - w_i + w_j) \leqslant C_{ij} - w_i + w_j$$

Hence, the dual constraints are satisfied.

CASE 3

$$i \in V_k - L_k, j \in L_{k+1}, (i, j) \in A$$

$$\hat{w}_i - \hat{w}_j = w_i - w_j - \varepsilon \leqslant w_i - w_j \leqslant C_{ij}.$$

$\hat{w}$ so defined satisfies the dual constraints.

CASE 4

$$i \in V_k - L_k, j \in V_{k+1} - L_{k+1}, (i, j) \in A$$

$$\hat{w}_i - \hat{w}_j = w_i - w_j \leqslant C_{ij}.$$

The dual constraints are satisfied.

These four cases exhaust all possibilities; hence $\hat{\mathbf{w}}$ is feasible for the dual problem.

Also note that $\hat{w}_0 - \hat{w}_* > w_0 - w_*$. Node 0 is always labeled and $\varepsilon > 0$; therefore $\hat{w}_0 = w_0 + \varepsilon > w_0$, and Node * is not labeled—that is, $\hat{w}_* = w_*$.

Thus we have obtained an improved dual solution. With this new solution, we consider $A_w = A_{\hat{w}}$ and repeat the steps until we get a feasible solution to the primal, that is, we have a path from 0 to *.

EXAMPLE 7.6.1 Consider the network given in Fig. 7.6.1.

Suppose we start with $\mathbf{w} = (0, \ldots, 0)'$ which is feasible as $C_{ij} > 0$ for all $(i, j) \in A$. Now only Node 0 is labeled, as all $\Delta_{ij} = C_{ij} > 0$. We find

$$\varepsilon = \min_{0 < k < 2} \left[\underset{\substack{i \in L_k \\ j \in V_{k+1} - L_{k+1}}}{\text{minimum}} \Delta_{ij} \right]$$

$$= \min_{\substack{i \in L_0 \\ j \in V_1}} \left[\Delta_{ij} \right] = \min_{\substack{i \in L_0 \\ j \in V_1}} C_{ij} = 4.$$

We change $w_0 = 0$ to $\hat{w}_0 = 4$. Now we have $\mathbf{w} = (4, 0, \ldots, 0)'$. So $\Delta_{02} = 4 - 4 + 0$

$=0$ and the rest of the Δ_{ij} are positive. Thus we have the restricted network given in Fig. 7.6.3.

We label Node 2 from 0, with 0^+. As there is no path in this network from 0 to *, we change w_i's.

$$\begin{aligned}\varepsilon &= \min(\Delta_{01}, \Delta_{23}, \Delta_{24})\\ &= \min[5-4+0, 6-0+0, 5-0+0]\\ &= 1.\end{aligned}$$

We change **w** as follows: $w_0 = 4+1 = 5$, $w_2 = 0+1 = 1$. The rest of the w_i's are zeroes. The restricted network corresponding to this solution is shown in Fig. 7.6.4. We label Nodes 1 and 2 from 0. We find

$$\begin{aligned}\varepsilon &= \min[\Delta_{23}, \Delta_{24}, \Delta_{13}, \Delta_{14}]\\ &= \min[6-1+0, 5-1+0, 3-0+0, 4-0+0]\\ &= 3.\end{aligned}$$

Now the new solution to the dual is found as $w_0 = 5+3 = 8$, $w_2 = 1+3 = 4$, $w_1 = 0+3 = 3$, for other i, $w_i = 0$. The restricted network corresponding to this solution is shown in Fig. 7.6.5.

Now the nodes 1, 2, are labeled from 0 and 3 is labeled from 1. We find

$$\begin{aligned}\varepsilon &= \min[\Delta_{23}, \Delta_{24}, \Delta_{14}, \Delta_{35}, \Delta_{36}]\\ &= \min[2, 1, 1, 5, 4] = 1.\end{aligned}$$

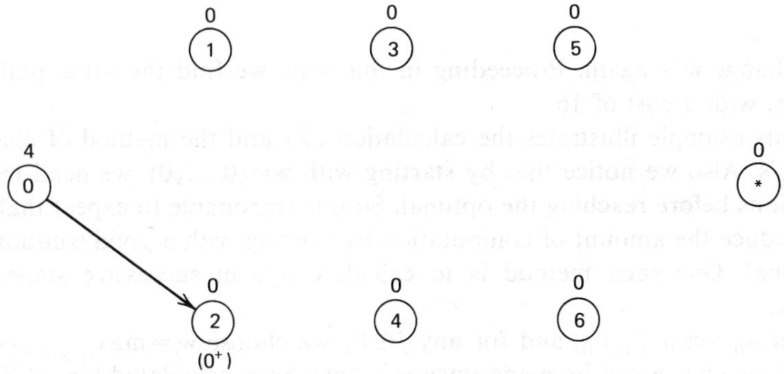

Figure 7.6.3 First restricted network discussed in Example 7.6.1.

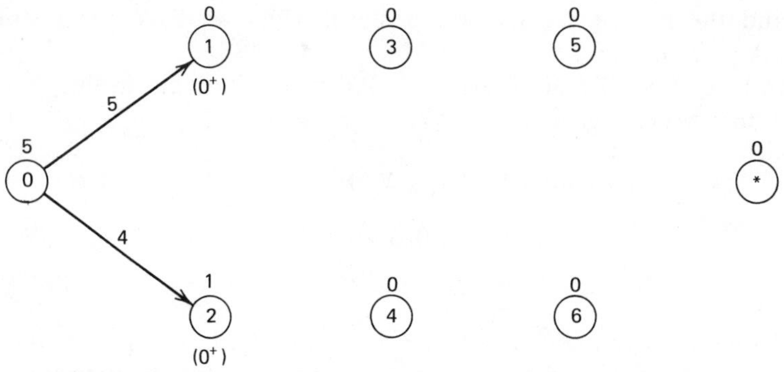

Figure 7.6.4 Second restricted network discussed in Example 7.6.1.

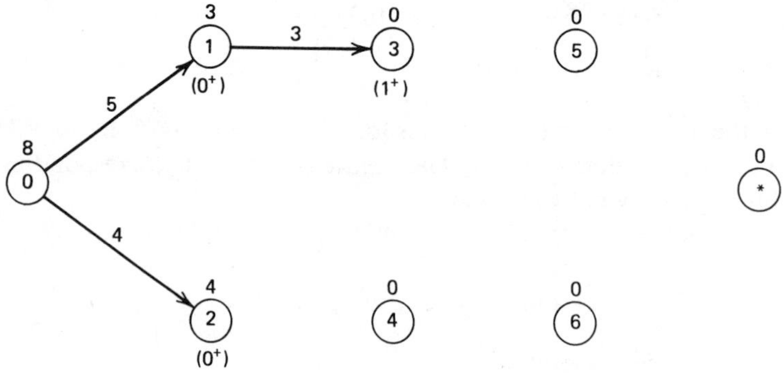

Figure 7.6.5 Third restricted network discussed in Example 7.6.1.

We change w_i's again. Proceeding in this way, we find the same path as earlier, with a cost of 16.

This example illustrates the calculation of ε and the method of changing w_i's. Also we notice that by starting with $\mathbf{w}=(0,\ldots,0)'$ we need many iterations before reaching the optimal. So it is reasonable to expect that we can reduce the amount of computation by starting with a good solution to the dual. One such method is to calculate w_i's in successive stages as follows:

Set $w_0=\min_{j\in V_1}C_{0j}$ and for any $j\in V_k$ we choose $w_j=\max_{(i,j)\in A}[w_i-C_{ij}]$. This choice can be made once w_i's have been calculated for $i\in V_{k-1}$.

The procedure can now be summarized as Algorithm 7.6.1.

Algorithm 7.6.1

Step 0 Form the m_0-stage network with cost C_{ij}.

Step 1 Find a feasible solution to the dual problem. That is, find w_i's satisfying $w_i - w_j \leqslant C_{ij}$, $(i, j) \in A$.

Step 2 Form the restricted network:

$$G_{\mathbf{w}} = (V, A_{\mathbf{w}}), \text{ where } A_{\mathbf{w}} = \{(i, j) | (i, j) \in A, \text{ and}$$

$$\Delta_{ij} = C_{ij} - w_i + w_j = 0\}.$$

Step 3 Use labeling procedure to find a path from 0 to *. If there is a path from 0 to *, stop; we have an optimal solution to the problem. Otherwise go to Step 4.

Step 4 Calculate

$$\varepsilon = \min_{0 \leqslant k \leqslant m_0 - 1} \left[\underset{\substack{i \in L_k \\ j \in V_{k+1} - L_{k+1}}}{\text{minimum}} [\Delta_{ij}] \right]$$

where L_k is the set of labeled nodes at Stage k. We label 0 always.

Step 5 Change $\mathbf{w}$ as follows:

$$\hat{w}_i = \begin{cases} w_i + \varepsilon & \text{if } i \in L_k \\ w_i & \text{if } i \in V_k - L_k. \end{cases}$$

Go to Step 2 with $\mathbf{w} = \hat{\mathbf{w}}$.

Notice that at each time we change w_i's we add some more arcs to the previous restricted network. But if we have an $(i, j) \in A_{\mathbf{w}}$ such that i is not labeled and j is labeled, then $\hat{w}_i = w_i$ and $\hat{w}_j = w_j + \varepsilon$

$$w_i - w_j - \varepsilon < w_i - w_j = C_{ij}.$$

Thus the new Δ_{ij} is greater than zero. So we may omit such arcs in the new network.

For instance, consider the network given in Example 7.6.1 with $C_{24} = 7$ and other C_{ij}'s unaltered, if $\mathbf{w}$ is as shown in Fig. 7.6.6 with the labeled nodes marked with ($\surd$) checks. In (4, 5), 5 is labeled as we can label 2 from 0, and 3 from 2, and 5 from 3. But 4 is not labeled. Since $C_{24} = 7$, $\Delta_{24} = 1$. So $\varepsilon = 1$. Therefore $\hat{w}_5 = -9 + 1 = -8$ and $\hat{w}_4 = -4$.

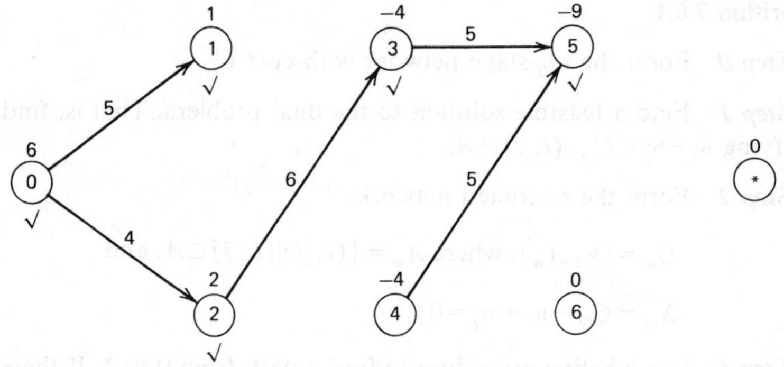

Figure 7.6.6 A restricted network to illustrate certain arcs may be dropped subsequently.

The new $\Delta_{45} = 5+4-8 = 1 > 0$. Thus (4, 5) does not appear in the next restricted network.

For this reason if the set

$$D = \{(i, j) | (i, j) \in A_{\mathbf{w}}, i \in V_k - L_k \text{ and } j \in L_{k+1}\}$$

is nonempty, we delete these arcs from $A_{\mathbf{w}}$ and add the new arcs to get $A_{\hat{\mathbf{w}}}$.

RESULT 7.6.1 If we choose $w_0 = 0$ and for $j \in V_{k+1}$

$$w_j = \max_{\substack{i \in V_k \\ (i, j) \in A}} [w_i - C_{ij}] \quad 0 \leqslant k \leqslant m_0 - 1$$

we indeed have an optimal solution to the dual problem given by this **w**.

Proof We consider Node *. We are choosing w_* such that there is a $j_{m_0-1} \in V_{m_0-1}$ such that $w_{j_{m_0-1}} - w_* = C_{j_{m_0-1*}}$ or $\Delta_{j_{m_0-1*}} = 0$. Therefore, there is an arc from j_{m_0-1} to * in the restricted network. Now suppose we have found a path from some node j_k in Stage k to * in the restricted network.

Now consider j_k. We shall show that there is a node $j_{k-1} \in V_{k-1}$ such that there is a path from j_{k-1} to * in the restricted network.

$$w_{j_k} = \max_{\substack{i \in V_{k-1} \\ (i, j) \in A}} [w_i - C_{ij_k}].$$

Therefore $w_{j_k} = w_{j_{k-1}} - C_{j_{k-1}j_k}$ for some $i = j_{k-1}$. The implication is that there is an arc from j_{k-1} to j_k in the restricted network. Hence, with the use of the induction hypothesis there is a path from j_{k-1} to * in the restricted network. Thus, the result is true for $k=0$, that is, there is a path from $j_0=0$ to * in the restricted network. The implication is that we have a flow of 1 unit from 0 to *, and hence a feasible solution for the primal problem, which means $\mathbf{w}$ is optimal for the dual problem.

REMARK 7.6.1 This result shows that we can find an optimal solution to the dual problem in a recursive way, which gives a corresponding path from 0 to * in the restricted network that is optimal for Problem 7.6.1. Now the fact can be used to solve the shortest-route problem arising in the m_0-stage network corresponding to a clustering problem. Also this result eliminates the trouble of solving the problem with Algorithm 7.6.1.

Notice the similarity between the procedure given by Result 7.6.1 for computing $\mathbf{w}$ and the dynamic-programming recursive equation.

Algorithm 7.6.1 considers a special case of the minimum-cost-flow problem. By modifying the labeling procedure, the definition, of ε, and the dual-variable-changing process, we can suitably make this algorithm solve problems in which we may not get such a simple solution as that given by Result 7.6.1. Algorithm 7.6.1 is discussed with a view to giving a basis for understanding the primal-dual method of solving minimal-cost-flow problems, that has application elsewhere. For instance, we have encountered such problems in the area of sampling and design of experiment.

7.7 BRANCH-AND-BOUND APPROACH

We have earlier seen the application of branch-and-bound methods for solving stepwise regression problems. In this section we are developing a branch-and-bound scheme for solving the clustering problem with the criterion of minimizing the total within-cluster sum of squares.

Let $N=\{1,2,\ldots,n\}$ be the set of n elements to be clustered. Let $\mathbf{X}=(\mathbf{X}_1,\ldots,\mathbf{X}_n)$, $\mathbf{X}_i \in R^z$ be the data on the n elements corresponding to the characteristics under consideration. For any subset $A \subset N$ we define

$$\tau(A)=\begin{cases} \dfrac{1}{l} \displaystyle\sum_{\substack{i<j \\ i,j\in A}} d_{ij}^2 & \text{for } \; l \geqslant 0 \\ 0 & \text{otherwise,} \end{cases}$$

where l is the cardinality of A.

Consider the problem of partitioning N into $J=(J_1, J_2,\ldots, J_m)$ such that

$$C'(J)=\sum_{i=1}^{m}\tau(J_i)$$

is minimized. (This problem was considered earlier.) We designate this problem the *main* problem.
To develop a branch-and-bound scheme to solve the problem we require certain results which are obtained in what follows.

First we state the following well-known result.

RESULT 7.7.1 Let $\mathbf{X}\in R^z$, then

$$\sum_{i=1}^{n}|\mathbf{X}-\mathbf{X}_i|^2 \geqslant \sum_{i=1}^{n}|\overline{\mathbf{X}}-\mathbf{X}_i|^2=\frac{1}{n}\sum_{1<i<j<n}|\mathbf{X}_i-\mathbf{X}_j|^2$$

where $\overline{\mathbf{X}}=1/n(\mathbf{X}_1+\mathbf{X}_2+\cdots+\mathbf{X}_n)$.

RESULT 7.7.2 For any two disjointed subsets A_1 and A_2 of N we have,

$$\tau(A_1\cup A_2)\geqslant\tau(A_1)+\tau(A_2).$$

This result follows from Result 7.7.1.

Before proceeding further with the *main* problem, consider the Problem 7.7.1. For notational convenience we denote C_2^n by $\bar{n}$.

Let $I=\{1,2,\ldots,\bar{n}\}$. In what follows, G denotes $(G_1,\ldots,G_m)$ where $G_i\subseteq I$ for all i and $G_i\cap G_j=\varnothing$ if $i\neq j$. Let n, m, and nonnegative real numbers $a_1\leqslant a_2\leqslant\cdots\leqslant a_{\bar{n}}$ be given.

Problem 7.7.1

Minimize $\quad C_2(G)=\sum_{\substack{i=1\\ n_i\geqslant 2}}^{m}\frac{1}{n_i}\sum_{j\in G_i}a_j$

subject to $\quad \sum_{i=1}^{m}n_i=n$

G_i has $\bar{n}_i$ elements for i such that $n_i\geqslant 2$

n_i positive integer

We shall connect the solutions to this problem with that of the main problem.

DEFINITION 7.7.1 An *alternative*, denoted by $\mathbf{P}$, is a vector $(n_1, n_2, \ldots, n_m)$ of positive integers such that $\Sigma_{i=1}^{m} n_i = n$ and $n_1 \leqslant n_2 \cdots \leqslant n_m$.

Let $\mathcal{P}_{n,m} = \{\mathbf{P} | \mathbf{P}$ is an alternative corresponding to n and $m\}$. Notice that an alternative is just a distribution form as defined in Section 7.5 but for the fact that the n_i's are ordered in a nondecreasing manner.

DEFINITION 7.7.2 G is called a *clustering induced* by $\mathbf{P} \in \mathcal{P}_{n,m}$ if the number of elements in G_i is $\bar{n}_i$ for all i, where $\mathbf{P} = (n_1, n_2, \ldots, n_m)$.

Let $\mathcal{G}(\mathbf{P}) = \{G | G$ is induced by $\mathbf{P}\}$ for a given $\mathbf{P} \in \mathcal{P}_{n,m}$. Clearly Problem 7.7.1 is equivalent to

$$\underset{\mathbf{P} \in \mathcal{P}_{n,m}}{\text{Minimize}} \left[\min_{G \in \mathcal{G}(\mathbf{P})} C_2(G) \right]. \tag{7.7.1}$$

First we consider $\min_{G \in \mathcal{G}(\mathbf{P})} C_2(G)$ and give an optimal solution for a given $\mathbf{P}$, and then consider Problem 7.7.1. In the following we write $\mathcal{P}$ for $\mathcal{P}_{n,m}$ when n and m are fixed.

RESULT 7.7.3 Given $\mathbf{P} \in \mathcal{P}$, $G^*(\mathbf{P})$ defined by

$$G_i^*(\mathbf{P}) = \left\{ j \,\middle|\, \sum_{k=1}^{i-1} \bar{n}_k + 1 \leqslant j \leqslant \sum_{k=1}^{i} \bar{n}_k \right\} \qquad \text{for} \quad 1 \leqslant i \leqslant m \tag{7.7.2}$$

minimizes $C_2(G)$ over $\mathcal{G}(\mathbf{P})$.

Proof The result follows from the fact that, given $a_1, a_2, \ldots, a_k$ and $b_1, b_2, \ldots, b_k$ nonnegative, $\Sigma_{i=1}^{k} a_{j_i} b_{j_i}$ is minimized when $j_1, \ldots, j_k$ is such that $a_{j_1} \leqslant \cdots \leqslant a_{j_k}$ and $b_{j_1} \geqslant \cdots \geqslant b_{j_k}$.

EXAMPLE 7.7.1 Consider $n=6$, $m=2$; so $\bar{n}=15$ with a_i's as given by $\mathbf{a}$.

$$\mathbf{a} = (1, 1, 2, 2, 2, 4, 4, 5, 5, 6, 8, 10, 11, 11, 12).$$

Then for the alternative $\mathbf{P} = (2,4)$, we have $\bar{n}_1 = C_2^2 = 1$; $\bar{n}_2 = C_2^4 = 6$. Using (7.7.2), $G^*(\mathbf{P})$, the optimal clustering induced by $\mathbf{P}$, given by

$$G_1^*(\mathbf{P}) = \{j | 1 \leqslant j \leqslant 1\} = \{1\},$$

$$G_2^*(\mathbf{P}) = \{j | 1+1 \leqslant j \leqslant 7\} = \{2,3,4,5,6,7\} \quad \text{and} \quad C_2(G^*(\mathbf{P}))$$

$$= \tfrac{1}{2}(1) + \tfrac{1}{4}(1+2+2+2+4+4) = 4.25.$$

We adopt the convention that $\Sigma_{k=1}^{i-1}\bar{n}_k=0$ if $i=1$, $\Sigma_{j\in G_i}a_j=0$ if $n_i=1$, and if $\Sigma_{k=1}^{i}\bar{n}_k<\Sigma_{k=1}^{i-1}\bar{n}_k+1$, we assume the cluster G_i has no elements, something that happens when the first few n_i's are 1's. In fact we need to consider only the $G_i^*(\mathbf{P})$ for i such that $n_i>1$. For instance, if $\mathbf{P}=(1,5)$, then $G_1^*(\mathbf{P})=\varnothing$ and $G_2^*(\mathbf{P})=\{1,2,\ldots,10\}$ and $C_2(G^*(\mathbf{P}))=0+\frac{1}{5}(1+1+\cdots+6)=6.4$.

Next we prove a result which brings out certain dominance among the alternatives.

RESULT 7.7.4 Let $\mathbf{P}=(n_1,n_2,\ldots,n_m)$ and $\mathbf{P}'=(n_1',n_2',\ldots,n_m')$ be alternatives such that

1 $1=n_k\leqslant n_k'\leqslant n_{k+1}$ and $n_s'\geqslant n_{s-1}$ for some k and s with $1\leqslant k\leqslant s\leqslant m$

and

2 $n_j'=n_j$ for $s\neq j\neq k$.

Then we have

$$C_2(G^*(\mathbf{P}))\geqslant C_2(G^*(\mathbf{P}')).$$

Proof Notice that $n_k'-n_k=n_s-n_s'$. Let $l=n_k'-n_k=n_s-n_s'$. Consider $G\in\mathcal{G}(\mathbf{P}')$, defined as follows:

$$G_i=G_i^*(\mathbf{P})\qquad \text{if}\quad s\neq i\neq k,$$

$$G_k=G_k^*(\mathbf{P})\cup H$$

where $H=$ the set of first $\bar{l}+n_k l$ terms from $G_s^*(\mathbf{P})$ and $G_s=$ the set of first $\bar{n}_s'$ terms from $G_s^*(\mathbf{P})-H$. Note that this formulation is made possible by the properties of n_k, n_k', n_s, and n_s' given in the hypothesis.

Let $F=G_s^*(\mathbf{P})-(G_s\cup H)$. Now

$$C_2(G^*(\mathbf{P}))-C_2(G)=\frac{1}{n_k}\sum_{j\in G_k^*(\mathbf{P})}a_j-\frac{1}{n_k'}\sum_{j\in G_k}a_j+\frac{1}{n_s}\sum_{j\in G_s^*(\mathbf{P})}a_j-\frac{1}{n_s'}\sum_{j\in G_s}a_j$$

$$=\left(\frac{1}{n_k}-\frac{1}{n_k'}\right)\sum_{j\in G_k^*(\mathbf{P})}a_j+\left(\frac{1}{n_s}-\frac{1}{n_k'}\right)\sum_{j\in H}a_j$$

$$+\left(\frac{1}{n_s}-\frac{1}{n_s'}\right)\sum_{j\in G_s}a_j+\frac{1}{n_s}\sum_{j\in F}a_j.$$

Let $\max_{j\in G_s}a_j=a'$. Then we have $\min_{j\in F}a_j\geqslant a'\geqslant\max_{j\in H}a_j$. Thus as the first term is zero in the expression for $C_2(G^*(\mathbf{P}))-C_2(G)$, we get

$$C_2(G^*(\mathbf{P}))-C_2(G)\geqslant\left(\frac{1}{n_s}-\frac{1}{n_k+l}\right)\left[\bar{l}+n_k l\right]a'+\left(\frac{1}{n_s}-\frac{1}{n'_s}\right)\bar{n}'_s a'$$

$$+\frac{1}{n_s}\left[\bar{n}_s-\bar{n}'_s-(l+n_k l)\right]a'$$

$$=\frac{a'l}{2}\left[1-\frac{(l-1+2n'_k)}{n_k+l}\right]$$

$$=0 \text{ after simplification.}$$

Thus $C_2(G^*(\mathbf{P}))\geqslant C_2(G)$. By definition, $C_2(G^*(\mathbf{P}'))\leqslant C_2(G)$ for any $G\in \mathcal{G}(\mathbf{P}')$. Hence the result follows.

REMARK 7.7.1 This result shows that when $n>2m$, if any $\mathbf{P}\in\mathcal{P}$ has some n_i equal to 1 then there exists a $\mathbf{P}'\in\mathcal{P}$ such that $C_2(G^*(\mathbf{P}'))\leqslant C_2(G^*(\mathbf{P}))$. Further, when $n\leqslant 2m$, $\mathbf{P}^*$ given by

$$n_i=\begin{cases}\left[\dfrac{n}{m}\right] & \text{for } 1\leqslant i\leqslant m-s\\[2ex] \left[\dfrac{n}{m}\right]+1 & \text{for } m-s+1\leqslant i\leqslant m\end{cases}$$

where $s=n-[n/m]m$, and $[n/m]$ is the integer part of n/m, is such that $G^*(\mathbf{P}^*)$ is optimal for Problem 7.7.1.

Going back to the main problem, consider any subset $A\subset N$ having at least 2 elements. Consider d_{ij}^2 for $i, j\in A$, $i<j$. Arrange them in nondecreasing order and rename them as

$$a_1, a_2, \ldots \text{that is, } a_k=d_{i_k j_k}^2$$

for some i_k and $j_k\in A$ for $k=1,2,\ldots,\bar{l}$ where l is the number of elements in A $(l\geqslant 2)$.

Let $\mathbf{P}^*$be such that

$$C_2(G^*(\mathbf{P}^*))\leqslant C_2(G^*(\mathbf{P}))\quad\text{for all}\quad \mathbf{P}\in\mathcal{P}_{l,m}.$$

Then we have the following two results.

RESULT 7.7.5 $C'(J^*)\geqslant C_2(G^*(\mathbf{P}^*))$ for any optimal partition J^* of A.

Proof We have

$$C'(J^*) = \sum_{k \ni n_k \geq 2} \frac{1}{n_k} \left[\sum_{\substack{i,j \in J_k^* \\ i<j}} d_{ij}^2 \right]$$

where n_k is the number of elements in J_k^*.

Also, without loss of generality, we assume that

$$n_1 \leq n_2 \leq \cdots \leq n_m.$$

Note that we can choose J^* such that no $n_i = 0$ (by Result 7.7.2).

Let $\mathbf{P} = (n_1, n_2, \ldots, n_m) \in \mathcal{P}_{l,m}$. Let $G \in \mathcal{G}(\mathbf{P})$ be defined by putting r in G_k such that $a_r = d_{ij}^2$ if and only if $i, j \in J_k^*$. Then,

$$C'(J^*) = C_2(G).$$

But $C_2(G^*(\mathbf{P}^*)) \leq C_2(G^*(\mathbf{P})) \leq C_2(G)$ and the result follows.

RESULT 7.7.6 $C_2(G^*(\mathbf{P}^*)) \geq (b-1)/2(\Sigma_{i=1}^{m-r} a_i) + (b/a)\Sigma_{i=m-r+1}^{m} a_i$ where $b = [l/m]$, $r = l - [l/m]m$.

Proof First let $l > 2m$. We may take it that no n_i is equal to 1 in $\mathbf{P}^*$ since otherwise we can get an optimal $\mathbf{P}'$ satisfying this condition. Therefore, $G_i^*(\mathbf{P}^*)$ will have at least one element, for all i.

Now

$$C_2(G^*(\mathbf{P}^*)) \geq \sum_{i=1}^{m} \frac{n_i(n_i - 1)}{2n_i} a_i \tag{7.7.3}$$

(since all elements in $G_i^*(\mathbf{P}^*) \geq 1$). The right-hand side of (7.7.3) is equal to $\frac{1}{2}\Sigma_{i=1}^{m}(n_i - 1)a_i$. Since $\Sigma_{i=1}^{m} n_i a_i$ is minimized subject to $\Sigma_{i=1}^{m} n_i = l$ and $n_1 \leq n_2 \leq \cdots \leq n_m$ when n_i's differ by at most 1 (note that $a_1 \leq a_2 \leq \cdots \leq a_m$), we have the required result.

Now let $l \leq 2m$. From Remark 7.7.1 we have the required result and in fact equality holds.

Let $\delta(A)$ be the lower bound for $C(G^*(\mathbf{P}^*))$ given by Result 7.7.6. Call any partition J of the set $N_q = \{1, 2, \ldots, q\} \subset N$ a *partial clustering* when $q < n$. Call any partition $J' = \{J_1', \ldots, J_m'\}$ of N a *completion* of any partial clustering J in case $J_i \subset J_i'$ for all $i = 1, 2, \ldots, m$. Let $C'(J) = \Sigma_{i=1}^{m} \tau(J_i)$ as defined earlier for complete clustering.

The following result provides a lower bound on the minimum total within-cluster sum of squares over all completions of a given partial clustering.

RESULT 7.7.7 Let J be a partition of N_q, the set of the first q elements of N. Let $\bar{C}(J)$ be the minimum of $C'(J')$ over all completions J' of J. Then

$$\bar{C}(J) \geqslant C'(J) + \delta(\bar{N}_q)$$

where $\delta(\bar{N}_q)$ is as given by Result 7.7.6, $\bar{N}_q = N - N_q$.

This result follows easily from Results 7.7.5 and 7.7.6.

Another lower bound on $\bar{C}(J)$ is obtained in Result 7.7.8.

RESULT 7.7.8 Let $J = (J_1, \ldots, J_m)$ be a partial partition of N_q with the number of elements in $J_k = q_k$. Let $S(J_k) = q_k \tau(J_k)$ and $S(J_k; j) = \Sigma_{i \in J_k} d_{ij}^2$. Let

$$\mu_j = \max\left\{0, \min_{1<k<n}\left[\frac{S(J_k; j) + S(J_k)}{j - q + q_k} - \frac{S(J_k)}{q_k}\right]\right\}$$

and

$$\nu(\bar{N}_q) = \sum_{j \in \bar{N}_q} \mu_j.$$

Then

$$\bar{C}(J) \geqslant C'(J) + \nu(\bar{N}_q).$$

Proof Observe that μ_j is the minimum contribution from j to $C'(J')$ for any completion, J' of J (when j is clustered with any of the J_k).

So, $\nu(\bar{N}_q) = \Sigma_{j \in \bar{N}_q} \mu_j$ gives a lower bound on the contribution to $C'(J')$ from the elements in $\bar{N}_q$ in any completion of J. Therefore, using Result 7.7.2 we get,

$$\bar{C}(J) \geqslant C'(J) + \nu(\bar{N}_q)$$

as required.

Combining Results 7.7.7 and 7.7.8, we can state the following result.

RESULT 7.7.9 $\bar{C}(J) \geqslant C'(J) + \max[\delta(\bar{N}_q), \nu(\bar{N}_q)] \triangleq \alpha(J).$

REMARK 7.7.2 As we have noticed earlier in branch-and-bound procedures, we require a certain lower bound corresponding to the candidate problem. Such a lower bound is provided in Result 7.7.9.

The branch-and-bound algorithm is explained briefly. We choose at first the empty clustering [.] represented by Node 0. We branch this node and find the lower bound given in Result 7.7.9. Choose the one of these descendants with least lower bound and branch. Further calculate the lower bounds for these descendants, and so forth. Whenever we obtain a complete clustering we update the incumbent if the newly generated complete cluster is better than the so-far-best clustering designated as the incumbent. As soon as the lower bounds for all the unbranched nodes are greater than or equal to the objective function value for the incumbent, we stop. Otherwise we backtrack, choose the unbranched node with the smallest lower bound, and proceed further.

We have already outlined the general branch-and-bound scheme while discussing stepwise regression.

Branching Scheme

All partitions, partial or complete clusterings, are represented by the nodes of a tree with a root at the empty clustering denoted by 0. Any node in the tree is called a node of order (q, s) if the number of elements clustered is q and the number of nonempty J_i's is s; in the corresponding partial or complete clustering $J=\{J_i, i=1,2,\ldots, m\}$. We take the s nonempty J_i's to be $\{J_1, J_2,\ldots, J_s\}$ without loss of generality.

A node σ of order (q, s), $0 \leqslant q \leqslant n-m+s$, $0 \leqslant s < m$ is connected by an arc to a node σ_{q+1} of order $(q+1, s+1)$, called a *descendant* of σ if the clustering of the first q elements in σ_{q+1} is identical with that of σ and the $(q+1)$st element is assigned to the $(s+1)$st cluster. Also a node σ of order (q, s), $0 \leqslant q < n-m+s$, $0 \leqslant s < m$, is connected by an arc to each of s nodes σ_k, $1 \leqslant k \leqslant s$, of order $(q+1, s)$, called the descendants of σ where the assignment of the first q elements in σ_k is identical with that of σ and the $q+1$st element is assigned to the kth nonempty cluster. If $q=n-m+s$ and $0 \leqslant s \leqslant m$, then a (q, s)-order node σ is connected by an arc to a node σ_1 of order (n, m) that represents a complete clustering, where the first q elements are assigned as in σ and the remaining $m-s$ elements are clustered into $(m-s)$ single-element clusters.

EXAMPLE 7.7.2 Let $n=8$, $m=3$. Suppose we have the partial cluster of $N_q=\{1,2,3,4,5\}$, that is, $q=5$, given by $J_1=\{1,3\}$, $J_2=\{2,4,5\}$, $J_3=\varnothing$.

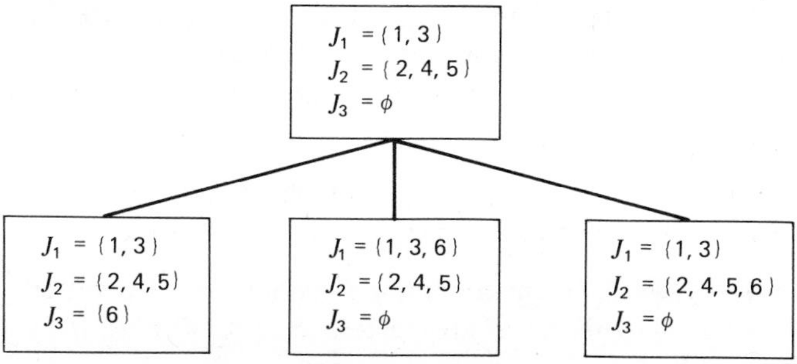

Figure 7.7.1 Branching as in Example 7.7.2.

The order of this partial cluster is (5,2), now $0 \leqslant 5 < 8-3+2$ and $2<3$. Therefore, we obtain as descendants of this partial cluster, $J_1 = \{1,3\}$, $J_2 = \{2,4,5\}$, and $J_3 = \{6\}$. Also we obtain the descendants

$$J_1 = \{1,3,6\}, \qquad J_2 = \{2,4,5\}, \qquad J_3 = \varnothing,$$

and

$$J_1 = \{1,3\}, \qquad J_2 = \{2,4,5,6\}, \qquad J_3 = \varnothing,$$

as shown in the tree given in Fig. 7.7.1.

EXAMPLE 7.7.3 Let $n=8$, $m=3$. Suppose instead we consider the partial cluster of order (6, 1) with $J_1 = \{1,2,3,4,5,6\}$, $J_2 = J_3 = \varnothing$. Then we have the descendant $J_1 = \{1,2,3,4,5,6\}$, $J_2 = \{7\}$, and $J_3 = \{8\}$. This branching is shown in Fig. 7.7.2.

Thus we can summarize the branching scheme by branching steps.

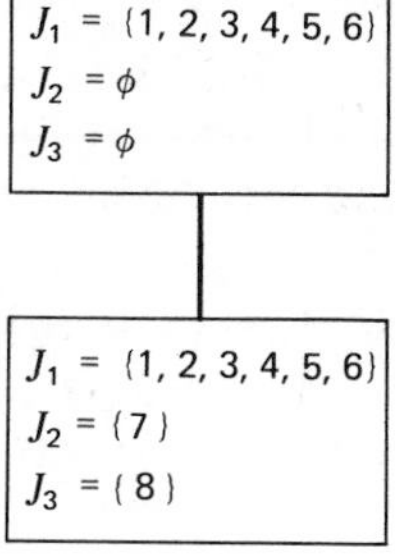

Figure 7.7.2 Branching as in Example 7.7.3.

Let the order of the node σ selected for branching be (q, s) and delete this node from further consideration. If $0 \leqslant q < n-m+s$ and $0 \leqslant s < m$, use Branching Rule 1. If $q = n-m+s$ and $s < m$, then use Branching Rule 2. Otherwise, use Branching Rule 3.

Branching Rule 1 Generate the $s+1$ descendants $\sigma_1, \sigma_2, \ldots, \sigma_{s+1}$ of σ. We get s descendants of order $(q+1, s)$ and one of order $(q+1, s+1)$.

Branching Rule 2 Generate the descendant σ_1 of σ by assigning the $(q+i)$th element to the $(s+i)$th cluster, for $1 \leqslant i \leqslant m-s$. σ_1 is a node of order (n, m) which is a complete clustering.

Branching Rule 3 Generate the m descendants of σ. Each of them is of order $(q+1, m)$.

Let $L(J)$ denote the best lower bound known for the partial clustering J: We take $L(J)$ to be equal to the maximum of the lower bounds for the predecessors of J and the lower bound $\alpha(J)$.

We now work out an example to illustrate the branch-and-bound scheme.

EXAMPLE 7.7.4 The following table gives the d_{ij}^2 for $1 \leqslant i < j \leqslant n$.

	j					
i	1	2	3	4	5	6
1	—	8	9	4	7	7
2		—	3	7	4	5
3			—	8	3	5
4				—	7	6
5					—	6
6						—

We are interested in clustering these six items into two clusters so as to minimize the total within-cluster sum of squares. Let the incumbent be $J_1 = \{1,4\}$, $J_2 = \{2,3,5,6\}$ with $C'(J) = 8.5$.

The calculation of $\delta(\overline{N}_q)$ is illustrated with $q=2$. Now, $\overline{N}_q = \{3,4,5,6\}$. Arranging the corresponding d_{ij}^2 in nondecreasing order, we get a_j's as $(3,5,6,6,7,8)$. As $b = [n-q/m] = [4/2] = 2$ and $r=0$, $\delta(\overline{N}_q) = \frac{1}{2}(3+5) = 4.0$.

Similarly $\delta(\bar{N}_q)$ is calculated for other values of q, and they are given below:

q	$\delta(\bar{N}_q)$
0	8.00
1	5.55
2	4.00
3	3.00
4,5,6	0.00

Next we illustrate how $\nu(\bar{N}_q)$ is calculated with the partial partition $J=(\{1,3\}, \{2,4\})$. We find that $S(J_1)=9$ and $S(J_2)=7$, $C'(J)=4.5+3.5=8.0$. Now

$$\nu\left(\bar{N}_q\right)=\mu_5+\mu_6 \text{ is obtained as follows:}$$

$$\mu_5=\max\left[0;\ \min\left(\frac{10+9}{3}-4.5;\ \frac{11+7}{3}-3.5\right)\right]$$

$$=1.83$$

and

$$\mu_6=\max\left[0;\ \min\left(\frac{12+9}{4}-4.5;\ \frac{11+7}{4}-3.5\right)\right]$$

$$=0.75.$$

Thus, $\nu(\bar{N}_q)=2.58$.

So

$$\bar{C}(J)\geqslant 8.0+\max[0,2.58]$$

$$=8.0+2.58=10.58.$$

With the algorithm, 14 nodes were generated to obtain an optimal complete clustering. The tree generated is given in Fig. 7.7.3. Optimal clustering is given by

$$J_1=\{1,4\}, \quad \text{and } J_2=\{2,3,5,6\};$$

the total within-cluster sum of squares is equal to 8.5.

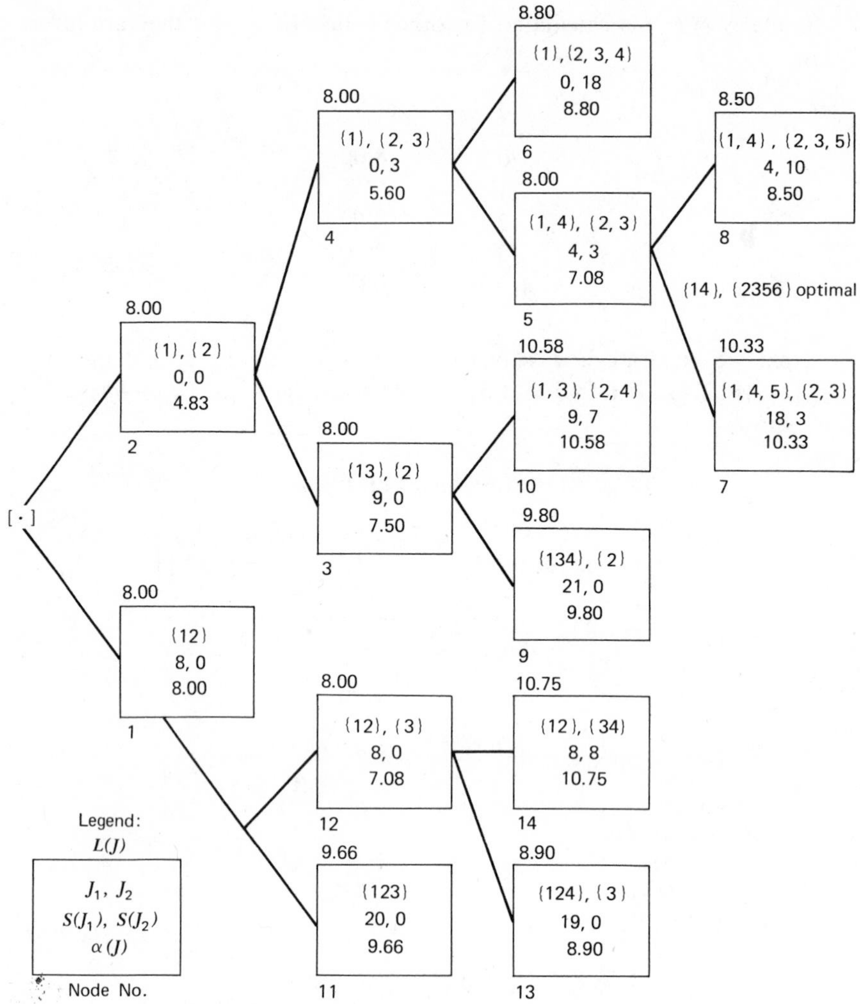

Figure 7.7.3 Tree generated for Example 7.7.4.

BIBLIOGRAPHICAL NOTES

7.1 There have been several books written on cluster analysis, from different points of view. For a quick introduction to the subject see Everitt (1974). Extensive references on this topic can be found in Duran and Odell (1974), Hartigan (1975), and Anderberg (1973).

7.2 Hierarchical clustering procedures can be found in Ward (1963), Everitt (1974), and

Sokal and Sneath (1963). Computer programs for these methods and also for computing different distances can be found in Hartigan (1975), Anderberg (1973), and Lance and Williams (1966a, 1966b). For graph-theoretic approach refer to Matula (1977). For distribution-based classification methods, books on multivariate analysis such as those by Anderson (1958), Morrison (1967), and Srivastava and Khatri (1979) may be referred to. Some definitions of cluster and an application of some methods to the study of race mixtures can be found in Rao (1977).

7.3 Hueristic methods for the problem discussed in this section can be developed analogous to those available in Fisher (1958), and MacQueen (1967). Problem 7.3.1 was formulated by Vinod (1969) for solving the homogenous clustering problem. The material presented in this section closely follows that of Mulvey and Crowder (1979). For results on Lagrangian relaxation and subgradient optimization see Agmon (1954), Motzkin and Schoenberg (1954), and Held, Wolfe, and Crowder (1974), Geoffrion (1974). Computational complexity of the problem is discussed in Cornuejols, Fisher, and Nemhauser (1977), who analyze the problem from an algorithmic point of view, through Integer programming for classification.

7.4 The problem considered in this section appears in Vinod (1969). String property was first observed by Fisher (1958). Rao (1971) considers the problem in higher dimensions. The criterion used was also considered by Edwards and Cavalli-Sfarza (1965), and they give a method for forming two clusters.

7.5 The formulation of clustering problems as dynamic-programming problems was proposed by Jensen (1969). For material on dynamic programming refer to Bellman and Dreyfus (1962). Jensen (1969) also provides formulas for the number of states in each stage and the number of nonredundant arcs from Stage k to Stage $k+1$. Example 7.5.1 is from Jensen (1969). Methods for solving shortest-route problems can be found in Garfinkel and Nemhauser (1969). The one-dimensional-problem with different criteria for clustering appears in Rao (1971).

7.6 An excellent pioneering work on network flow problems is the book by Ford and Fulkerson (1962). They consider the problem of finding maximal flow through network, and provide a labeling procedure for solving the problem. The out-of-kilter algorithm for the minimal-cost-flow problem is also developed in that book.

7.7 The material in this section is from Arthanari (1974). For an application of the branch-and-bound method in classification problems see Liittschwager and Wang (1978). For the cluster-median problem such an approach is used in Jarvinen, Rajala, and Sinervo (1972). Also see Spielberg (1969) for a related application.

REFERENCES

Agmon, S. (1954). "The Relaxation Method for Linear Inequalities." *Canad. J. Math.* **6**, 382.

Anderberg, M. R. (1973). *Cluster Analysis for Applications*. Academic Press, New York.

Anderson, T. W. (1958). *An Introduction to Multivariate Statistical Analysis*. Wiley, New York.

Arthanari, T. S. (1974). "On Some Problems in Sequencing and Grouping." Unpublished doctoral dissertation, Indian Statistical Institute, Calcutta.

Beale, E. M. L. (1969). "Euclidean Cluster Analysis." *Bull. I.S.I.* **43**, 92.

Bellman, R. E., and Dreyfus, S. E. (1962). *Applied Dynamic Programming*. Princeton Univ. Press, Princeton, N.J.

Cornuejols, G., Fisher, M. L., and Nemhauser, G. L. (1977). "Location of Bank Accounts to

Optimize Float: An Analytic Study of Exact and Approximate Algorithms." *Manage. Sci.* **23**, 789.

Cox, D. R. (1957). "Note on Grouping." *J. Am. Stat. Assoc.* **52**, 543.

Duran, B. S., and Odell, P. L. (1974). "Cluster Analysis: A Survey." In *Lecture Notes in Economics and Mathematical Systems*. Springer-Verlag, New York.

Edwards, A. W. F., and Cavalli-Sfarza, L. L. (1965). "Methods for Cluster Analysis." *Biometrics*, **21**, 362.

Everitt, B. S. (1974). *Cluster Analysis*. Halstead Press, London.

Fisher, W. D. (1958). "On Grouping for Maximum Homogeneity." *J. Am. Stat. Assoc.* **53**, 789.

Ford, L. R., and Fulkerson, D. R. (1962). *Flows in Networks*. Princeton Univ. Press, Princeton, N.J.

Garfinkel, R. S., and Nemhauser, G. L. (1972). *Integer Programming*. Wiley, New York.

Geoffrion, A. M. (1974). "Lagrangian Relaxation in Integer Programming." *Math. Program. Stud.* **2**, 82.

Hansen, P., and Delattre, M. (1978). "Complete-Link Cluster Analysis by Graph Coloring." *J. Am. Stat. Assoc.* **73**, 397.

Hartigan, J. A. (1975). *Clustering Algorithms*. Wiley, New York.

Held, M., Wolfe, P., and Crowder, H. (1974). "Validation of Subgradient Optimization." *Math. Program.* **6**, 62.

Jensen, R. E. (1969). "A Dynamic Programming Algorithm for Cluster Analysis." *Oper. Res.* **17**, 1034.

Jarvinen, P., Rajala, J., and Sinervo, H. (1972). "A Branch and Bound Algorithm for Seeking the p-Median." *Oper. Res.* **20**, 173.

Kendal, M. G. (1968). *A Course in Multivariate Analysis* (ed. 4). Hafner, New York.

Lance, G. N., and Williams, W. T. (1966a). "Computer Programmes for Hierarchical Polythetic Classification—Similarity Analysis." *Comput. J.* **9**, 60.

Lance, G. N., and Williams, W. T. (1966b). "A General Theory of Classificatory Sorting Strategies. 1. Hierarchical Systems." *Comput. J.* **9**, 373.

Liittschwager, J. M., and Wang, C. (1978). "Integer Programming Solution of a Classification Problem." *Manage. Sci.* **24**, 1515.

Mahalanobis, P. C. (1936). "On the Generalized Distance in Statistics." *Proc. Nat. Inst. Sci. India* **2**, 49.

Matula, D. W. (1977). "Graph Theoretic Techniques for Cluster Analysis Algorithms." In *Classification and Clustering*. J. V. Ryzin, Ed., Academic Press, New York, pp. 95–129.

Morrison, D. F. (1967). *Multivariate Statistical Methods*. McGraw-Hill, New York.

Motzkin, T., and Schoenberg, I. J. (1954). "The Relaxation Method for Linear Inequalities." *Canad. J. Math.* **6**, 393.

Mulvey, J. M., and Crowder, H. P. (1979). "Cluster Analysis: An Application of Lagrangian Relaxation." *Manage. Sci.* **25**, 329.

Rao, C. R. (1977). "Cluster Analysis Applied to a Study of Race Mixture in Human Population." In *Classification and Clustering*. J. V. Ryzin, Ed., Academic Press, New York, pp. 175–97.

Rao, M. R. (1971). "Cluster Analysis and Mathematical Programming." *J. Am. Stat. Assoc.* **66**, 622.

Ryzin, I. V. (1977). *Classification and Clustering*. Academic Press, New York.

Schrage, L. (1975). "Implicit Representation of Variable Upper Bounds in Linear Programming." *Math. Program. Stud.* **4**, 118.

Spielberg, K. (1969). "An Algorithm for the Simple Plant Location Problem with Some Side Constraints." *Oper. Res.* **17**, 85.

Sokal, R. R., and Sneath, P. H. (1963). *Principles of Numerical Taxonomy*. W. H. Freeman and Company, San Francisco.

Srivastava, M. S., and Khatri, C. G. (1979). *An Introduction to Multivariate Statistics*. Elsevier, New York.

Vinod, H. D. (1969). "Integer Programming and Theory of Grouping." *J. Am. Stat. Assoc.* **64**, 506.

Ward, J. H. Jr. (1963). "Hierarchical Grouping to Optimize an Objective Function." *J. Am. Stat. Assoc.* **58**, 236.

CHAPTER 8

Epilogue

8.1 INTRODUCTION

In this book so far we have seen how mathematical programming can be applied to problems arising in statistics. Encouragingly, there are other applications that unfortunately could not be discussed for various reasons. Some were in areas of special interest, so that a discussion of the problems would require elaborate description of the problem area itself, before we could go on to describe the application of mathematical programming to them. Some would have required much abstract discussion of the concepts. Some are similar to those discussed herein. Finally, there may be other applications that we were not aware of.

In this epilogue we briefly review some of these problems and applications, and provide a concluding remark.

8.2 SELECTION PROBLEMS AMENABLE TO MATHEMATICAL-PROGRAMMING APPLICATION

The problem of selecting individuals for a specific purpose so as to insure that the selected individuals are the best in some sense for serving that purpose arise in several contexts. However, in many situations measurements on the criterion variables are not available at the time of selection. We may have measurements only on some related variables, called "predictor variables," on which to base selection. Also, the predictor variables may not be available all at one time. In such cases selection is made in stages. There are different ways of selecting the individuals in multistage selection.

The objective of a selection program may be maximization of (1) the average value of a measurement in the selected population subject to a given frequency of selection; or (2) the frequency of selection for a specified average value of a measurement; or (3) the proportion of "prospective candidates," subject to some restriction on the set of individuals chosen.

Rao (1964) has considered such problems and has discussed the difficulties involved in obtaining optimal solutions. Rao emphasizes the need for mathematical programming for solving these problems. These problems can be related to the problem of testing of hypotheses. Here we are seeking for a "best" region, such that if the observed measurements for the predictor variables fall in this region we accept the individual or else we reject the individual. Interestingly, Rao poses a problem which resembles the generalized Neyman-Pearson problem. He also obtains solutions in special cases, where the objective function is a linear function and the constraints are nonlinear. Computational procedures to solve the problems formulated by Rao (1964) offer a possible area of research, which involves application of mathematical programming.

8.3 OPTIMAL DESIGNS

Different optimality criteria are considered for finding optimal designs. For an extensive bibliography on the theory of optimal designs see Ash and Hedayat (1978). Gribik and Kortanek (1977) identify the mathematical-programming problems arising in such situations and provide a cutting-plane algorithm to solve the problems.

Consider the model,

$$y|x=\sum_{r=1}^{p}\theta_r f_r(x)+\varepsilon(x)$$

where x belongs to the factor space, $E(\varepsilon(x))=0$, and $\mathrm{Var}(\varepsilon(x))=\sigma_x^2$.

Let the n_i independent observations of y be made at x_i, $i=1,\ldots,m$. The observations are made independently at x_i and x_j, $i\neq j$. Let y_{ij} denotes the value of the jth observation made at point x_i. Let

$$n=\sum_{i=1}^{m} n_i \qquad \text{and} \qquad M=\sum_{i=1}^{m}\frac{n_i}{n}\mathbf{f}(x_i)\mathbf{f}(x_i)'$$

where

$$\mathbf{f}(x_i)=(f_1(x_i),\ldots,f_p(x_i)).$$

M corresponds to the information matrix $(X'X)$ in the linear model.

Assume that M is nonsingular. Then, because $1/nM^{-1}$ is the covariance matrix of the least-squares estimator $\hat{\theta}$ of θ, it would be desirable to choose M for a given n, n_i, x_i, so that the covariance matrix of $\hat{\theta}$ is "best" in terms of some criterion.

Suppose the criterion function is $C(M^{-1})$. Then the problem can be formulated as

$$
\begin{aligned}
&\text{Minimize} \quad C(M^{-1})\\
&\text{subject to} \quad M=\sum_{i=1}^{m} p_i \mathbf{f}(x_i)\mathbf{f}(x_i)', \text{ which is nonsingular,}\\
&\qquad\qquad \sum_{i=1}^{m} p_i = 1\\
&\qquad\qquad p_i \geq 0, \qquad i=1,\ldots,m\\
&\qquad\qquad p_i n \quad \text{is an integer}, \qquad i=1,\ldots,m\\
&\qquad\qquad m \quad \text{is an arbitrary positive integer.}
\end{aligned}
$$

This problem corresponds to finding an *exact* design. An *approximate design*, $\xi(\cdot)$, is a probability measure over the factor space $\mathscr{F}$. A similar problem for approximate designs can be stated as:

$$
\begin{aligned}
&\text{Minimize} \quad C(M^{-1})\\
&\text{subject to} \quad M\in\Omega=\{M(\xi)\in R^{p\times p}\mid\\
&\qquad\qquad M(\xi)=\int \mathbf{f}(x)\mathbf{f}'(x)\xi(dx) \qquad\qquad (8.3.1)\\
&\qquad\qquad \text{for } \xi(\cdot), \text{ a probability measure on the}\\
&\qquad\qquad \text{factor space } \mathscr{F}\}\\
&\text{and} \qquad M \text{ nonsingular.}
\end{aligned}
$$

Choosing $C(\cdot)$ differently, we get different optimality criterion. When $C(M^{-1})$ is (1) the determinant of M^{-1}, we have the D-optimality criterion; (2) a linear function of M^{-1}, we have the L-optimality criterion; (3) the trace of M^{-1}, we have the A-optimality criterion; and (4) the modulus of the largest eigenvalue of M^{-1}, we have the E-optimality criterion.

Considering approximate designs, Keifer and Wolfowitz (1960) have proved the equivalence between some of these optimality criteria.

Gribik and Kortanek (1977) consider problems with general criterion functions satisfying certain assumptions. They characterize the optimal solution to the problems. And their results provide a system of linear inequalities on Ω in (8.3.1). The set of solutions to this system provides the set of optimal information matrices. Gribik and Kortanek also provide a cutting-plane algorithm specifically developed for such problems.

Pukelsheim (1978) also considers design optimality problems from a mathematical-programming point of view, and characterizes optimal designs through abstract duality.

Another optimality criterion is cost. Neuhardt and Mount-Campbell (1978) consider the problem of minimizing a cost function that reflects the cost of selecting certain treatment combinations in a factorial layout. They give a partial search procedure for the case of additive cost function. They also consider a parametric version of the problem, under some assumptions.

8.4 OBTAINING NONNEGATIVE ESTIMATES OF VARIANCE COMPONENTS

Obtaining negative estimates of the parameters that are expected to be nonnegative estimates creates considerable discomfort for an experimenter using analysis of variance. As discussed in Chapter 2, nonnegativity restrictions can be introduced into the optimization problem under consideration to avoid this situation. Thompson (1962) considers one such situation when negative estimates are obtained in variance-components analysis. He approaches the problem from the mathematical-programming point of view, exploiting the special structure of constraints of the problem, and obtains a special algorithm.

Consider the following random-effect two-way classification model with interaction

$$y_{ijk}=\mu+\alpha_i+\beta_j+(\alpha\beta)_{ij}+\varepsilon_{ijk} \tag{8.4.1}$$

where y_{ijk} is the response variable; μ represents a grand mean; ε_{ijk} is the random error with mean 0 and variance σ^2; and α_i, β_j, $(\alpha\beta)_{ij}$ are uncorrelated random variables, with zero mean and variances σ_α^2, σ_β^2, and $\sigma_{\alpha\beta}^2$, respectively. Let a, b denote the number of levels of α and β, respectively. In this model the levels of both factors are considered to be random. There are n observations per cell.

Let S_A, S_B, S_{AB}, and S_E denote the sum of squares due to Factor A, Factor B, Interaction AB, and error, respectively. Then the expected values of S_A, S_B, S_{AB}, and S_E are obtained, via traditional analysis of the model, as

$$\begin{aligned} E(S_A) &= (a-1)(\sigma^2 + n\sigma_{AB}^2 + bn\sigma_A^2) \\ E(S_B) &= (b-1)(\sigma^2 + n\sigma_{AB}^2 + an\sigma_B^2) \\ E(S_{AB}) &= (a-1)(b-1)(\sigma^2 + n\sigma_{AB}^2) \end{aligned} \tag{8.4.2}$$

and

$$E(S_E) = ab(n-1)\sigma^2.$$

The estimates of the components of variance are needed in the related hypothesis-testing problem. For this reason we assume normality and independence of the variables. Let

$$\sigma_1^2 = \sigma^2;\ \sigma_2^2 = \sigma_{AB}^2;\ \sigma_3^2 = an\sigma_B^2 \quad \text{and} \quad \sigma_4^2 = bn\sigma_A^2. \tag{8.4.3}$$

Let $\omega_1, \ldots, \omega_4$ denote the expected mean squares of A, B, AB and error respectively. We have

$$\begin{aligned} \omega_1 &= \sigma_1^2 + \sigma_2^2 + \sigma_4^2 \\ \omega_2 &= \sigma_1^2 + \sigma_2^2 + \sigma_3^2 \\ \omega_3 &= \sigma_1^2 + \sigma_2^2 \\ \omega_4 &= \sigma_1^2. \end{aligned} \tag{8.4.4}$$

Usually these mean squares are shown in the analysis-of-variance table, as in Table 8.4.1. The equations of 8.4.4 imply $\boldsymbol{\omega} = \mathbf{A}\boldsymbol{\sigma}^2$ where $\mathbf{A}$ is nonsingu-

Table 8.4.1

Source	Mean Square	Expected Mean Square
Rows	w_1	$\omega_1 = \sigma_1^2 + \sigma_2^2 + \sigma_4^2$
Columns	w_2	$\omega_2 = \sigma_1^2 + \sigma_2^2 + \sigma_3^2$
Interaction	w_3	$\omega_3 = \sigma_1^2 + \sigma_2^2$
Error	w_4	$\omega_4 = \sigma_1^2$

lar, and $\mathbf{A}$ has a special structure: it is triangular. Therefore, we have

$$\boldsymbol{\sigma}^2 = \mathbf{A}^{-1}\boldsymbol{\omega}. \tag{8.4.5}$$

Conventional maximum-likelihood estimates of $\omega_1, \ldots, \omega_4$ are $w_1, \ldots, w_4$. So it is easy to notice that there is a possibility of getting negative estimates of the variance components.

Let us consider the general model, with multiple classification. We have correspondingly $\boldsymbol{\omega} = \mathbf{A}\boldsymbol{\sigma}^2$, where $\boldsymbol{\omega} = (\omega_1, \ldots, \omega_p)'$; $\boldsymbol{\sigma}^2 = (\sigma_1^2, \ldots, \sigma_p^2)'$ and $\mathbf{A}$ is a nonsingular square matrix, with the property of triangularity. From the results available in the area of sampling distributions, we have the density function of the random variables W_i corresponding to the ith mean square in the table, following a gamma distribution.

$$f(w_i;\, \omega_i) = \left[\Gamma(n_i/2)(2\omega_i/n_i)^{n_i/2}\right]^{-1} w_i^{n_i/2-1} e^{-n_i w_i/2\omega_i} \qquad w_i \geqslant 0;\ \omega_i > 0 \tag{8.4.6}$$

where w_i is a realization of W_i, n_i is the corresponding degrees of freedom, and $\omega_i = E(W_i)$.

Consider the likelihood of $(w_1, \ldots, w_p)$, given by $L(\boldsymbol{\omega}) = \prod_{i=1}^{p} f(w_i;\, \omega_i)$. Instead of maximizing the likelihood function $L(\boldsymbol{\omega})$ we can minimize twice the log of $L(\boldsymbol{\omega})$. That is, we maximize $g(\boldsymbol{\omega})$ with respect to $\boldsymbol{\omega}$, where

$$g(\boldsymbol{\omega}) = K - \sum_{i=1}^{p} n_i(\log w_i + w_i/\omega_i) \tag{8.4.7}$$

where K is a constant.

However, we require $\boldsymbol{\omega}$ to satisfy

$$\boldsymbol{\sigma}^2 = \mathbf{A}^{-1}\boldsymbol{\omega} \geqslant \mathbf{0}$$

to obtain nonnegative estimates of $\sigma_1^2, \ldots, \sigma_p^2$.

Thus we have the following problem:

Problem 8.4.1

$$\text{Maximize} \quad g(\boldsymbol{\omega})$$

$$\text{subject to} \quad \mathbf{A}^{-1}\boldsymbol{\omega} \geqslant \mathbf{0}.$$

Problem 8.4.1 is a nonlinear-programming problem with a differentiable, strictly concave function as the objective function, and the constraints are linear.

Necessary conditions, drawn from Kuhn and Tucker as discussed in Section 2.19, can be stated for a ω^* to be optimal for Problem 8.4.1. However, in Section 2.19 we discussed a minimization problem and so the convexity of the objective function was used to prove that any local minimum is also global minimum for the problem. Here, we have a strictly concave objective function to be maximized, so the results go through similarly; we also have an unique solution, implied by the strict concavity of the objective function. We state, slightly differently from Kuhn and Tucker, necessary and sufficient conditions for ω^* to be optimal for Problem 8.4.1 in Result 8.4.1. Let

$$\nabla g(\omega^*) = \left(\frac{\partial g(\omega)}{\partial \omega_1}, \ldots, \frac{\partial g(\omega)}{\partial \omega_p} \right)\Bigg|_{\omega=\omega^*}, \text{ the gradient vector at } \omega^*.$$

and

$$g_{\omega_i}(\omega^*) \text{ denotes } \partial g(\omega)/\partial \omega_i \big|_{\omega=\omega^*}.$$

Let $\mathbf{x}_s$ be the surplus variable vector corresponding to $A^{-1}\omega \geqslant 0$, that is, $\mathbf{A}^{-1}\omega - \mathbf{I}\mathbf{x}_s = \mathbf{0}$.

RESULT 8.4.1 A set of necessary and sufficient conditions for ω^* to be optimal for Problem 8.4.1 is

$$(1) \quad \mathbf{A}^{-1}\omega^* \geqslant \mathbf{0}$$

$$(2) \quad \mathbf{A}'\nabla \mathrm{g}(\omega^*) \leqslant \mathbf{0} \tag{8.4.8}$$

and

$$(3) \quad \omega^* \nabla \mathrm{g}(\omega^*) = \mathbf{0} \tag{8.4.9}$$

From the strict concavity of $g(\omega)$, we also have a unique optimal solution to (8.4.1).

Thompson connects this problem to an optimization problem involving the node potentials of a tree.

Let the graph

$$\mathcal{G} = (\mathcal{N}, \mathcal{A}) \text{ be a rooted tree, where } \mathcal{N} = \{1, 2, \ldots, p\}$$

with the root designed as r.

There is a unique path $P(r,s)=r, (r;\, i_1), i_1, (i_1;\, i_2), i_2, \ldots, (i_k;\, s), s$, from r to s, for any $s \in \mathfrak{N}$. Let

$$\mathcal{P}(s)=\{i_l \mid i_l \in p(r,s)\}$$

$\mathcal{P}(s)$ is called the set of predecessors of s, and $\mathcal{P}(r)=\varnothing$. The immediate predecessor of s denoted by $P(s)$ is the node i_k in the unique path $p(r,s)$.

The concept of predecessors corresponding to the tree $\mathcal{G}$ introduces a partial order in the set $\mathfrak{N}$.

Let us write $t<s$ if $t \in \mathcal{P}(s)$; $t=s$ implies that t and s denote the same node in $\mathfrak{N}$.

Let

$$\omega=(\omega(1),\ldots,\omega(p))' \in R^p$$

$\omega(s)$ is called the potential of $s \in \mathfrak{N}$.

Consider a differentiable function

$$g(\omega)=g(\omega(1),\ldots,\omega(p)).$$

Problem 8.4.2

$$\begin{aligned} &\text{Maximize} && g(\omega) \\ &\text{subject to} && \omega(t) \geqslant \omega(P(t)) \qquad\qquad (8.4.10) \\ &\text{for all} && t \in \mathfrak{N}. \end{aligned}$$

Assume $\omega(P(r))=0$ for all ω.

Problem 8.4.1 is reformulated as a special case of Problem 8.4.2, with the observation that the constraint matrix has a tree structure in Problem 8.4.1. Thompson develops a recursive algorithm based on the graph-related results obtained for Problem 8.4.2, using Result 8.4.1.

Applications of this kind bring out an important message: Those who wish to apply mathematical programming in these areas have to exploit fully the structural and other properties of the problem under consideration.

Rao (1979) identifies five different types of optimal allocation problems in stratified sampling. Some of these problems have constraint matrices similar to that of Problem 8.4.1. Hence he suggests the use of Thompson's result for these problems.

McLean and Anderson (1966) use the programming approach to find the constrained optimal treatment combination for the extreme-vertices design-of-mixture experiments.

Rao (1971) has developed MINQUE (Minimum Norm Quadratic Unbiased Estimator) for estimation of variance components. However, the problem of getting negative estimates while using MINQUE still exists. Pukelsheim (1977, 1978) has modified MINQUE to obtain nonnegative estimates. His approach is through abstract duality. He also proves sufficient optimality conditions for this kind of estimators. Considering the problem of estimation of heteroscedastic variances, Pukelsheim (1978) shows that the sample variance of the kth group provides the nonnegative MINQUE. Pukelsheim (1977) proves the existence and uniqueness of the nonnegative MINQUE.

Optimization problems in statistics involving variables that are probability measures, or elements of some abstract spaces, require the knowledge of abstract mathematical programming as a necessary tool for applying programming methods in these areas. (See Rockafellar (1974), Luenberger (1969), Luenberger (1973), and Gol'steĭn (1972) for treatment of programming in abstract spaces).

8.5 CONCLUDING REMARK

It does not seem to end here.

REFERENCES

Ash, A., and Hedayat, A. (1978). "An Introduction to Design Optimality with an Overview of the Literature." *Commun. Stat. Theor. Math.* **A7**, 1295.

Cochran, W. G. (1951). "Improvement by Means of Selection." In *Proceedings of the Second Berkeley Symposium on Mathematical Statistics and Probability*. Univ. California Press, Berkeley, Calif., pp. 449–70.

Federov, V. V. (1972). *Theory of Optimal Experiments*. Academic Press, New York.

Gribik, P. R., and Kortanek, K. O. (1977). "Equivalence Theorems and Cutting Plane Algorithms for a Class of Experimental Design Problems." *SIAM J. Appl. Math.* **32**, 232.

Kempthorne, O., and Nordskog, A. W. (1959). "Restricted Selection Indices." *Biometrics* **15**, 10.

Kiefer, J. (1959). "Optimum Experimental Designs." *J. R. Stat. Soc.* **B21**, 272.

Kiefer, J., and Wolfowitz, J. (1959). "Optimum Designs in Regression Problems." *Ann. Math. Stat.* **30**, 271.

Kiefer, J., and Wolfowitz, J. (1960). "The Equivalence of Two Extremum Problems." *Canad. J. Math.* **12**, 363.

Luenberger, D. G. (1969). *Optimization by Vector Space Methods*. Wiley, New York.

Luenberger, D. G. (1973). *Introduction to Linear and Nonlinear Programming*. Addison-Wesley, Reading, Mass.

McLean, R. A., and Anderson, V. L. (1966). "Extreme Vertices Design of Mixture Experiments." *Technometrics* **8**, 447.

Neuhardt, J. B., and Mount-Campbell, C. A. (1978). "Selection of Cost-Optimal 2^{k-p} Fractional Factorials." *Commun. Stat. Simula. Comput.* **B7**, 369.

Pukelsheim, F. (1977). *Linear Models and Convex Programs: Unbiased Nonnegative Estimation in Variance Component Models*. Technical Report 104, Department of Statistics, Stanford Univ. Press, Palo Alto, Calif.

Pukelsheim, F. (1978). *A Quick Introduction to Mathematical Programming with Applications to Most Powerful Tests, Nonnegative Variance Estimation, and Optimal Design Theory*. Technical Report 128, Department of Statistics, Stanford Univ. Press, Palo Alto, Calif.

Rao, C. R. (1964). *Problems of Selection Involving Programming Techniques*. Proceedings of the IBM Scientific Computing Symposium on Statistics, pp. 29–51.

Rao, C. R. (1971a). "Estimation of Variance and Covariance Components MINQUE Theory." *J. Multivar. Anal.* **1**, 257.

Rao, C. R. (1971b). "Minimum Variance Quadratic Unbiased Estimation of Variance Components." *J. Multivar. Anal.* **1**, 445.

Rao, J. N. K. (1973). "On the Estimation of Heteroscedastic Variances." *Biometrics* **29**, 11.

Rao, J. N. K. (1979). "Optimization in the Design of Sample Surveys." In *Optimizing Methods in Statistics*. J. S. Rustagi, Ed. Academic Press, New York, pp. 419–34.

Rockafellar, R. T. (1974). "Conjugate Duality and Optimization." Philadelphia, Pa. Society for Industrial and Applied Mathematics.

Sibson, R. (1974). "D_A-Optimality and Duality." In *Progress in Statistics*. J. Gani, K. Sarkadi, and I. Vincze, Eds. North Holland Publishing, Amsterdam, pp. 677–92.

Thompson, W. A. Jr. (1962). "The Problem of Negative Estimates of Variance Components." *Ann. Math. Stat.* **33**, 273.

Index

See bibliography/references at the end of chapter for names that are not cited in the index.

Applied Probability and Statistics (Continued)

FLEISS • Statistical Methods for Rates and Proportions, *Second Edition*
GALAMBOS • The Asymptotic Theory of Extreme Order Statistics
GIBBONS, OLKIN, and SOBEL • Selecting and Ordering Populations: A New Statistical Methodology
GNANADESIKAN • Methods for Statistical Data Analysis of Multivariate Observations
GOLDBERGER • Econometric Theory
GOLDSTEIN and DILLON • Discrete Discriminant Analysis
GROSS and CLARK • Survival Distributions: Reliability Applications in the Biomedical Sciences
GROSS and HARRIS • Fundamentals of Queueing Theory
GUPTA and PANCHAPAKESAN • Multiple Decision Procedures: Theory and Methodology of Selecting and Ranking Populations
GUTTMAN, WILKS, and HUNTER • Introductory Engineering Statistics, *Second Edition*
HAHN and SHAPIRO • Statistical Models in Engineering
HALD • Statistical Tables and Formulas
HALD • Statistical Theory with Engineering Applications
HARTIGAN • Clustering Algorithms
HILDEBRAND, LAING, and ROSENTHAL • Prediction Analysis of Cross Classifications
HOEL • Elementary Statistics, *Fourth Edition*
HOLLANDER and WOLFE • Nonparametric Statistical Methods
JAGERS • Branching Processes with Biological Applications
JESSEN • Statistical Survey Techniques
JOHNSON and KOTZ • Distributions in Statistics
Discrete Distributions
Continuous Univariate Distributions—1
Continuous Univariate Distributions—2
Continuous Multivariate Distributions
JOHNSON and KOTZ • Urn Models and Their Application: An Approach to Modern Discrete Probability Theory
JOHNSON and LEONE • Statistics and Experimental Design in Engineering and the Physical Sciences, Volumes I and II, *Second Edition*
JUDGE, GRIFFITHS, HILL and LEE • The Theory and Practice of Econometrics
KALBFLEISCH and PRENTICE • The Statistical Analysis of Failure Time Data
KEENEY and RAIFFA • Decisions with Multiple Objectives
LANCASTER • An Introduction to Medical Statistics
LEAMER • Specification Searches: Ad Hoc Inference with Nonexperimental Data
McNEIL • Interactive Data Analysis
MANN, SCHAFER and SINGPURWALLA • Methods for Statistical Analysis of Reliability and Life Data
MEYER • Data Analysis for Scientists and Engineers
MILLER, EFRON, BROWN, and MOSES • Biostatistics Casebook
OTNES and ENOCHSON • Applied Time Series Analysis: Volume I, Basic Techniques
OTNES and ENOCHSON • Digital Time Series Analysis
POLLOCK • The Algebra of Econometrics
PRENTER • Splines and Variational Methods
RAO and MITRA • Generalized Inverse of Matrices and Its Applications
RIPLEY • Spatial Statistics
SCHUSS • Theory and Applications of Stochastic Differential Equations
SEAL • Survival Probabilities: The Goal of Risk Theory
SEARLE • Linear Models
SPRINGER • The Algebra of Random Variables
UPTON • The Analysis of Cross-Tabulated Data
WEISBERG • Applied Linear Regression
WHITTLE • Optimization Under Constraints